Phytoremediation

METHODS IN BIOTECHNOLOGY™

John M. Walker, SERIES EDITOR

24. **Animal Cell Biotechnology,** *Methods and Protocols, Second Edition,* edited by *Ralf Pörtner, 2007*
23. **Phytoremediation,** *Methods and Reviews,* edited by *Neil Willey, 2007*
22. **Immobilization of Enzymes and Cells,** *Second Edition,* edited by *Jose M. Guisan, 2006*
21. **Food-Borne Pathogens,** *Methods and Protocols,* edited by *Catherine Adley, 2006*
20. **Natural Products Isolation,** *Second Edition,* edited by *Satyajit D. Sarker, Zahid Latif, and Alexander I. Gray, 2005*
19. **Pesticide Protocols,** edited by *José L. Martínez Vidal and Antonia Garrido Frenich, 2005*
18. **Microbial Processes and Products,** edited by *Jose Luis Barredo, 2005*
17. **Microbial Enzymes and Biotransformations,** edited by *Jose Luis Barredo, 2005*
16. **Environmental Microbiology:** *Methods and Protocols,* edited by *John F. T. Spencer and Alicia L. Ragout de Spencer, 2004*
15. **Enzymes in Nonaqueous Solvents:** *Methods and Protocols,* edited by *Evgeny N. Vulfson, Peter J. Halling, and Herbert L. Holland, 2001*
14. **Food Microbiology Protocols,** edited by *J. F. T. Spencer and Alicia Leonor Ragout de Spencer, 2000*
13. **Supercritical Fluid Methods and Protocols,** edited by *John R. Williams and Anthony A. Clifford, 2000*
12. **Environmental Monitoring of Bacteria,** edited by *Clive Edwards, 1999*
11. **Aqueous Two-Phase Systems,** edited by *Rajni Hatti-Kaul, 2000*
10. **Carbohydrate Biotechnology Protocols,** edited by *Christopher Bucke, 1999*
9. **Downstream Processing Methods,** edited by *Mohamed A. Desai, 2000*
8. **Animal Cell Biotechnology,** edited by *Nigel Jenkins, 1999*
7. **Affinity Biosensors:** *Techniques and Protocols,* edited by *Kim R. Rogers and Ashok Mulchandani, 1998*
6. **Enzyme and Microbial Biosensors:** *Techniques and Protocols,* edited by *Ashok Mulchandani and Kim R. Rogers, 1998*
5. **Biopesticides:** *Use and Delivery,* edited by *Franklin R. Hall and Julius J. Menn, 1999*
4. **Natural Products Isolation,** edited by *Richard J. P. Cannell, 1998*
3. **Recombinant Proteins from Plants:** *Production and Isolation of Clinically Useful Compounds,* edited by *Charles Cunningham and Andrew J. R. Porter, 1998*
2. **Bioremediation Protocols,** edited by *David Sheehan, 1997*
1. **Immobilization of Enzymes and Cells,** edited by *Gordon F. Bickerstaff, 1997*

METHODS IN BIOTECHNOLOGY™

Phytoremediation

Methods and Reviews

Edited by

Neil Willey

*Center for Research in Plant Science,
University of the West of England, Bristol, UK*

HUMANA PRESS ✳ TOTOWA, NEW JERSEY

© 2007 Humana Press Inc.
999 Riverview Drive, Suite 208
Totowa, New Jersey 07512

www.humanapress.com

All rights reserved. No part of this book may be reproduced, stored in a retrieval system, or transmitted in any form or by any means, electronic, mechanical, photocopying, microfilming, recording, or otherwise without written permission from the Publisher. Methods in Biotechnology™ is a trademark of The Humana Press Inc.

All papers, comments, opinions, conclusions, or recommendations are those of the author(s), and do not necessarily reflect the views of the publisher.

This publication is printed on acid-free paper. ∞

ANSI Z39.48-1984 (American Standards Institute) Permanence of Paper for Printed Library Materials.

Production editor: Rhukea J. Hussain

Cover design by Carlotta L. C. Craig

Cover illustration: Asiatic Dayflower (*Commelina communis* L) growing on copper-mining spoil in China (courtesy Dr. Shirong Tang).

For additional copies, pricing for bulk purchases, and/or information about other Humana titles, contact Humana at the above address or at any of the following numbers: Tel.: 973-256-1699; Fax: 973-256-8341; E-mail: orders@humanapr.com; or visit our Website: www.humanapress.com

Photocopy Authorization Policy:
Authorization to photocopy items for internal or personal use, or the internal or personal use of specific clients, is granted by Humana Press Inc., provided that the base fee of US $30.00 per copy is paid directly to the Copyright Clearance Center at 222 Rosewood Drive, Danvers, MA 01923. For those organizations that have been granted a photocopy license from the CCC, a separate system of payment has been arranged and is acceptable to Humana Press Inc. The fee code for users of the Transactional Reporting Service is: [978-1-58829-541-5 • 1-58829-541-9/07 $30.00].

Printed in the United States of America. 10 9 8 7 6 5 4 3 2 1

eISBN: 1-59745-098-7

ISBN 13: 978-1-58829-541-5

Library of Congress Cataloging-in-Publication Data

Phytoremediation : methods and reviews / edited by Neil Willey.
    p. cm. -- (Methods in biotechnology ; 23)
  Includes bibliographical references and index.
  ISBN 1-58829-541-9 (alk. paper)
  1. Phytoremediation. I. Willey, Neil. II. Series.
  TD192.75.P47 2006
  628.5--dc22
                            2006015505

Preface

The term phytoremediation was coined in the 1980s to describe the use of plants to ameliorate degraded or polluted substrates. Utilizing plants to control soil and water degradation has a long history. Many early agriculturalists developed plant-based systems to minimize soil erosion, and the use of plants to restore disturbed environments and cleanse water is well established. In the 20th century, the potential of plants as extractors of pollutants began to be explored, for example during investigations of radionuclide-contaminated soils by Nishita and colleagues in the 1950s. In the last decade or so there has been rapid development of scientific methods relevant to phytoremediation. It is, therefore, an appropriate time to consider methodological developments in phytoremediation and to review their current use. In Parts I and II of *Phytoremediation: Methods and Reviews*, methods are described for enhancing contaminant degradation, uptake, and tolerance by plants, for exploiting plant biodiversity for phytoremediation, for modifying contaminant availability, and for experimentally analyzing phytoremediation potential. Then, in Parts III and IV, a variety of phytoremediation technologies and their use around the world is reviewed.

The ability of plants to degrade, take up, or tolerate the effects of pollutants is the *sine qua non* of phytoremediation. Plants that have an innate ability to degrade organics or accumulate heavy metals were the focus of the first phase of phytoremediation research. A full understanding of the mechanisms underpinning these processes in plants is now, at least in theory, possible. The transformation of the life sciences in the second half of the 20th century not only provided the potential for full mechanistic understanding, but also the opportunity to engineer plant properties for phytoremediation. Knowledge and manipulation of fundamental properties might facilitate the wide development of phytoremediation. It is easy to be overly optimistic in our hopes for the use of molecular manipulation in phytoremediation, but it is equally easy to underestimate its possibilities. By the time the 50th anniversary of elucidating the structure of DNA was approaching, plants had been genetically engineered to take up and tolerate arsenic and mercury, to degrade residues of synthetic explosives, and to solubilize and take up some of the most insoluble inorganic nutrients in soil, such as iron and phosphate. The production of plants with these properties became a reality in what was approximately the working life of a generation of scientists. This book, therefore, includes a number of chapters, spread across Parts I, II, and III, devoted to recent developments in the manipulation of pollutant degradation, uptake, and tolerance by plants.

If pollutants are not available to plants, plant abilities are irrelevant to phytoremediation. Paradoxically, pollutant availability to plants can, therefore, be singled out as the *sine qua non* of phytoremediation just as can the ability of plants to deal with pollutants. The allocation to Part II of most chapters focused on pollutant availability in the soil risks perpetuating an unhelpful dichotomy in the study of soil–plant systems between "uptake" and "availability" in the soil. Ultimately, only the ability of plants to deal with pollutants from real soils or effluents can be meaningfully used to infer phytoremediation potential, and availability for plant uptake can only be defined in plant terms. "Uptake" and "availability" are less easily disentangled, and more frequently interact, than we often assume, and neither is more important. I hope that spreading chapters on degradation, uptake, and tolerance by plants across sections, including Part II, which has chapters on manipulating soil availability, avoids emphasizing the soil–plant dichotomy. I have, however, kept most chapters on the topic of availability together in Part II to emphasize the importance of overcoming the barrier of availability and the significance of recent research into it.

Biodiversity is a raw material of biotechnology, but there are few established techniques for mining its potential. This is unfortunate, particularly for phytoremediation, in which the identification of wild plants with unusual metal uptake patterns was instrumental in establishing the discipline. This book includes methods that might be useful for charting and exploiting what might be called the "biodiversity landscape" of phytoremediation potential. Presently, almost all the organisms actually used in bioremediation were identified at contaminated sites. Such sites clearly provide a useful screen for plants that can degrade, take up, or tolerate pollutants. However, the performance of species can be site specific and some of the species with useful phenotypes for phytoremediation might not occur at contaminated sites. In fact, the species is a reproductive unit and there is no reason at present, other than convenience, for believing it to be the most useful taxonomic unit on which to focus phytoremediation efforts. In reality very little of the biodiversity that might be useful for phytoremediation, either as suitable taxa at any level of the taxonomic hierarchy or as useful genes, has so far been investigated. There has been much recent advance in understanding how biodiversity arises, the phylogenies that constrain plant phenotype, and in defining plant functional types. Therefore, I have included in this book research focused on exploiting biodiversity for improving phytoremediation.

The efficacy and utility of technologies, including environmental biotechnologies like phytoremediation, is affected by context. Socioeconomic and ecological contexts will determine, at least partly, the successful application of phytoremediation. In Parts III and IV, phytoremediation in a variety of contexts is reviewed. The insights this provides will be useful, I hope, not only for the development of phytoremediation, but also to remind us that the success of the technology depends not only on the novelty of the underlying science but also on its suitability to context. Phytoremediation will probably always be a relatively slow, solar-powered, low-technology fix to problems of soil and water degradation or pollution. However, this type of technology is suitable to a surprising variety of contexts and might have particular resonance with 21st-century demands. I hope that the inclusions of chapters outlining the use of phytoremediation in a variety of contexts emphasizes this.

Scientific investigations of phytoremediation might have revealed insurmountable limitations by the end of the 20th century, but they did not. The limitations of phytoremediation were brought sharply into focus, but the number of end-users, publications, conferences, and grants that continue to focus on phytoremediation indicate that it has survived its first phase of testing. The chapters for this book demonstrate, I believe, some of the advances possible in phytoremediation and some of the contexts for them that the 21st century might provide. It has strengthened my belief that phytoremediation can become a more widely used environmental biotechnology. If it does become more widely used, it will confirm the foresight of the researchers who developed the fundamental concepts and the determination of those who carried out the first phase of experiments. It seems to me that the intensity of the scrutiny and the effort required during the first phase of tests of a technology can lead to despondency amongst its advocates. Continued optimism and obstinacy are required to develop technologies beyond their first phase. Young researchers often, but not exclusively, possess these qualities. In *Phytoremediation: Methods and Reviews*, I have endeavored to include many younger researchers who were not necessarily involved with the first investigations into phytoremediation, because it is they who will carry phytoremediation through its next phases of development.

Soil and water are clearly two of the most important natural resources sustaining terrestrial life on Earth. Degradation and pollution of them can have adverse consequences for ecosystems and hence, from an anthropocentric viewpoint, human food security and health. If terrestrial ecosystems of which humans are part are to be sustained, the degradation and pollution of soil and water have to be kept, at the very least, to a level that does not have devastating consequences. I believe that the research recounted in this book by a wide variety of authors shows that if we have the perseverance to improve and implement phytoremediation of soil and water, the technology has the potential to play a more important role in the sustainable use of these resources than is currently the case. Overall, therefore, I hope this book is helpful to those tackling the formidable challenge of realizing this potential.

Neil Willey

Contents

Preface v
Contributors xiii

PART I MANIPULATING PHENOTYPES AND EXPLOITING BIODIVERSITY

1. Genetic Engineering of Plants for Phytoremediation of Polychlorinated Biphenyls
   Shigenori Sonoki, Satoru Fujihiro, and Shin Hisamatsu 3

2. Increasing Plant Tolerance to Metals in the Environment
   Jennifer C. Stearns, Saleh Shah, and Bernard R. Glick 15

3. Using Quantitative Trait Loci Analysis to Select Plants for Altered Radionuclide Accumulation
   Katharine A. Payne, Helen C. Bowen, John P. Hammond, Corrina R. Hampton, Philip J. White, and Martin R. Broadley 27

4. Detoxification of Soil Phenolic Pollutants by Plant Secretory Enzyme
   Guo-Dong Wang and Xiao-Ya Chen 49

5. Using Real-Time Polymerase Chain Reaction to Quantify Gene Expression in Plants Exposed to Radioactivity
   Yu-Jin Heinekamp and Neil Willey 59

6. Plant Phylogeny and the Remediation of Persistent Organic Pollutants
   Jason C. White and Barbara A. Zeeb 71

7. Producing Mycorrhizal Inoculum for Phytoremediation
   Abdul G. Khan 89

8. Implementing Phytoremediation of Petroleum Hydrocarbons
   Chris D. Collins 99

9. Uptake, Assimilation, and Novel Metabolism of Nitrogen Dioxide in Plants
   Misa Takahashi, Toshiyuki Matsubara, Atsushi Sakamoto, and Hiromichi Morikawa 109

PART II MANIPULATING CONTAMINANT AVAILABILITY AND DEVELOPING RESEARCH TOOLS

10. Testing the Manipulation of Soil Availability of Metals
    Fernando Madrid Diaz and M. B. Kirkham 121

11 Testing Amendments for Increasing Soil Availability
    of Radionuclides
    Nicholas R. Watt *131*

12 Using Electrodics to Aid Mobilization of Lead in Soil
    David J. Butcher and Jae-Min Lim *139*

13 Stable Isotope Methods for Estimating the Labile Metal Content
    of Soils
    Andrew J. Midwood *149*

14 In Vitro Hairy Root Cultures as a Tool
    for Phytoremediation Research
    Cecilia G. Flocco and Ana M. Giulietti *161*

15 Sectored Planters for Phytoremediation Studies
    Chung-Shih Tang *175*

16 Phytoremediation With Living Aquatic Plants:
    Development and Modeling of Experimental Observations
    Steven P. K. Sternberg *185*

17 Near-Infrared Reflectance Spectroscopy: *Methodology
    and Potential for Predicting Trace Elements in Plants*
    **Rafael Font, Mercedes del Río-Celestino,
    and Antonio de Haro-Bailón** *205*

PART III CURRENT RESEARCH TOPICS IN PHYTOREMEDIATION

18 Using Hydroponic Bioreactors to Assess Phytoremediation
    Potential of Perchlorate
    Valentine Nzengung *221*

19 Using Plant Phylogeny to Predict Detoxification
    of Triazine Herbicides
    Sylvie Marcacci and Jean-Paul Schwitzguébel *233*

20 Exploiting Plant Metabolism for the Phytoremediation
    of Organic Xenobiotics
    Peter Schröder *251*

21 Searching for Genes Involved in Metal Tolerance, Uptake,
    and Transport
    **Viivi H. Hassinen, Arja I. Tervahauta,
    and Sirpa O. Kärenlampi** *265*

22 Manipulating Soil Metal Availability Using EDTA
    and Low-Molecular-Weight Organic Acids
    Longhua Wu, Yongming Luo, and Jing Song *291*

Contents

23 Soils Contaminated With Radionuclides: *Some Insights for Phytoextraction of Inorganic Contaminants*
Neil Willey 305

24 Assessing Plants for Phytoremediation of Arsenic-Contaminated Soils
Nandita Singh and Lena Q. Ma 319

PART IV CONTEXTS AND UTILIZATION OF PHYTOREMEDIATION

25 Phytoremediation in China: *Inorganics*
Shirong Tang 351

26 Phytoremediation in China: *Organics*
Shirong Tang and Cehui Mo 381

27 Phytoremediation of Arsenic-Contaminated Soil in China
Chen Tong-Bin, Liao Xiao-Yong, Huang Ze-Chun, Lei Mei, Li Wen-Xue, Mo Liang-Yu, An Zhi-Zhuang, Wei Chao-Yang, Xiao Xi-Yuan, and Xie Hua 393

28 Phytoremediation in Portugal: *Present and Future*
Cristina Nabais, Susana C. Gonçalves, and Helena Freitas 405

29 Phytoremediation in Russia
Yelena V. Lyubun and Dmitry N. Tychinin 423

30 Phytoremediation in India
M. N. V. Prasad 435

31 Phytoremediation in New Zealand and Australia
Brett Robinson and Chris Anderson 455

Index 469

Contributors

CHRIS ANDERSON • *Soil and Earth Sciences, Massey University, Palmerston North, New Zealand*
HELEN C. BOWEN • *Department of Plant Science, Warwick HRI, Wellesbourne, Warwick, UK*
MARTIN R. BROADLEY • *Division of Plant Sciences, University of Nottingham, Sutton Bonnington, UK*
DAVID J. BUTCHER • *Department of Chemistry and Physics, Western Carolina University, Cullowhee, NC*
WEI CHAO-YANG • *Institute of Geographic Sciences and Natural Resources Research, Chinese of Academy of Sciences, Beijing, China*
XIAO-YA CHEN • *Institute of Plant Physiology and Ecology, Chinese Academy of Sciences, Shanghai, China*
CHRIS D. COLLINS • *Reader in Soil Science, Department of Soil Science, School of Human and Evironmental Sciences, The University of Reading, Whiteknights, Reading, UK*
CECILIA G. FLOCCO • *Cátedra de Microbiología Industrial y Biotecnología, Facultad de Farmacia y Bioquímica, Universidad de Buenos Aires, Buenos Aires, Argentina*
RAFAEL FONT • *Department of Agronomy and Plant Breeding, Institute for Sustainable Agriculture, Córdoba, Spain*
HELENA FREITAS • *Department of Botany, University of Coimbra, Coimbra, Portugal*
SATORU FUJIHIRO • *Graduate School of Environmental Health, Azabu University, Kanagawa, Japan*
ANA M. GIULIETTI • *Cátedra de Microbiología Industrial y Biotecnología, Facultad de Farmacia y Bioquímica, Universidad de Buenos Aires, Buenos Aires, Argentina*
BERNARD R. GLICK • *Department of Biology, University of Waterloo, Waterloo, Ontario, Canada*
SUSANA C. GONÇALVES • *Department of Botany, University of Coimbra, Coimbra, Portugal*
JOHN P. HAMMOND • *Department of Plant Science, Warwick HRI, Wellesbourne, Warwick, UK*
CORRINA R. HAMPTON • *Department of Plant Science, Warwick HRI, Wellesbourne, Warwick, UK*

ANTONIO DE HARO-BAILÓN • *Department of Agronomy and Plant Breeding, Institute for Sustainable Agriculture, Córdoba, Spain*

VIIVI H. HASSINEN • *Institute of Applied Biotechnology, University of Kuopio, Kuopio, Finland*

YU-JIN HEINEKAMP • *Center for Research in Plant Science, University of the West of England, Bristol, UK*

SHIN HISAMATSU • *Graduate School of Environmental Health, Azabu University, Kanagawa, Japan*

XIE HUA • *Institute of Geographic Sciences and Natural Resources Research, Chinese of Academy of Sciences, Beijing, China*

SIRPA O. KÄRENLAMPI • *Institute of Applied Biotechnology, University of Kuopio, Kuopio, Finland*

ABDUL G. KHAN • *Department of Microbiology, Quaid-i-Azam University, Islamabad, Pakistan*

M. B. KIRKHAM • *Department of Agronomy, Kansas State University, Manhattan, KS*

MO LIANG-YU • *Institute of Geographic Sciences and Natural Resources Research, Chinese of Academy of Sciences, Beijing, China*

JAE-MIN LIM • *Department of Chemistry and Physics, Western Carolina University, Cullowhee, NC*

YONGMING LUO • *Soil and Environment Bioremediation Research Center, Institute of Soil Science, Chinese Academy of Sciences, Nanjing, China*

YELENA V. LYUBUN • *Institute of Biochemistry and Physiology of Plants and Microorganisms, Russian Academy of Sciences, Saratov, Russia*

LENA Q. MA • *Soil and Water Science Department, University of Florida, Gainesville, FL*

FERNANDO MADRID DIAZ • *Instituto de Recursos Naturales y Agrobiología de Sevilla, Consejo Superior de Investigaciones Científicas, Sevilla, Spain*

SYLVIE MARCACCI • *Laboratory for Environmental Biotechnology, Swiss Federal Institute of Technology, Lausanne, Switzerland*

TOSHIYUKI MATSUBARA • *Core Research for Evolutional Science and Technology, Japan Science and Technology Agency, Kawaguchi, Japan*

LEI MEI • *Institute of Geographic Sciences and Natural Resources Research, Chinese of Academy of Sciences, Beijing, China*

ANDREW J. MIDWOOD • *Analytical Group, The Macaulay Institute, Aberdeen, Scotland*

CEHUI MO • *Department of Environmental Engineering, Jinan University, Guangzhou, China*

HIROMICHI MORIKAWA • *Department of Mathematical and Life Sciences, Graduate School of Science, Hiroshima University, Higashi-Hiroshima, Japan*

Contributors

CRISTINA NABAIS • *Department of Botany, University of Coimbra, Coimbra, Portugal*
VALENTINE NZENGUNG • *Department of Geology, University of Georgia, Athens, GA*
KATHARINE A. PAYNE • *Department of Plant Science, Warwick HRI, Wellesbourne, Warwick, UK*
M. N. V. PRASAD • *Department of Plant Sciences, University of Hyderabad, Hyderabad, Andhra Pradesh, India*
MERCEDES DEL RÍO-CELESTINO • *Department of Agronomy and Plant Breeding, Institute for Sustainable Agriculture, Córdoba, Spain*
BRETT ROBINSON • *Swiss Federal Institute of Technology, Institute of Terrestrial Ecology, Zurich, Switzerland*
ATSUSHI SAKAMOTO • *Department of Mathematical and Life Sciences, Graduate School of Science, Hiroshima University, Higashi-Hiroshima, Japan*
PETER SCHRÖDER • *Department of Rhizosphere Biology, Institute for Soil Ecology, National Research Center for Environment and Health, Neuherberg, Germany*
JEAN-PAUL SCHWITZGUÉBEL • *Laboratory for Environmental Biotechnology, Swiss Federal Institute of Technology, Lausanne, Switzerland*
SALEH SHAH • *Alberta Research Council, Vegreville, Alberta, Canada*
NANDITA SINGH • *Department of Biomass Biology and Environmental Sciences, National Botanical Research Institute, Lucknow, India*
JING SONG • *Soil and Environment Bioremediation Research Center, Institute of Soil Science, Chinese Academy of Sciences, Nanjing, China*
SHIGENORI SONOKI • *Graduate School of Environmental Health, Azabu University, Kanagawa, Japan*
JENNIFER C. STEARNS • *Department of Biology, University of Waterloo, Waterloo, Ontario, Canada*
STEVEN P. K. STERNBERG • *Chemical Engineering Department, University of Minnesota, Duluth, MN*
MISA TAKAHASHI • *Department of Mathematical and Life Sciences, Graduate School of Science, Hiroshima University, Higashi-Hiroshima, Japan*
CHUNG-SHIH TANG • *Department of Molecular Biosciences and Bioengineering, University of Hawaii, Honolulu, HI*
SHIRONG TANG • *College of Environmental Sciences and Technology, Guangzhou University, Guangzhou, China*
ARJA I. TERVAHAUTA • *Institute of Applied Biotechnology, University of Kuopio, Kuopio, Finland*
CHEN TONG-BIN • *Institute of Geographic Sciences and Natural Resources Research, Chinese of Academy of Sciences, Beijing, China*
DMITRY TYCHININ • *Institute of Biochemistry and Physiology of Plants and Microorganisms, Russian Academy of Sciences, Saratov, Russia*

GUO-DONG WANG • *Institute for Plant Physiology and Ecology, Chinese Academy of Sciences, Shanghai, China*
NICHOLAS R. WATT • *British Nuclear Group, Berkeley Center, Gloucestershire, UK*
LI WEN-XUE • *Institute of Geographic Sciences and Natural Resources Research, Chinese of Academy of Sciences, Beijing, China*
JASON C. WHITE • *Department of Soil and Water, The Connecticut Agricultural Experiment Station, New Haven, CT*
PHILIP J. WHITE • *Environment-Plant Interactions Program, Scottish Crops Research Institute, Dundee, Scotland*
NEIL WILLEY • *Center for Research in Plant Science, University of the West of England, Bristol, UK*
LONGHUA WU • *Soil and Environment Bioremediation Research Center, Institute of Soil Science, Chinese Academy of Sciences, Nanjing, China*
LIAO XIAO-YONG • *Institute of Geographic Sciences and Natural Resources Research, Chinese of Academy of Sciences, Beijing, China*
XIAO XI-YUAN • *Institute of Geographic Sciences and Natural Resources Research, Chinese of Academy of Sciences, Beijing, China*
HUANG ZE-CHUN • *Institute of Geographic Sciences and Natural Resources Research, Chinese of Academy of Sciences, Beijing, China*
BARBARA A. ZEEB • *Department of Chemistry and Chemical Engineering, Royal Military College of Canada, Ontario, Canada*
AN ZHI-ZHUANG • *Institute of Geographic Sciences and Natural Resources Research, Chinese of Academy of Sciences, Beijing, China*

I

MANIPULATING PHENOTYPES AND EXPLOITING BIODIVERSITY

1

Genetic Engineering of Plants for Phytoremediation of Polychlorinated Biphenyls

Shigenori Sonoki, Satoru Fujihiro, and Shin Hisamatsu

Summary

Phytoremediation is an emerging technology that uses certain plants to clean up soil, water, and air contaminated with environmental pollutants such as polychlorinated biphenyls (PCBs) through degradation, extraction, or immobilization of contaminants. This technology has been receiving attention lately as an innovative, cost-effective, and long-term alternative to the more established engineering methods used at hazardous waste sites. This chapter describes methods for the construction of two kinds of transgenic plants useful for the remediation of environments polluted by PCBs. The first one is an enhancer-trap Ac/Ds transposon-tagging transformant of *Arabidopsis thaliana*. This contains the nonautonomous mobile transposable element (Ds transposon) into which a β-glucuronidase (*GUS*) reporter gene is inserted. The miniature promoter fused to a *GUS* reporter gene can drive gene expression only when the Ds transposon-containing *GUS* reporter gene is moved near the enhancer region of the gene in the plant genome. Thus, the gene(s) involved with the catabolism of PCBs are expected to be found through monitoring the change of reporter gene expression. The second transgenic *A. thaliana* has the gene for a lignin-degrading enzyme from white-rot fungi in the genomic DNA and is expected to be useful for direct degradation of PCBs.

Key Words: PCBs; *Agrobacterium tumefaciens*; *Arabidopsis thaliana*; transgenic plant; Ac/Ds transposon; *GUS* reporter gene; white-rot fungil lignin-degrading enzyme.

1. Introduction

It has been of great concern that halogenated aromatic hydrocarbons can disrupt the endocrine system of animals. They include numerous polychlorinated biphenyls (PCBs), which, because of their nonflammability, chemical stability, high boiling point, and electrical insulating properties, were used for industrial and commercial applications from the 1960s. Although the use of PCBs was widely prohibited more than 39 yr ago because of their high toxicity, a significant amount of them are still detected even now in almost all environments

(1–4). Among the possible 209 PCB congeners, the environmental toxicity of coplanar PCBs (Co-PCBs) is becoming more severe, especially in Japan. To clean up the polluted environment, biological remediation systems using plants, "phytoremediation," is expected to solve the environmental pollution problem. This chapter introduces novel methods for the construction of two kinds of transgenic plants engineered to remediate environments polluted by PCBs *(5)*. They both use *Arabidopsis thaliana*, a widely used model organism in plant biology. *A. thaliana* is a small, flowering plant that is a member of the mustard (Brassicaceae) family and it offers important advantages for basic research in genetics and molecular biology because of its small, sequenced genome (125 megabases/5 chromosomes) estimated to have about 26,000 genes (at sequence completion in 2000). A rapid life cycle (about 6 wk from germination to mature seed) and easy cultivation in restricted space are further advantages.

Because, in contrast to mobile animals, plants cannot move, they possess a unique genetic inheritance, gained in the long process of evolution. This inheritance consists especially of characters adapting plants for suboptimal environmental conditions including high temperature, cold, dehydration, high salt concentration, xenobiotics, and so on. This led us to suspect the existence of special gene(s) in the genome of *A. thaliana*, particularly impacting on the catabolism of PCBs. To look for such gene(s), the first transgenic plant described here, an enhancer-trap Ac/Ds transposon-tagging line of *A. thaliana*, was used. This kind of transgenic plant contains the nonautonomous mobile transposable element (Ds transposon) into which the reporter gene for β-glucuronidase (*GUS*) is inserted. This *GUS* reporter gene is connected with the miniature 35S promoter from the cauliflower mosaic virus and is driven with the aid of an enhancer region of the gene only when the Ds transposon containing the *GUS* reporter gene is moved in the vicinity of the gene in the plant genome. Insertional mutagenesis using these "enhancer traps" involves generating a large number of lines of *A. thaliana* that have the reporter gene integrated at different sites throughout the genome. The gene(s) that respond to the stress of PCBs are, therefore, expected to be found through monitoring the change of reporter gene expression using a large number of enhancer trap lines of *A. thaliana*.

The second novel transgenic plant described here is constructed using a different concept. The degradation of a variety of environmentally persistent pollutants such as chlorinated aromatic compounds, polyaromatic hydrocarbons, and synthetic high polymers by Basidiomycetes, such as white-rot fungi, has been extensively studied in the process of lignin degradation *(6–9)*. As a result, unique extracellular oxidative enzymes, namely, lignin peroxidase (LiP), manganese-dependent peroxidase (MnP) and laccase (Lac) produced by white-rot fungi were found to be responsible for degrading a wide variety of organic recalcitrants in addition to lignin. These oxidative enzymes are potentially useful in the bioremediation of PCBs. Actually, the lignin-degrading enzymes Lip, Mnp, and

Lac produced by the white-rot fungus *Phanerochaete chrysosporium* have been well examined for their ability to degrade PCB congeners *(10–12)*. Here, a method is described for the construction of transgenic *A. thaliana* that involves the genes for each of the lignin-degrading enzymes Lip, Mnp, and Lac from white-rot fungi, being introduced into the genomic DNA to make transgenic plants that might be utilized in remediating environments polluted by harmful chemicals including PCBs.

2. Materials
2.1. Plants, Bacteria, Fungi, Transposon, and Plasmid

1. Seeds of *A. thaliana* (ecotype: Columbia) are obtained from the SENDAI Arabidopsis Seed Stock Center, Japan.
2. The competent cell of *A. tumefaciens* LBA 4404 is purchased from Life Technologies, MD, and also the competent cell of *Escherichia coli* DH5α from Invitrogen Japan K. K., Tokyo, Japan (*see* **Note 1**).
3. *P. chrysosporium* (UAMH 3641) and *Trametes versicolor* (UAMH 8272) are purchased from University of Alberta Microfungus Collection, Canada (www.devonian.ualberta.ca/uamh).
4. Ds-GUS-T-DNA and Ac-T-DNA, both of which are individually integrated in pCGN binary vector plasmid for plant transformation, are kindly gifted by Dr. Nina V. Fedoroff, The Pennsylvania State University *(13,14)*. Ds-GUS-T-DNA/pCGN includes the Hm-resistant (Hm^r) gene and *GUS* gene connected with the miniature 35S promoter in the nucleotide sequence of the Cs-resistant (Cs^r) gene, and also carries the Km-resistant (Km^r) gene. Ac-T-DNA/pCGN harbors a transposase gene in addition to Km^r gene (*see* **Note 2**).
5. pEGAD, the binary vector plasmid for plant transformation, is obtained from the Arabidopsis Biological Resource Center (ABRC) through TAIR (http://www.arabidopsis.org/) (*see* **Note 3**).

2.2. Culture Media

All culture media are autoclaved for 20 min at 121°C and stored at room temperature or 4°C for up to 1 mo without change.

1. SOC medium: 2% (w/v) tryptone, 0.5% (w/v) yeast extract, 10 mM NaCl, 2.5 mM KCl, 10 mM MgCl$_2$, 10 mM MgSO$_4$, and 20 mM glucose (pH 7.0).
2. LB (Luria-Bertani) medium: 1% (w/v) tryptone, 0.5% (w/v) yeast extract, and 10 mM NaCl (pH 7.0).
3. YM (yeast mannitol) medium: 0.04% (w/v) yeast extract, 1% (w/v) mannitol, 1.7 mM NaCl, 0.8 mM MgSO$_4$, and 2.2 mM K$_2$HPO$_4$ (pH 7.0).
4. YEB (yeast beef) medium: 0.5% (w/v) sucrose, 0.5% (w/v) peptone, 0.5% (w/v) beef extract, 0.1% (w/v) yeast extract, and 0.2 mM MgSO$_4$ (pH 7.2).
5. MS (Murashige-Skoog) medium: 4.41 g/L basal MS medium *(15)* (ICN Biomedicals, Inc., OH; cat. no. 26-100-22), 1% (w/v) sucrose (pH 5.6–5.8). When necessary, 0.2 g Gelrite (Wako Pure Chemical Industries, LTD, Osaka, Japan) is added for 100 mL solid plate medium.

6. Infiltration MS medium, solution I: 2.2 g/L basal MS medium, 5% (w/v) sucrose, 0.05% (w/v) MES (2-morpholinoethanesulfonic acid monohydrate), 112 mg/L Gamborg's B5 vitamins containing 100 mg myo-inositol, 10 mg thiamine-HCl, 1 mg nicotinic acid, and 1 mg pyridoxine-HCl (pH 5.7). Solution I is autoclaved first, and then other filter-sterilized components are added to 1000 mL solution I as follows: 10 µL of 1 mg/mL benzylaminopurine dissolved in dimethylsulfoxide and 0.02% Silwet L-77 (Osi Specialties, Inc., A Witco Company, NY).
7. Kirk culture medium for white-rot fungi: 1% (w/v) glucose, 1 g/L KH_2PO_4, 1 g/L $Ca(H_2PO_4)_2$, 221 mg/L ammonium tartrate, 500 mg/L $MgSO_4·7H_2O$, 1 mg/L thiamine-HCl, and 10 mL Kirk mineral solution *(16)* (pH 4.5).

2.3. Chemicals and Reagents

All chemicals used are of analytical grade from commercial sources, unless otherwise stated.

1. Antibiotics: all antibiotic solutions such as kanamycin (Km), hygromycin (Hm), chlorsulfuron (Cs), and gentamycin (Gm) are dissolved in sterile distilled water at the 10^3-fold concentration, filter-sterilized, divided into portions for one use, and then stored under −20°C for up to 6 mo without change.
2. GUS active-staining reagent mix: 0.1 mol/L sodium phosphate, 0.01 *M* EDTA, 0.5 *M* each potassium ferri- and ferrocyanide, 1 m*M* X-Gluc (5-bromo-4-chloro-3-indolyl-β-glucuronide), and 0.1% Triton X-100 (optional) (pH 7.0).
3. Co-PCBs: in this study 12 congeners of Co-PCB are used among the 209 PCB congeners. The Co-PCBs are purchased from AccuStandard Inc., New Haven, CT under the restriction of law. The commercial product contains 5 µg/mL each of 12 congeners dissolved in dimethylsulfoxide and the components are as follows: 3,3′,4,4′- and 3,4,4′,5-tetrachlorobiphenyl, 2,3,3′,4,4′-, 2,3,4,4′,5-, 2,3′,4,4′,5-, 2′,3,4,4′,5- and 3,3′,4,4′,5-pentachlorobiphenyl, 2,3,3′,4,4′,5-, 2,3,3′,4,4′,5′-, 2,3′,4,4′,5,5′- and 3,3′,4,4′,5,5′-hexachlorobiphenyl and 2,3,3′,4,4′,5,5′-heptachlorobiphenyl. Because PCBs are harmful compounds, be sure to handle them very carefully with some effective protection. When not in use keep them at 4°C in tightly closed container.

3. Methods

3.1. Searching for Gene(s) Involved With Catabolism of PCBs Using Enhancer-Trap Transposon-Tagging Lines of A. thaliana

3.1.1. Construction of Enhancer-Trap Ac/Ds Transposon-Tagging Lines of A. thaliana

3.1.1.1. PREPARATION OF pCGN BINARY VECTOR PLASMID INVOLVING Ds-GUS-T-DNA OR Ac-T-DNA

1. Mix Ds-Gus-T-DNA/pCGN or Ac-T-DNA/pCGN and 1 µL competent cell solution of *E. coli* DH5α in a microcentrifuge tube.
2. Pipet the mixture briefly and then leave it for 30 min on ice.
3. Incubate the mixture for 40 s at 42°C, quickly after that cool it for 2 min on ice.
4. Add 90 µL SOC medium to the mixture, and then incubate the mixture for 1 h at 37°C.

Genetic Engineering of Plants

5. Spread the mixture over the LB plate medium containing 10 µg/mL Gm, and then grow bacterial cells for 16 h at 37°C.
6. Select the proper Gm-resistant colony, transfer it to 5 mL LB liquid medium, and then incubate overnight at 37°C.
7. Prepare and purify the pCGN binary vector plasmid, and then confirm the plasmid with 0.7% agarose gel electrophoresis at 100 V for 1 h. (According to the QIAGEN manufacturer's instructions, the preparation and purification of pCGN binary vector plasmid were carried out using QIAprep Spin Miniprep Kit.)
8. Precipitate the vector plasmid with 70% ethanol, and then wash the precipitated plasmid with 70% ethanol three times.
9. Dry the plasmid under vacuum at room temperature, and then resuspend it with sterile distilled water at the final concentration of 0.1 µg/µL.

3.1.1.2. INSERTION OF Ds-GUS-T-DNA OR Ac-T-DNA INTO THE GENOME OF A. TUMEFACIENS USING AN ELECTROPORATION SYSTEM (SEE NOTE 4)

1. Mix 1 µL of 0.1 µg/µL pCGN binary vector plasmid involving Ds-GUS-T-DNA or Ac-T-DNA and 20 µL competent cell solution of A. *tumefaciens* in a well-cooled microcentrifuge tube on ice.
2. Pipet the mixture thoroughly, and then transfer it to a cuvet that has been well chilled at –20°C overnight.
3. Place the cuvet on a Gene Pulser apparatus for electroporation (Bio-Rad Japan Laboratories, Tokyo, Japan).
4. Perform electroporation at the electrical field of 2.5 kV/cm with pulse lengths of 5 ms.
5. Add 1 mL YM medium to the electroporated solution, and leave it for 3 h at 28°C (*see* **Note 5**).
6. Spread the solution over the YM plate medium containing 10 µg/mL Gm for the selection of transformed A. *tumefaciens*, and then grow cells for 2 d at 28°C.
7. Select some Gm-resistant colonies, and transfer each of them to 5 mL YEB liquid medium and incubate overnight at 28°C.
8. Keep each culture individually with glycerol at –80°C until use.

3.1.1.3. TRANSFORMATION OF A. THALIANA USING A VACUUM INFILTRATION SYSTEM (SEE NOTE 6)

1. Culture the transformed A. *tumefaciens* harboring Ds-GUS-T-DNA/pCGN or Ac-T-DNA/pCGN individually in 5 mL YEB liquid medium at 28°C.
2. Dilute 5 mL cultures 100-fold into YEB medium containing 10 µg/mL Gm for large-scale culture, and then grow cultures overnight at 28°C with shaking until culture OD_{600} becomes over 2.0.
3. Harvest cells by centrifugation at 5000 rpm (approx 3000g) for 15 min at room temperature.
4. Resuspend cells in 1 L infiltration MS medium to an OD_{600} of 0.8.
5. Place resuspended culture in a beaker inside a vacuum desiccator.
6. Invert 4- to 6-wk-old plants, grown in soil-mounded pots at 26°C on a 16 h light/8 h dark cycle, into the suspension culture so that the entire plant is immersed, including rosette, but not too much of the soil is submerged (*see* **Note 7**).

7. Draw a vacuum of 400 mmHg, and then let the plants stay under this vacuum level for 10 min.
8. Release the vacuum quickly, and then remove the pots from the suspension.
9. Place the pots horizontally in a tray lined with a paper pad for drainage, cover the tray with plastic wrap to maintain high humidity, and then place the tray back in a growth chamber.
10. Uncover the pots the next day and set them upright, allow plants to grow to maturity, and then harvest seeds from each plant individually.
11. Sterilize seeds with 70% ethanol, sodium hypochlorite solution (available chlorine: 2%), and then rinse them in sterile distilled water.
12. Sow seeds of each plant infiltrated by *A. tumefaciens* harboring Ds-GUS-T-DNA/pCGN or Ac-T-DNA/pCGN on MS solid medium containing 20 μg/mL Hm and 50 μg/mL Km, or 50 μg/mL Km.
13. Select the Hm^r and Km^r transformants involving single-copy Ds-Gus-T-DNA or Km^r transformants involving single-copy Ac-T-DNA, respectively.

3.1.1.4. CROSSING BETWEEN DS-GUS-T-DNA AND AC-T-DNA TRANSFORMANTS AND SELECTION OF GUS TRANSPOSON-TAGGING LINES

1. Sow seeds of Ds-GUS-T-DNA or Ac-T-DNA transformants as the pollen parent or egg parent, respectively.
2. Choose an inflorescence in 4- to 6-wk-old plants, and then remove all the flowers that are too young (too small) and the ones that already show white petals (*see* **Note 8**).
3. Remove all the flower parts except the ovary using very fine forceps cleaned by dipping them in 95% ethanol followed by distilled water in between flowers.
4. Choose fully mature flowers, remove the stamens, and then brush the prepared ovaries several times with these stamen to complete pollination on top of the ovary.
5. Cover the ovaries with a plastic wrap to avoid cross-contamination.
6. Leave ovaries developing until they start browning, and then harvest seeds carefully before shedding seeds.
7. Sow F_1 seeds on MS solid medium containing 20 μg/mL Hm to select Hm^r F_1 plants, grow Hm^r F_1 plants, and then allow them to self-pollinate to collect F_2 seeds.
8. Sow F_2 seeds on MS solid medium containing 20 μg/mL Hm and 6 μg/mL Cs to select Hm^r and Cs^r F_2 plants in which the occurrence of excision and transposition of GUS transposon on the genome is expected.
9. Score the plants for acquisition of resistance to Hm and Cs.

3.1.1.5. SELECTION OF ENHANCER-TRAP AC/DS-GUS TRANSPOSON-TAGGING LINES USING A GUS HISTOCHEMICAL-STAINING METHOD

1. Sow F_3 seeds harvested from Hm^r and Cs^r F_2 plants.
2. Place 2-wk-old plants to be stained in a test tube filled with enough GUS active-staining reagent mix to completely cover the whole plants.
3. Wrap the whole test tube with an aluminum foil to shade it, and then draw a vacuum of 400 mmHg for about 5 min until air bubbles are created in a vacuum desiccator.
4. Release the vacuum, cover the test tube to prevent evaporation of the reagent mix, and then incubate at 37°C overnight.

Genetic Engineering of Plants 9

5. Transfer the stained plants to a test tube filled with 70% ethanol, and then incubate overnight to fix GUS-stained parts in the plants and to decolorize chlorophyll to distinguishing blue GUS color from the green color of chlorophyll.
6. Score the plants that are stained blue in the Hm^r and Cs^r F_3 plants.

3.1.2. Searching for the Gene(s) Responding to the Stress of PCBs

3.1.2.1. PCB EXPOSURE AND SCREENING OF ENHANCER-TRAP TRANSPOSON LINES RESPONDING TO PCBS (*SEE* **NOTE 9**)

1. Prepare two kinds of sterile MS solid media in the Petri dish containing 5 ng/mL PCBs or 1 µL/mL DMSO as a control, respectively.
2. Sterilize each seed of enhancer-trap transposon lines that are obtained as in **Subheading 3.1.1.5.** with 70% ethanol, sodium hypochlorite solution (available chlorine: 2%), and then rinse them in sterile distilled water.
3. Sow 10 seeds of each enhancer-trap transposon line on MS solid medium containing PCBs or DMSO, respectively.
4. Use 2-wk-old plants grown on each PCBs/MS or DMSO/MS medium at 26°C on a 16 h light/8 h dark cycle for histochemical staining of GUS-active sites with the same method as in **Subheading 3.1.1.5.**
5. Monitor the change of GUS expression between PCBs exposed and control plants to capture enhancer-trap transposon line(s) responds to the stress of PCBs.

3.1.2.2. IDENTIFICATION OF THE GENE(S) RESPONDING TO THE STRESS OF PCBS

1. Prepare the genomic DNA from enhancer-trap transposon line(s) responding to the stress of PCBs using cetyltrimethylammonium bromide *(17)*.
2. Prepare two sets of inverse PCR primers in accordance with the nucleotide sequence of GUS, digest the genomic DNA by restriction enzyme, *Sna*BI (which has one cutting site in the GUS sequence), and then perform inverse PCR after the self-ligation of DNA fragments *(18)*.
3. Perform cloning and sequencing of inverse PCR products using *Taq* DyeDeoxy terminator cycle-sequencing reactions (*see* **Note 10**).
4. Identify the genes using the sequence database of The Arabidopsis Information Resource (TAIR) (http://www.arabidopsis.org/) on the basis of the nucleotide sequence determined at **step 3**.

3.2. Construction of Transgenic A. thaliana With the Gene for Lignin-Degrading Enzyme From White-Rot Fungi

3.2.1. Preparation of Full-Length cDNA for Lignin-Degrading Enzymes

1. Culture each of the white-rot fungi *P. chrysosporium* (UAMH 3641) and *T. versicolor* (UAMH 8272) in 50 mL Kirk culture medium for 5 d at 37°C without shaking.
2. Prepare the total RNA from the culture of *P. chrysosporium* for cDNA of Lip and MnP, and from *T. versicolor* for cDNA of Lac with the intact total RNA isolation system using guanidine thiocyanate as the potent inhibitors of RNase and denaturing reagent for nucleoprotein complexes, and lithium chloride to preferentially precipitate RNA *(19,20)*.

3. Prepare the total cDNA from the total RNA using a reverse transcriptase with an oligo-dT primer only for the transcription of mRNA.
4. Design forward and reverse primers for PCR of full-length cDNA of each Lip, Map, and Lac according to the nucleotide sequence of full-length mRNA derived from DNA Data Bank of Japan (http://www.ddbj.nig.ac.jp/Welcome.html) as follows:

    Lip: Forward/5'-ATGGCCTTCAAGCAGCTCTTCGCA-3'
           Reverse/5'-TCATTTAAGCACCCGGAGGCGGA-3'
    Mnp: Forward/5'-ATGGCCTTCGGTTCTCTCCTCG-3'
           Reverse/5'-TTAGGCAGGGCCATCGAACTGAACA-3'
    Lac: Forward/5'-ATGGGCAAGTTTCACTCTTTTGTGAACGTC-3'
           Reverse/5'-TCAGAGGTCGGACGAGTCCAAAG-3'

5. Perform PCR in the following way: preheating at 98°C for 3 min; first cycle reactions at 98°C for 10 s, 60°C for 10 s, 72°C for 90 s/5 cycles; second cycle reactions at 98°C for 10 s, 72°C for 90 s/30 cycles; extension at 72°C for 7 min.
6. Purify the PCR products, and then confirm the PCR bands with 0.7% agarose gel electrophoresis at 100 V for 1 h (*see* **Note 11**). (According to the QIAGEN manufacturer's instructions, the purification of PCR products was carried out using QIAquick PCR Purification Kit.)
7. Determine the nucleotide sequence of PCR products using the DNA sequencing system described in **Subheading 3.1.2.2.**
8. Measure the concentration of PCR products and save them at –20°C until use.

3.2.2. Cloning of the cDNA for Each Lip, Mnp, and Lac in the Genome of A. tumefaciens

3.2.2.1. CONSTRUCTION OF pEGAD INVOLVING THE GENE FOR EACH LIP, MNP, AND LAC

A universal cloning method based on the site-specific recombination system of bacteriophage λ reported by Landy is carried out using Gateway Cloning Technology of Invitrogen in accordance with the manufacturer's instructions (http://www.invitrogen.com/content/sfs/manuals/gatewayman.pdf) *(21)*.

3.2.2.2. INSERTION OF EACH GENE FOR LIP, MNP, AND LAC INTO THE GENOME OF A. TUMEFACIENS

1. Mix 1 µL of 0.1 µg/µL pEGAD involving the gene for each Lip, Mnp, and Lac and 20 µL competent cell solution of *A. tumefaciens* in a well-cooled microcentrifuge tube on the ice.
2. Follow the procedures described in **Subheading 3.1.1.2.** except that BASTA is used instead of Gm for the selection of transformed *A. tumefaciens*.

3.2.2.3. TRANSFORMATION OF A. THALIANA

1. Culture the transformed *A. tumefaciens* prepared as in **Subheading 3.2.2.2.** in 5 mL YEB liquid medium at 28°C.

Genetic Engineering of Plants

2. Dilute 5 mL cultures 100-fold into YEB medium containing 10 µg/mL BASTA for large-scale overnight culture.
3. Take the same procedures as in **Subheading 3.1.1.3.** except that BASTA is used as a substitute for some antibiotics for the selection of transformants.

4. Notes

1. *A. tumefaciens,* which causes crown gall disease of a wide range of dicotyledonous (broad-leaved) plants, has been used extensively for genetic engineering of plants. The basis of *Agrobacterium*-mediated genetic engineering is that the part of the DNA (T-DNA) on a large plasmid (Ti plasmid) of *A. tumefaciens* is excised and integrated into the plant genome through the infection process by this bacterium. Thus, any foreign DNA inserted into the T-DNA region will also be cointegrated *(22)*.
2. Reporter genes of enhancer-trap lines have a minimal promoter that is only expressed when inserted near the cis-acting element of chromosomal enhancer regions, on the other hand, reporter genes of gene trap lines have no promoter, so that reporter gene expression can occur only when the reporter gene inserts within a transcribed chromosomal gene, creating a transcriptional fusion. Thus, it is expected that the enhancer-trap lines work with great efficiency to capture many chromosomal genes compared with the gene trap lines, and furthermore transposons have an important advantage in that their genome integration typically occurs in a simple and single-copy insertion manner. T-DNA, on the other hand, tends to be integrated in multiple arrays. The Ds-GUS transposon-tagging lines are now made available from the seed stocks of ABRC through TAIR (http://www.arabidopsis.org/).
3. pEGAD includes the gene for green fluorescent protein from luminescent jellyfish as a reporter gene and the gene for BASTA that is a herbicide for selection of transformants.
4. A freeze–thaw method in the presence of calcium chloride for *A. tumefaciens* transformation is also very convenient and efficient compared with the electroporation system because it does not require the special apparatus for electroporation; however, chemical transformation seems to be less efficient than the electroporation system *(23)*.
5. Use of YM medium in electroporation of *A. tumefaciens* results in an increase in the transformation efficiency when compared with the use of LB medium.
6. There are some other methods available to deliver foreign genes into the plant genome apart from the vacuum infiltration system used in this study, such as an *Agrobacterium*-mediated leaf disk transformation *(24)* and a particle bombardment delivery system *(25)*; however, the vacuum infiltration system has the advantage of working well to gain the transformant seeds directly and quickly from vacuum-infiltrated plants. The *Agrobacterium*-mediated leaf disk or particle bombardment transformation method, on the other hand, requires a regeneration step through undifferentiated embryogenic calli tissue formation to get the transformed seeds.
7. To improve the efficiency for transformation, be sure to water plants well the day before infiltration so that the stomata will be open that day and make sure no large bubbles are trapped under the plant when in filtration.

8. A pair of sharp forceps is an indispensable tool to remove the flower parts carefully and ×3.5 headband magnifiers (Lehkle Seeds, TX) are successfully used to remove parts without fail. If the flowers are open and the stamen are already mature with pollen, do not use that flower because the opening flowers will tend to have started self-fertilization.
9. Because PCBs are significant health hazards, wear the appropriate protective clothing—gloves made of chemically resistant materials and an organic vapor mask to ensure that researcher's exposure to PCBs is minimized. In case of contact, immediately wash skin with plenty of soap and water for at least 15 min.
10. In this study the sequencing is carried out by the Thermo Sequenase Cycle Sequencing Kit (Shimadzu, Tokyo, Japan), following the protocol of the manufacturer, using a DNA sequencer (model DSQ-1000L, Shimadzu).
11. If the PCR products can not be seen by agarose gel electophoresis after performing PCR because of small amounts of the mRNA of interest, try the second PCR using 100-fold diluted first PCR products as a template to successfully detect the PCR bands.

References

1. Ohsaki, Y., Matsueda, T., and Ohno, K. (1995) Levels and source of non-ortho coplanar polychlorinated biphenyls, polychlorinated dibenzo-p-dioxins and polychlorinated dibenzofurans in pond sediments and paddy field soil. *Water Res.* **29,** 1379–1385.
2. Ohsaki, Y., Matsueda, T., and Kurokawa, Y. (1997) Distribution of polychlorinated dibenzo-p-dioxins, polychlorinated dibenzofurans and non-ortho coplanar polychlorinated biphenyls in river and offshore sediments. *Environ. Pollut.* **96,** 79–88.
3. Kurokawa, Y., Matsueda, T., Nakamura, M., Takada, S., and Fukamachi, K. (1996) Characterization of non-ortho coplanar PCBs, polychlorinated dibenzo-p-dioxins and dibenzofurans in the atmosphere. *Chemosphere* **32,** 491–500.
4. Soong, D. K. and Ling, Y. C. (1997) Reassessment of PCDD/DFs and Co-PCBs toxicity in contaminated rice-bran oil responsible for the disease "Yu-Cheng". *Chemosphere* **34,** 1579–1586.
5. Sonoki, S., Kobayashi, A., Matsumoto, S., Nitta, S., and Hisamatsu, S. (2000) Analysis of plant gene response to the stress of coplanar PCB using transgenic *Arabidopsis thaliana. Organohal. Comp.* **49,** 450–454.
6. Higson, F. K. (1991) Degradation of xenobiotics by white rot fungi. *Rev. Environ. Contam. Toxicol.* **122,** 111–152.
7. Aust, S. D. and Benson, J. T. (1993) The fungus among us: use of white rot fungi to biodegrade environmental pollutants. *Environ. Health Perspect.* **101,** 232–233.
8. Barr, D. P. and Aust, S. D. (1994) Pollutant degradation by white rot fungi. *Rev. Environ. Contam. Toxicol.* **138,** 49–72.
9. Levin, L., Jordan, A., Forchiassin, F., and Viale, A. (2001) Degradation of anthraquinone blue by *Trametes trogii. Rev. Argent Microbiol.* **33,** 223–228.
10. Novotny, C., Vyas, B. R., Erbanova, P., Kubatova, A., and Sasek, V. (1997) Removal of PCBs by various white rot fungi in liquid cultures. *Folia Microbiol (Praha).* **42,** 136–140.

11. Krcmar, P. and Ulrich, R. (1998) Degradation of polychlorinated biphenyl mixtures by the lignin-degrading fungus *Phanerochaete chrysosporium*. *Folia Microbiol (Praha)*. **43,** 79–84.
12. Beaudette, L. A., Davies, S., Fedorak, P. M., Ward, O. P., and Pickard, M. A. (1998) Comparison of gas chromatography and mineralization experiments for measuring loss of selected polychlorinated biphenyl congeners in cultures of white rot fungi. *Appl. Environ. Microbiol.* **64,** 2020–2025.
13. Fedoroff, N. V. and Smith, D. L. (1993) A versatile system for detecting transposition in *Arabidopsis*. *Plant J.* **3,** 273–289.
14. Smith, D., Yanai, Y., Liu, Y. G., et al. (1996) Characterization and mapping of Ds-GUS-T-DNA lines for targeted insertional mutagenesis. *Plant J.* **10,** 721–732.
15. Murashige, T. and Skoog, F. (1962) A revised medium for rapid growth and bioassays with tobacco tissue cultures. *Physiol. Plant.* **15,** 473–497.
16. Kirk, T. K., Schultz, E., Connors, W. J., Lorenz, L. F., and Zeikus, J. G. (1978) Influence of culture parameters on lignin metabolism by *Phanerochaete chrysosporium*. *Arch. Microbiol.* **117,** 277–285.
17. Rogers, S. O. and Bendich, A. J. (1985) Extraction of DNA from milligram amounts of fresh, herbarium and mummified plant tissues. *Plant Mol. Biol.* **5,** 69–76.
18. Ochman, H., Medhora, M. M., Garza, D., and Hartle, D. L. (1990) Amplification of flanking sequences by inverse PCR. In: *PCR Protocols: A Guide to Methods and Application* (Innis, M. A., Gelfand, D. H., Sninsky, J. J., and White, T. J., eds.), Academic Press Inc., London, pp. 219–227.
19. Chirgwin, J. M., Przybyla, A. E., MacDondald, R. J., and Rutter, W. J. (1979) Isolation of biologically active ribonucleic acid from sources enriched in ribonuclease. *Biochem.* **24,** 5294–5299.
20. Cathala, G., Savouret, J., Mendez, B., et al. (1983) A method for isolation of intact, translationally active ribonucleic acid. *DNA* **2,** 329–335.
21. Landy, A. (1989) Dynamic, structural and regulatory aspects of lambda site-specific recombination. *Ann. Rev. Biochem.* **58,** 913–949.
22. Van Montagu, M. and Schell, J. (1982) The Ti plasmids of *Agrobacterium*. *Curr. Top. Microbiol. Immunol.* **96,** 237–254.
23. Cui, W., Liu, W., and Wu, G. (1995) A simple method for the transformation of *Agrobacterium tumefaciens* by foreign DNA. *Chin. J. Biotechnol.* **11,** 267–274.
24. Curtis, I. S., Davey, M. R., and Power, J. B. (1995) Leaf disk transformation. *Methods Mol. Biol.* **44,** 59–70.
25. Sagi, L., Panis, B., Remy, S., et al. (1995) Genetic transformation of banana and plantain (*Musa* spp.) via particle bombardment. *Biotechnol.* **13,** 481–485.

2

Increasing Plant Tolerance to Metals in the Environment

Jennifer C. Stearns, Saleh Shah, and Bernard R. Glick

Summary

An effective metal phytoremediation strategy depends on the ability of plants to tolerate and accumulate metals from the environment. Metals in soil and water exert a stress on plants that is detectable at the organismic and cellular level, and as a consequence of this stress, plant ethylene levels increase. The increased levels of ethylene typically exacerbate stress symptoms such that reducing this excess ethylene improves plant survival. Plant growth-promoting bacteria that express the enzyme 1-aminocylcopropane-1-carboxylic acid (ACC) deaminase have been shown to lower plant ethylene levels and enhance a plant's ability to proliferate in metal-contaminated soil. In this chapter, the isolation of ACC deaminase-containing soil bacteria, the isolation of the ACC deaminase gene, and its use in constructing transgenic plants is described. These transgenic plants are more tolerant of high levels of metals, flooding, pathogen attack, and salt stress, making them excellent candidates for phytoremediation strategies.

Key Words: Ethylene; ACC deaminase; transgenic plants; stress resistance; metals; plant growth-promoting bacteria.

1. Introduction

Water and soil may be contaminated with heavy metals from various sources. Among the metals that are most toxic to plant and animal life are those that displace essential metal ions in biological processes, including cadmium, zinc, mercury, copper, lead, and nickel (*1*). These metals can cause both the repression of some cellular activities and the activation of others, as well as overall plant stress. When an essential metal ion is out-competed or when metal ions bind to proteins that do not require a metal cofactor, inactivation of the proteins results. In plants, activities that are turned on by high metal concentrations encompass those that scavenge inactive proteins and free-radical species, as well as processes for sequestering and detoxifying contaminants.

Enzymes such as superoxide dismutase, metallothionins, and phytochelatins are among the most common over-expressed proteins seen in plants exposed to heavy metals *(2)*.

1.1. Plant Stress

To effectively utilize plants to remove metals from the environment, plants must first be able to tolerate and then accumulate metals into their tissues, the stress that the metal imposes on the plant notwithstanding. Environmental stresses induce a number of different reactions from plants depending on the type of stress and the species of plant. One common theme to many stress responses in plants is the increased production of the gaseous hormone ethylene, also termed stress ethylene *(3)*. Increased ethylene levels have been observed in plants exposed to chilling, heat, wounding, pathogen infection, and high levels of metals *(3–6)*. Van Loon et al. *(7)* showed that upon infection with a pathogen, plant damage was caused by ethylene and not by the direct action of the pathogen. Exogenous applications of ethylene has been shown to increase the severity of damage in many cases of stress and, thus, lowering ethylene levels has been effective in decreasing the amount of damage to cells *(6)*.

1.2. Ethylene

The synthesis of ethylene involves two main enzymes: 1-aminocylcopropane-1-carboxylic acid (ACC) synthase, which converts *S*-adenosylmethionine to ACC, and ACC oxidase, that converts ACC to ethylene *(8)*. The enzymes ACC synthase and ACC oxidase are both rate limiting to the synthesis of ethylene *(9)* and both are difficult to purify from plant tissues. In any particular plant, both enzymes are encoded by a multigene family, suggesting that different isoforms are active under different conditions *(10)*. The polygenic nature of these enzymes makes it difficult to limit their expression, and hence ethylene production, through the use of antisense versions of the genes.

1.3. ACC Deaminase

The enzyme ACC deaminase is thought to be specific to micro-organisms; no plant version of this enzyme has been isolated to date. It has been associated with plant growth-promoting bacteria that reduce ethylene levels by degrading ACC from plant tissues *(11)* and promote plant proliferation, especially under stressful conditions *(12–16)*. A model has been proposed to explain the promotion of plant growth by ACC deaminase-containing bacteria *(17)*.

Transgenic plants that express this gene have been shown to be more tolerant to a variety of stresses than their nontransformed parent plants. For instance, transgenic canola grown in soil containing arsenate had both a higher germination rate and a significantly higher biomass than nontransformed canola seedlings *(18)*. Similarly, transgenic tomato plants containing the ACC deaminase gene under the control of different promoters, were more resistant than nontransformed plants to the toxic effects of a variety of metals *(4)*. The transgenic tomato plants generally had a better germination percentage, greater biomass, and higher accumulation of some metals than nontransformed tomato plants *(4)*. In the face of other stresses, such as flooding and pathogen attack, transgenic plants containing the ACC deaminase gene also fared better *(5,6)*.

Here, we describe a method for decreasing the amount of ethylene produced in a plant through insertion of a bacterial ACC deaminase gene under the control of different promoters. Such transgenic plants may then be used as a component of an effective metal phytoremediation strategy. The steps involved are: (1) isolation of bacteria that contain ACC deaminase, (2) isolation of the ACC deaminase gene, (3) introduction of ACC deaminase gene into a binary vector, (4) transformation of canola, (5) selection of homozygous plants, and (6) measurement of ACC deaminase activity in plant tissues.

2. Materials
2.1. Isolation of Bacteria that Contain ACC Deaminase

1. PAF media: per 1 L: 10 g each proteose peptone and casein hydrolysate, 1.5 g each anhydrous $MgSO_4$ and K_2HPO_4, and 10 mL glycerol.
2. Minimal medium with DF salts (based on **ref. *19***).
   a. Per 980 mL: 4 g KH_2PO_4, 6 g Na_2HPO_4, 0.20 g $MgSO_4·7H_2O$, 2 g gluconic acid, 2 g citric acid, 2 g $(NH_4)_2SO_4$ (this is omitted for nitrogen-free DF medium), 0.10 mL each of stock solution of trace elements (*see* **item 2b**) and $FeSO_4·7H_2O$ solution (*see* **item 2c**), 20 mL sterile stock solution of glucose (*see* **item 2d**) added after autoclaving (*see* **Note 1**).
   b. Stock solution of trace elements (per 100 mL): 10 mg H_3BO_3, 0.012 g $MnSO_4·H_2O$, 0.126 g $ZnSO_4·H_2O$, 0.0782 g $CuSO_4·5H_2O$, and 0.010 g MoO_3, (*see* **Note 2**).
   c. $FeSO_4·7H_2O$ solution: 0.100 g $FeSO_4·7H_2O$ is dissolved in 10 mL water.
   d. 0.05 *M* Sterile stock solution of glucose: 9 g glucose is added to 100 mL distilled water and filter-sterilized.
3. ACC (Calbiochem, San Diego, CA).
4. Nitrogen-free Bacto agar (BD, Franklin Lakes, NJ).

2.2. Isolation of the ACC Deaminase Gene

1. Agarose.
2. *Escherichia coli* DH5α cells.
3. Tryptic soybean broth medium (BD).
4. Isopropyl-β-D-thiogalactopyranoside (IPTG) (Fermentas, Hanover, MD).

5. 5-bromo-4-chloro-3-indolyl-β-D-galactopyranoside (Fermentas).
6. ^{32}P (GE Healthcare Biosciences AB, Upsala, Sweden) or digoxigenin (Boehringer Ingelheim, Ridgefield, CT).
7. Chloroform.
8. Sodium acetate.

2.3. Introduction of ACC Deaminase Gene Into a Binary Vector

1. *Taq* DNA polymerase.
2. Restriction enzymes and buffers.

2.4. Canola Transformation

1. Nutrient broth liquid medium (BD).
2. Murashige minimal organics cocultivation medium (Sigma-Aldrich, St. Louis, Mo).
3. 6-Benzylaminopurine (Sigma-Aldrich).
4. Timentin (GlaxoSmithKline, Philadelphia, PA).
5. Kanamycin.
6. 1-Naphthalene acetic acid (Sigma-Aldrich).
7. Murashige and Skoog medium (Sigma-Aldrich).
8. Potting soil such as Promix-BX (Premier Horticulture, Riviere-du-loup, PQ, Canada).

2.5. Selecting Plants Homozygous for the Transgene

1. Kanamycin.
2. Murashige minimal organics cocultivation medium (Sigma-Aldrich).

2.6. Measuring ACC Deaminase Activity in Plant Tissues

2.6.1. Making a Plant Extract

1. Polyvinylpolypyrrolidone (polyclar AT) (Sigma-Aldrich).
2. 100 mM Tris-HCl, pH 7.4.
3. β-Mercaptoethanol.
4. Glycerol.
5. Cheesecloth or Mira cloth (Calbiochem).
6. Bio-Rad reagents (for total protein measurements).

2.6.2. Measuring ACC Deaminase Activity From Cell Extracts

1. α-Ketobutyrate.
2. ACC (Calbiochem).
3. 0.56 M HCl.
4. 2 M HCl.
5. 0.2% 2,4-Dinitophenylhydrazine (2,4-DNP) (Sigma-Aldrich).
6. 2 M NaOH.
7. 100 mM Tris, pH 8.5.

3. Methods
3.1. Isolation of Bacteria that Contain ACC Deaminase

ACC deaminase activity and plant-growth promotion are reasonably correlated, therefore, the development of an ACC deaminase activity screening method for bacteria has greatly accelerated the identification of new plant growth-promoting bacteria (20). Using the following procedure, and soil from around plant roots, many novel bacteria with ACC deaminase activity have been isolated. The method for isolating ACC deaminase-containing bacteria is from Penrose and Glick (21) and relies on a bacterium's ability to use ACC as its sole nitrogen source.

1. In a 250-mL flask add 50 mL PAF media and 1 g of soil from the plant rhizosphere.
2. Incubate in a shaking water bath overnight at 200 rpm at a temperature of between 25 and 35°C (*see* **Note 3**).
3. After incubation add a 1-mL aliquot of the bacterial suspension to 50 mL of fresh PAF media and incubate overnight at the same conditions as in **step 2**; this enriches for pseudomonads and similar bacteria and discourages fungal growth.
4. Add a 1-mL aliquot of this culture to 50 mL minimal medium with DF salts with $(NH_4)_2SO_4$ and incubate overnight at the same conditions as previously listed in **step 2**.
5. After incubation, add a 1-mL aliquot to 50 mL minimal medium with DF salts containing 3.0 mM ACC instead of $(NH_4)_2SO_4$ and incubate overnight at the same conditions as above. It is important to add ACC to the growth medium just prior to use, because it is quite labile in solution (*see* **Note 4**) (21).
6. Plate dilutions of this culture onto solid nitrogen-free minimal medium with DF salts (i.e., without $(NH_4)_2SO_4$), containing 1.8% nitrogen-free Bacto agar that has been spread with ACC (*see* **Note 4**), at a final concentration of 1.5 mM.
7. Incubate plates at the same temperature used for the liquid cultures, for a maximum of 3 d with most growth occurring within 48 h. Single isolated colonies then represent soil bacteria with the ability to use ACC as the sole source of nitrogen.

3.2. Isolation of ACC Deaminase Genes

According to the method from Shah et al. (14,15,22) ACC deaminase genes may be isolated by PCR using synthetic oligonucleotide primers (5'-TA(CT)GC (CG)AA(AG)CG(ACGT)GA(AG)GA(CT)TGCAA-3' and 5'-CCAT(CT)TC (AGT)ATCAT(ACGT)CC(GA)TGCAT-3') designed from the DNA sequences of the known ACC deaminase genes. PCR products representing about three-quarters of the ACC deaminase structural gene are excised and recovered from an agarose gel, then cloned into pGEM-T. *E. coli* DH5α cells are transformed with the reaction mixture and recombinant plasmids containing the ACC deaminase gene fragment can be isolated from white colonies on tryptic soybean broth medium containing IPTG (*see* **Note 5**) and 5-bromo-4-chloro-3-indolyl-β-D-galactopyranoside. The plasmids are then digested with *Nco*I and *Sal*I to obtain the insert. The ACC deaminase gene fragment is then used as a DNA hybridization probe, and is labeled with either ^{32}P or digoxigenin via the random primer labeling method *(23)*.

1. Isolate chromosomal DNA from an ACC deaminase-containing bacterium using a phenol:chloroform (1:1) method *(23)* and precipitate with chloroform and sodium acetate.
2. Partially digest DNA with *Sau*3AI so that an average-sized fragment is approx 2–6 kb.
3. Ligate into the pUC18 plasmid that has been digested with *Bam*HI.
4. Transform *E. coli* DH5α with the reaction mixture and plate on selective media.
5. Probe the resultant colonies for the ACC deaminase gene with ^{32}P or digoxigenin-labeled probe.
6. Isolate plasmids from positive clones (rescreened several times to avoid both false-positive and false-negatives) and digest with *Nco*I and *Sal*I to obtain the DNA inserts, which can be sent to a commercial laboratory for DNA sequence analysis.

3.3. Introduction of ACC Deaminase Gene Into a Binary Vector

1. Using PCR amplify the 1.0-kb open reading frame, encoding the ACC deaminase structural gene, from either the cloned gene or the chromosomal DNA of the ACC deaminase-containing bacterium.
2. Clone PCR product downstream of the root specific *rolD* promoter (P_{rolD}) from *Agrobacterium rhizogenes* *(24)* (*see* **Note 6**).

3.4. Canola Transformation

Canola (*Brassica napus*) cultivar "Westar" can be transformed with the ACC deaminase gene construct using a protocol developed by Moloney et al. *(25)*. Briefly:

1. Cut off fully unfolded cotyledons from 5-d-old seedlings including the petiole as close to the apical meristem as possible, without including it, with a sharp blade.
2. Dip the cut end of the petiole briefly into a 1-mL nutrient broth liquid medium (NB), with an optical density of approx 0.5 at 600 nm, of *Agrobacterium tumefaciens* harboring the previously mentioned ACC gene construct.
3. Embed the petioles into Murashige minimal organics cocultivation medium with 4.5 mg/L benzyl adenine (MMO-BA) in Petri plates so that explants stand up vertically.
4. Seal plates with surgical tape and keep in a growth room at 25°C with a photoperiod of 16 h of light and 8 h of darkness and a light intensity of 70–80 µE m^{-2} s^{-1} for 2–3 d.
5. Induce callus by transferring the explants into MMO-BA medium containing 300 mg/L Timentin.
6. Induce shoot formation from the callus by transferring the explants into plates of MMO-BA medium containing 300 mg/L Timentin and 20 mg/L kanamycin.

Increasing Plant Tolerance to Metals

7. Cut off these shoots from the explants and put into magenta vessels containing MMO medium with antibiotics (but without benzyl adenine) for shoot development.
8. When the shoots with normal morphology and apical dominance develop, transfer them to a root-induction medium: Murashige and Skoog medium containing antibiotics and 0.1 mg/L naphthalene acetic acid.
9. Once a good root system forms (approx 3–4 wk) remove the plants from the vessel, remove attached agar under running water, transfer the plants to moist potting soil (such as Promix-BX greenhouse mix) and cover with jars to avoid dehydration. Place the plants into a humidity chamber and harden off the plants by slowly removing the cover (i.e., allowing air in at intervals for 2 wk).

3.5. Selecting Plants Homozygous for the Transgene

1. Collect T1 seeds from regenerated plants before they are fully mature, remove seed coat, chop embryo into smaller pieces, and place on MMO-BA medium containing 20 mg/L kanamycin (*see* **Note 7**).
2. Obtain homozygous lines by growing seeds from plants that produce dark-green embryos on MMO-BA medium for successive generations until there is no further segregation of genes.
3. Test these lines by Southern hybridization for the presence of the transgene.

3.6. Measuring ACC Deaminase Activity in Plant Tissues

3.6.1. Making a Plant Extract

A crude plant extract can be used for the measurement of ACC deaminase activity.

1. In a chilled mortar, or for tough tissues a chilled blender, homogenize 10 g of plant tissue with 40 mL of extraction buffer, then quickly add a polyvinylpolypyrrolidone (PVP) mixture (1 g PVP in 10 mL extraction buffer) (*see* **Note 8**).
2. Filter the sample through either several layers of prewet cheesecloth or two layers of Mira cloth.
3. Centrifuge filtrate for 30 min at 100,000g and 4°C.
4. Recentrifuge the supernatant at 100,000g and 4°C for 15 min to remove any remaining particulate matter. The supernatant can be stored at –20°C or assayed immediately for ACC deaminase activity.
5. Use a 100-µL aliquot of the supernatant to calculate total protein using the Bradford method, which is most easily done using Bio-Rad reagents and following the protocol provided by the company.

3.6.2. Measuring ACC Deaminase Activity From Cell Extracts

ACC deaminase activity is calculated by measuring the concentration of α-ketobutyrate (α-KB), produced by the hydrolysis of ACC, through a change in absorbance at a wavelength of 540 nm.

1. In a 1.5-mL tube, mix 200 µL of plant extract with 20 µL of a 0.5-M stock solution of ACC, briefly vortex, and incubate for 15 min.
2. Add 1 mL 0.56 M HCl, vortex the mixture to stop the reaction, and then centrifuge for 5 min at 16,000g at room temperature.
3. Add a 1-mL aliquot of the supernatant to 800 µL 0.56 M HCl and vortex.
4. Add a 300-µL aliquot of the 2,4-DNP reagent, vortex the mixture, and incubate for 30 min at 30°C. The 2,4-DNP reagent is made by dissolving 0.2% 2,4-DNP in 2 M HCl.
5. Add 2 mL 2 M NaOH, vortex the mixture, and measure the absorbance at 540 nm. As a blank use a mixture of all assay reagents including ACC, without the plant extract. The background can be calculated by measuring the amount of α-KB in the plant extract without the addition of ACC.
6. Use a standard curve of known concentrations of α-KB in the range of 0.1 to 1.0 µM to calculate the concentration of α-KB produced, and the ACC deaminase activity is expressed in nanomoles of α-KB/mg/h. To make a standard curve, a 0.1 M stock solution of α-KB is made in 100 mM Tris pH 8.5 buffer then diluted to 10 mM prior to use, from which standards between 0.1 and 1 µM are made in a volume of 200 µL of the previously described buffer (*see* **Note 9**). To these are added 300 µL of the 2,4-DNP reagent, the mixtures are vortexed and then incubated for 30 min at 30°C. Two milliliters of 2 M NaOH are then added and the absorbance of each standard measured at 540 nm. A mixture of the 2,4-DNP reagent, 2 M NaOH, and 200 µL buffer is used as a blank.

4. Notes

1. To make minimal media with DF salts, dissolve the listed compounds in distilled water and autoclave for 20 min. Top this solution to 980 mL with sterile distilled water (to account for evaporation during autoclaving) then add 20 mL of the sterile glucose stock solution (0.05 M). When making solid DF medium, use Bacto agar from Difco Laboratories because it contains the least amount of impurities, which can lead to unwanted bacterial growth.
2. To avoid the formation of a precipitate in the stock solution of trace elements, stir the solution continuously during preparation and allow each compound to dissolve before adding the next. A precipitate may form after several months of storage of this solution.
3. All of the currently known ACC deaminase enzymes are inactivated above about 37°C, therefore, choose a temperature between 25 and 35°C and keep it constant throughout.
4. When ACC is stored in solution at 4°C, breakdown occurs at a rate of approx 5% per day. When stored at –20°C, without thawing, the breakdown is limited to approx 10% over a 2-mo period. To avoid problems that result from its lability, ACC solutions should be made, stored, and used in the following manner: to make a 0.5 M stock solution, solid ACC is dissolved in water, filter-sterilized, aliquoted, and frozen at –20°C. ACC stock solution is thawed and added to the liquid medium just prior to use, or it can be spread on solid medium and allowed to dry

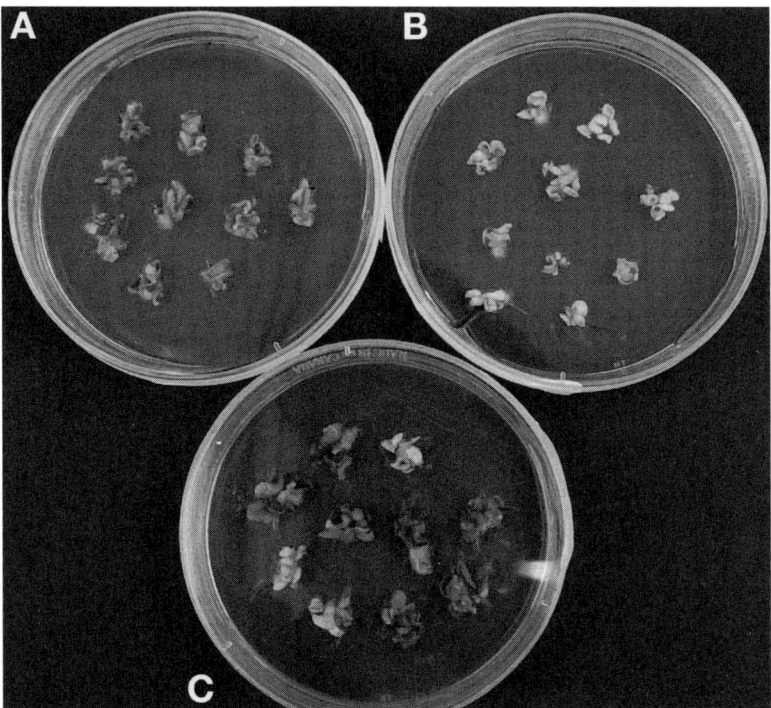

Fig. 1. Selecting plants homozygous for the transgene by visually choosing dark green from pale green T1 generation embryos (*see* **Note 7**). **(A)** The embryo is homozygous for the transgene, **(B)** the embryo does not contain the transgene, and **(C)** the embryo is heterozygous for the transgene.

  completely before a bacterial culture is added. Solid medium containing ACC should be stored at 4°C and used within 1 wk.
5. Typically, 0.01–0.10 mM IPTG not only fully induces the *lac* promoter but also yields the highest level of foreign gene expression *(26)*. The higher levels of IPTG that are used by some researchers may be deleterious to transformed *E. coli* cells.
6. In this case, the PCR primers should be designed based on the precise DNA sequence of the ACC deaminase gene. These primers should also incorporate restriction enzyme cut sites, taking care to avoid any cut sites that are contained within the ACC deaminase structural gene. A binary vector carries a neomycin phosphotransferase gene that confers kanamycin resistance (Km^R) as a selectable marker for transformed plant cells, a tetracycline resistance gene (Tc^R) as a selectable marker for bacterial cells, DNA sequence from the right border and left border of the Ti plasmid T-DNA, a transcription terminator sequence from the pea *rbcS*-E9 gene, a multiple cloning site, and a broad-host-range bacterial origin of replication from plasmid RK2. One such binary vector that we have used to introduce ACC deaminase genes into plants is pKYLX71 *(27)*.

7. When grown on this medium, tissue containing the transgene appears dark green in color, whereas tissue from nontransgenic plants is pale green (**Fig. 1**). This is a consequence of the neomycin phosphotransferase gene in tandem with the ACC deaminase gene on the binary vector. Thus, embryos from transgenic plants are either all dark green or a combination of dark green and pale green, the ratio dependent on the number of transgene integrations, whereas seeds from nontransgenic plants are completely pale green.
8. For the extraction buffer, a solution of 100 mM Tris (pH 7.4), 100 mM β-mercaptoethanol, and 20% glycerol (v/v) is presparged with nitrogen to reduce its oxidizing potential. When measuring the activity of sulfhydryl enzymes (such as ACC deaminase) it is important to remove all oxidizing agents from buffers and solutions because these will inactivate the enzyme. It is also important to remove phenolic compounds from the plant homogenate as soon as possible. Sparging with nitrogen will remove oxidizing activity from the extraction buffer and addition of PVP to the homogenate will eliminate phenolics. If the supernatant is yellow, some of the phenolic compounds have not been removed from the solution and will increase background levels of α-KB measured. This can usually be remedied by re-extracting the phenolic compounds from the extract.
9. Solutions of α-KB with concentrations in the range of 0.1 and 1 μM have an absorbance between 1 and 1.6 at 540 nm and the lower limit of measurement of α-KB spectrophotometrically is 0.1 μM. Stock solutions can be stored at 4°C.

Acknowledgments

The work described here was supported by a Strategic Grant from the Natural Science and Engineering Research Council of Canada to B. R. Glick and by funds from the Alberta Research Council to S. Shah.

References

1. Prasad, M. N. V. and Strzalka, K. (eds.) (2002) *Physiology and Biochemistry of Metal Toxicity and Tolerance in Plants.* Kluwer Academic Publishers, Boston, MA.
2. Robinson, N. J., Urwin, P. E., Robinson, P. J., and Jackson, P. J. (1994) Gene expression in relation to metal toxicity and tolerance. In: *Stress-Induced Gene Expression in Plants,* (Basra, A. S., ed.), Harwood Academic Publishers, Reading, UK, pp. 217–237.
3. Hyodo, H. (1991) Stress/wound ethylene. In : *The Plant Hormone Ethylene,* (Mattoo, A. K. and Shuttle, J. C., eds.), CRC Press, Boca Raton, FL, pp. 65–80.
4. Grichko, V. P., Filby, B., and Glick, B. R. (2000) Increased ability of transgenic plants expressing the bacterial enzyme ACC deaminase to accumulate Cd, Co, Cu, Ni, Pb, and Zn. *J. Biotechnol.* **81**, 45–53.
5. Grichko, V. P. and Glick, B. R. (2001) Flooding tolerance of transgenic tomato plants expressing the bacterial enzyme ACC deaminase controlled by the *35S*, *rolD* or PRB-1*b* promoter. *Plant Physiol. Biochem.* **39**, 19–25.
6. Robison, M. M., Shah, S., Tamot, B., Pauls, K. P., Moffatt, B. A., and Glick, B. R. (2001) Reduced symptoms of *Verticillium* wilt in transgenic tomato expressing a bacterial ACC deaminase. *Mol. Plant Path.* **2**, 135–145.

7. Van Loon, L. C. (1984) Regulation of pathogenesis and symptom expression in diseased plants by ethylene. In : *Ethylene: Biochemical, Physiological and Applied Aspects,* (Fuchs, Y. and Chalutz, E. eds.), Martinus Nijhoff/Dr. W. Junk, The Hague, The Netherlands, pp. 171–180.
8. Yang, S. F. and Hoffman, N. E. (1984) Ethylene biosynthesis and its regulation in higher plants. *Annu. Rev. Plant Physiol.* **35,** 155–189.
9. Fluhr, R. and Mattoo, A. K. (1996) Ethylene: biosynthesis and perception. *Crit. Rev. Plant Sci.* **15,** 479–523.
10. Abeles, F. B., Morgan, P. W., and Saltveit, Jr M. E. (eds.) (1992) *Ethylene in Plant Biology.* Academic Press, NY, NY.
11. Penrose, D. M. and Glick, B. R. (2001) Levels of ACC and related compounds in exudates and extracts of canola seeds treated with ACC deaminase-containing plant growth promoting bacteria. *Can. J. Microbiol.* **47,** 368–372.
12. Burd, G. I., Dixon, D. G., and Glick, B. R. (1998) A plant growth promoting bacterium that decreases nickel toxicity in seedlings. *Appl. Enviro. Microbiol.* **64,** 3663–3668.
13. Ghosh, S., Penterman, J. N., Little, R. D., Chavez, R., and Glick, B. R. (2003) Three newly isolated plant growth promoting bacilli facilitate the seedling growth of canola, *Brassica campestris. Plant Physiol. Biochem.* **41,** 277–281.
14. Ma, W., Sebestianova, S. B., Sebestian, J., Burd, G. I., Guinel, F. C., and Glick, B. R. (2003) Prevalence of 1-aminocyclopropane-1-carboxylic acid deaminase in *Rhizobium* spp. *A. Van Leeuw. J Microb.* **83,** 285–291.
15. Ma, W., Guinel, F. C., and Glick, B. R. (2003) *Rhizobium leguminosarum* biovar viciae 1-aminocyclopropane-1-carboxylate deaminase promotes nodulation of pea plants. *Appl. Enviro. Microbiol.* **69,** 4396–4402.
16. Van Loon, L. C. and Glick, B. R. (2004) Increased plant fitness by *Rhizobacteria.* In : *Molecular Ecotoxicology of Plants,* (Sandermann, H., ed.), Springer-Verlag, Berlin, Germany, pp. 177–205.
17. Glick, B. R., Li, J., and Penrose, D. M. (1998) A model for the lowering of plant ethylene concentrations by plant-growth-promoting bacteria. *J. Theor. Biol.* **190,** 63–68.
18. Nie, L., Shah, S., Rashid, A., Burd, G. I., Dixon, D. G., and Glick, B. R. (2002) Phytoremediation of arsenate contaminated soil by transgenic canola and the plant growth-promoting bacterium *Enterobacter cloacae* CAL2. *Plant Physiol. Biochem.* **40,** 355–361.
19. Dworkin, M. and Foster, J. (1958) Experiments with some microorganisms which utilize ethane and hydrogen. *J. Bacteriol.* **75,** 592–601.
20. Glick, B. R., Karaturovíc, D., and Newell, P. (1995) A novel procedure for rapid isolation of plant growth-promoting rhizobacteria. *Can. J. Microbiol.* **41,** 533–536.
21. Penrose, D. and Glick, B. R. (2003) Methods for isolating and characterizing ACC deaminase containing plant growth-promoting rhizobacteria. *Physiol. Plant.* **118,** 10–15.
22. Shah S., Li J., Moffatt, B. A., and Glick, B. R. (1998) Isolation and characterization of ACC deaminase genes from two different plant growth-promoting rhizobacteria. *Can. J. Microbiol.* **44,** 833–843.

23. Sambrook, J. and Russell, D. W. (eds.) (2001) *Molecular Cloning: a Laboratory Manual*. Cold Spring Harbor Laboratory Press, Cold Spring Harbor, NY.
24. Elmayan, T. and Tepfer, M. (1995) Evaluation in tobacco of the organ specificity and strength of the *rolD* promoter, domain A of the 35S promoter and the 35S^2 promoter. *Transgenic Res.* **4,** 388–396.
25. Moloney, M., Walker, J. M., and Sharma, K. K. (1989) High efficiency transformation of *Brassica napus* using *Agrobacterium* vectors. *Plant Cell Rpts.* **8,** 238–242.
26. Donovan, R. S., Robinson, C. W., and Glick, B. R. (2000) Optimizing the expression in *Escherichia coli* of a monoclonal antibody fragment from the *Escherichia coli lac* promoter. *Can. J. Microbiol.* **46,** 532–541.
27. Schardl, C. L., Byrd, A. D., Benzion, G., Altschuler, M. A., Mildebrand, D. F., and Hunt, A. G. (1987) Design and construction of a versatile system for the expression of foreign genes in plants. *Gene* **61,** 1–11.

3

Using Quantitative Trait Loci Analysis to Select Plants for Altered Radionuclide Accumulation

Katharine A. Payne, Helen C. Bowen, John P. Hammond, Corrina R. Hampton, Philip J. White, and Martin R. Broadley

Summary

The uptake and accumulation of toxic cations, including radionuclides, by plants growing on contaminated soils can adversely affect the health of humans and livestock. Using natural genetic variation and molecular-based quantitative genetic approaches, it is possible to identify chromosomal loci that underpin genetic variation in plant shoot radionuclide accumulation. Resolving these loci could allow the identification of candidate genes impacting on shoot radionuclide accumulation. Such methods enable gene-based crop selection or improvement strategies to be contemplated to either (1) exclude radionuclides from the food chain to minimize health risks or (2) enhance radionuclide phytoextraction. Using radiocesium (^{137}Cs) as a case study, this chapter provides an overview of how natural genetic variation and quantitative trait loci approaches in a model plant species, *Arabidopsis thaliana*, can be used to identify candidate genes/genetic loci impacting on radionuclide accumulation by plants.

Key Words: Gene-based selection; mapping populations; natural genetic variation; phytoextraction; quantitative trait loci; QTL; radiocaesium.

1. Introduction

Fundamental research to dissect mechanisms of mineral uptake and accumulation by plants is underpinning technological advances to: (1) improve fertilizer-use efficiency (i.e., maximize crop quality, and minimize production costs and environmental pollution), (2) optimize the delivery of minerals to the diets of humans and livestock, and (3) allow crops to be cultivated on marginal (e.g., nutrient poor or contaminated) land. Daar et al. *(1)* suggest that altering the accumulation of cations by crops for bioremediation and for nutritional enhancement are 2 of the top 10 biotechnological targets for improving human health in developing countries.

In this chapter, we illustrate how knowledge of candidate genetic loci, based on quantitative genetical or functional analyses, can provide information to underpin gene-based selection or crop-improvement strategies to support technological advances in altering the cation composition of the shoot. We use radionuclide accumulation in plants, and specifically shoot ^{137}Cs accumulation, as a case study. Shoot radionuclide accumulation is an appropriate model for cation accumulation traits for two reasons. First, radionuclide accumulation in crops is a serious socio-economic issue for states of the former Soviet Union and "safe" crops are a potential countermeasure to reduce radiological doses to human populations. In a recent international review of 130 countermeasures for managing radiological contamination, selective crop breeding was one of only six countermeasures deemed worthy of further exploratory experimentation *(2)*. Second, novel techniques for managing radionuclide accumulation in crop plants could underpin new technologies for cleansing soils through phytoextraction. Within this framework, we describe how quantitative trait loci (QTL) impacting on shoot ^{137}Cs accumulation can be identified in a model plant species, *Arabidopsis thaliana*. Further, we provide notes on how the establishment of a link between genotype and ^{137}Cs-accumulation phenotypes could underpin gene-based selection or crop-improvement strategies for managing cation accumulation in crops in a postgenomic context, using *Brassica oleracea* as an example crop.

1.1. ^{137}Cs Accumulation by Plants

Radioactive ^{137}Cs is a persistent contaminant of soils (half-life = 30.2 yr) arising from industrial discharges. Cs is an alkali metal that exists in the environment predominantly as Cs^+ and has similar chemical properties to K^+—an ion accumulated in large amounts by plants. Plant accumulation of ^{137}Cs, following its uptake by roots, impacts severely on the health of humans and livestock, and still affects the health and economy of some populations following the accident at Chernobyl, Ukraine, in 1986 *(3)*. The management of soils contaminated with ^{137}Cs usually relies on either the disposal of soil in repositories, or on minimizing its transfer into the food chain using agricultural countermeasures such as heavy fertilization, deep-ploughing, or mulching *(2,4,5)*. However, soil disposal is costly and can only be justified for remediating small areas of high-value land (e.g., in urban environments). Countermeasures also involve high infrastructure and labor costs and, although countermeasures can reduce ^{137}Cs transfer to crops, they are not fully effective. This is witnessed by the continued restrictions on the movement of produce from areas contaminated following the Chernobyl accident *(6)*.

Traditional constraints on managing ^{137}Cs-contaminated soils led, in the 1990s, to the suggestion that crop plants could be used to manage ^{137}Cs-contaminated sites, either to exclude ^{137}Cs from the food chain ("safe" crops to minimize health risks), or to "phytoextract" ^{137}Cs to cleanse soils *(3,7)*. However, crop plants have not yet been screened systematically for an ability to exclude, i.e., *hypo*accumulate, ^{137}Cs in their edible tissues and thus safe crop strategies have not yet been exploited as a countermeasure. Further, because no high biomass ^{137}Cs *hyper*accumulating plant species have yet been identified, phytoextraction strategies may not be realistic in the short-term (i.e., <10 yr) *(7–9)*.

One method to alter the accumulation of ^{137}Cs in crop shoot (or other edible) tissues is to alter the expression or selectivity of those transport proteins that determine shoot ^{137}Cs concentration. One such scheme has been proposed by White et al. *(3)*. In this scheme, Cs$^+$ reaching the shoot via the xylem must be transported across the plasma membranes of root cells at least twice (i.e., in and out of at least one cell) through plasma membrane-bound proteins. In the plasma membrane of root cells, inward-rectifying K$^+$ channels (KIRCs), outward-rectifying cation channels (KORCs and NORCs), voltage-insensitive cation channels and voltage-dependent Ca^{2+} channels (HACCs and DACCs) are all permeable to Cs$^+$ and K$^+$ *(10–14)*. Further, it is predicted that additional Cs$^+$ influx is catalyzed by Cs$^+$/H$^+$ symporters encoded by the *KUP/HAK* gene family *(15)*. Therefore, by altering the selectivity or expression of a subset of one or more of these transporters, Cs hypo- or hyper-accumulation phenotypes may be achievable *(3)*. Alternatively, identifying candidate genes through functional studies could allow the development of gene-based selection strategies for screening crop germplasm. However, each transporter mechanism identified to date represents at least one gene family containing many members *(16)*. For example, in *A. thaliana*, the model plant species whose genome was fully sequenced in 2000 *(17)*, 13 genes encoding KUP/HAK transporters, 8 genes encoding KORC, and tens of voltage-insensitive cation channels have been identified *(14)*. Gene redundancy or functional compensation may therefore hinder the application of functional approaches to developing reliable gene-based selection or crop improvement techniques *(3,18)*.

1.2. Natural Genetic Variation in ^{137}Cs Accumulation by Plants

Two approaches have been proposed, based on exploiting natural genetic variation, to select plants with altered ^{137}Cs-accumulation characteristics, which could circumvent complexities of gene redundancy or functional compensation (reviewed in **ref.** *3*). In the first approach, natural genetic variation in ^{137}Cs concentration between plant species is quantified, based on, for example, (1) soil-to-plant transfer data compiled by radioecologists, or (2) the laboratory screening of available genotypes. Certain functionally (e.g., green leafy vegetables) or phylogenetically (e.g., the angiosperm order Caryophyllales, which contains beet and amaranth species) defined groups of plants have high potential for ^{137}Cs accumulation *(19,20)*. Other groups of plants (e.g., cereals) have low potential for ^{137}Cs accumulation. Within groups of plants with extreme ^{137}Cs-accumulation characteristics, existing crop varieties suitable for ^{137}Cs phytoextraction, or "safe" crops, could subsequently be identified. If natural genetic variation is present within a crop species with extreme ^{137}Cs-accumulation characteristics, phenotypes could be further improved through iterative breeding and selection programs. However, there is little information on within-species variation in the uptake and accumulation of ^{137}Cs by plants, although preliminary data in certain cereal and grass species have been reported *(21)*.

In the second approach, ^{137}Cs accumulation is quantified in different accessions or ecotypes of a species for which genetic maps are available. Inbred progeny from crosses between two accessions are then used to map quantitative gene effects to specific chromosomal loci. This technique is referred to as QTL analysis *(22)*. Following the identification of QTL impacting on shoot ^{137}Cs accumulation, the underlying genes involved in the phenotype can be resolved, or sequence information from this chromosomal region used directly, to develop locus-specific molecular markers for gene-based selection or crop-improvement strategies. Where significant genetic variation is identified at candidate genetic loci of a target species, the contribution of particular alleles to a desired trait can be subsequently quantified and exploited. For example, knowledge of the genes and/or chromosomal loci controlling ^{137}Cs accumulation in one plant species can be used directly in a different target species.

1.3. A QTL–Based Approach to Dissect Natural Genetic Variation in ^{137}Cs Accumulation by Plants

The use of QTL analyses can proceed in several ways if natural genetic variation is present within a species *(23)*. *Arabidopsis* has been used extensively as a model plant for laboratory-based genetic studies because of its small stature, short life-cycle (<6 wk), self-compatibility, and high seed production. Further, there is considerable natural genetic variation in *Arabidopsis* because it has a wide geographical distribution *(22)*. Many thousands of accessions (ecotypes) and mutants with distinctive phenotypes have thus been collected or produced in laboratories around the world. Seeds of these lines are curated at the Nottingham *Arabidopsis* Stock Centre (NASC) in Europe, and at the *Arabidopsis* Biological Resource Center (ABRC) in the United States, and they are available for distribution to the scientific and commercial sectors. As a result of its use as a model genetic organism, and because it has a small genome compared to other flowering plants (n = five chromosomes totalling 125 Mb, as compared to 2500 and 16,000 Mb for maize and wheat, respectively), *Arabidopsis*

(Columbia ecotype) was the first plant genome to be fully sequenced and this was completed in 2000 *(17)*.

In addition to large collections of accessions and mutants, several mapping populations of *Arabidopsis* have been generated from crosses between selected accessions. Mapping populations of *Arabidopsis* represent several hundred lines, or progeny, derived from a single cross (F2), which have been driven to homozygosity through self-pollination and single-seed decent for more than eight generations *(22)*. In *Arabidopsis*, several mapping populations of immortalized progeny, called recombinant inbred lines (RILs), are available. These include a population of RILs from a cross between the ecotypes Landsberg *erecta* (L*er*-0) × Columbia (Col-4) *(24)*. Single seed from this cross were driven through single-seed descent for eight generations until loci were almost homozygous (residual heterozygosity is 0.42%). Other mapping populations have been derived from crosses between Wassilewskija × W100F *(25)*, Niederzenz (Nd-1) × Columbia *(26,27)*, Bay-0 × Shahdara *(28)*, and Landsberg *erecta* × Cape Verde Islands *(29)*. Full details of available mapping populations in *Arabidopsis* are given at NASC (http://nasc.nott.ac.uk/) and ABRC (http://www.arabidopsis.org/abrc/).

The chromosomes of these RILs are populated with molecular markers that are polymorphic between parents. By screening the phenotypes (traits) of the lines of the mapping population, the effect of alleles contributed by each parent at each of the loci, as defined by the molecular markers, can be determined through marker mean analysis. For each marker, the mean trait score for all accessions homozygous for parent one is compared to the mean trait score for accessions homozygous for parent two, and to the mean trait score for accessions that are heterozygous. Thus, marker means define the individual associations between markers and phenotype. Once allelic effects at each locus are identified, different techniques can be used to position precise loci (i.e., QTL) influencing the trait. These techniques include marker regression *(30)*, interval mapping *(31)*, and multiple mapping strategies *(32)*. Marker regression locates QTL with respect to all markers simultaneously by regression onto the marker means. It also estimates the additive (and dominance) effects, tests their significance, and tests for the presence of more than one QTL *(30,33)*. Interval mapping tests for the presence of a QTL between two markers using either maximum likelihood or regression-based algorithms to test for goodness-of-fit *(31)*. Multiple mapping uses markers as cofactors to reduce the genetic background noise *(32,34)*. Different cofactor markers are tested around the putative QTL and a selection of cofactors are used to narrow the putative QTL region.

Using *Arabidopsis* RILs to identify QTL also allows the heritability of the trait to be determined. Heritability is a measure of the contribution of genetics to the total phenotypic variation and it thus provides an indication of the potential ease of conventional directed-breeding efforts to deliver new crop phenotypes. A detailed discussion of calculating heritability is given in Kearsey and Pooni *(23)*. Heritability can lie between one, if all phenotypic variation is from additive genetic factors, or zero when all phenotypic variation is from nongenetic (i.e., environmental) factors. In *Arabidopsis*, heritabilities of approx 0.5 for nitrogen-use efficiency *(35)*, between 0.07 and 0.48 for flowering time *(36)* and between 0.14 and 0.41 for phosphate-use efficiency (JP Hammond, unpublished observations) have been reported. From heritability estimates, it is possible to quantify the contribution of each QTL to the genetic, and thus to the phenotypic variation. Further, if experiments are conducted on the same population under different environmental conditions, the robustness of QTL, and the heritability of the trait in its broadest sense can be determined, to include estimates of genetic (QTL) × environmental (G × E) interactions.

Overall, therefore, screening populations of *Arabidopsis* RILs allows the QTL impacting on the trait, and the contribution of genetic and environmental factors impacting on the trait (i.e., the heritability of the trait), to be determined. If traits are analyzed on several plants per line, this minimizes environmental variation and improves the accuracy of the QTL mapping and estimates of heritability. Populations of *Arabidopsis* RILs have been used in several mineral nutrition-related studies to map mineral nutrition traits, including nitrogen-use efficiency *(35,37)*, aluminium tolerance *(38,39)*, phosphate accumulation *(40)*, and Cs accumulation *(41)*. A methodology that demonstrates how QTL analysis can be used to resolve genetic loci impacting on shoot ^{137}Cs accumulation is outlined here.

2. Materials

1. Seeds of RILs, e.g., for *A. thaliana* L*er* × Cvi, or *A. thaliana* L*er* × Columbia, obtainable form the NASC, Nottingham, UK.
2. Vegetative-phase plants of RILs grown on nutrient agar with 1 µ*M* Cs or radionuclide of interest.
3. β-Counter and/or γ-counter/spectrometer for analysis of radionuclide of interest.
4. QTL-mapping software such as QTL Café or MAPQTL.

3. Methods

There are several ways to conduct QTL experiments; next is an example using *A. thaliana* to determine QTL impacting on shoot Cs concentration and shoot Cs extraction, which is based on a study of Payne et al. *(41)*.

1. Determine appropriate experimental conditions needed to assay large numbers of accessions, for example, find appropriate harvest times, a suitable external radionuclide concentration (and stable carrier isotope concentration, if appropriate), suitable growth media, and determine if there is discrimination between isotopes (*see* **Note 1**). Experimental design strongly affects the quality of QTL analysis, including

QTL Analysis for Altered Radionuclide Accumulation 33

the number of QTL detected, the accuracy of their map positions, and their phenotypic effect *(22)*. Each experimental design will vary depending on species, resources, and local radiological rules and it is best to gain advice from a statistician.

2. Analyze radioactive samples in accordance to their principle decay (α, β, or γ) in a relevant alpha-, liquid scintillation-, or auto-γ counter, respectively. First, normalize the counter for the isotope in use. Calibration optimizes the detector window to recognize each isotope. Background normalization will need to be made at least once for each instrument; follow manufacturer's instructions. In most cases the instrument will automatically subtract the background. However, a background reading may still appear when counting an empty tube. This background must be subtracted manually from the counts. Samples should be counted for the same length of time or to a certain "error" (*see* **Note 2**).

3. Undertake a varietal screen to quantify if there is natural genetic variation within the species for the trait in question (*see* **Note 3**). If the trait shows a continuous phenotypic and an approximate normal distribution within the species, then the trait is controlled by more than one gene. This natural variation can then be used to identify the loci impacting on the trait using a QTL approach. A varietal screen helps decide on the appropriate mapping population to assay. To gain the most variation within a species the parents of the mapping population should have the largest segregation possible. However, because of time constraints it is often more practicable to choose an appropriate mapping population that is already available through the current stock centers. Plant species chosen for study will depend largely on the availability of suitable plant resources.

4. Obtain appropriate mapping population information to include information on markers/genotypes (*see* **Note 4**). A marker is an identifying factor; a gene or other DNA of known location that is used to track the inheritance and so on of other genes whose exact location is not yet known. Markers must be generated or known *a priori* for the mapping population prior to QTL analysis.

5. Assay mapping population for trait.

6. Statistically analyze results (*see* **Notes 5** and **6**). Evaluation of data will depend on initial experimental design.

7. Conduct QTL analysis by mapping accession mean data to the genotype of each accession (*see* **Note 7**); when analyzing QTL data it is important to choose the correct computer package. There are many QTL programs varying in interface and algorithms used, e.g., R/qtl, BQTL (runs in R), WebQTL, Pseudomarker (runs in MATLAB®), Multimapper, Mapmaker/QTL, MultiQTL, Map manager QTX, PLABQTL, QTL Cartographer, Epistat, HAPPY, The QTL Café, MAPL, QTL Express, MapQTL, QU-GENE, Bmapqtl, MCQTL, and MQTL/NQTL. Choice of program will depend on crop type, marker information available, the availability of software, and the platform in which the program will run. All packages have help information available. The following data files are needed for QTL analysis (these are applicable to the two most commonly used QTL packages for plant species; QTL Café and MAPQTL). Data files should all have the same file name but with different extensions (*see* **Note 7**).

a. Genotype data: columns indicate marker names, chromosome, and distance along chromosome (QTL Café; *filename.mrk* or MAPQTL; *filename.map*).
b. Loci information: the first row always contains three numbers representing the number of individuals (n-ind), the number of markers (n-mrk), and the number of traits (n-tra). Then follows n-ind rows of individuals (lines), each row containing the genotype at each marker locus of the individual. The genotypes are coded: 1, homozygous for parent one; 2, heterozygous; 3, homozygous for parent two (QTL Café; *filename.gen*; *33*) or A, homozygous for parent one; H, heterozygous; and B, homozygous for parent two (MAPQTL; *filename.loc*; see **Note 7**).
c. Trait information: each column represents a single trait and contains the trait values for all the individuals; names of the individuals do not need to be included (QTL Café; *filename.tra* or MAPQTL; *filename.qua*; see **Note 7**).
   Load data (see **Note 7**) and run the QTL program. There are several QTL analysis options (see **Notes 8–10**):
d. Interval mapping: this option tests for the presence of a QTL between two markers using a goodness-of-fit test either using maximum likelihood or regression (*[31]*; see **Note 8**).
e. Marker means: this option assesses all individual associations between markers and phenotype. For each marker, the mean of the trait scores for all those lines homozygous for the allele from parent one is compared with the mean trait score for all those lines homozygous for the allele from parent two and to the mean of those heterozygous for this marker (*[33]*; see **Note 8**).
f. Marker regression: this option locates the QTL with respect to all markers simultaneously by regression onto the marker means. It also estimates the additive (and dominance) effects, tests their significance, and tests for more than one QTL (*[30,33]*; see **Note 8**).
g. The multiple mapping option ("MQM") within the MAPQTL program *(32)* uses markers as cofactors to reduce the genetic background noise *(34,42)*. Different cofactor markers are tested around the putative QTL with the final selection of cofactors maximizing the LOD score *(43)*. This statistical analysis narrows the putative QTL region to approx 10 cM (see **Note 8**).
h. Composite interval mapping; this option analyzes for QTL using simple-interval mapping but also taking into consideration other forms of variance and can thus identify more than one QTL per location (see **Note 9**).

8. Interpret results.
   a. Threshold values: the likelihood of a QTL occurring at a particular locus vs no QTL is calculated using the LOD score. If the LOD exceeds a threshold LOD score then a putative QTL is likely at that locus *(44)*. For a 5% LOD threshold for suggestive linkage LOD values are fixed, for example, a suggestive LOD threshold for *A. thaliana* recombinant inbred populations is 2.4 *(44,45)*.
   b. Confidence intervals: one difficulty associated with QTL mapping is the quantification of accurate confidence interval size. A confidence interval is important as it determines the strategies for further experimentation. The size of the average confidence interval depends heavily on the population size and the effect of the

QTL. All QTL-mapping methods whether based on maximum likelihood *(34,47)* or on regression *(31,48)* do not lend themselves to straightforward calculation of confidence intervals for the QTL location. Currently QTL-mapping methods can only locate QTL within approximately one-third of a chromosome and calculations of confidence intervals can vary with each method. The maximum likelihood approach calculates CIs using LOD scores; one or two LOD score intervals either side of each QTL equals 90 or 95% CI, respectively *(49,50)*. A bootstrap method *(51)* has also been proposed to determine confidence intervals for QTL *(46)*. Bootstrapping can be calculated for each simulated backcross population or each trait and allows the empirical calculations of 90 and 95% CIs. This is calculated by ordering the samples and taking the bottom fifth and top 2.5th percentile. The simulated QTL lies either within or outside this empirical CI. Averaged over replicate populations, the proportion of empirical bootstrap CI that contain the QTL is calculated. If the method works perfectly, this proportion would be 0.90 or 0.95 when the 90 or 95% empirical bootstrap CI was determined *(46)*.
9. Calculate, for each mapping population and each trait, the heritability (ease of conventional breeding) defined as the contribution of the accession variance component, as a proportion of the sum of all variance components compared with the contribution to phenotypic variation contributed by other variance components (*see* **Note 10**).
10. Calculate the proportion of the genetic variation contributed by each QTL as the additive effect of an allele at a QTL squared, divided by heritability (*see* **Note 11**).
11. QTL can be fine-mapped using, for example in *Arabidopsis,* heterozygous inbred families *(52)*, near-isogenic lines *(40,53)*, stepped aligned inbred recombinant strains *(54)*, and single-nucleotide polymorphism analyses *(55)*. Fine-mapping of QTL can be integrated with *in silico* searches of the *Arabidopsis* databases and functional studies of candidate genes to resolve the location of important genes that impact on the trait of interest (reviewed in **ref. 56**; *see* **Note 12**).
12. In the absence of precise positional cloning or functional confirmation of candidate genes, QTL identified in *Arabidopsis* can still be used to direct crop improvement by exploiting conservation of gene order (collinearity) between *Arabidopsis* and a target crop species *(57,58)*.

4. Notes

1. QTL analyses are dependant on environment and thus experimental design. Payne et al. *(41)* assayed *Arabidopsis* in agar radiolabeled with 1 µM Cs, and identified a harvest time of 18 d after sowing as a suitable sampling point to assay *Arabidopsis* for shoot Cs concentrations because plants were still in their vegetative stages, were growing most rapidly, and shoot Cs concentration differed between the two accessions (**Figs. 1** and **2**). Prior to this, Payne et al. *(41)* determined that there was no discrimination between the uptake of ^{133}Cs and ^{134}Cs by *Arabidopsis* and thus the amount of ^{134}Cs in a sample was used to quantify the total amount of Cs each sample.
2. Longer counting time usually equals better accuracy. Payne et al. *(41)* used 900 s as a pragmatic optimum to count each sample.

Fig. 1. *Arabidopsis thaliana* grown for 18 d in nutrient-replete agar supplemented with 1 µM Cs inside a polycarbonate box (dimensions ca. 10 × 10 × 10 cm). Cape Verde Island accessions are on the left, Landsberg *erecta* on the right.

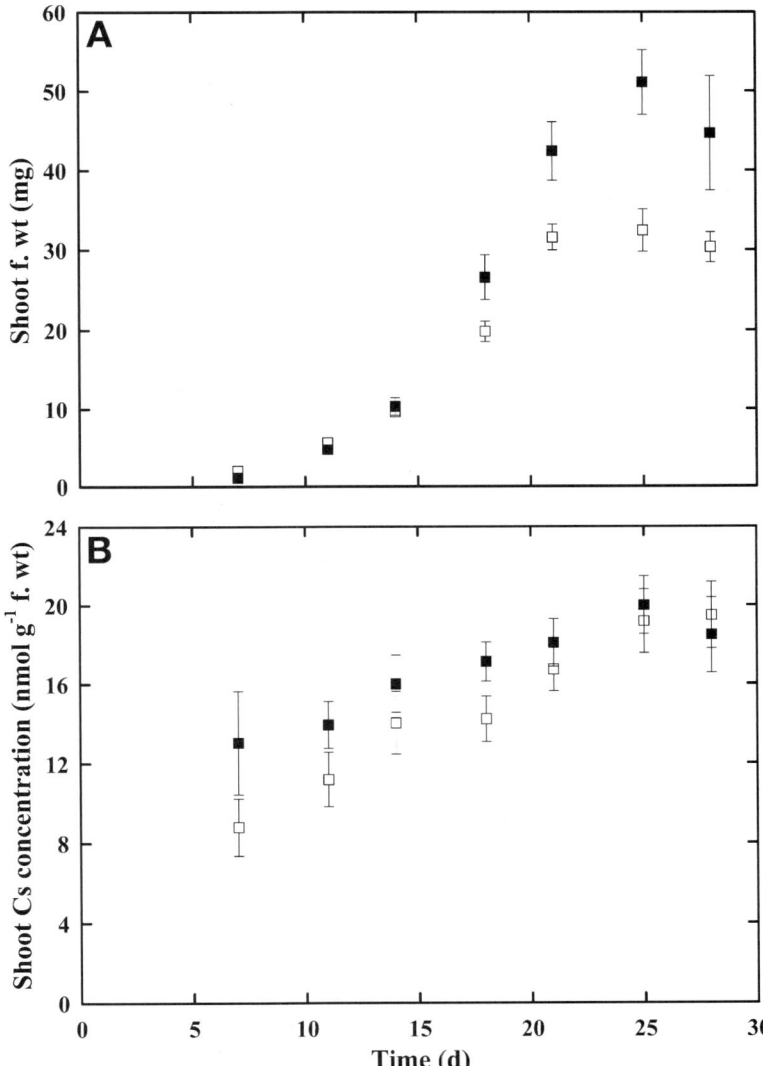

Fig. 2. (A) Shoot fresh weight and **(B)** shoot Cs concentration of *Arabidopsis thaliana* grown for 28 d in nutrient-replete agar supplemented with 1 µM Cs. Filled squares represent Landsberg *erecta* and unfilled squares represent Cape Verde Island accessions (mean ± SEM, $n = 12$). (Modified from **ref. 41**, with the kind permission of Blackwell Publishing Ltd, Oxford, UK, on behalf of the *New Phytologist* Trust.)

3. Payne et al. *(41)* for example, assayed 56 accessions representing the worldwide distribution of *Arabidopsis*.
4. Payne et al. *(41)*, for example, used a Ler × Cvi population of 162 RILs and an Ler (Ler-0; NW20) × Columbia (Col-4; N933) population of 100 RILs obtainable from the NASC. Genotypic data for the Ler × Cvi and Ler × Col RILs were taken as a set of 99 and 100 markers, respectively, covering most of the *Arabidopsis* genome *(29)*.
5. Payne et al. *(41)*, for example, harvested 159 RILs for Ler × Cvi RILs, and 88 RILs for Ler × Col RILs and determined shoot fresh weight, shoot Cs concentration, and shoot Cs extraction for each line. In both populations, shoot Cs concentration and shoot Cs extraction showed an approximately normal distribution (**Fig. 3**). Transgressive segregation occurred for shoot Cs concentration and shoot Cs extraction indicating that a quantitative gene effect controlled these traits.
6. Among the Ler × Cvi RILs of Payne et al. (*[41]*; **Fig. 3B**) there was a significant threefold variation in shoot Cs concentration. Accessions N22146 and N22024 had the lowest and highest shoot Cs concentration, respectively. Among Ler × Col RILs (**Fig. 3D**) there was a significant twofold variation in shoot Cs concentration. Accessions N1992 and N1990 had the lowest and highest shoot Cs concentration, respectively.
7. Each file must be tab-delimited and have exactly the same filename to function when using QTL packages such as QTL Café and MAPQTL. If data is missing, define within the package the value to be used (i.e., default settings; -999 in QTL Café or a dash (-) in MAPQTL). The order of lines in the trait file must be the same as in the loci file (as some mapping files do not run sequentially). QTL packages are sensitive to units with more than 2 d.p.; multiply data if necessary. QTL Café specific: before loading data check that the correct settings are made before analysis, i.e., if the map distances between the markers have been calculated using the Kosambi-mapping function; the Kosambi mapping function option should be selected (*see* program manual).
8. In Payne et al. *(41)*, marker mean values of shoot fresh weight, shoot Cs concentration, and Cs extraction were calculated. Further, marker regressions *(30)* were used to test for the location and effect of one QTL on each chromosome using 5000 simulations and the regression mean square (for 1 d.f.) and significance of the fit were calculated. If, upon visual inspection, the distribution of markers indicated the presence of >1 QTL per chromosome, and if the significance of fitting one QTL was <0.2, the possible location and effect of 2 QTL per chromosome were explored using 5000 simulations; regression mean squares were calculated for 2 d.f. These QTL analyses were supplemented using the interval mapping and multiple mapping (MQM; *[34]*) options of the MAPQTL program *(32)* to confirm the likelihood of the presence of QTL.
9. A disadvantage of interval mapping is that the model only fits a QTL at one location. Therefore if there is more than one QTL the combined effects of them will cause biased results. Jansen and Stam *(42)* suggested composite interval mapping to map more than one QTL. Composite interval mapping is reliant on the correct number of markers. Beyond a certain point, additional background markers will

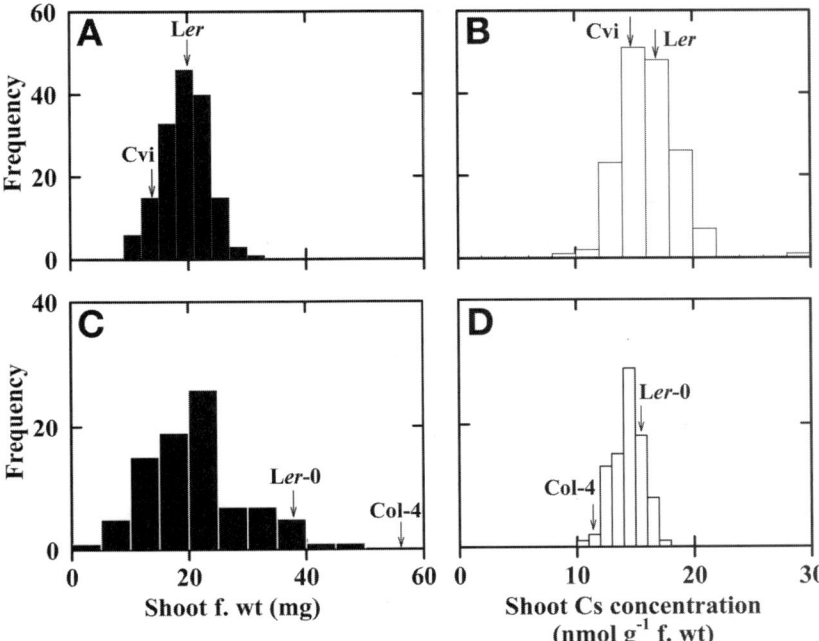

Fig. 3. Frequency distributions of shoot fresh weight (filled bars; **A,C**) and shoot Cs concentration (unfilled bars; **B,D**) of (**A,B**) 159 Landsberg *erecta* (L*er*) × Cape Verde Island recombinant inbred lines (RILs) and (**C,D**) 88 L*er* × Col RILs of *Arabidopsis thaliana* grown for 18 d in nutrient-replete agar supplemented with 1 µM Cs (mean ± SEM, n = 12 or 36). Arrows depict the means of parental accessions. (Modified from **ref. 41**, with the kind permission of Blackwell Publishing Ltd, Oxford, UK, on behalf of the *New Phytologist* Trust.)

severely weaken the ability to detect QTL because (1) they use up the finite number of degrees of freedom and (2) if background markers are too close to the position being tested, they will absorb much of the effects of the target QTL *(59)*.

10. In Payne et al. *(41)* the heritability of shoot fresh weight, shoot Cs concentration, and shoot Cs extraction were estimated and compared to the phenotypic variation contributed by other nongenetic components. The heritability of shoot fresh weight was 0.18 and 0.28, and the heritability of shoot Cs concentration was 0.08 and 0.06, in L*er* × Cvi and L*er* × Col respectively (**Table 1**). The heritability of shoot Cs extraction was 0.17 and 0.16 in L*er* × Cvi and L*er* × Col, respectively.

11. In the results of Payne et al. *(41)* for example, for shoot Cs concentration, significant allelic effects of L*er* were observed on chromosome I and chromosome V within the L*er* × Cvi cross (**Fig. 4A**). Putative QTL for shoot Cs concentration were subsequently identified on chromosome I and on chromosome V (**Table 2**). Each shoot Cs concentration QTL accounted for up to 7.7% of the genetic contribution to the phenotypic variation. For shoot Cs concentration, significant allelic effects of

Table 1
The Proportion of the Phenotypic Variation in Shoot Fresh Weight, Shoot Cs Concentration, and Cs Extraction Contributed by Each of the Variance Component Nested Within a Residual Maximum Likelihood (REML) Model[a]

Variance component	Ler × Cvi			Ler × Col		
	Shoot f. wt shoot	Cs concentration	Cs extraction	Shoot f. wt	Shoot Cs concentration	Cs extraction
Replicate	0.16	0.13	0.02	0.09	0.70	0.41
Run	0.02	0.01	0.00	–	–	–
Box	0.10	0.59	0.49	0.04	0.06	0.04
Accession (heritability)	0.18	0.08	0.17	0.28	0.06	0.16
Plant	0.53	0.20	0.32	0.59	0.18	0.39

[a]Modified from **ref. 41**, with the kind permission of Blackwell Publishing Ltd, Oxford, UK, on behalf of the *New Phytologist* Trust.

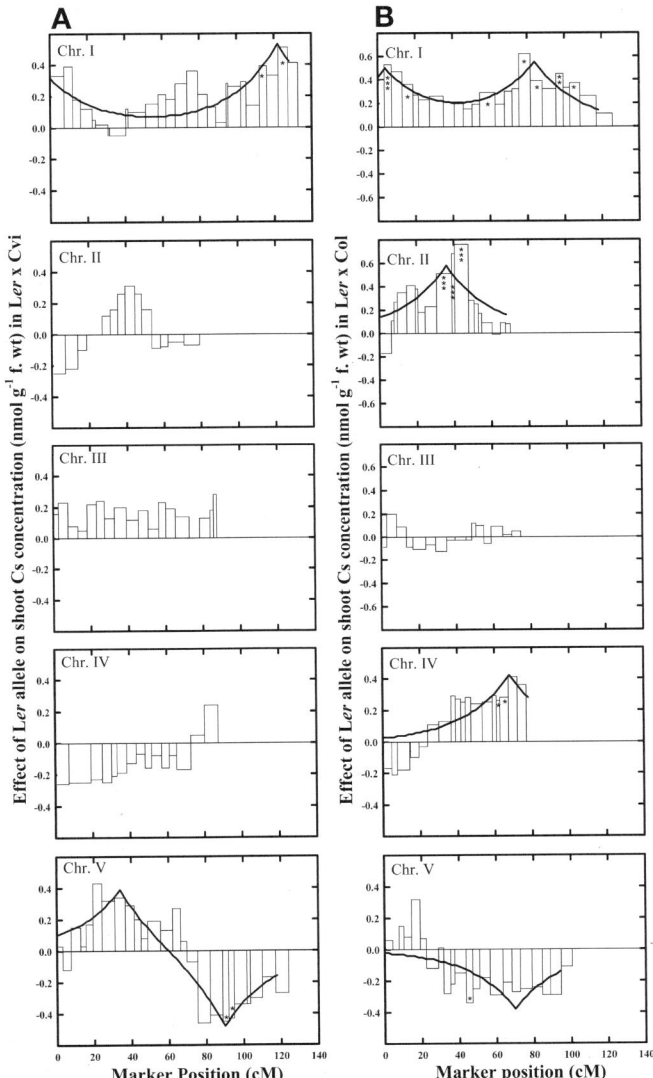

Fig. 4. The allelic effect of markers, homozygous for Landsberg *erecta* (L*er*), on shoot Cs concentration according to their position on the five *Arabidopsis thaliana* chromosomes, in plants grown for 18 d in nutrient-replete agar supplemented with 1 μ*M* Cs. Marker means were calculated from phenotypic and genotypic data derived from (**A**, left column) 159 L*er* × Cape Verde Island recombinant inbred lines (RILs) and (**B**, right column) 88 L*er* × Col RILs. Significant allelic effects of L*er* are indicated by asterisks (*, **, *** represent $p = 0.05, 0.01, 0.001$, respectively). The left-hand side of each bar represents the position of the marker used in the analyses. (Modified from **ref. *41***, with the kind permission of Blackwell Publishing Ltd, Oxford, UK, on behalf of the *New Phytologist* Trust.)

Table 2
The QTL Location, the Ler Allelic Effect, and the Percentage of the Genetic Variance Assigned to the QTL, for Shoot Fresh Weight and Shoot Cs Concentration in Ler × Cvi and Ler × Col Mapping Populations of Recombinant Inbred Lines of *Arabidopsis thaliana*

Population	Trait		1 QTL simulated per chromosome				
		Chromosome	Position	Ler effect	Percentage genetic variance	Regression mean square (1 d.f.)	P
Ler × Cvi	Shoot f. wt (mg)	II	68.0	1.52	21.5	1065	0.002
		III	20.0	−1.33	16.5	1286	0.001
		V	48.0	0.76	5.4	471	0.110
	Shoot Cs concentration (nmol g-1 f. wt)	I	120.0	0.54	7.7	169	0.100
Ler × Col	Shoot f. wt (mg)	V	96.0	−0.40	4.2	121	0.157
		IV	72.0	2.57	12.6	2916	0.041
		I	78.0	−0.61	30.4	146	0.036
	Shoot Cs concentration (nmol g-1 f. wt)	II	36.0	−0.58	27.4	149	0.040
		IV	68.0	−0.42	14.4	83	0.035
		V	70.0	0.38	11.7	56	0.089

(*Continued*)

Table 2 (*Continued*)

	QTL 1	Percentage genetic	QTL simulated per chromosome		QTL 2% genetic	Regression
Position	Ler effect	variance	position	Ler effect	variance	mean square (2 d.f.)
14.0	0.69	4.42	70.0	1.19	13.09	658
50.0	0.872	7.07	116.0	−0.43	1.7	271
0.0	0.312	2.61	122.0	0.53	7.4	106
34.0	0.445	5.30	90.0	−0.53	7.6	124
4.0	−0.476	18.69	84.0	−0.531	23.3	108

*QTL were simulated using marker regression analyses in QTL Café. Regression mean squares and significance of simulating one QTL per chromosome are presented for all chromosomes where $p < 0.2$; simulations of two QTL per chromosome are presented for selected chromosomes. (Modified from **ref. 41**, with the kind permission of Blackwell Publishing Ltd, Oxford, UK, on behalf of the *New Phytologist* Trust.)

Ler were observed on chromosomes I, II and IV and V within the Ler × Col cross (**Fig. 4B**). Significant QTLs for shoot Cs concentration were identified on chromosomes I, II, and IV (**Table 2**). A putative QTL for shoot Cs concentration was identified on chromosome V. Within Ler × Col, QTL accounted for over 80% of the genetic contribution to the phenotypic variation in shoot Cs concentration.

12. Although the locations of QTL identified by Payne et al. *(41)* represent more than 10 cM, there are genes for at least three proteins that are likely to catalyze Cs^+ transport within a genomic region of 100,000 bp (approx 25 genes) on either side of a marker where a significant allelic effect on shoot Cs concentration. These are a plasma membrane glutamate receptor channel on chromosome 2, a vacuolar K^+-channel, and proton-coupled cation transporter on chromosome 1.

Acknowledgments

Modified figures and tables have been adapted from an original paper with the kind permission of Blackwell Publishing Ltd, Oxford, UK, on behalf of the *New Phytologist* Trust.

References

1. Daar, A. S., Thorsteinsdóttir, H., Martin, D. K., Smith, A. C., Nast, S., and Singer, P. A. (2002) Top ten biotechnologies for improving health in developing countries. *Nat. Genet.* **32,** 229–232.
2. http://www.strategy-ec.org.uk. Last accessed on.
3. White, P. J., Swarup, K., Escobar-Gutiérrez, A. J., Bowen, H. C., Willey, N. J., and Broadley, M. R. (2003) Selecting plants to minimise radiocaesium in the food chain. *Plant Soil* **249,** 177–186.
4. Yera, T. S., Vallejo, R., Tent, J., Rauret, G., Omelyanenko, N., and Ivanov, Y. (1999) Mulching as a countermeasure for crop contamination within the 30 km zone of Chernobyl nuclear power plant. *Environ. Sci. Technol.* **33,** 882–886.
5. Zhu, Y. G. and Shaw, G. (2000) Soil contamination with radionuclides and potential remediation. *Chemosphere* **41,** 121–128.
6. Commission of the European Communities (2000) Council Regulations (EC) No 616/2000 of 20 March 2000 amending Regulation (EEC) No 737/90 on the conditions governing imports of agricultural products originating in third countries following the accident at the Chernobyl nuclear power station. *Official Journal of the European Communities,* **L 075,** 1–2.
7. Dushenkov, S. (2003) Trends in phytoremediation of radionuclides. *Plant Soil* **249,** 167–175.
8. Willey, N., Hall, S., and Mudiganti, A. (2001) Assessing the potential of phytoremediation at a site in the U.K. contaminated with ^{137}Cs. *Int. J. Phytorem.* **3,** 321–333.
9. Watt, N. R., Willey, N. J., Hall, S. C., and Cobb, A. (2002) Phytoextraction of ^{137}Cs: the effect of soil ^{137}Cs concentration on ^{137}Cs uptake by *Beta vulgaris*. *Acta Biotechnol.* **22,** 183–188.

10. Gaymard, F., Pilot, G., Lacombe, B., et al. (1998) Identification and disruption of a plant Shaker-like outward channel involved in K$^+$ release into the xylem sap. *Cell* **94**, 647–655.
11. White, P. J. and Broadley, M. R. (2000) Mechanisms of caesium uptake by plants. *New Phytol.* **147**, 241–256.
12. Broadley, M. R., Escobar-Gutiérrez, A. J., Bowen, H. C., Willey, N. J., and White, P. J. (2001) Influx and accumulation of Cs$^+$ by the *akt1* mutant of *Arabidopsis thaliana* (L.) Heynh. lacking a dominant K$^+$ transport system. *J. Exp. Bot.* **52**, 839–844.
13. White, P. J. and Davenport, R. J. (2002) The voltage-independent cation channel in the plasma membrane of wheat roots is permeable to divalent cations and may be involved in cytosolic Ca^{2+} homeostasis. *Plant Phys.* **130**, 1386–1395.
14. White, P., Bowen, H., Broadley, M., Hammond, J., Hampton, C., and Payne, K. (2004) The mechanisms of caesium uptake by plants. In: *Proceedings of the International Symposium on Radioecology and Environmental Dosimetry,* Institute of Environmental Sciences, Rokkasho, Aomori, Japan, 22–24 October 2003, pp. 255–262.
15. Rubio, F., Santa-María, G. E., and Rodríguez-Navarro, A. (2000) Cloning of *Arabidopsis* and barley cDNAs encoding HAK potassium transporters in root and shoot cells. *Phys. Plant.* **109**, 34–43.
16. Mäser, P., Thomine, S., Schroeder, J. I., et al. (2001) Phylogenetic relationships within cation transporter families of *Arabidopsis*. *Plant Physiol.* **126**, 1646–1667.
17. The Arabidopsis Genome Initiative. (2000) Analysis of the genome sequence of the flowering plant *Arabidopsis thaliana*. *Nature* **408**, 796–815.
18. Hirschi, K. D. (2003) Insertional mutants: a foundation for assessing gene function. *Trends Plant Sci.* **8**, 205–207.
19. Broadley, M. R., Willey, N .J., and Mead, A. (1999) A method to assess taxonomic variation in shoot caesium concentration among flowering plants. *Environ. Poll.* **106**, 341–349.
20. Frissel, M. J., Deb, D. L., Fathony, M., et al. (2002) Generic values for soil-to-plant transfer factors of radiocesium. *J. Environ. Radioac.* **58**, 113–128.
21. Øhlenschlæger, M. and Gissel-Nielsen, G. (1991) Differences in the ability for barley and rye grass varieties to absorb cesium through the roots. *Acta Agric. Scand.* **41**, 321–328.
22. Alonso-Blanco, C. and Koornneef, M. (2000) Naturally occurring variation in *Arabidopsis*: an underexploited resource for plant genetics. *Trends Plant Sci.* **5**, 22–29.
23. Kearsey, M. J. and Pooni, H. S. (1996) *The Genetical Analysis of Quantitative Traits.* Chapman and Hall, London, UK.
24. Lister, C. and Dean, C. (1993) Recombinant inbred lines for mapping RFLP and phenotypic markers in *Arabidopsis thaliana*. *Plant J.* **4**, 745–750.
25. Reiter, R. S., Williams, J. G. K., Feldmann, K. A., Antoni Rafalski, J., Tingey, S. V., and Scolnik, P. A. (1992) Global and local genome mapping in *Arabidopsis thaliana* by using recombinant inbred lines and random amplified polymorphic DNAs. *Proc. Natl. Acad. Sci. USA* **89**, 1477–1481.

26. Holub, E. B. and Beynon, J. L. (1997) Symbiology of mouse-ear cress (*Arabidopsis thaliana*) and oomycetes. *Advan. Bot. Res.* **24,** 227–273.
27. Deslandes, L., Pileur, F., Liaubet, L., et al. (1998) Genetic characterization of RRS1, a recessive locus in *Arabidopsis thaliana* that confers resistance to the bacterial soilborne pathogen *Ralstonia solanacearum*. *Mol. Plant-Micro Interact.* **11,** 659–667.
28. Loudet, O., Chaillou, S., Camilleri, C., Bouchez, D., and Daniel-Vedele, F. (2002) Bay-0 x Shahdara recombinant inbred line population: a powerful tool for the genetic dissection of complex traits in *Arabidopsis*. *Theor. App. Gen.* **104,** 1173–1184.
29. Alonso-Blanco, C., Peeters, A. J. M., Koornneef, M., et al. (1998) Development of an AFLP based linkage map of L*er*, Col and Cvi *Arabidopsis thaliana* ecotypes and construction of a L*er*/Cvi recombinant inbred line population. *Plant J.* **14,** 259–271.
30. Kearsey, M. J. and Hyne, V. (1994) QTL analysis: a simple 'marker-regression' approach. *Theor. App. Gen.* **89,** 698–702.
31. Haley, C. S. and Knott, S. A. (1992) A simple regression method for mapping quantitative trait loci in line crosses using flanking markers. *Hered.* **69,** 315–324.
32. van Ooijen, J. and Maliepaard, C. (1996) MAPQTL™, Version 3.0: software for the calculation of QTL positions on genetic maps. CPRO-DLO, Wageningen, The Netherlands.
33. Seaton, G. (2000) *The QTL Café*. http://www.bham.ac.uk/g.g.seaton. Last accessed.
34. Jansen, R. C. (1993) Interval mapping of multiple quantitative trait loci. *Gen.* **135,** 205–211.
35. Loudet, O., Chaillou, S., Merigout, P., Talbotec, J., and Daniel-Vedele, F. (2003) Quantitative trait loci analysis of nitrogen use efficiency in *Arabidopsis*. *Plant Physiol.* **131,** 345–358.
36. Van Berloo, R. and Stam, P. (1999) Comparison between marker-assisted selection and phenotypical selection in a set of *Arabidopsis thaliana* recombinant inbred lines. *Theor. App. Gen.* **98,** 113–118.
37. Rauh, B. L., Basten, C., and Buckler, E. S. (2002) Quantitative trait loci analysis of growth response to varying nitrogen sources in *Arabidopsis thaliana*. *Theor. App. Gen.* **104,** 743–750.
38. Kobayashi, Y. and Koyama, H. (2002) QTL analysis of Al tolerance in recombinant inbred lines of *Arabidopsis thaliana*. *Plant Cell Physiol.* **43,** 1526–1533.
39. Hoekenga, O. A., Vision, T. J., Shaff, J. E., et al. (2003) Identification and characterization of aluminium tolerance loci in *Arabidopsis* (Landsberg *erecta* x Columbia) by quantitative trait locus mapping. A physiologically simple but genetically complex trait. *Plant Physiol.* **132,** 936–948.
40. Bentsink, L., Yuan, K., Koornneef, M., and Vreugdenhil, D. (2003) The genetics of phytate and phosphate accumulation in seeds and leaves of *Arabidopsis thaliana*, using natural variation. *Theor. App. Gen.* **106,** 1234–1243.
41. Payne, K. A., Bowen, H. C., Hammond, J. P., et al (2004) Natural genetic variation in caesium (Cs) accumulation by *Arabidopsis thaliana*. *New Phytol.* **162,** 535–548.
42. Jansen, R. C. and Stam, P. (1994) High resolution of quantitative traits into multiple loci via interval mapping. *Gen.* **136,** 1447–1455.

43. Alonso-Blanco, C., El Assal, S. E., Coupland, G., and Koornneef, M. (1998) Analysis of natural allelic variation at flowering time loci in the Landsberg *erecta* and Cape Verde Islands ecotypes of *Arabidopsis thaliana*. *Gen.* **149,** 749–764.
44. van Ooijen, J. (1999) LOD significance thresholds for QTL analysis in experimental populations of diploid species. *Hered.* **83,** 613–624.
45. Jansen, R. C., van Ooijen, J. W., Stam, P., Lister, C., and Dean, C. (1995) Genotype-by environment interaction in genetic mapping of multiple quantitative trait loci. *Theor. App. Gen.* **91,** 33–37.
46. Visscher, P. M., Thompson, R., and Haley, C. S. (1996) Confidence intervals in QTL mapping by bootstrapping. *Gen.* **143,** 1013–1020.
47. Lander, E. S. and Botstein, D. (1989) Mapping Mendelian factors underlying quantitative traits using RFLP linkage maps. *Gen.* **121,** 185–199.
48. Martinez, O. and Curnow, R. N. (1992) Estimating the locations and the sizes of the effects of quantitative trait loci using flanking markers. *Theor. App. Gen.* **85,** 480.
49. van Ooijen, J. W. (1992) Accuracy of mapping quantitative trait loci in autogamous species. *Theor. App. Gen.* **84,** 803.
50. Mangin, B., Goffinet, B., and Rebai, A. (1994) Constructing confidence intervals for QTL location. *Gen.* **138,** 1301–1308.
51. Efron, B. (1979) Bootstrap methods; another look at the jacknife. *Ann. Statist.* **7,** 1–26.
52. Tuinstra, M. R., Ejeta, G., and Goldsbrough, P. B. (1997) Heterogeneous inbred family (HIF) analysis: a method for developing near-isogenic lines that differ at quantitative trait loci. *Theor. App. Gen.* **95,** 1005–1011.
53. Swarup, K., Alonso-Blanco, C., Lynn, J. R., et al. (1999) Natural allelic variation identifies new genes in the *Arabidopsis* circadian system. *Plant J.* **20,** 67–77.
54. Koumproglou, R., Wilkes, T. M., Townson, P., et al. (2002) STAIRS: a new genetic resource for functional genomic studies of *Arabidopsis*. *Plant J.* **31,** 355–364.
55. El-Assal, S. E.-D., Alonso-Blanco, C., Peeters, A. J. M., Raz, V., and Koornneef, M. (2001) A QTL for flowering time in *Arabidopsis* reveals a novel allele of CRY2. *Nature Gen.* **29,** 435–440.
56. Borevitz, J. O. and Nordborg, M. (2003) The impact of genomics on the study of natural variation in *Arabidopsis*. *Plant Physiol.* **132,** 718–725.
57. King, G. J. (2002) Through a genome darkly: comparative analysis of plant chromosomal DNA. *Plant Mol. Biol.* **48,** 5–20.
58. Li, G., Gao, M., Yang, B., and Quiros, C. F. (2003) Gene for gene alignment between the Brassica and Arabidopsis genomes by direct transcriptome mapping. *Theor. App. Gen.* **107,** 168–180.
59. http://gnome.agrenv.mcgill.ca/tinker/pgiv/qtltitle.htm. Last accessed.

4

Detoxification of Soil Phenolic Pollutants by Plant Secretory Enzyme

Guo-Dong Wang and Xiao-Ya Chen

Summary

The enormous growth of industrialization and agriculture has resulted in serious environmental pollution, and polychlorophenols are among the most hazardous pollutants. Because of the large investment required for traditional physical and chemical detoxification methods, engineering transgenic plants for phytoremediaton of polluted soil has received more and more attention. In most cases, however, the plants are employed to take up the pollutants and the toxic compounds are often not destroyed but merely accumulated or displaced. Recently, we developed a novel system of phytoremediation *ex planta* based on the overexpression of a secretory laccase that catalyzes the oxidation of various aromatic compounds, including 2,4,6-trichlorophenol. Because there are rich sources of various detoxifying enzymes, the technique of using plant exudation machinery for phytoremediation should be applicable to other types of pollutants in soil.

Key Words: Phytoremediation; laccase; secretory enzyme; polychlorophenol; allelochemicals.

1. Introduction

The contamination of soil by organic pollutants is a major environmental problem and a threat to human health. Polychlorinated phenols, such as pentachlorophenol and 2,4,6-trichlorophenol (TCP), have found wide use in pesticides, herbicides, wood/glue preservatives, and the antimildew agents for textiles. These polychlorophenols are highly toxic to organisms and persistent in the environment, and are among the most hazardous and recalcitrant organic pollutants in soil and water *(1)*. Environmental recovery by traditional physical and chemical methods demands large investments of economy and technology. This problem can be solved, at least partly, by phytoremediation, a cost-effective and

From: *Methods in Biotechnology, vol. 23: Phytoremediation: Methods and Reviews*
Edited by: N. Willey © Humana Press Inc., Totowa, NJ

plant-based technology, which has gained more and more attention in recent years *(2,3)*. Currently, phytoremediation can be divided into phytoextraction, phytodegradation, rhizofiltration, phytostabilization, and phytovolatilization, and various physiological mechanisms are involved in each of these processes *(4)*. However, most phytoremediation strategies are designed on the basis of plant uptake of toxins. This may reduce the viability of the plants in contaminated sites because many pollutants are also toxic to plant cells. Furthermore, for phytoremediation to be a viable alternative, plants need to destroy the contaminants rather than simply accumulate or displace them.

1.1. Phytoremediation and Allelopathy

Recently we developed a system of phytoremediation *ex planta* by engineering plant-secretory enzymes to transform or degrade organic contaminants in soil *(5)*. It is well known that plants release phytotoxic compounds (allelochemicals) into soil to inhibit growth of neighboring plants *(6–8)*. In wheat root exudates, for example, a number of phenolic acids with allelopathic activities have been identified *(9)*. In addition to secondary metabolites, plant roots also secrete enzymes that are capable of degrading allelochemicals and organic pollutants in soil. Such soil enzymes of plant origin include laccase, dehalogenase, and peroxidase *(4)*. These secretory enzymes and the plant secretion machinery may be employed for detoxifying pollutants in soil.

1.2. The Selection of Plants

For phytoremediation *ex planta* to be practical one needs to consider two factors: the enzyme used for transformation of the pollutants, and the plant used for expression and release of the enzyme. The ideal plant for use in phytoremediation *ex planta* should have the following traits: (1) a highly developed root system that has the ability to secret a substantial amount of the enzyme that can render the pollutants harmless; (2) tolerance to the pollutants at a concentration found in soil; and (3) fast growth and a relatively high biomass (large size). Although poplar trees are now considered one of the most useful potential candidates for phytoremediation *(10)*, most researchers are using the model plant *Arabidopsis thaliana* for pilot research. In the laboratory this plant offers several advantages over other plants, including the small size, fast growth, and available molecular and genetic tools that make research less time-consuming and more informative. The complete genome sequences of *A. thaliana* are known and accessible, and there are several methods available for producing mutants and identifying mutant genes *(11)*. In recent years some encouraging results have been produced by using *Arabidopsis* as a tool. For example, Song et al. *(12)* reported that transgenic *Arabidopsis* plants overexpressing a yeast protein YCF1, which detoxifies cadmium by transporting it into vacuoles, exhibited enhanced tolerance of Pb(II) and Cd(II) and accumulated a greater amount of these metals. Similar results were obtained with phytoremediation of other pollutants *(13–16)*. The technology we have developed with *Arabidopsis* will eventually be transferred to more suitable plants, such as poplar and tobacco.

1.3. The Use of Laccase for Phytoremediation

There is no doubt that the enzyme used for phytoremediation depends on the pollutant of concern. The activity of the candidate enzyme should be catalyzing the pollutants into a harmless form, and the enzyme should be expected to maintain a high activity in soil for a relatively long duration. A specific criterion of the phytoremediation *ex planta* system is that the enzyme should be secreted into soil by the plant root system. A portion of both prokaryotic and eukaryotic proteins have a signal domain that directs protein secretion into extracellular space. Even if the enzyme to be used is originally intracellular, secretion can be easily achieved by fusing the protein with a signal peptide, thus theoretically all enzymes can be engineered for plant root secretion.

Laccase (*p*-diphenol:dioxygen oxidoreductase, EC 1.10.3.2), a multicopper-containing oxidase, has been studied for more than 120 yr after it was first described by Yoshida in 1893 *(17,18)*. Laccases are widespread in plants and fungi. In most of the wood-rotting fungi, laccases are involved in the lignin-degrading procedure *(19)*. Fungal laccases are also important in physiological and developmental processes such as pigment formation *(20)*, spore formation *(21,22)*, and pathogenesis *(23)*. In plants, laccases are proposed to play a role in phenolic metabolism, lignin polymerization, and cell wall biosynthesis *(24,25)*. The range of substrates that laccases catalyze is rather broad—any substrates similar to *p*-diphenol will be oxidized by laccase. The general catalytical mechanism is that laccase has the ability to produce a free radical from a suitable substrate, the ensuing secondary reactions are responsible for producing many varied products. During this procedure, laccase needs to take up O_2. Because of its stability and broad spectrum of substrates, laccase has a wide application in industry and research, such as woody tissue delignification *(26)*, ethanol production *(27)*, biosensors *(28)*, and bioremediation.

Recently, laccases from fungi have been shown to be useful for the degradation of a variety of persistent environmental pollutants, including alkenes, bisphenol A, and industrial synthetic dyes *(18)*. Leontievsky et al. *(29)* found that immobilized laccase from the white-rot fungus *Coriolus versicolor* can degrade TCP; the degradation products include 3,5-dichlorocatechol, 2,6-dichloro-1,4-benzoquinone, and 2,6-dichloro-1,4-hydroquinone, along with polymers. We found that a plant-secretory laccase from cotton (*Gossypium arboreum*) also has the activity of transforming TCP *(5)*. It is therefore worth using plant-secretory laccase to attempt to detoxify polychlorophenols in soil.

The approach illustrated in this chapter includes isolation of cDNAs, assay of enzyme activities, transferring foreign genes into plants, testing the tolerance of transgenic plants to various phenolic acids and TCP, and chemical analysis of plant phenolic components and enzymatic products.

2. Materials

2.1. cDNA Isolation, Expression, and Enzyme Assay

1. Plants of cotton cultivars (*G. hirsutum* L. cv. Zhong-12 and *G. hirsutum* L. cv. Hai-1) were grown in the field, and those of *A. thaliana* (Col-0, transgenic and wild type) in the greenhouse (22–25°C, continuous illumination or 16 h light/8 h dark).
2. *Pichia pastoris* GS115 or similar yeast-expression system.
3. PCR, sodium dodecyl sulfate-polyacrylamide gel electrophoresis, and laccase-assay protocols.

2.2. Transformation of Arabidopsis and Selection of Putative Transformants

1. *A. tumefaciens* strain GV3101.
2. 30% Bleach containing 0.05% Tween-20.
3. 1/2 Murashige and Skoog (MS) selection agar plates containing antibiotics appropriate to the binary vector used.

2.3. Analysis of Enzyme Activities Secreted From Transgenic Plants

1. 1/2 MS liquid medium.
2. Laccase-assay protocols (*see* **Subheading 2.1.**).

2.4. Plant-Resistance Assays

1. 30% Bleach containing 0.05% Tween-20.
2. 1/2 MS plates.
3. Solutions of test pollutant(s).
4. Pots of soil soaked in pollutant(s).

2.5. Analysis of Arabidopsis Soluble and Cell Wall-Bound Phenolic Compounds

1. 50% Methanol containing 1.5% (v/v).
2. Folin–Denis reagent. To prepare add 100 g sodium tungstate and 20 g phosphomolybdic acid to 750 mL water, and add 50 mL orthophosphoric acid. Reflux for 2 h, allow to cool, make up to 1 L, and store in dark *(30)*.
3. Saturated sodium carbonate solution.
4. Gallic acid.
5. 1 N NaOH.
6. 2 N HCl.

Detoxification of Soil Phenolic Pollutants

3. Methods
3.1. cDNA Isolation, Expression, and Enzyme Assay

Before designing and generating the transgenic plants, we suggest testing the properties of the enzyme to make sure that it fits the criteria of phytoremediation *ex planta*.

1. Isolate the full-length cDNA of target enzyme(s) using different methods. The cDNA can be isolated by screening the cDNA library, or by amplification of the reverse transcripts (RT) by PCR *(31)* (*see* **Note 1**).
2. After DNA sequencing to confirm the cDNA identity, subclone the open reading frame into a suitable expression vector, and express the candidate cDNA in a suitable host. For plant-secretory enzyme, the yeast-expression systems of *Pichia pastoris* GS115 (Invitrogen, Carlsbad, CA) may be used for protein production and secretion.
3. Harvest yeast cells by centrifugation, and extract proteins from the yeast cells. Collect and concentrate the culture medium to 1/10 ~1/50 of the original volume. The proteins may be checked by sodium dodecyl sulfate-polyacrylamide gel electrophoresis.
4. Assay the enzyme activities present in cellular proteins and in concentrated medium with appropriate substrates and the pollutant under concern. For assaying laccase activities, protocols can be found in literature *(32)* (*see* **Note 2**).

3.2. Transformation of Arabidopsis and Selection of Putative Transformants

1. Insert the candidate gene (or cDNA) into a binary vector and introduce the plasmid into *A. tumefaciens* strain GV3101. For the purpose of phytoremediation of soil pollutants, a constitutive promoter or a root-specific promoter may be used to drive the transgene expression.
2. Grow the *Arabidopsis* plants (Col-0) in a greenhouse, and select the healthy plants for transformation.
3. Transform the plants by using the floral dip method *(33)*, and harvest the seeds after the plants are fully mature.
4. Sterilize the seeds in the 30% bleach for 5 min, followed by five rinses with sterile water. Plate the seeds on 1/2 MS selection agar plates containing antibiotics (depending on the binary vector used).
5. Place the plates in a cold room (4°C) for 2 d to synchronize germination (vernalization); then place the plate in the greenhouse (22–25°C) for seed germination and seedling growth.
6. After approx 10 d, select the dark-green seedlings with roots that extend over and into the selection agar, and transfer them into pots of soil.
7. Harvest the seeds from T1 plants. Repeat **steps 4–6** to grow T2 plants. Select the homozygous transgenic lines on the basis of segregation rates of antibiotic resistance. Check the number of T-DNA copies inserted into the plant genome by Southern blotting, and select the single-copy insertion lines for further investigation.

8. Check the transgene expression levels by RT-PCR and/or Northern blot. Transgenic lines with different expression levels of the transgene can be used for phytoremediation assays.

3.3. Analysis of Enzyme Activities Secreted From Transgenic Plants

1. Surface-sterilize seeds as described in **Subheading 3.2.**
2. Put the seeds in a sterilized flask with one-third volume of 1/2 MS liquid medium. Put the flask into a shaker at 100 rpm, 22°C, 12 h light/12 h dark, for 1–2 wk.
3. Collect the medium and concentrate the liquid by more than 100-fold. Assay the enzyme activity as described in **Subheading 3.1.**

3.4. Plant-Resistance Assay

3.4.1. On Plates

1. Surface sterilize and vernalize the seeds as described in **Subheading 3.2.** Place the seeds on the agar-gelled 1/2 MS plate without antibiotics, and place in a greenhouse for 4 d for seed germination and seedling growth.
2. Transfer those seedlings with about 1-cm long root to a fresh plate supplemented with the pollutants or other testing compounds at different concentrations (*see* **Note 3**). Record the root length accurately at the day of transfer.
3. Place the plates in the greenhouse and allow the seedlings to grow for 10 d.
4. Measure the root length of the seedlings and calculate the relative root growth, which is expressed by the ratio of the root length at the day of measurement to the original root length at the day of transfer.

3.4.2. In Soil

1. Plant the seeds in soil (pots) that has been soaked with the solutions of the pollutants, such as TCP or phenolic acids, at a given concentration.
2. Keep the pots in cold room (4°C) for 2 d to synchronize germination.
3. Transfer pots with cold-treated seeds into the greenhouse and allow the plants to grow for 2 wk.
4. The 2-wk-old seedlings may be further treated by spraying with solutions of the chemicals (pollutants).
5. Record the plant growth and the phenotypic changes; measure the biomass when the plants are completely mature.

3.5. Analysis of Arabidopsis Soluble and Cell Wall-Bound Phenolic Compounds

1. Grind 2 g of 4-wk-old plants of *Arabidopsis* to fine powder in liquid nitrogen, extract with 20 mL of 50% methanol containing 1.5% (v/v) acetic acid, at 4°C for 4 h.
2. Centrifuge at 10,000g for 10 min. Keep the residue for analysis of the cell wall-bound phenolics.
3. Take 20 µL of supernatant for HPLC or HPLC/mass spectrometric analysis.

4. The total soluble phenolics can be quantitatively determined with the Folin–Denis method (see **ref. 25**).
5. To 25 mL of water add 200 µL of initial extract, followed by 2.5 mL of Folin–Denise reagent; mix well. After 3 min add 5 mL of saturated sodium carbonate and bring to volume with water; mix well.
6. After 20 min measure the absorbance at 760 nm, use gallic acid as a phenolic standard, and express the results as milligrams of gallic acid equivalents per g tissue.
7. For analysis of cell wall-bound phenolics, rinse the residue with ethanol three times and dry it with nitrogen gas.
8. Hydrolyze the residue with 10 mL of 1 N NaOH at 37°C for 24 h.
9. Centrifuge at 10,000g for 10 min, take 60% of the supernatant, and adjust the solution with 2 N HCl to pH 1.0–2.0.
10. Extract with equal volume of ethyl acetate three times and combine the upper phase.
11. Concentrate to 1 mL and take 50 µL for HPLC analysis.

4. Notes

1. We isolated a cotton laccase cDNA, *LAC1*, from a library constructed from suspension-cultured cells of *G. arboreum* by using a PCR 96-well plate method *(5)*.
2. For example, we determined the laccase (LAC1 of *G. arboreum*) activity at 30°C, using 2,2′-azino-*bis*-ethylbenthiazoline (ABTS) as a standard substrate *(5)*. The reaction mixture of 0.5 mL contains 50 mM sodium acetate buffer (pH 4.5), 1.8 mM ABTS, 100 µM CuCl$_2$, and 10 µL crude protein. One unit of enzyme activity is defined as the amount of enzyme required to oxidize 1 µmol ABTS/min using an absorption value at 420 nm for oxidized ABTS (3.6×10^4/mol/cm). For other substrates, determine the laccase activity by using a spectrometer at different wavelengths: sinapic acid (376 nm), syringic acid (272 nm), ferulic acid (318 nm), and catechol (450 nm).
3. For example, we performed the plate assay of seedling resistance to the following chemicals at different concentrations and combinations: 1, 0.5 mM syringic acid only; 2, 0.5 mM sinapic acid only; 3, 0.3 mM vanillic acid only; 4, 0.15 mM syringic acid + 0.15 mM sinapic acid; 5, 0.25 mM syringic acid + 0.25 mM sinapic acid; 6, 0.3 mM vanillic acid + 0.2 mM sinapic acid. We also examined the resistance of transgenic seedlings to TCP at 5, 10, 20, and 40 µM *(5)*.

References

1. Gupta, S. S., Stadler, M., Noser, C. A., et al. (2002) Rapid total destruction of chlorophenols by activated hydrogen peroxide. *Science* **296,** 326–328.
2. Boyajian, G. E. and Carreira, L. H. (1997) Phytoremediation: a clean transition from laboratory to marketplace? *Nat. Biotechnol.* **15,** 127–128.
3. Watanabe, M. E. (2001) Can bioremediation bounce back? *Nat. Biotechnol.* **19,** 1111–1115.
4. Salt, D. E., Smith, R. D., and Raskin, I. (1998) Phytoremediation. *Annu. Rev. Plant Physiol. Plant Mol. Biol.* **49,** 643–668.

5. Wang, G. D., Li, Q. J., Luo, B., and Chen, X. Y. (2004) *Ex planta* phytoremediation of trichlorophenol and phenolic allelochemicals via an engineered secretory laccase. *Nat. Biotechnol.* **22,** 893–897.
6. Blum, U., Shafer, S. R., and Lehman, M. E. (1999) Evidence for inhibitory allelopathic interactions involving phenolic acids in field soil: concepts vs. an experimental model. *Crit. Rev. Plant Sci.* **18,** 673–693.
7. Reigosa, M. J., Sanchez-Moreiras, A., and Gonzalez, L. (1999) Ecophysiological approach in allelopathy. *Crit. Rev. Plant Sci.* **18,** 577–608.
8. Bais, H. P., Vepachedu, R., Gilroy, S., Callaway, R. M., and Vivanco, J. M. (2003) Allelopathy and exotic plant invasion: from molecules and genes to species interactions. *Science* **301,** 1377–1380.
9. Wu, H., Haig, T., Pratley, J., Lemerle, D., and An, M. (2002) Biochemical basis for wheat seedling allelopathy on the suppression of annual ryegrass (*Lolium rigidum*). *J. Agric. Food Chem.* **50,** 4567–4571.
10. Brunner, A. M., Busov, V. B., and Strauss, S. H. (2004) Poplar genome sequence: functional genomics in an ecologically dominant plant species. *Trends Plant Sci.* **9,** 49–56.
11. Arabidopsis Genome Initiative (2000) Analysis of the genome sequence of the flowering plant *Arabidopsis thaliana. Nature* **408,** 796–815.
12. Song, W. Y., Sohn, E. J., Martinoia, E., et al. (2003) Engineering tolerance and accumulation of lead and cadmium in transgenic plants. *Nat. Biotechnol.* **21,** 914–919.
13. Bizily, S. P., Rugh, C. L., Summers, A. O., and Meagher, R. B. (1999) Phytoremediation of methylmercury pollution: *merB* expression in *Arabidopsis thaliana* confers resistance to organomercurials. *Proc. Natl. Acad. Sci. USA* **96,** 6808–6813.
14. Dhankher, O. P., Li, Y., Rosen, B. P., et al. (2002). Engineering tolerance and hyperaccumulation of arsenic in plants by combining arsenate reductase and γ-glutamylcysteine synthetase expression. *Nat. Biotechnol.* **20,** 1140–1145.
15. Lee, J., Bae, H., Jeong, J., et al. (2003) Functional expression of a bacterial heavy metal transporter in Arabidopsis enhances resistance to and decreases uptake of heavy metals. *Plant Physiol.* **133,** 589–596.
16. LeDuc, D. L., Tarun, A. S., Montes-Bayon, M., et al. (2004) Overexpression of selenocysteine methyltransferase in Arabidopsis and Indian mustard increases selenium tolerance and accumulation. *Plant Physiol.* **135,** 377–383.
17. Yoshida, H. (1883) Chemistry of lacquer (urushi). *J. Chem. Soc.* **43,** 472–486.
18. Mayer, A. M. and Staples, R. C. (2002) Laccase: new functions for an old enzyme. *Phytochemistry* **60,** 551–565.
19. Heinzkill, M., Bech, L., Halkier, T., Schneider, P., and Anke, T. (1998) Characterization of laccases and peroxidases from wood-rotting fungi (family Coprinaceae). *Appl. Environ. Microbiol.* **64,** 1601–1606.
20. Cardenas, W. and Dankert, J. R. (2000) Cresolase, catecholase and laccase activities in haemocytes of the red swamp crayfish. *Fish Shellfish Immunol.* **10,** 33–46.
21. Kurtz, M. B. and Champe, S. P. (1981) Dominant spore color mutants of *Aspergillus nidulans* defective in germination and sexual development. *J. Bacteriol.* **148,** 629–638.

22. Kurtz, M. B. and Champe, S. P. (1982) Purification and characterization of the conidial laccase of *Aspergillus nidulans*. *J. Bacteriol.* **151,** 1338–1345.
23. Choi, G. H., Larson, T. G., and Nuss, D. L. (1992) Molecular analysis of the laccase gene from the chestnut blight fungus and selective suppression of its expression in an isogenic hypovirulent strain. *Mol. Plant Microb. Interact.* **5,** 119–128.
24. Bao, W., O'Mally, D. M., Whetten, R., and Sedero, R. R. (1993) A laccase associated with lignification in loblolly pine xylem. *Science* **260,** 672–674.
25. Ranocha, P., Chabannes, M., Chamayou, S., et al. (2002) Laccase down-regulation causes alterations in phenolic metabolism and cell wall structure in poplar. *Plant Physiol.* **129,** 145–155.
26. Argyropoulos, D. S. (2001) *Oxidative Delignification Chemistry.* ACS Symposium Series 785. American Chemical Society, Washington, DC.
27. Larson, T. G., Cassland, P., and Jonsson, L. J. (2001) Development of a *Saccharomyces cerevisiae* strain with enhanced resistance to phenolic fermentation inhibitors in lignocellulose hydrolysates by heterologous expression of laccase. *Appl. Environ. Microbiol.* **67,** 1163–1170.
28. Bauer, C. G., Kuehn, A., Gajovic, N., et al. (1999) New enzyme sensors for morphine and codeine based on morphine dehydrogenase and laccase. *Fres. J. Anal. Chem.* **364,** 179–183.
29. Leontievsky, A. A., Myasoedova, N. M., Baskunov, B. P., Golovleva, L. A., Bucke, C., and Evans, C. S. (2001) Transformation of 2,4,6-trichlorophenol by free and immobilized fungal laccase. *Appl. Microbiol. Biotechnol.* **57,** 85–91.
30. Pritchard, S., Peterson, C., Runion, G. B., Prior, S., and Rogers, H. (1997) Atmospheric CO_2 concentration, N availability, and water status affect patterns of ergastic substance deposition in longleaf pine (*Pinus palustris* Mill.) foliage. *Trees* **11,** 494–503.
31. Sambrook, J., Fritsch, E. F., and Maniatis, T. (2001) *Molecular Cloning: A Laboratory Manual, 3rd ed.* Cold Spring Harbor Laboratory, Cold Spring Harbor, NY.
32. Min, K. L., Kim, Y. H., Kim, Y. W., Jung, H. S., and Hah, Y. C. (2001) Characterization of a novel laccase produced by the wood-rotting fungus *Phellinus ribis*. *Arch. Biochem. Biophys.* **392,** 279–286.
33. Clough, S. J. and Bent, A. F. (1998) Floral dip: a simplified method for *Agrobacterium*-mediated transformation of *Arabidopsis thaliana*. *Plant J.* **16,** 735–743.

5

Using Real-Time Polymerase Chain Reaction to Quantify Gene Expression in Plants Exposed to Radioactivity

Yu-Jin Heinekamp and Neil Willey

Summary

Plants have the ability to take up harmful substances and then store or metabolize them. This suggests the possibility of phytoremediation of soil contaminated with organic, inorganic, and radioactive substances, all of which are difficult to remove from soil with conventional methods. Phytoremediation is becoming an important tool for decontaminating soil, water, and air by detoxifying, extracting, hyperaccumulating, and/or sequestering contaminants, especially at low levels where, using current methods, costs exceed effectiveness. The genome of *Arabidopsis thaliana* has recently been sequenced and it has been shown that various of its genes can have an impact on its uptake of harmful substances. To identify and analyze all the genes that might be useful for phytoremediation is desirable because plants might then be engineered for faster and more precise phytoremediation. Here, a method is described for using real time-PCR to monitor gene expression in plants exposed to radioactivity. This method can monitor the expression of phytoremediation-related genes under a variety of conditions. The experiments described here focus on DNA-repair genes and show how analyses can be extended to whole-genome studies of phytoremediation-related genes using DNA microarrays. A method to grow *A. thaliana* in hydroponic cultures is also outlined.

Key Words: Radioactivity; *Arabidopsis thaliana*; real-time PCR; phytoremediation; Cs-137; hydroponics.

1. Introduction

Human activities have increased the exposure of the environment to radioactivity. Weapons and power stations, and particularly the Chernobyl accident in 1986, are the biggest contributors to radioactive contamination at present *(1)*. Because of the long-lived nature of some artificially created isotopes, the consequences of their presence in the environment will have to be mitigated.

Conventional decontamination methods are costly and might harm the environment further through use of chemicals and agents used to remove the radioactivity *(2)*. Currently, there is research into the use of plants and micro-organisms to decontaminate radioactivity, especially at low levels at which, using currently available methods, the expense often exceeds the benefit.

Current techniques in molecular biology enable the study of single genes and gene clusters associated with plant stress and the ability to tolerate stress. These techniques are important for unravelling the pathways in plants and microorganisms that underpin the resistance to toxic substances that is essential for bioremediators. Furthermore, such knowledge might guide the enhancement of resistance. Real time (RT)-PCR and DNA microarrays together form a powerful tool not only to scan and analyze the genome of *Arabidopsis thaliana* for candidate genes associated with phytoremediation, but also to analyze the expression of phytoremediation-related genes.

1.1. Phytoremediation

Plants can provide an alternative to, for example, excavation, in the decontamination of soils with low levels of organic and inorganic substances *(3)*. They provide an eco-friendly tool for decontamination but there are obstacles, such as how to dispose of contaminated plants, especially with radioactivity. Problems such as how to increase the biomass, and therefore the total uptake of pollutants, in the most efficient species have also to be addressed in the near future. It is known that certain families and species are more efficient than others in the uptake of selected pollutants *(4)*. Some of the plants, which have a high uptake of contaminants, do not, however, develop a high biomass and therefore the total uptake of non-nutrient substances is low *(5)*. So combining uptake and tolerance with increased biomass of plants is a focus of ongoing research. Another focus is the resistance to and detoxification of the pollutants accumulated.

1.2. DNA-Repair Genes and Phytoremediation-Related Genes

Traits useful for phytoremediation are clearly linked to certain types of gene categories, e.g., heavy metal accumulators activate redox catalysts, transporters/pumps, and receptors when exposed to heavy metals *(6,7)*. Genes induced by exposure of plants to radioactivity include a range of cell-cycle maintenance and DNA-repair genes, e.g., *AtPARP-1*, *AtLIG4*, *AtRAD51*, and *AtGR1*. These genes enable plants to deal with potentially lethal stress, e.g., DNA breaks and reactive oxygen species, and therefore increase resistance to a range of pollutants when overexpressed. Such genes, and more pollutant-specific genes, are candidates for the engineering of plants for phytoremediation.

Here, a method is outlined based on the DNA repair and cell-cycle maintenance genes *AtRAD51*, *AtLIG4*, and *AtPARP-1* that monitors gene expression after exposure to cesium-137, with *AtACT2* as an internal housekeeping gene. *AtRAD51*, *AtLIG4*, *AtGR1*, and *AtPARP-1* are known to be inducible by irradiation with gamma sources such as cesium-137 and cobalt-60 *(8–11)*. *AtACT2* has been shown to be expressed strongly and constitutively in vegetative tissue *(12)* and is therefore suitable as a housekeeping control gene.

1.3. RT-PCR

RT-PCR is a very sensitive method to detect subtle changes in gene expression. It is, in theory, possible to detect single copies of sequences and to quantify changes in gene expression between control and induced samples, as well as during development. The two key factors for a successful RT-PCR are high-quality cDNA and equally good primers. Using the right primers ensures that only expression of the target genes will be recorded and that results will not be falsified by primer dimers. Different approaches are available but only three will be briefly mentioned.

The use of SYBR green fluorescence shows the lowest specificity toward the amplified products and does not distinguish between primer dimers and PCR products because SYBR green binds to all double-stranded DNA (**Fig. 1A**). TaqMan and Molecular Beacon probes are highly specific toward the target sequence and the monitored fluorescence is a more accurate picture of the amplification process. Both probes bind to a complementary sequence between the primer binding sites. The TaqMan probe has a fluorophore attached to one end and a quencher to the other end. The fluorophore is excited by the machine and passes its energy, via fluorescence resonance energy transfer, to the quencher. The quencher is excited by the energy from the fluorophore and fluoresces in a different wavelength to the fluorophore and, therefore, emission of the fluorophore is low. In the process of amplification the *Taq* polymerase degrades the 5′-end of the primer, and fluorophore and quencher are then separated. This leads to an irreversible increase in fluorescence from the fluorophore and to a decrease from the quencher (**Fig. 1B**). Molecular Beacon probes consist of a hairpin structure where the loop is complementary to the target sequence. The stem of Cs and Gs holds it in the hairpin structure, on one end of the stem is an attached fluorophore and on the other a quencher. With the fluorophore and the quencher brought so close together, collisional quenching occurs. If the Molecular Beacon probe binds to the target the structure formed is thermodynamically more stable than the hairpin structure at the fluorescent-acquisition temperature. Once the probe binds to the target the hairpin is opened and fluorophore and quencher are separated. The fluorescence that occurs is reversible because the probe dissociates at high temperatures and again assumes the hairpin form (**Fig. 1C**).

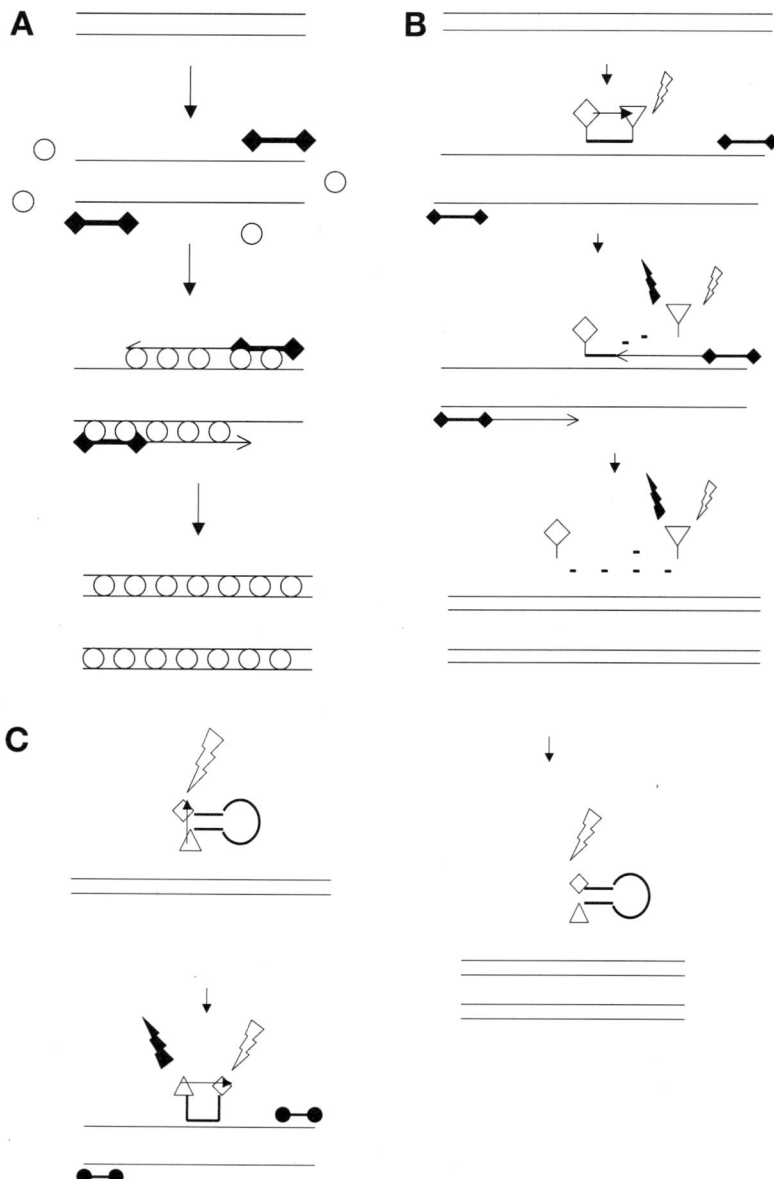

Fig. 1. (A) The mechanism by which SYBR green is bound to double-stranded DNA (↔ = primers, ○ = SYBR green). **(B)** The binding of a TaqMan probe to DNA via the target sequence, and the release of flourophore and quencher owing to action of *Taq* polymerase. **(C)** The binding of Molecular Beacon probes to DNA via the target sequence showing quenched flourophore before binding, flourescence on binding, and dissociation and reformation of probe at high temperatures.

A relative expression level = $2^{(CT\ sample1 - CT\ sample2)}$

B relative expression level = $E_{target}^{(Ct\ sample\ 1 - Ct\ sample\ 2)}$

C relative expression level = $\dfrac{E_{target}^{delta\ Ct\ target\ gene\ (Ct\ sample\ 1 - Ct\ sample\ 2)}}{E_{housekeeping}^{delta\ Ct\ housekeeping\ gene\ (Ct\ sample\ 1 - Ct\ sample\ 2)}}$

Fig. 2. Formulas for calculating the x-fold difference with the ΔCT values obtained by RT-PCR: (**A**) efficiency is assumed to be 100% and expression of housekeeping gene between two samples is identical; (**B**) efficiency of target gene is taken into account and ΔCt should be ≤1 for the housekeeping genes; (**C**) normalized level of relative gene expression in relation to the housekeeping gene *(15)*.

Choosing the right probe is dependent on the experimental setup. TaqMan probes can be used for quantification and mutation detection, and Molecular Beacon probes are convenient to use for allelic discrimination or gene expression. Analyses of RT-PCR results depend on how the efficiency is taken into account (**Fig. 2**).

1.4. Growing A. thaliana Hydroponically

Most experiments involving *A. thaliana* and exposure to radioactivity have used cell cultures and distant high-dose sources of radioactivity. Many experiments with other pollutants have used similar experimental setups. To imitate the real conditions for plants and the effect of radioactivity, a system that allows the use of whole plants is useful. Toquin et al. *(13)* describe a low-maintenance system to grow *A. thaliana* in hydroponic systems in which no aeration is needed to maintain healthy plants *(13)*. The growth medium does not need to be changed if the plants are grown for 3–4 wk. If, however, plants are grown for their whole life cycle then replacement of growth medium is required. The system permits use of whole plants for irradiation experiments, and radioactivity can be taken up through the roots as it would be from contaminated soil. Plants grow normally and **Fig. 3** shows their development (**B, C**) approx 10 and 29 d after planting seeds onto the agar filled tubes.

2. Materials

Everything used with RNA should be sterile, autoclaved, and RNAase-free, e.g., pipet tips, water, and Eppendorf tubes. If necessary, equipment (e.g., pipets) can be wiped with 75% ethanol before use.

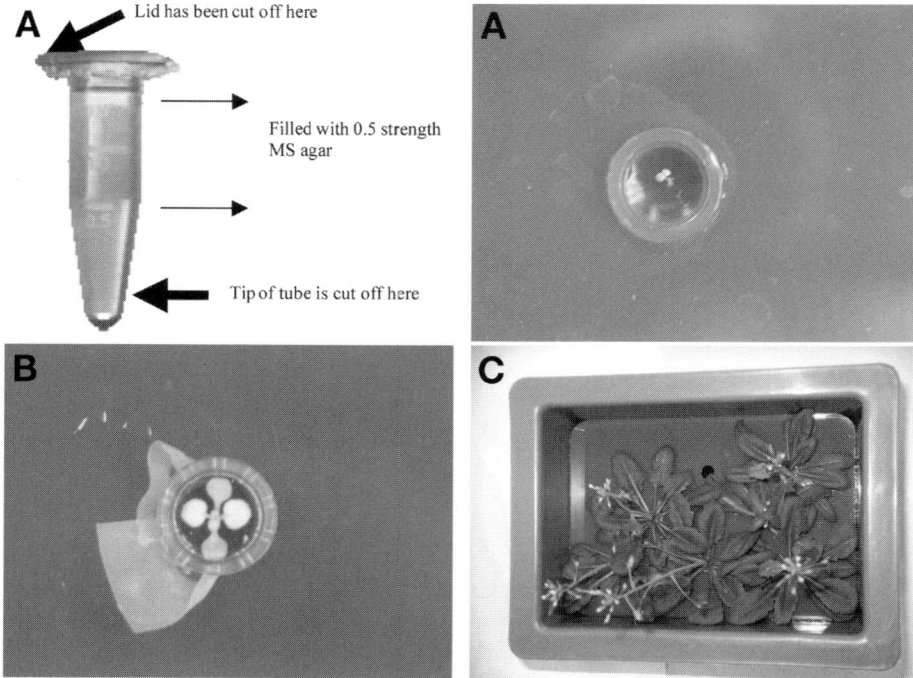

Fig. 3. (**A**) The hydroponic system with *Arabidopsis thaliana* plants 2–3 d after germination. A 0.5-mL Eppendorf tube has been customized by cutting off the lid and the tip, smaller arrows indicate the level of 0.5 strength MS agar filled into the tube. (**B**) The plants are now approx 10 d old. (**C**) Plants have been kept in the containers for 29 d without aeration or change of growth medium. They have been allowed to mature and the seeds were then collected. Prior to that stage the growth medium had to be supplemented as a result of complete uptake by the plants.

2.1. Sterilizing A. thaliana Seeds

1. 10% Ethanol.
2. 10% Bleach.
3. Distilled water.
4. *A. thaliana* seeds (Col-0, N1092).
5. Sterile 1.5-mL Eppendorf tubes.
6. Centrifuge.

2.2. Growing A. thaliana Hydroponically

1. 0.5-mL Eppendorf tubes (**Fig. 2**).
2. Murashige and Skoog (MS) agar (8 g Bacto agar/L).
3. Distilled water.
4. Appropriate containers to hold tubes and growth medium.

5. MS plant medium (half strength).
  6. Environmental chamber.

2.3. Exposure of A. thaliana Plants to Cesium-137

  1. Cesium-137 (Cs-137).
  2. 10 µM CsCl solution.
  3. 3-wk *A. thaliana* plants.

2.4. RNA Isolation

  1. TRI REAGENT (Sigma, Dorset, UK).
  2. Pestle and mortar.
  3. Liquid nitrogen.
  4. Cool centrifuge.
  5. Chloroform.
  6. Isopropanol.
  7. 75% Ethanol.
  8. 1.5-mL Screw-cap Eppendorf tubes (sterile).
  9. 2.0-mL Eppendorf tubes (sterile).
  10. Sterile Rnase-free distilled water.
  11. Ultraviolet (UV) spectrometer (CE2021, Cecil, Cambridge, UK).

2.5. DNase Treatment

  1. Sterile RNase-free distilled water.
  2. Deoxyribonuclease I (DNase I), amplification grade (Sigma).
  3. Heating block.

2.6. cDNA Synthesis

  1. iScript reverse transcriptase (iScript kit from Bio-Rad, Hercules, CA).
  2. iScript 5X reaction mix.
  3. RNase-free water.
  4. PCR machine (PTC-200, MJ Research, Waltham, MA).

2.7. Primer Design

  1. Sequence of target gene.
  2. Primer design program (Beacon Designer 2, Bio-Rad).

2.8. RT-PCR

  1. iQ SYBR green supermix (Bio-Rad).
  2. Set of gene specific primers.
  3. Distilled RNase-free water.
  4. iCycler iQ Multicolor real time PCR detection system (Bio-Rad).
  5. 1.5-mL Eppendorf tubes (sterile).
  6. 96-Well plate.
  7. Optical clear sealing tape.

3. Methods

3.1. Sterilizing A. thaliana Seeds

1. Seeds are first washed for 5 min in an Eppendorf tube in 1 mL 10% ethanol and centrifuged for 30 s to spin the seeds down. The ethanol is decanted and 1 mL 10% bleach is added to the tube, vortexed, and then left for 5 min. The seeds are centrifuged again for 30 s, the bleach removed, and then the seeds are washed five times with distilled water with centrifuging for 30 s between each wash. The seeds can be stored up to 1 wk at 4°C in distilled water.

3.2. Growing A. thaliana Plants Hydroponically

A. thaliana plants can be grown in the hydroponic system through their entire life cycle without aeration but growth medium might have to be added or replaced (*see* **Note 1**).

1. To grow the plants, prepare 0.5-mL Eppendorf tubes by cutting off the lid and the tip of the tubes (**Fig. 3**). Place upside down on masking tape and fill with MS agar. After the agar has set, place the tubes in the containers, which are filled with 0.5X MS medium (*see* **Note 2**).
2. Place sterilized A. thaliana seeds in the middle of each 0.5-mL Eppendorf tube on the MS agar (8 g Bacto agar/L). Put plants in an environment chamber to grow at 23°C with 16 h light and 8 h dark with 85% humidity.

3.3. Exposure of A. thaliana Plants to Cesium-137

1. A. thaliana plants (3 wk old) in the 0.5-mL Eppendorf tubes are placed in an Eppendorf holder with the tubes standing in a Petri dish.
2. 10 mL of a 10 μM CsCl solution containing 187.5 kBq/L Cs-137 is added to the plate and the plants are returned to the environmental chamber for 4 h. The control is placed in the same way in a 10 μM CsCl solution.

3.4. RNA Isolation

1. The plants are harvested by gently pulling out the whole plant from the agar. The roots are rinsed briefly with water, dried carefully with tissue, wrapped in foil, and shock-frozen in liquid nitrogen. The plants are ground using a pestle and mortar using liquid nitrogen to prevent thawing.
2. The sample is carefully transferred to a sterile 2.0-mL Eppendorf tube and, per 100 mg plant tissues, 1 mL of TRI REAGENT (Sigma) added to the tube, vortexed, and left at room temperature for 5 min. The whole mixture is transferred to a 1.5-mL Eppendorf screw-cap tube and 200 μL chloroform is added per 1 mL TRI REAGENT, vortexed, and left at room temperature for 15 min.
3. The tubes are centrifuged at 4°C for 15 min at 12,000g. Three different layers are formed in the process. The upper aqueous top layer contains the RNA and is carefully pipetted into a new 1.5-mL screw-cap tube. 500 μL isopropanol/mL TRI REAGENT are added and left for 10 min at room temperature. The RNA sample

Gene Expression in Plants Exposed to Radioactivity

is centrifuged at 4°C for 10 min at 12,000g. Pellets containing RNA are formed. The supernatant is removed and the pellet washed with 1 mL 75% ethanol/mL TRI REAGENT. The sample is further centrifuged at 4°C at 7500g for 5 min. If the pellet floats after washing then washing should be repeated and the sample centrifuged at 12,000g. The RNA obtained is briefly air-dried for 5–10 min and resuspended in sterile RNase-free water. To dissolve the pellet completely it is heated up to 55–60°C while occasionally pipetting up and down.

4. RNA purity and concentration are determined using an UV spectrometer at 260- and 280-nm wavelengths.

3.5. DNase Treatment

1. 1 µg RNA is pipetted into a 0.5-mL Eppendorf tube in a maximum of 8 µL of RNase-free distilled water. 1 µL of DNase reaction buffer (10X) and DNase (1 U/µL) are added to the tube and incubated at room temperature for 15 min.
2. 1 µL of the stop solution (EDTA) is added to a total of 11 µL in the tube and placed on a heating block for 10 min at 70°C and afterwards immediately placed on ice.

3.6. cDNA Synthesis

1. To the RNA treated with DNase, the following components are added: 4 µL distilled RNase-free water, 4 µL 5X reaction mix, and 1 µL iScript reverse transcriptase.
2. The cDNA synthesis is carried out in a thermal cycler using the program: 5 min at 25°C, 35 min at 42°C, 5 min at 85°C and hold at 4°C (optional). The cDNA can then be frozen overnight at –20°C (*see* **Note 3**).

3.7. Primer Design

To design the primers a program supplied for the iCycler (Bio-Rad) is used (*see* **Note 4**).

1. Parameters are set as low as possible for –ΔG values and so stringent for the other values that no more then a couple of primer sets are given. The target T_a is set to 55°C ± 5°C, and the lengths of the primer are set to between 17 and 25 bp. The amplicon length itself is set to be 80–150-bp long.
2. The maximum primer pair T_m mismatch temperature default setting is 3°C, maximum cross dimer ΔG value (3'-end) 3 –kcal/mol and the same for the internal cross dimer maximum ΔG value. The maximum ambiguous bases in the amplicon were defined as 0. If the default setting does not result in any primer pairs then stringency is reversed, first on settings such as primer length or amplicon length and then on the other settings.

3.8. RT-PCR

1. The total amount of all components used is calculated before the run. It is recommended to use 50-µL samples but we find that 20 µL of total sample volume gives equally reliable results. For every gene triplicate, or at least duplicate, analysis is carried out along with water controls. A mix with SYBR green supermix and

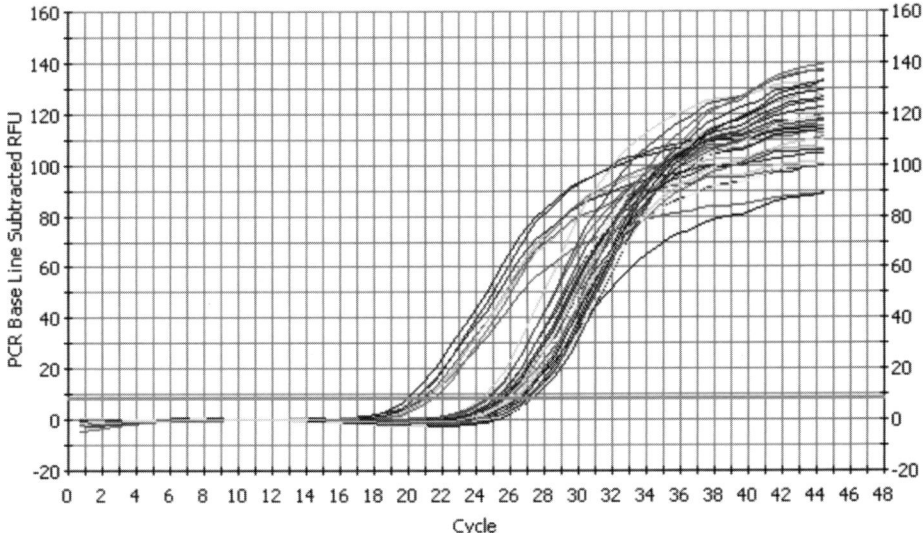

Fig. 4. Amplification curves obtained by RT-PCR showing the threshold cycle values as obtained using the Bio-Rad iCycler. (Data analysis parameters: calculated threshold has been replaced by the user-selected threshold 8.1; selected baseline cycles are 10–13; data analysis window is set at 95.00% of a cycle, centered at end of the cycle; weighted mean digital filtering has been applied; global filtering is off.)

water is prepared before being divided to create mastermixes for every set of primers used. The primers are then added in proportional amount to the number of samples in the mastermix and kept on ice.

2. To calculate the efficiency for every set of primers it is necessary to perform a dilution series with 10-fold-diluted cDNA starting with 1 µL of cDNA and dilute to 1×10^{-5} or 10^{-6} if enough wells are available. For every gene, therefore, efficiency analysis is needed in each RNA sample. The total concentration of SYBR green fluorescence, and any other fluorescence dye used, must be the same across the samples. (Each well contained 10 µL SYBR green, 1 µL cDNA or diluted cDNA or water, 1 µL of each primer [10 µM], and 7 µL of RNAse-free distilled water in our setup.) Once the efficiency is established no further analysis by dilution series is required.

3. The threshold cycle line shows the amount of cDNA optimal for the RT-PCR: if the exponential phase starts at ± 20 in the run-time central window then cDNA concentration is considered to be optimal. An example of ct-value diagram after a completed RT-PCR run is shown in **Fig. 4** (*see* **Note 5**).

4. Notes

1. It is important that the seeds are kept at high humidity for germination. If humidity levels are too low then the agar in the tubes will dry out and no germination will occur. Therefore it helps, after sterilizing, to soak the seeds in distilled water

and keep them at 4°C, and to put a cover over the containers holding the agar tubes. The cover has to be removed 2–3 d after germination, otherwise the plants will suffer from chlorosis and fungal infections. If problems with high contamination with fungi occur then rinsing and wiping every container and lid with 70% ethanol is recommended.
2. The medium used to grow the plants can be bought or made up but it has been shown that one-fourth or one-eighth strength medium shows the best effects on growth if used in hydroponics *(14)*. If plants show signs of chlorosis after germination then addition of ammonium and iron in the form of ammonium nitrate (NH_4NO_3, 30 mM) and ferric citrate (iron[III] citrate, 90 mM) into the MS agar can help. Aeration is not necessary.
3. cDNA synthesis is best done the same day as the RT-PCR and should, if possible, not be stored and used later. DNase treatment should always be done before cDNA synthesis is carried out to avoid contamination with genomic DNA. Furthermore, it should be ensured that the reverse transcriptase performs well because cDNA quality and primers have to be optimal to perform a satisfactory RT-PCR run. Only high-quality reverse transcriptase should be used and if possible the reverse transcriptase should come in a kit with ready-to-use solutions because these exclude fluctuations in the concentration of the reagents and are optimized for use. The cDNA for the dilution series for the RT-PCR should not be mixed by vortexing. Gently mixing by flicking is recommended. If a good pipet is used then reproducibility of pipetting is guaranteed with amounts as low as 1 µL for the dilution series.
4. Designing the right primers is crucial to a successful RT-PCR because few parameters beside cDNA quality cannot be changed. Guidelines recommending the length of the amplicon to be between 50 and 175 bp should be followed as well as the suggestions for the annealing temperature being around 55 ± 5°C for all primers used in the same run.
5. Contamination of the negative controls in the RT-PCR can be reduced significantly by wiping all items such as pipets and Eppendorf holders with 70% ethanol as well as changing gloves between the major steps and after handling the different sets of primers. If possible, leaving wells between the samples and negative controls decreases the risk of further contamination.

Acknowledgments

The work described here was supported by a bursary of the University of The West of England. John Hancock, Mrs. Judy Brown, and Mr Arthur Gough have given technical support and valuable advice.

References

1. Zhu, Y. G. and Shaw, G. (2000) Soil contamination with radionuclides and potential remediation. *Chemosphere* **41,** 121–128.
2. Cunningham, S. D., Berti, W. R., and Huang, J. W. (1995) Phytoremediation of contaminated soils. *Trends Biotechnol.* **13,** 393–397.

3. Cunningham, S. D. and Ow, D. W. (1996) Promises and prospects of phytoremediation, *Plant Physiol.* **110**, 715–719.
4. Broadley, M. R. and Willey, N. J. (1997) Differences in root uptake of radiocaesium by 30 plant taxa, *Environ. Pollut.* **97**, 11–15.
5. McGrath, S. P. and Zhao, F. J. (2003) Phytoextraction of metals and metalloids from contaminated soils. *Curr. Opin. Biotechnol.* **14**, 277–282.
6. Assunção, A. G. L., da Costa Martins, P., de Folter, S., Vooijs, R., Schat, H., and Aarts, M. G. M. (2001) Elevated expression of metal transporter genes in three accessions of the metal hyperaccumulator *Thlaspi caerulescens. Plant Cell Env.* **24**, 217–226.
7. Cobbett, C. S. and Meagher, R. B. (2002) Arabidopsis and the genetic potential for the phytoremediation of toxic elemental and organic pollutants. In: *The Arabidopsis On-Line Textbook,* (Somerville, C. R. and Meyerowitz, E. M., eds.), American Society of Plant Biologists, Rockville, MD, doi. 10.1199/tab. 0032.
8. Doucet-Chabeaud, G., Godon, C., Brutesco, C., de Murcia, G., and Kazmaier M. (2001) Ionising radiation induces the expression of *PARP-1* and *PARP-2* genes in *Arabidopsis. Mol. Genet. Genomics* **265**, 954–963.
9. West, C. E., Waterworth, W. M., Jiang, Q., and Bray, C. M. (2000) *Arabidopsis* DNA ligase IV is induced by gamma-irradiation and interacts with an *Arabidopsis* homologue of the double strand break repair protein XRCC4. *Plant J.* **24**, 67–78.
10. Doutriaux, M. P., Couteau, F., Bergounioux, C., and White C. (1998) Isolation and characterisation of the *RAD51* and *DMC1* homologs from *Arabidopsis thaliana, Mol. Gen. Genomics* **257**, 283–291.
11. Deveaux, Y., Alonso, B., Pierrugues, O., Godon, C., and Kazmaier, M. (2000) Molecular cloning and developmental expression of *AtGR1*, a new growth-related *Arabidopsis* gene strongly induced by ionizing radiation. *Radiat Res.* **154**, 355–364.
12. An, Y. Q., McDowell, J. M., Huang, S., McKinney, E. C., Chambliss, S., and Meagher, R. B. (1996) Strong, constitutive expression of the *Arabidopsis ACT2/ACT8* actin subclass in vegetative tissues, *Plant J.* **10**, 107–121.
13. Tocquin, P., Corbesier, L., Havelange, A., et al. (2003) A novel high efficiency, low maintenance, hydroponic system for synchronous growth and flowering of *Arabidopsis thaliana,* BMC *Plant Biol.* **30**, 2.
14. Arteca, R. N. and Arteca, J. M. (2000) A novel method for growing *Arabidopsis thaliana* plants hydroponically. *Physiol. Plant.* **108**, 188–193.
15. Pfaffl, M. W. (2001) A new mathematical model for relative quantification in real-time RT–PCR. *Nucl. Acids. Res.* **29**, 2002–2007.

6

Plant Phylogeny and the Remediation of Persistent Organic Pollutants

Jason C. White and Barbara A. Zeeb

Summary

The purpose of this chapter is twofold. First, we review the relevant literature regarding the phytoremediation of persistent organic pollutants (POPs) as it relates to plant phylogeny. The phytoremediation of POPs is a rapidly developing field of research and as such, a review of the literature is important for understanding the development of experimental methodologies. Second, we present in-depth coverage of the methods we have used for the analysis of weathered POPs in both soil and vegetative tissues. Although the protocols described pertain to the specific POPs cited, the techniques could readily be used for many other contaminants.

Key Words: Phytoremediation; persistent organic pollutants; bioremediation; plant uptake; PCBs; DDT; dioxin; furans; phytoextraction.

1. Introduction

Persistent organic pollutants (POPs) are a group of compounds singled out because of their long-term persistence in the environment, toxicity, and global distribution *(1,2)*. Extensive reviews on POPs and individual contaminants within that group are available; we here briefly touch on the basic properties as they relate to phytoremediation. There has been a significant international effort to minimize or eliminate the usage of a specific subgroup of POPs known as the "dirty dozen" and as such, when many people refer to POPs, they are actually referring to this smaller group of compounds. In fact, much of the literature review here will apply to the dirty dozen, although the methodologies utilized would likely cover the more comprehensive list of contaminants as well. The dirty dozen are split into three groups of compounds based on their intended usage; insecticides, industrial chemicals, and chemical byproducts *(3)*. The insecticides include aldrin, chlordane, DDT, mirex, toxaphene, hexachlorobenzene, dieldrin, endrin, and heptachlor. With the exception of hexachlorobenzene (a fungicide), these compounds are all organochlorine insecticides used to control a range of insect species through a variety of toxic modes of action. The industrial POPs include polychlorinated biphenyls (PCBs). PCBs are complex mixtures of chlorinated hydrocarbons that were extensively used in industry for purposes such as dielectrics in transformers, heat exchange fluids, and paint additives. Polychlorinated dibenzo-*p*-dioxins and polychlorinated dibenzo-*p*-furans (often simply referred to as dioxins and furans) are not produced commercially but are byproducts from the production of other chemicals. For example, dioxins have been detected during the incineration of coal, hospital waste, and car emissions, whereas furans are the main contaminant in PCBs *(3)*. This chapter will also mention polycyclic aromatic hydrocarbons (PAHs). Although not part of the dirty dozen, most people consider these compounds as POPs and a large amount of research on the phytoremediation of these compounds has been done over the last decade. As such, a brief discussion of that literature and the relevant methodologies will be included.

With regard to persistence, half-lives of POPs in the environment are frequently measured in years or decades *(4,5)*. This general resistance to microbial attack is the result of their unique molecular structure that has no analogs in the natural world and, thus, no enzyme systems have evolved to facilitate their breakdown. These compounds are also characterized as being highly hydrophobic, with log K_{ow} values (octanol–water coefficients) of 3.0–8.3. This hydrophobicity leads to several problems of environmental concern. First, in a soil or sediment, POPs will bind very strongly to the organic fraction of the soil, and as the compounds persist over the decades, they become weathered or sequestered within the organic matter *(6–9)*. This sequestration generally leads to reduced bioavailability as measured by a number of assays and renders many *in situ*-remediation strategies ineffective. Second, the hydrophobicity can result in contaminant bioaccumulation in the fatty tissues of living organisms *(10,11)*, and, potentially, biomagnification within food chains. Compounding the issue of bioaccumulation is the fact that some of the POPs are highly toxic, including having known or suspected carcinogenic and mutagenic properties *(12–14)*. The final problematic feature of POPs is that many of them are semivolatile and have achieved global distribution *(1,2,15)*. This feature is perhaps best represented by studies reporting bioaccumulation of various POPs within arctic food chains *(11,16–19)*.

The literature pertaining to the phytoremediation of POPs is not extensive. It is noteworthy that much of the work done with plants and POPs relates to uptake of the unaltered parent compound; for most other organic contaminants, degradation either outside or inside the plant tends to be the fate most observed. Lichtenstein in the late 1950s and early 1960s published extensively on the uptake of several chlorinated hydrocarbon insecticides now considered to be POPs. Although the pesticides had only been present in the soil for a few years, he was able to measure varying amounts of uptake and translocation of aldrin and heptachlor into the roots and shoots of species such as carrots (*Daucus carota*), peas (*Pisum sativum*), cucumber (*Cucumis sativum*), lettuce (*Lactuca sativa*), alfalfa (*Medicago sativa*), and soybeans (*Glycine max*) *(20–23)*. The Analytical Chemistry department at the Connecticut Agricultural Experiment Station (CAES), as part of its marketbasket survey of pesticide residues in produce, reported variable levels of POPs such as chlordane and heptachlor epoxide in the fruit of various plant species including squash (summer and winter; *Cucurbita pepo*), cantaloupe (*Cucumis melo*), cucumber (*C. sativum*), pumpkin (*C. pepo*), and sweet potato (*Ipomoea batatas*) *(24,25)*.

Later research projects at CAES focused specifically on two of the pesticide members of the POPs group, *p,p'*-DDE and chlordane. The list of species investigated for uptake and translocation of POPs under both field and greenhouse conditions includes lettuce (*L. sativa*), spinach (*Spinacia oleracea*), peppers (*Capsicum annuum*), tomatoes (*Lycopersicon esculentum*), carrots (*D. carota*), potatoes (*Solanum tuberosum*), beans (*Phaseolus vulgaris*), beets (*Beta vulgaris*), eggplant (*Solanum melongena*), corn (*Zea mays*), dandelion (*Taraxacum officinale*), mustard (*Brassica juncea*), vetch (*Vicia villosa*), ryegrass (*Lolium perenne*), lupins (*Lupinus albus* and *Lupinus angustifolius*), canola (*Brassica napus*), peanut (*Arachis hypogaea*), pigeonpea (*Cajanus cajan*), pumpkin (*C. pepo*), cucumber (*C. sativus*), squash (*C. pepo*), clover (*Trifolium pratense*), and alfalfa (*M. sativa*) *(26–32)*. In many cases, more than one cultivar of a species has been investigated. In all cases, measurable levels of either DDE or chlordane are evident in the root tissue, although the precise concentrations vary by well over an order of magnitude. However, significant translocation of the contaminants from the roots to the shoots has been observed for only one species; *C. pepo* (pumpkin and zucchini). Similar findings have been reported by researchers at the Royal Military College of Canada (RMC). One study compared the ability of five plant species (zucchini—*C. pepo* L. cv. summer squash, tall fescue—*Festuca arundinacea* Schreb., alfalfa—*M. sativa*, rye grass—*Lolium multiflorum*, and pumkin—*C. pepo* cv. Howden) to mobilize and translocate DDT under greenhouse conditions. Because of low ambient temperatures, the DDT mixture in soil had degraded very little, and consisted of largely 4,4′-DDT. Although DDT was found in the roots and shoots of all five plant species, *C. pepo* (pumpkin and zucchini) were the only plants to achieve significant extraction and translocation, with little or no preferential translocation or transformation of DDT compounds observed within the plant tissues *(33)*. Recent work at CAES has shown that two distinct subspecies of *C. pepo* phytoextract have significantly *different* quantities of the residues; ssp *texana* is characterized by minimal-to-low quantities of DDE in only its roots, whereas ssp *pepo* has high concentrations of contaminant throughout the plant *(32)*. Similar data is available for the second group of POPs, dioxins and furans. Hulster et al. *(34)* described the "exceptionally high" uptake of these two weathered contaminants by zucchini (*C. pepo* ssp *pepo*) from soil. Interestingly, through the use of congener profiles, the much lower levels of POPs in plants such as cucumber (*C. sativum*) were attributed to aerial deposition and not soil-to-plant transfer. Campanella and Paul *(35)* also report significant phytoextraction of weathered dioxins and furans by certain Cucurbitaceae.

There has been some work on the phytoremediation of the third group of POPs, PCBs, but it is again noteworthy that much of the effort has focused on dechlorination and/or degradation of the congeners outside of the plant. The traditional assumption has been that these contaminants, as with other POPs, are too hydrophobic to accumulate in plant tissues *(36,37)*. Puri et al. *(38)* produced a comprehensive database on the possible uptake and metabolism of PCBs in plants using a prepared PCB mixture containing Aroclors 1221, 1242, and 1260 dissolved in hexane and clean-sieved soil. To study the translocation of PCBs, the prepared PCB mixture in dimethyl sulfoxide was injected directly into plants grown in uncontaminated soil. They concluded that plants take up PCBs to a very limited extent from their environment and do not modify the extractable congeners appreciably.

John Fletcher's group at the University of Oklahoma has published extensively on the use of mulberry (*Morus rubra*) to facilitate the remediation of PCBs through a process they term "rhizoremediation." The mechanism of removal focuses on the release of certain phenolic constituents from mulberry roots that in turn support the growth of bacteria that degrade PCBs co-metabolically *(39)*. Similar rhizosphere effects on PCBs have been reported elsewhere *(40)*. A group at Michigan State University is involved in a project assessing the likelihood of PCB co-metabolism within the rhizsophere of alfalfa through the inoculation of bacteria engineered with PCB-degrading genes. There have been a number of studies on the metabolism of PCBs by cell cultures of various plant species, including wheat (*Triticum aestivum*), black nightshade (*Solanum nigrum*), tomato (*L. esculentum*), soybean (*G. max*), barley (*Hordeum brachyantherum*), birch (*Betula uliginosa)*, and mulberry (*M. rubra*) *(41–43)*. Perhaps more analogous to earlier studies, Pier et al. *(44)* examined PCB concentrations in more than 1000 vascular plant specimens and the soils in which they grew. All plants (but not all soils) contained detectable levels of PCBs. Bioaccumulation factors ([PCB] tissues/[PCB] soil) decreased with increasing soil concentrations, suggesting that at higher levels of exposure, accumulation of PCBs may be kinetically limited within the plant. Total PCB root:shoot ratios approached unity in all plant specimens leading to the hypothesis that PCB redistribution was occurring within the plants. Recent work by Zeeb et al. *(45)* examined nine plant species (*F. arundinacea, Gycine max, M. sativa, Phalaris arundinacea, L. multiflorum, Carex normalis*, and three varieties of *C. pepo*) grown in three concentrations of PCB-contaminated soil (~70 µg/g, ~150 µg/g, and ~4500 µg/g) under greenhouse conditions designed to control for volatilization. Overall, results indicated that varieties of *C. pepo* were the most effective at removing PCBs from soils, and the translocation of highly chlorinated congeners was observed.

As previously mentioned, although not included as a member of the dirty dozen, most people consider PAHs to be POPs and this is the group of organic contaminants on which much phytoremediation effort has focused. The research has included not only individual PAH contaminants such as phenanthrene, pyrene, chrysene, benzo(a)pyrene, benzo(a)anthracene, and dibenz(a)anthracene *(46–49)*, but also on more complex mixtures of materials that contain PAHs; namely gasoline, various fuel oils, diesel, and aged petroleum sludge *(50–54)*. Similar to some of the work previously described for PCBs, the focus on PAHs has been degradation of the contaminants within the rhizospheres of various plant species *(55)*. The list of vegetation shown to be somewhat successful in facilitating PAH removal is extensive and has generally focused on various species of perennial grasses because of their dense fibrous root system. Some of the species investigated include big bluestem (*Andropogon gerardii*), switch grass (*Panicum virgatum*), Indian grass (*Sorghastrum nutans*), western wheatgrass (*Agropyron smithii*), Canada wild rye (*Elymus canadensis*), side oats grams (*Bouteloua curtipendula*), blue grams (*Bouteloua gracilis*), little bluestem (*Schizachyrium scoparius*), alfalfa (*M. sativa*), fescue (*F. arundinacea*), sudan grass (*Sorghum vulgare* var. *Sudanese*), willow (*Salix babylonica*), hybrid willow (*Prairie cascade*), alpine bluegrass (*Poa alpina*), ryegrass (*L. perene*), and Bermuda grass (*Cynodon dactylon*).

2. Materials

The precise list of materials required for the extraction and recovery of persistent organic pollutants from vegetation (and soil) is somewhat dependent on which of the analytical methodologies are followed in the next section. The vendor is not specified if the material is generic and available through large scientific suppliers.

1. Potential solvents needed are methanol, acetone, trimethylpentane, dichloromethane, petroleum ether, diethyl ether, hexanes, and 2-propanol. Required reagents include anhydrous sodium sulfate and PR-grade 60/100-mesh Florisil (US Silica, Berkeley Springs, WV). An explosion-proof blender or soxhlet extraction system may be required for the solvent extraction of vegetation.
2. Basic laboratory supplies include 500-mL separatory funnels with glass stoppers and Teflon stopcocks, glass funnels, blender jars, 500-mL Kuderna–Danish flask

fitted with a 10-mL concentrator tube, 100-mL graduated cylinders with glass stoppers, aluminum foil, aluminum weigh dishes, an oven capable of 100°C, and 10- to 1000-µL syringes.
3. If available, a CEM MES-1000 microwave solvent extraction system (CEM Corporation, Mathews, NC) may be used for soil extractions. Analyte standards may be purchased through Chem Service (West Chester, PA) or acquired from the US Environmental Protection Agency National Pesticide Standard Repository (Fort Meade, MD).

3. Methods

The methodology is split into two separate sections based on the matrix to be analyzed, soil or vegetation. Within each matrix, methods utilized by the CAES, United States, and the RMC Canada are described. A third brief section on instrument conditions for analysis will conclude the methodology.

3.1. POPs Extraction From Soil

Although this chapter focuses on the relationship between plant phylogeny and POPs remediation, some mention of procedures related to soil preparation and analysis is necessary both to establish methodologies for determining the initial contaminant concentration or load in the source compartment, but also as a means to assess phytoremediation efficiency via postharvest contaminant quantitation and to support mass balance studies.

3.1.1. Soil Collection and Preparation

1. In terms of soil collection, it is often necessary to consider three separate soil fractions. The *bulk soil* consists of nonvegetated control soil and is routinely collected via soil cores from the top 15-cm of soil prior to planting or at harvest. The *near-root zone* soil is operationally defined as both the soil that falls off the roots at harvest and as the soil within the area encompassed by the plant roots. The *rhizosphere* soil is operationally defined as the soil retained by the roots at harvest.
2. Root samples can be placed on aluminum foil for approx 1–2 h to allow the attached soil to air-dry, and the rhizosphere soil can then removed with a fine-bristle toothbrush or by vigorously shaking the roots.
3. All soil fractions should be sieved to 0.5 mm to remove nonsoil debris and to provide sample homogeneity. The soil fractions should be stored in 250-mL amber glass bottles and with Teflon-lined screw caps at 4°C.
4. A representative 3.0-g portion of each of the soil treatments can be placed in an oven at 100°C for 24 h for moisture determination.

3.1.2. POPs Extraction

At CAES, two procedures have been used to extract weathered POPs from soil; first is a microwave-assisted extraction (MAE), the second a hot solvent soak (*see* **Note 1**).

3.1.2.1. MAE of POPs

1. Portions (3.0 g) of the soils are transferred to a Teflon PFA-lined digestion vessel from the CEM MES-1000 microwave solvent extraction system (CEM). 50 mL of 2:3 (v/v) hexane/acetone is added to the vessel, followed by 500–1000 ng of transnonachlor (TN) as an internal standard. For chlordane analysis, ^{13}-C TN, ^{13}-C *cis*-chlordane (CC), and ^{13}-C *trans*-chlordane (TC) are used.
2. The vessels are sealed and placed in the CEM MES-1000 oven and extracted using the following program: 100% power, 7 min ramp to 120°C; 20 min hold time.
3. The liquid phase is decanted into Kuderna–Danish flasks (Organomation, Berlin, MA) fitted with 10-mL concentrator tubes containing a boiling chip. The residual soil in the extraction vessels is rinsed twice with 15-mL portions of 2:3 hexane/acetone, and the solvent is combined with the original extract.
4. A Snyder column is fitted to the flask and the solvent is reduced to less than 1 mL in a 95°C water bath. Three milliliters of 2,2,4-trimethylpentane are added through the Snyder column and the volume is again reduced to 1 mL. The flask is then removed and the column is rinsed with an additional 3 mL of trimethylpentane. Upon cooling, the final volume of solvent is adjusted to 10 mL. Prior to analysis, the soil extract is passed through a glass microfiber filter (0.2 µm; Laboratory Science Inc., Sparks, NV) to ensure complete removal of particulates.

3.1.2.2. Hot Solvent Soak Extraction

1. Portions (3.0 g) of the soil fractions are weighed into 40-mL amber vials containing 15 mL of hexanes and 500–1000 ng of TN (^{13}-C TN, ^{13}-C CC, and ^{13}-C TC are used for chlordane analysis). The vials are sealed with Teflon-lined caps and placed in an oven at 70°C for 4–6 h.
2. After cooling for 10 min, a 1-mL aliquot of the supernatant is passed through a glass microfiber filter for particulate removal prior to analysis. For a discussion of the differences between these extraction procedures, *see* **Note 1**.

3.1.2.3. EPA 3540C Method of Extraction

For phytoremediation studies completed at RMC, one of two laboratories are utilized; the Analytical Services Group (ASG) on the RMC campus, or the Analytical Services Unit (ASU) at nearby Queen's University.

1. In both cases, soil samples are extracted according to the industry standard methods (EPA 3540C) in a soxhlet apparatus for 4–6 h at 4–6 cycles/h with 250 mL of dichloromethane (*see* **Note 2**). All samples are spiked with an internal standard; decachlorobiphenyl for DDT and ^{13}C-labeled PCB congener 2,2′,4,4′,5,5′-hexachlorobiphenyl for PCB congener analysis.
2. The extract is concentrated by roto-evaporation to 1 mL and the solvent exchanged to hexanes using a Büchi Rotavapor R-114. The extract is applied to a Florisil column, and diluted with hexanes to 10 mL. A fraction of the eluant (1 mL) is transferred to a vial in preparation for analysis. If necessary, samples are concentrated before analysis by blowing down with nitrogen.

3.2. POPs Extraction From Vegetation

At CAES, the method used for extracting POP residues from vegetation is that of Pylypiw *(56)* (*see* **Note 3**).

3.2.1. The CAES Method

1. Samples of vegetation are collected from the field and where appropriate, split into roots, stems, leaves, and fruit or simply roots and shoots, and the biomass of individual tissue compartments is determined. The vegetation should be thoroughly rinsed with tap water to remove attached soil particles. Residual soil on the surface of the root samples could confound results obtained from vegetative extracts of this plant tissue. Therefore, after rinsing, the root samples may be added to 300 mL of distilled water and sonicated for 15 min (FS30, Fisher Scientific, Springfield, NJ). All vegetative samples are finely chopped with a knife or commercial food chopper and are either extracted immediately or are stored in 250- to 500-mL amber glass bottles with Teflon-lined caps in a freezer.
2. 25-g portions of the vegetation are weighed into a 1-qt blender container with 25-mL of 2-propanol and 500–3000 ng of TN as an internal standard (^{13}-C TN, ^{13}-C CC, and ^{13}-C TC are used for chlordane analysis). The sample is blended at high speed in an explosion-proof blender (Fisher Scientific) for 30 s and 50 mL of petroleum ether is added. Smaller quantities of vegetation may be extracted if necessary but the ratio of biomass to 2-propanol (1:1) and 2-propanol to petroleum ether (1:2) should remain constant. The sample is blended at 40% of full speed for 4–5 min and then the sample is allowed to settle for 30 s.
3. The extract is decanted into a glass funnel packed with glass wool and collected in a 500-mL glass separatory funnel with a Teflon stopcock. After complete draining of the solids (approx 20 min), 100 mL of distilled water and 10 mL of saturated sodium sulfate solution are added separately to each funnel. The funnel is capped and swirled gently for 10 s. The funnel should sit for 15–20 min for phase separation. The water can then be drawn off and the petroleum ether should be rinsed two additional times with distilled water.
4. The final extract (approx 30–50 mL) is then amended with 15 g of anhydrous sodium sulfate. For analysis by gas chromatography with electron capture detection (GC/ECD), the extracts are allowed to sit for 2–3 h prior to analysis. However, analysis of chlordane is by chiral-based mass spectrometry and an additional extract cleanup step may used to promote GC column life.
5. For additional extract clean up, a fritted chromatography column (22-mm internal diameter by 34-cm long) is packed with a 12-cm layer of PR-grade 60/100-mesh Florisil (US Silica) on top of which was added 2 cm anhydrous sodium sulfate. The column should be conditioned with 50 mL of petroleum ether, which is discarded. The vegetation extract in petroleum ether is added to the column, and the graduated cylinder should be rinsed two times with 10˙mL of petroleum ether. These washings are transferred to the column and the liquid is drained, but not to dryness, into the column packing. The column is first eluted with 100 mL of

petroleum ether, which is discarded, and then rinsed with 100 mL of 6% diethyl ether in petroleum ether. This solvent fraction is collected in a 500-mL Kuderna–Danish flask (Organomation) fitted with a 10-mL concentrator tube and a three-ball Snyder column, and is concentrated to 1 mL in a hot water bath at 95°C and solvent exchanged to trimethylpentane as previously described. The trimethylpentane is concentrated to 0.5 mL with gentle heat under a stream of nitrogen and the final volume is adjusted to 2 mL with trimethylpentane.

3.2.2. The RMC Method

1. At RMC, vegetation is washed by immersion in a water bath, and then under running tap water until visually no solids remain. It is patted dry using absorbent Kim towels, split into compartments (i.e., roots, lower stem, upper stem, leaves, and flowers), and then air-dried and weighed for moisture determination. Dried plant samples are combined with anhydrous sodium sulfate and ground to a fine powder with a mortar and pestle.
2. Samples are extracted in a soxhlet apparatus for 4–6 h at 4-to-6 cycles/h using 250 mL of dichloromethane. The extract is concentrated to approx 10 mL by roto-evaporation. The concentrated extract is then filtered through a 0.45-µm syringe filter into a 50-mL vial and is rinsed thoroughly with dichloromethane. The filtered extract is then concentrated to 1 mL by blowing down with nitrogen gas.
3. The extract is run through a gel permeation cleanup column to separate the pesticides or PCBs from the plant lipids. Two fractions are collected. The lipid fraction is evaporated to determine the lipid content of the plants. The organochlorine fraction collected from the column is solvent exchanged with hexanes and concentrated to 1 mL by roto-evaporation. The concentrated extract is applied to a Florisil column and is flushed through with 10 mL of hexanes and then concentrated down to 0.5 mL by blowing down with nitrogen gas. A sub-sample (~50 µL) is transferred to a vial for analysis.
4. For quality assurance/quality control, one control, one sample duplicate, and one blank are prepared and extracted with every batch of 10 samples, and all values are reported as µg/g dry weight.

3.3. Analytical Procedures

3.3.1. CAES Procedures

3.3.1.1. DDE Quantitation

1. The amount of p,p'-DDE in the hexane or trimethylpentane soil extracts and the petroleum ether vegetative extracts is determined on a Agilent (Agilent, Avondale, PA) 5890 or 6890 GC with a ^{63}Ni electron-capture detector (ECD). The column is a 30 m × 0.53 mm 0.5 µm SPB(-1 film (Supelco, Inc., Bellefonte, PA).
2. The GC program is as follows: 175°C initial temperature; ramped at 1°C/min to 205°C; then ramped at 15°C/min to 250°C; with a hold time of 5 min. The total run time is 38 min. A 2-µL splitless injection is used, and the injection port is

maintained at 250°C. The carrier gas is He, and the make-up gas is 5% CH_4 in Ar at 20 mL/min. The ECD was maintained at 325°C.
3. A stock standard of crystalline p,p'-DDE can be purchased from Chem Service or acquired from the US Environmental Protection Agency National Pesticide Standard Repository. A portion of the stock is dissolved in either petroleum ether (for vegetative extracts), trimethylpentane (for soil extracts), or hexane (for soil extracts) and diluted to prepare a series of calibration standards of p,p'-DDE at 10, 25, 50, 100, 150, 250, and 500 ng/mL for each solvent. Each calibration level contained 100 ng/mL TN as an internal standard. The retention times of TN and p,p'-DDE were 20.95 and 23.95 min, respectively.

3.3.1.2. CHLORDANE QUANTITATION

1. Extracts are analyzed on a Saturn 2000 Ion Trap GC/MS system (Varian, Sugar Land, TX) equipped with a 30 m × 0.25-mm I.D. × 0.25-μm film thickness γ-DEX-120 chiral column (Supelco). A deactivated silica-guard column (0.5 m × 0.25 mm) is attached before and after the analytical column with press-tight connectors (Restek).
2. The GC oven is programmed as follows: initial temperature 120°C, hold 1 min; ramped at 20°C/min to 155°C; ramped at 0.5°C/min to 195°C; ramped at 20°C/min to 230°C, hold for 21.6 min. The injection port is maintained at 230°C, and a 3-μL splitless injection is used. The mass spectrometer conditions are: a 38-min filament delay, emission current 60 μA, target total ion current 5000 counts, maximum ionization time 25,000 μs, multiplier offset 200 V, and scan range m/z 345 to 425.
3. Chlordane is a multicomponent compound and is quantitated on the basis of the three main constituents that persist: TN, CC, and TC. In addition, TC and CC exist as +/− enantiomers. A set of calibration standards containing TC, CC, and TN are prepared in trimethypentane at the following levels: 10, 25, 50, 100, 250, 500, 1000 ng/mL. Each solution contains racemic TC and CC at the cited concentration, and TN at one-half the cited amount. The calibration solutions may also contain oxychlordane (metabolite of chlordane). Every calibration solution contains 50 ng/mL of each labeled component: $(+)-{}^{13}C_{10}$–TC, $(-)-{}^{13}C_{10}$–CC, and ${}^{13}C_{10}$–TN as internal standards.

3.3.2. RMC Procedures

1. GC/ECD is used to determine the composition and quantity of ΣDDT and Aroclors in soils. Standards are run with each batch of samples and GC/ECD analysis is done using an Agilent 6890 gas chromatograph equipped with a ^{63}Ni GC/ECD, an SPB™-1 fused-silica capillary column (30 m, 0.25 mm I.D. × 0.25-mm film thickness).
2. The chromatographic conditions are as follows: sample volume 2 μL; splitless injection; initial temperature 100°C held for 2 min; ramp 10°C/min to 150°C, 5°C/min to 300°C; final time 5 min. Carrier gas is helium at a flow rate of 2 mL/min. Nitrogen is used as a makeup gas for the ECD. Results obtained for

4,4'-DDT, 2,4-DDT, 4,4'-DDE, 2,4-DDE, 4,4'-dichlorodiphenyl-dichloroethane (DDD), and 2,4-DDD are expressed as nanograms of pesticide per gram dry weight of soil (ng/g). The analytical detection limit for this method is 10 ng/g in the original sample (soil or plant) for all of the DDT, DDE, and DDD compounds and 0.1 ppm Aroclor.

3. GC with mass selective detection is used to determine the composition and quantity of PCB congeners in soils. Two systems are utilized. Most samples are analyzed using an HP 6890 series gas chromatograph equipped with a HP 5973 mass selective detection (GC/MS), and a SPB-1 fused-silica capillary column (30 m, 0.25 mm I.D. × 0.25-mm film thickness).
4. The chromatographic conditions are as follows: sample volume 1 µL; splitless injection; initial temperature 70°C for 4 min; ramp 8°C/min to 180°C hold for 1 min; 3°C/min to 240°C hold for 1 min; 20°C/min to 300°C; final time 10 min. For Aroclor 1260 samples, the oven ramp was modified slightly to improve separation, and the temperature ramp from 180 to 240°C was 1.5°C/min. The carrier gas used here is helium at a flow rate of 1 mL/min.
5. A Varian 3400 CX GC equipped with a Saturn 4D mass selective detection is used for some congener analyses. Chromatographic conditions are similar to those described above. All values are reported as ppm (mg/g) on a dry weight basis.
6. Calibration standards containing all 209 PCB congeners are diluted and run with each batch of samples. Data are collected in single ion-monitoring mode with one primary and one secondary ion. Data-selection criteria are based on compound retention time and on the relative intensity of primary and secondary ions for standard reference congeners and extracted samples.
7. All quality control samples (controls, duplicates, and blanks) are also analyzed. All blanks should be below sample detection limits and controls should be within acceptable limits of the control target, as set by EPA Method 8250A *(57)* (*see* **Notes 4** and **5**).

4. Notes

1. Soils. The protocols described in this chapter for the extraction of POPs from soil have been used successfully in our laboratories and include standard US EPA methods. The MAE procedure was employed first at CAES but is rather labor intensive and the equipment is costly. The hot solvent soak was selected based on roughly equivalent performance with the MAE but involves considerably less resources. However, we do have unpublished evidence that as the contaminant concentration in soil declines, the MAE will outperform the hot solvent extraction. At soil concentrations of 100 ng/g or higher, a hot solvent soak likely performs adequately. At concentrations less than that, we would recommend a more rigorous extraction protocol such as MAE or a soxhlet extraction.
2. One final consideration involves a study conducted by Anderson et al. *(58)*; the authors reviewed a variety of experimental apparatuses and methods to quantify uptake of PAHs from soil by vegetation. They concluded that in general, soxhlet extractions yielded higher percent recovery of PAHs than sonication and

homogenization, especially for aged PAHs, soils with high organic carbon (OC) content, and high molecular weight PAHs. They determined that the primary disadvantage of soxhlet extraction was the time required to exhaustively extract the sample. Sonication and homogenization were quicker, but extraction efficiencies for these two methods were lower than that achieved by soxhlet extraction. Differences in extraction efficiencies for the three techniques were, however, less pronounced for two- and three-ringed PAHs in soils with low (<1%) OC content. Their final analysis was that soxhlet extraction was the method of choice for PAHs with four or more rings in soils with greater than 1% OC or when PAHs had aged with the soil.

3. Vegetation. As previously mentioned, the multiresidue-screening procedure used at CAES was developed to provide simple and straightforward extraction of organochlorine and organophosphate pesticides from various vegetative matrices, but the protocol will likely work well for most hydrophobic organic contaminants. Certain vegetative matrices may be problematic regardless of the contaminant being analyzed. For example, canola (*B. napus*) tissues have large amounts of oils that confound analysis by GC/ECD and thus the florisil cleanup procedure is required. One should take care to cover the blender tightly; we recommend using aluminum foil with the dull side down. The shiny side of the foil contains waxes that may interfere with analysis of the contaminants. In addition, the organic solvents used may extract certain constituents from the rubber and/or plastic tops that are routinely sold with the blender jars, further confounding analysis. Another potential drawback of this procedure is the significant cost of the explosion-proof blender. There are alternative blenders that have the motor encased to avoid sparking in the presence of the organic solvents, but significant quantities of compressed air are required to cool and vent the motor. The RMC soxhlet-extraction protocol is an exhaustive standard extraction method. Although the cleanup steps are labor intensive and the extraction is more time consuming than the CAES method, excellent detection limits are achieved. It is quite likely that procedures such as finely chopping vegetation and shaking vigorously with organic solvents, such as propanol and petroleum ether, would produce adequate recovery of POPs and may prove to be viable options for laboratories with limited resources. However, such methods would need to be validated prior to use.

4. Overall analytical considerations. The analytical methods previously described should work very well for PAHs and for analysis we highly recommend GC/MS. Although GC with flame ionization detection (FID) can provide useful information for standards and matrices containing relatively high concentrations of PAHs, confirmation of individual PAH components should be done via mass spectrometry. Finally, we have found that the use of internal standards for calibration is critical in accurate quantitation and analysis of persistent organic pollutants. The internal standard eliminates much of the typical variability and error associated with instrumentation and experimental protocols.

5. The US EPA has a RAM index for pesticides in food, feed, and animal commodities and an environmental chemistry methods index for pesticides in soil, sediment, and water at http://www.epa.gov/oppbead1/methods/index.htm. Within the RAM index, the listing is by contaminant and many of the methods have been verified

and/or modified by EPA laboratories. Within the environmental chemistry methods index, the listing is also by contaminant, but only 25% of the methods have been verified by EPA laboratories. These two resources focus on currently registered products but some of the methods listed may be of interest to readers. Further methods specifically for food products can be found within the US Food and Drug Administration (FDA) Pesticide Analytical Manual *(59)* or at the FDA website (http://www.fda.gov/). Within the manual, there are a variety of multiresidue-screening methods listed, usually involving some variation of organic solvent extraction, followed by preparatory chromatography or solvent partitioning with GC analysis by a range of detectors. The CAES method described in the current chapter was developed because many of the previous methods were matrix specific. The currently described method eliminates much matrix variability and although specifically designed for organochlorines and organophosphates, will likely be effective at extracting most hydrophobic organic pollutants. Also of interest may be information on tolerances or limits for specific persistent organic pollutants in environmental compartments or food crops. Title 40 of the Code of Federal Regulations is entitled "Protecting the Environment" and is available at http://www.access.gpo.gov/ecfr/. In addition, the FDA lists tolerances for PCBs and pesticides at http://www.cfsan.fda.gov/~lrd/fdaact.html. No such tolerances are available for dioxins and furans.

References

1. Ray, A. J. S. (2002) Pollutants without borders. *The Scientist*, Sept. 2, 16–18.
2. Wania, F. and Mackay, D. (1996) Tracking the distribution of persistent organic pollutants. *Environ. Sci. Technol.* **30,** 390A–396A.
3. Ritter, L., Solomon, K. R., Forget, J., Stemeroff, M., and O'Leary, C. (1995) *Persistent Organic Pollutants.* Prepared for the International Programme on Chemical Safety (IPCS) within the framework of the Inter-Organization Programme for the Sound Management of Chemicals (IOMC), United Nations Environment Program.
4. Nash, R. G. and Woolson, E. A. (1967) Persistence of chlorinated hydrocarbon insecticides in soil. *Science* **157,** 924–927.
5. Mattina, M. J. I., Iannucci-Berger, W., Dykas, L., and Pardus, J. (1999) Impact of long-term weathering, mobility, and land use on chlordane residues in soil. *Environ. Sci. Technol.* **33,** 2425–2431.
6. Alexander, M. (1995) How toxic are chemicals in soil? *Environ. Sci. Technol.* **29,** 2713–2717.
7. Alexander, M. (2000) Aging, bioavailability, and overestimation of risk from environmental pollutants. *Environ. Sci. Technol.* **34,** 4259–4265.
8. Robertson, B. K. and Alexander, M. (1998) Sequestration of DDT and dieldrin in soil: Disappearance of acute toxicity but not the compounds. *Environ. Toxicol. Chem.* **17,** 1034–1038.
9. Loehr, R. C. and Webster, M. T. (1997) Effect of treatment on contaminant availability, mobility, and toxicity. In: *Environmentally Acceptable Endpoints in Soil,*

(Linz, D. G. and Nakles, D. V., eds.), American Academy of Environmental Engineers, Annapolis, MD, pp. 137–386.
10. Fraser, A. J., Burkow, I. C., Wolker, H., and Mackay, D. (2002) Modeling biomagnification and metabolism of contaminants in harp seals of the Barents Sea. *Environ. Toxicol. Chem.* **21,** 55–61.
11. Kelly, B. C. and Gobas, F. A. P. C. (2001) Bioaccumulation of persistent organic pollutants in lichen-caribou-wolf food chains of Canada's central and western arctic. *Environ. Sci. Technol.* **35,** 325–334.
12. Gosselin, R. E., Smith, R. P., and Hodge, H. C. (1984) In: *Clinical Toxicology of Commercial Products, 5th ed.,* (Williams, A. and Wilkens, A., eds.), Section III, Baltimore, MD, 3 pp. 108–109.
13. McKinney, J. J. and Waller, C. L. (1994) Polychlorinated biphenyls as hormonally active structural analogues. *Environ. Health Perspect.* **102,** 290–297.
14. Grasman, K. A., Scanlon, P. F., and Fox, G. A. (1998) Reproductive and physiological effects of environmental contaminants in fish-eating birds of the Great Lakes: a review of historical trends. *Environ. Monit. Assess.* **53,** 117–145.
15. Meijer, S. N., Steinnes, E., Ockenden, W. A., and Jones, K. C. (2002) Influence of environmental variables on the spatial distribution of PCB's in Norwegian and U.K. soils: implications for global cycling. *Environ. Sci. Technol.* **36,** 2146–2153.
16. Arctic Monitoring and Assessment Programme (AMAP) (1997) *Arctic Pollution Issues: A State of the Arctic Environment Report.* AMAP, Oslo, Norway, pp. 188.
17. Canadian Arctic Contaminants Assessment Report (CACAR) (1997) (Jensen, J., Adare, K., and Shearer, R., eds.). Published under the authority of the Department of Indian Affairs and Northern Development, Ottawa, Canada, pp. 460.
18. Canadian Arctic Contaminants Assessment Report II (CACAR II) (2003a) Contaminant levels, trends and effects in the biological environment. (Fisk, A. T., Hobbs, K., and Muir, D. C., eds.). Published under the authority of the Department of Indian Affairs and Northern Development, Ottawa, Canada, pp. 175.
19. Canadian Arctic Contaminants Assessment Report II (CACAR II) (2003b) *Human Health.* (Van Oostdam, J., Donaldson, S., Feeley, M., and Tremblay, N., eds.). Published under the authority of the Department of Indian Affairs and Northern Development, Ottawa, Canada. pp. 127.
20. Lichtenstein, E. P. (1959) Absorption of some chlorinated hydrocarbon insecticides from soils into various crops. *J. Agric. Food Chem.* **7,** 430–433.
21. Lichtenstein, E. P. (1960) Insecticidal residues in various crops grown in soils treated with abnormal rates of aldrin and heptachlor. *J. Agric. Food Chem.* **8,** 448–451.
22. Lichtenstein, E. P. and Schulz, K. R. (1960) Translocation of some chlorinated hydrocarbon insecticides into the aerial parts of pea plants. *J. Agric. Food Chem.* **8,** 452–456.
23. Lichtenstein, E. P., Schulz, K. R., Skrentny, R. F., and Stitt, P. A. (1965) Insecticidal residues in cucumbers and alfalfa grown on aldrin- or heptachlor-treated soils. *J. Econ. Entomol.* **58,** 742–746.
24. Pylypiw, H., Naughton, E., and Hankin, L. (1991) DDT persists in soil: uptake by squash plants. *Dairy Food Environ. Sanit.* **11,** 200–201.

25. Pylypiw, H. M., Misenti, T., and Mattina, M. J. I. (1996) Pesticide residues in produce sold in Connecticut. The Connecticut Agricultural Experiment Station, Bulletin 940, New Haven, CT.
26. Mattina, J. J. I., Iannucci-Berger, W., and Dykas, L. J. (2000) Chlordane uptake and its translocation in food crops. *J. Agric. Food Chem.* **48,** 1909–1915.
27. White, J. C. (2001) Plant-facilitated mobilization and translocation of weathered 2,2-bis(p-chlorophenyl)-1, 1-dichloroethylene (p,p'-DDE) from and agricultural soil. *Environ. Toxicol. Chem.* **20,** 2047–2052.
28. Mattina, M. J. I., White, J. C., Eitzer, B. D., and Iannucci-Berger, W. (2002) Cycling of weathered chlordane residues in the environment: compositional and chiral profiles in contiguous soil, vegetation, and air compartments. *Environ. Toxicol. Chem.* **21,** 281–288.
29. White, J. C. (2002) Differential bioavailability of field-weathered p,p'-DDE to plants of the *Cucurbita* and *Cucumis* genera. *Chemosphere* **49,** 143–152.
30. White, J. C. and Kottler, B. D. (2002) Citrate-mediated increase in the uptake of weathered p,p'-DDE residues by plants. *Environ. Toxicol. Chem.* **21,** 550–556.
31. White, J. C., Mattina, M. J. I., Eitzer, B. D., and Iannucci-Berger, W. (2002) Tracking chlordane compositional and chiral profiles in soil and vegetation. *Chemosphere* **47,** 639–646.
32. White, J. C., Wang, X., Gent, M. P. N., et al. (2003) Subspecies-level variation in the phytoextraction of weathered p,p'-DDE by *Cucurbita pepo*. *Environ. Sci. Technol.* **37,** 4368–4373.
33. Lunney, A., Zeeb, B. A., and Reimer, K. J. (2004) Uptake of weathered DDT in vascular plants: Potential for phytoremediation. *Environ. Sci. Technol.* **38,** 6147–6154.
34. Hülster, A., Muller, J. F., and Marschner, H. (1994) Soil-plant transfer of polychlorinated dibenzo-p-dioxins and dibenzofurans to vegetables of the cucumber family (Cucurbitaceae) *Environ. Sci. Technol.* **28,** 1110–1115.
35. Campanella, B. and Paul, R. (2000) Presence, in the rhizosphere and leaf extracts of zucchini (*Curcurbita pepo L.*) and melon *(Cucumis melo L.)*, of molecules capable of increasing the apparent aqueous solubility of hydrophobic pollutants. *Int. J. Phytoremed.* **2,** 145–158.
36. Briggs, G. G., Bromilow, R. H., and Evan, A. A. (1982) Relationships between lipophilicity and root uptake and translocation of non-ionized chemicals by barley. *Pestic. Sci.* **13,** 495–504.
37. Sicbaldi, F., Sacchi, G. A., Trevisan, M., Attilio, A. M., and Del Re, A. M. (1997) Root uptake and xylem translocation of pesticides from different chemical classes. *Pestic. Sci.* **50,** 111–119.
38. Puri, R. K., Qiuping, Y., Kapila, S., Lower, W. R., and Puri, V. (1997) Plant uptake and metabolism of polychlorinated biphenyls (PCBs). In: *Plants for Environmental Studies*, (Wang, W., Gorsuch, J. W., and Hughes, J. S., eds.), CRC Press LLC, pp. 481–513. Boca Raton, Florida.
39. Leigh, M. B., Fletcher, J. S., Fu, X., and Schmitz F. J. (2002) Root turnover: An important source of microbial substrates in rhizosphere remediation of recalcitrant contaminants. *Environ. Sci. Technol.* **28,** 1110–1115.

40. Mehmannavaz, R., Prasher, S. O., and Ahmad, D. (2002) Rhizospheric effects of alfalfa on biotransformation of polychlorinated biphenyls in a contaminated soil augmented with Sinorhizobium meliloti. *Proc. Biochem.* **37,** 955–963.
41. Mackova, M., Macek, T., Kucerova, P., Burkhard, J., Pazlarova, J., and Demnerova, K. (1997) Degradation of polychlorinated biphenyls by hairy root culture of *Solanum nigrum*. *Biotechnol. Let.* **19,** 787–790.
42. Burkard, J., Mackova, M., Macek, T., Kucerova, P., and Demnerova, K. (1997) Analytical procedure for the estimation of polychlorinated biphenyl transformation by plant tissue cultures. *Anal. Commun.* **34,** 287–290.
43. Kucerova, P., in der Weische, C., Wolter, M., Macek, T., Zadrazil, F., and Mackova, M. (2001) The ability of different plant species to remove polycyclic aromatic hydrocarbons and polychlorinated biphenyls from incubation media. *Biotechnol. Let.* **23,** 1355–1359.
44. Pier, M. D., Zeeb, B. Z., and Reimer, K. J. (2002) Patterns of contamination among vascular plants exposed to local sources of polychlorinated biphenyls in the Canadian Arctic and Subarctic. *Sci. Total Environ.* **297,** 215–227.
45. Zeeb, B. A., Amphlett, J. A., Rutter, A., and Reimer, K. J. (2006) Potential for phytoremediation of polychlorinated biphenyl (PCB)-contaminated soil. *Intl. J. Phytorem.* In Press.
46. Aprill, W. and Sims, R. C. (1990) Evaluation of the use of prairie grasses for stimulating polycyclic aromatic hydrocarbon treatment in soil. *Chemosphere* **20,** 253–265.
47. Liste, H. H. and Alexander, M. (2000) Accumulation of phenanthrene and pyrene in rhizosphere soil. *Chemosphere* **40,** 11–14.
48. Banks, M. K., Lee, E., and Schwab, A. P. (1999) Evaluation of the dissipation mechanisms for benzo[a]pyrene in the rhizosphere of tall fescue. *J. Environ. Qual.* **28,** 294–298.
49. Rogers, H. B., Beyrouty, C. A., Nichols, T. D., Wolf, D. C., and Reynolds, C. M. (1996) Selection of cold-tolerant plants for growth in soils contaminated with organics. *J. Soil Contam.* **5,** 171–186.
50. Hutchinson, S. L., Schwab, A. P., and Banks, M. K. (2001) Phytoremediation of aged petroleum sludge: effect of inorganic fertilizer. *J. Environ. Qual.* **30,** 395–403.
51. Hutchinson, S. L., Schwab, A. P., and Banks, M. K. (2001) Phytoremediation of aged petroleum sludge: effect of irrigation techniques and scheduling. *J. Environ. Qual.* **30,** 1516–1522.
52. Vouillamoz, J. and Milke, M. W. (2001) Effect of compost in phytoremediation of diesel-contaminated soils. *Water Sci. Technol.* **43,** 291–295.
53. Corseuil, H. X. and Netta Morena, F. (2001) Phytoremediation potential of willow trees for aquifers contaminated with ethanol-blended gasoline. *Water Res.* **35,** 3013–3017.
54. Carman, E. P., Crossman, T. L., and Gatliff, E. G. (1998) Phytoremediation of No. 2 fuel oil-contaminated soil. *J. Soil Contam.* **7,** 455–466.
55. Glick, B. R. (2003) Phytoremediation: synergistic use of plants and bacteria to clean up the environment. *Biotech. Adv.* **21,** 383–393.
56. Pylypiw, H. M. (1993) Rapid gas chromatographic method for the multiresidue screening of fruits and vegetables for organochlorine and organophosphate pesticides. *J. AOAC Int.* **76,** 1369–1373.

57. United States Environmental Protection Agency. Method 8250A. In: *USEPA Methods SW-846, 3rd ed.* Updates I, II, and IIA. September 1994.
58. Anderson, T. A., Hoylman, A. M., Edwards, N. T., and Walton, B. T. (1997) Uptake of polycyclic aromatic hydrocarbons by vegetation: a review of experimental methods. In: *Plants for Environmental Studies*, (Wang, W., Gorsuch, J. W., and Hughes, J. S., eds.). CRC Press LLC, Washington, DC. pp. 451–480.
59. *Pesticide Analytical Manual* (1967 Rev.) vol. 1, U.S. Food and Drug Administration, Washington, DC.

7

Producing Mycorrhizal Inoculum for Phytoremediation

Abdul G. Khan

Summary

This chapter describes the latest ultrasonic nebulizer technology for the production of inocula of arbuscular mycorrhizal fungi for the purpose of mycorrhizo-remediation of metal-contaminated soils. It is a superior alternative to common pot-culture and conventional atomizing disc or spray nozzle systems used for the production of arbuscular mycorrhizal fungi inocula. This technology employs high-frequency sound to nebulize nutrient solution into microdroplets 1 µm in diameter. Fast growth of roots of the test plant, pre-colonized with arbuscular mycorrhizal fungus obtained from the contaminated site, is achieved in the ultrasonic nebulizer chamber containing low-P nutrient solution. These roots can be sheared and used as inoculum in small doses to improve survival and growth of plants used for mycorrhizo-remediation of metal-contaminated soils, which is a prerequisite for phytoremediation.

Key Words: Mycorrhizo-remediation; arbuscular mycorrhizae; commercial inoculum; ultrasonic nebulizer; phytoremediation.

1. Introduction

Mycorrhizal fungi, especially arbuscular mycorrhizae (AM), are ubiquitous soil inhabitants forming symbioses with most naturally growing terrestrial *(1)* and aquatic *(2)* plants. In AM mycorrhizas, Glomeromycota fungi produce arbuscules, hyphae, and sometimes vesicles within the cortices of the roots *(3)*. A recent study highlighted the presence of AM fungal structures, including arbuscules, in the roots of nonmycorrhizal Brassicaceous plants in a heavy metal (HM)-contaminated soil *(4)*. Plant associations with AM fungi are the most ancient, common, and wide-spread symbiosis in higher plants.

In the past, there has been considerable interest in the potential use of AM fungi in agricultural, horticultural, and forestry practices, but neglect of their importance in disturbed and contaminated derelict lands. For example, there is limited knowledge concerning AM fungi in HM-contaminated soils. HM-contaminated soils support unique, diverse, and poor vegetation, which is an outcome of a long evolutionary process resulting in adaptation of plants to the extreme environments on such soils *(5)*. Re-establishment of these ecosystems is a challenge.

In HM-contaminated soils, AM symbiosis is likely to be a strategy adopted by plants growing there to improve nutrition and HM tolerance. During recent decades many studies have demonstrated beyond doubt that these symbionts, when compared with nonmycorrhizal controls, increase the ability of mycorrhizal plants to take up nutrients, including relatively mobile elements such as P, Cu, Ni, Pb, and Zn present in low levels *(6–8)*. Improved plant nutrition associated with AM is thought to be responsible for an increase in plant tolerance to HM *(5)*. The mycorrhizal effect is owing to enhanced absorbing area and accessible soil volume, decreased soil pH, a high affinity to P on the membranes of AM fungi (AMF), efficient hyphal transport of P as polyphosphate, and an increased capacity of mycorrhizal plants to utilize unavailable P *(9)*. They have potential in removing or detoxifying HMs from polluted soils. HM-contaminated lands often support characteristic plant species, the roots of which are often associated with AM fungi and N-fixing rhizobial bacteria *(10,11)*. Pioneer plants growing on nutrient-deficient derelict soils are mycorrhizal and their rhizospheres contain propagules of these symbionts *(12)*. These plants and their associated symbionts have evolved a tolerance to the environmental conditions *(13)*. The infection of roots of such plants with AM fungi leads to improved survival and plant growth in such soils *(14)*. But, many factors such as individual fungus–host combination compatibility between AM fungus and host, light intensity, temperature, soil nutrient levels especially P, root density, and so on, may affect the benefit derived from AM fungi *(9)*.

Many studies suggest that plants with AM are successful as primary colonizers of pioneer and derelict habitats, and that manipulation of and inoculation with appropriate and ecologically adapted AM fungi offer a challenging but not yet fully exploited opportunity *(15)*. This is partly a result of the obligate nature of these fungi and inadequate methods for their large-scale inoculum production. Because each AM fungal species and isolate has specific ecological requirements, their screening is necessary to select the most superior, effective, and efficient isolate for successful introduction into plantings *(16)*. Furthermore, sound experimental design is critical before embarking on field experiments as establishment of field experiments is very labor intensive and expensive. Excellent practical descriptions of planning and designing field experiment have been provided by Brundrett et al. *(17)*.

AM fungi produce large asexual spores and coenocytic hyphae distributed throughout the soil and, based on morphological characteristics of spores, approx 150 species have been identified belonging to six genera *(18)*. Characteristic colonizing patterns inside host roots can also be used to identify different genera of AM fungi in some cases *(17)*. However, difficulties exist in the identification of AM fungi from these morphological characteristics. Identification of AM fungi in the soil and in the plant roots is now possible with modern molecular-based methods such as PCR and restriction analysis of ribosomal DNA *(19–25)*. Based on molecular phylogenetic studies, AM fungi have been elevated to the phylum Glomeromycota *(26,27)*.

AM spores can be separated from soil samples collected from around roots of plants growing on contaminated sites by various techniques involving sucrose density layers or gradients or/and wet sieving and decanting techniques *(17)*. Healthy looking mature spores collected in this way can be used to start pot culture propagation to produce inoculum. But it is time consuming, bulky, and often not pathogen free.

To overcome these problems, soil-free methods such as axenic culturing of AM fungi *(26)* and soil-less growth media culture techniques such as hydroponics and aeroponics have been proposed *(28)*. Conventional aeroponics are a recent development in hydroponic methods that has gained much popularity over recent years (**Fig. 1**). Sylvia and Jarstfer *(29)* used a conventional atomizing disc and spray nozzle system to produce aeroponically grown roots colonized with AM fungi. Mycorrhizal roots so produced were sheared and used as commercial inoculum *(29)*, but this system does not produce fine mists of nutrient solution, which diminish rapid absorption of nutrients *(30)*. Sylvia and Jarstfer's *(29)* aeroponic system uses syphon tubes which may be blocked by roots, the drive shaft may become disconnected from the motor, or roots may become entangled and destroyed by the drive shaft or impeller, which also reduces the space for roots to expand. All these factors might limit growth and AM-fungal colonization of roots in the system. These limitations were overcome by the use of the most recent innovation in aeroponics, i.e., use of piezo ceramic element technology employing high-frequency sound to nebulize the nutrient solution (ultrasonic nebulizer technology). This provides a fog-like mist to the plant roots, rather than a spray to deliver the nutrient solution as in a conventional aeroponic system (**Fig. 2**). The mist is much finer microdroplets, 1 µm in diameter *(30)*, than is achievable in the conventional atomizing disc or pump and spray nozzle systems. This technology also improves the humidity inside the chamber. This improved aeroponic technique has been used by Mohammad et al. to produce concentrated high-quality AM isolates that can be used in small doses to produce a large response *(31–33)*, a prerequisite for commercialization of AM technology.

This technology is currently used as an alternative to hydroponics for growing plants *(30)*. This chapter describes this technology to further improve the

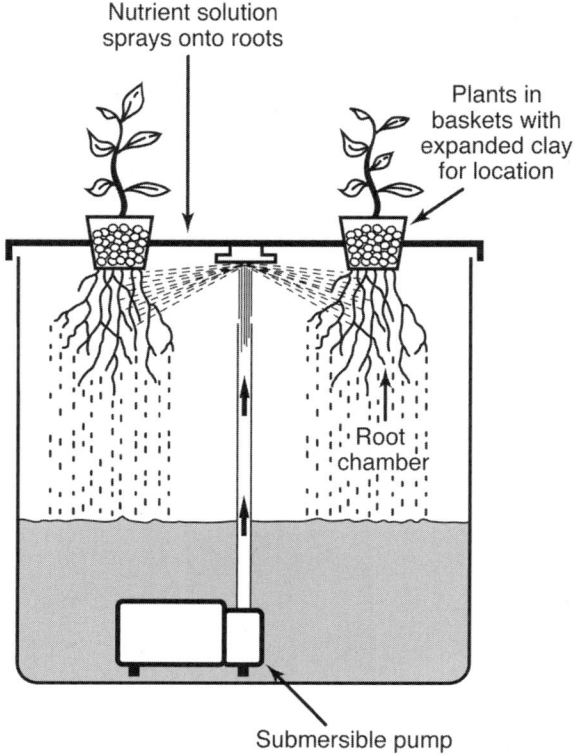

Fig. 1. Schematic view of conventional aeroponic system. Aeroponics is the most recent development in hydroponic methods, and one that has gained much publicity over recent years. (Modified from **ref. 30**.)

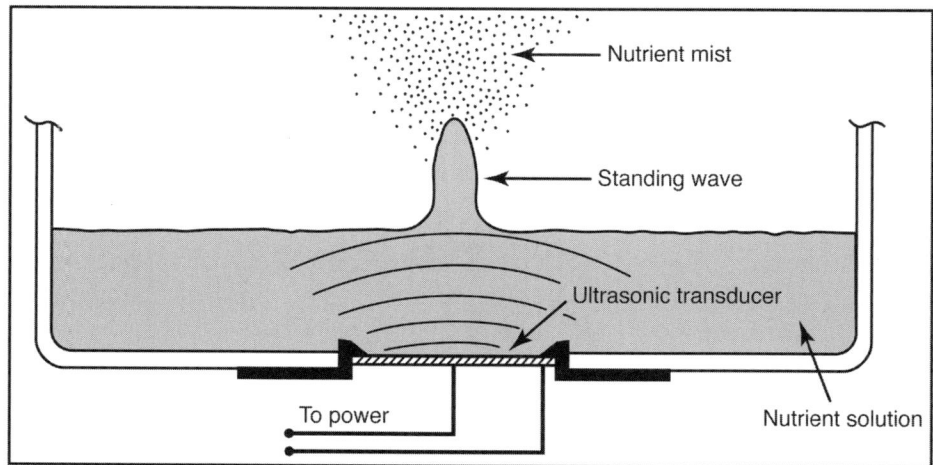

Fig. 2. Schematic view of an AGS ultrasonic nebulizer system. (Modified from **ref. 30**.)

efficiency of AM fungal biomass production for mycorrhizo-remediation (phytoremediation by using mycorrhizal plants) purposes.

2. Materials

2.1. Establishment of Pot Cultures

1. Rhizosphere soil from contaminated site.
2. Plastic bucket of 1-L capacity to mix rhizosphere soil and water.
3. Set of sieves for wet-sieving and spore extraction from soil suspension.
4. Binocular microscope to observe AM spore morphology.
5. Plastic pots (14 × 13 cm) to grow test plants.
6. Pasteurized coarse sand and vermiculite (3:1) to mix with the rhizosphere soil (1:1) as growth medium for the initiation of AM infection in Sudan grass seedlings.
7. pH meter to measure soil pH.
8. Sudan grass seeds as test plant.
9. 5% Sodium hypochlorite solution to sterilize seeds.
10. Sterilized plastic Petri dishes (90 × 15 mm) containing sterilized (by autoclaving) wet filter papers for seed germination.
11. Growth chamber.
12. Methylene stain (1% aquouse) for staining roots to observe AM fungal infection in the roots.
13. Microscopic glass slides and cover slips.
14. Compound microscope to observe AM fungal colonization and calculate percentage infection.

2.2. Improved Aeroponic Technique

1. Ultrasonic nebulizer aeroponic nutrient chamber (*see* **Fig. 2**).
2. Low-P nutrient solution for the nutrient chamber *(34)*.
3. Growth conditions.
4. Infective seedlings from pot cultures in **Subheading 2.1.**
5. Electric blender to shear roots.

2.3. Mycorrhizal Infectivity Assessment

1. Pasteurized sand to mix with dried sheared-root inoculum.
2. 50-mL Plastic vials.
3. Test plant such as onion seedlings.

3. Methods

3.1. Pot Cultures

3.1.1. Establishment and Maintenance of Pot Cultures

1. Collect 10 sub-samples of 1 kg each from top 10 cm of the soil profile at 2-m intervals on five transect lines 2-m apart at each site (contaminated or control). Mix

thoroughly sub-samples collected at each site, air-dry them, and then sieve through 2-mm sieve. Take five sub-samples from the composite sample per site at random and analyze its Electrical Conductivity (EC) and pH (assessed in 0.01 M $CaCl_2$ soil suspension using 1:5 [w/v] soil-to-solution ratio), total and DTPA-extractable HM by atomic absorption spectrophotometry.

2. Soak clover seeds in a 5% sodium hypochlorite for 5 min, followed by three washings with DI water. Place surface-sterilized seeds in a sterilized Petri dish on a filter paper moistened with DI water and incubate at 20°C in the dark until seeds germinate (3–4 d). Transfer five germinating seedlings into the pots containing 1 kg soil sample, with three replicates per site. Place the pots in a glasshouse (20–25°C; photoperiod 16 h) for 8 wk in a randomized fashion and water them with DI water to field capacity.

3. Fertilize seedlings with approx 100 mL low P Hoagland solution *(34)* on a weekly basis.

3.1.2. Observing Roots for AM Colonization

1. At harvest, wash soil off the roots over a sieve, collect a few root sub-samples from each seedling, cut them into 1-cm fragments, and store in 90% ethanol for assessing AM infection.

2. Select 50 root segments at random from the 1-cm root fragments collected from pot cultures for clearing, staining, and assessing percentage colonization by AM fungi. Clear the roots with 1% KOH solution by heating at 90°C for 1 h. Wash cleared roots three times with distilled water and acidify them once with 1% HCl for 5 min. Now stain the cleared and acidified root segments with 0.05% aniline blue in lactic acid for 2–4 h and store in lactic acid at room temperature until ready for microscopic observations by mounting in groups of 10 on a glass slide parallel to the short axis of the glass slide.

3. Examine each root segment on the slide for the presence of any mycorrhizal structure, i.e., inter- and intracellular hyphae, vesicles, arbuscules, and denote as positive (+) or negative (–). Determine the total percentage colonization by dividing the counts for number of AM positive root segments with the total number of segments scored across the slide and multiplying with 100.

3.2. Aeroponic Production of AM-Colonized Roots

1. Select the infected and healthy plants from the above pots and place them, supported by cottonwool plugs, into the holes of the lid of the ultrasonic nebulizer aeroponic nutrient chamber (**Fig. 2**) containing low P Hoagland nutrient solution *(34)* (*see* **Note 1**).

2. Check daily and maintain the nutrient solution level by adding additional amounts, if required. Replace the nutrient solution every 2 wk.

3. Monitor pH of the solution weekly and maintain it near neutral.

4. Remove the fast-growing roots from the chamber by trimming off with scissors above the solution at regular intervals to avoid anaerobiosis. Store them in a self-sealing polythene bag at 4°C until ready to shear.

5. Trim off the roots from each plant after 12 wk of growth in the chamber and assess them for mycorrhizal colonization and extra-matrical chlamydospore production by the method previously described (*see* **Subheading 3.1.2., steps 2 and 3**).
6. Combine all the root samples including those collected at intervals plus the root at harvest into one pool and mix with ionized water in a 1:10 ratio (w/v) and shear with a super blender for about 80 s to produce a slurry of the aeroponically grown mycorrhizal roots containing root fragments, interradical vesicles, mycelia, and spores. The air-dried slurry of sheared roots so produced contains mycorrhizal root fragments and AM fungal propagules.
7. Assess the inoculum potential of the AM inoculum so produced by using a Most Probable Number (MPN) bioassay as per method of Khan *(35)*. Prepare a 10-fold series of dilutions of the sheared root inoculum with acid-washed steamed sand as diluent and plant with 5-d-old onion seedlings in 50-mL plastic vials using one seedling per vial and five replicates per dilution. Maintain vials just below field capacity by frequent watering to a constant weight and arrange them in a randomized fashion in a glasshouse (temperature range 15 to 25°C and photoperiod 14 h). Remove roots after 6 wk, wash, clear, and stain in 0.05% trypan blue and score for the presence or absence of AM infection as described in **Subheading 3.1.2**. Calculate the MPN of AM propagules for each incculum using a MPN table *(36)*.

4. Notes

1. Ultrasonic nebulizer aeroponic chamber *(30)*. This nebulizes the nutrient solution into a fog-like mist, in much the same way that the application of medication is atomized and delivered to the alveoli in the lungs of asthma patients. It uses a high-frequency sound that blasts the nutrient solution into a fog-like mist, which is absorbed more rapidly by roots growing in the chamber. It consists of a root chamber, much like the conventional aeroponic chamber, which contains the nutrient solution. The ultrasonic transducer is embedded in polystyrene foam that floats on the solution, with the chamber itself aerated to circulate the mist and to remove exudates.

References

1. Jeffries, P. (1987) Use of mycorrhizae in agriculture. *Crit. Rev. Biotech.* **5,** 319–357.
2. Khan, A. G. (2004) Mycotrophy and its significance in wetland ecology and wetland management. In: M. H. Wong (ed.), *Wetland Ecosystems in Asia: Function and Management,* Volume 1, Chapter 7: pp. 95–114. Amsterdam, Oxford: Elsevier.
3. Schussler, A., Schwarzott, D., and Walker, C. (2001) A new fungal phylum, the Glomeromycota: phylogeny and evolution. *Mycol. Res.* **105,** 1413–1421.
4. Orlowska, E., Zubek, S., Jurkiewicz, A., Szarek-Lukaszewska, G., and Turnau, K. (2002) Influence of restoration on vesicular arbuscular mycorrhiza of *Biscutella laevigata* L. (Brassicaceae) and *Plantago lanceolata* L. (Plantaginaceae) from calamine spoil mounds. *Mycorr.* **12,** 153–160.
5. Meharg, A. A. and Cairney, J. W. G. (2000) Co-evolution of mycorrhizal symbionts and their hosts to metal-contaminated environments. *Adv. Ecol. Res.* **30,** 69–112.

6. Killham, K. and Firestone, M. K. (1986) Vesicular arbuscular mycorrhizal mediation of grass response to acid and heavy metal deposition. *Plant Soil* **72**, 39–48.
7. Jamal, A., Ayub, N., Usman, M., and Khan, A. G. (2002) Arbuscular mycorrhizal fungi enhance zinc and nickel uptake from contaminated soil by soybean and lentil. *Int. J. Phytorem.* **4**, 205–221.
8. Liu, A., Hamel, C., Hamilton, R. I., Ma, B. L., and Smith, D. L. (2000) Acquisition of Cu, Zn, Mn, and Fe by mycorrhizal maize (*Zea mays* L.) grown in soil at different P and micronutrient levels. *Mycorr.* **9**, 331–336.
9. Smith, S. E. and Read, D. J. (1997) *Mycorrhizal Symbiosis*. 2nd ed. Academic Press, London, UK.
10. Khan, A. G., Chaudhry, T. M., Hayes, W. J., et al. (1998) Physical, chemical, and biological characterization of a steel works waste site at Port Kembla, NSW, Australia. *J. Water, Air Soil Pollut.* **104**, 389–402.
11. Khan, A. G. (2001) Relationship between chromium biomagnification ratio, accumulation factor, and mycorrhizae in plants growing on tannery effluent polluted soils. *Environ. Int.* **26**, 417–423.
12. Chaudhry, T. M. and Khan, A. G. (2003) Plants growing on abandoned mine site and their root symbionts. In: *Proceedings 7th International Conference on Biogeochemistry of Trace Elements,* (Gobran, G. R. and Lepp, N., eds.), Swedish University of Agriculture Service, Uppsala, Sweden, pp. 134–135.
13. Khan, A. G. (2006) Mycorrhizoremediation–an enhanced form of phytoremediation. *Journal of Zhejiang Uni. SCIENCE B.* **7(7)**, 503–514.
14. Khan, A. G., Kuek, C., Chaudhry, T. M., Khoo, C., and Hayes, W. J. (2000) The role of plants, mycorrhizae, and phytochelators in heavy metal contaminated land remediation. *Chemosphere* **41**, 197–207.
15. Khan, A. G. (2002) The significance of microbes. In: *The Restoration and Management of Derelict Lands,* (Ming, M. H. and Bradshaw, A. D., eds.), World Scientific Publisher, Singapore, pp. 149–160.
16. Khan, A. G. (2002) The handling of microbes. In: *The Restoration and Management of Derelict Lands,* (Ming, M. H. and Bradshaw, A. D., eds.), World Scientific Publisher, Singapore, pp. 80–92.
17. Brundrett, M., Bougher, N., Dell, B., Grove, T., and Malajczuk, N. (1996) *Working With Mycorrhizas in Forestry and Agriculture*. Australian Centre of International Agricultural research, Canberra, Australia.
18. Morton, J. B. and Benny, G. L. (1990) Revised classification of arbuscular mycorrhizal fungi (Zygomycetes): a new order, Glomales, two new suborders, Glomineae and Gigasporineae, and two new families, Acaulosporaceae and Gigasporaceae, with an emendation of Glomaceae. *Mycotaxon* **37**, 471–491.
19. Redecker, D., Theirfelder, H., Walker, C., and Werne, D. (1997) Restriction analysis of PCR amplified internal transcribed spacers of ribosomal DNA as a tool for species identification in different genera of the order Glomales. *App. Environ. Microbiol.* **63**, 1765–1771.
20. Redecker, D. (2000) Specific PCR primers to identify arbuscular mycorrhizal fungi within colonized roots. *Mycorrhiza* **10**, 73–80.

21. DiBonito, R., Elliott, M. L., and Des Jardin, E. A. (1995) Detection of arbuscular mycorrhizal fungi in roots of different plant species with PCR. *App. Environ. Microbiol.* **61,** 2809–2810.
22. Clapp, J. P., Fitter, A. H., and Young, J. P. W. (1999) Ribosomal small subunit sequence variation within spores of an arbuscular mycorrhizal fungus, *Scutellospora* sp. *Mol. Ecol.* **8,** 915–921.
23. Daniell, T. J., Husband, R., Fitter, A. H., and Young, J. P. W. (2001) Molecular diversity of arbuscular mycorrhizal fungi colonizing arable crops. *FEMS Micro. Ecol.* **36,** 203–209.
24. Jacquote-Plumey, E., Tuinen, D. V., Chatagnier, O., Gianinazzi, S., and Gianinazzi-Pearson, V. (2001) 25S rDNA based molecular monitoring of Glomalean fungi in sewage sludge treated field plots. *Environ. Microbiol.* **3,** 525–531.
25. Renker, C., Heinriche, J., and Kaldorf, M. (2003) Combining nested PCR and restriction digest of the internal transcribed spacer region to characterize arbuscular mycorrhizal fungi on roots from the field. *Mycorr.* **13,** 191–198.
26. Schubler, A., Shwarzott, D., and Walker, C. (2001) A new fungal phylum, the Glomeromycota: phylogeny and evolution. *Mycol. Res.* **105,** 1413–1421.
27. Helgason, T., Watson, I. J., Peter, J., and Young, W. (2003) Phylogeny of the Glomerales and Diversisporales (Fungi: Glomeromycota) from actin and elongation factor 1-alpha sequences. *FEMS Microbiol. Lett.* **229,** 127–132.
28. Mohammad, A. and Khan, A. G. (2002) Monoxenic *in vitro* production and colonization potential of AM fungus *Glomus intraradices*. *Ind. J. Exp. Biol.* **40,** 1087–1091.
29. Sylvia, D. M. and Jarstfer, A. G. (1992) Sheared root inoculum of vesicular-arbuscular mycorrhizal fungi. *App. Environ. Microbiol.* **58,** 229–232.
30. Carruthers, S. (1992) Aeroponics systems review. *Practical Hydroponics* July/August 1992, pp. 18–21.
31. Mohammad, A., Mitra, B., and Khan, A. G. (2004) Effects of sheared-root inoculum of *Glomus intraradices* on wheat grown at different phosphorus levels in the field. *Agric. Ecosystems Environ.* **103,** 245–249.
32. Mohammad, A., Khan, A. G., and Kuek, C. (2000) Improved aeroponic culture technique for production of inocula of arbuscular mycorrhizal fungi. *Mycorr.* **9,** 337–339.
33. Asif, M., Khan, A. G., and Kuek, C. (1997) Growth responses of wheat to sheared-root and sand-culture inocula of arbuscular-mycorrhizal (AM) fungi at different phosphorus levels. *Kavaka* **25,** 71–78.
34. Hoagland, D. R. and Arnon, D. I. (1938) The water culture method for growing plants without soil. *Calif. Agric. Exp. Sta. Cir.* **347,** 1–39.
35. Khan, A. G. (1988) Inoculum density of *Glomus mosseae* and growth of onion plants in unsterilized bituminous coal soil. *Soil Biol. Biochem.* **20,** 749–753.
36. Alexander, M. (1965) Most probable number method for microbial population. In: *Methods of Soil Analysis, Part 2, Chemical and Microbiological Properties,* (C. A. Black, ed.), Am. Soc. Agron., Madison, WI, pp. 1467–1472.

8

Implementing Phytoremediation of Petroleum Hydrocarbons

Chris D. Collins

Summary

An evaluation of the current "state of the art" for the phytoremediation of total petroleum hydrocarbons (TPH) is given, which will allow for well-informed decisions to be made when the technology is being applied to this contamination problem. Information is provided on phytotoxicity, plant selection, and management as well as useful supplementary practical data sources. A management decision tree is presented to aid in the successful application of phytoremediation to TPH-contaminated sites. Finally, deficiencies in the current knowledge are identified, which need to be addressed to improve the effectiveness of phytoremediation to this problem.

Key Words: Total petroleum hydrocarbons; plant selection; field application; decision tree.

1. Introduction

1.1. Total Petroleum Hydrocarbons in the Environment

The petroleum hydrocarbons are some of the most universally detected organic pollutants in the environment because of the high industrial use of petroleum products world wide. The world petroleum consumption in 2001 was 77 million barrels per day *(1)*, this scale of use results in a high potential for contamination from both accidental and fugitive releases. For example, in the United Kingdom alone there are estimated to be 120,000 contaminated petrol station sites with an associated remediation cost of £2.5 billion. The remediation of contaminated oil terminals and refinery sites will increase this figure. The US petroleum industry alone spent $0.8 billion dollars in 2001 on remediation *(2)*. The large scale and economic importance of this contamination

Table 1
Carbon Number Ranges and Associated Boiling Point Ranges of Different Fuels and Oils

Fuel or oil	Carbon number range	b.p. range (°C)
Petroleum	C4–C12	40–200
Jet fuel	C5–C14	150–275
Kerosene	C6–C16	150–300
Diesel	C8–C21	200–325
Motor oil	C18–C34	325–600

has resulted in a significant effort in developing remediation technologies for its clean-up. Phytoremediation is one of the remediation technologies that has been developed to address this problem. Phytoremediation is advantageous when a low-cost solution is needed that can be easily applied to diffuse sources of contamination. This is typical of the large sites used by the petrochemical industry. Such sites are usually in places that do not have significant value for redevelopment (where economic pressures require a rapid clean up) so it is usually a matter of reducing liability. Phytoremediation can also be used to reduce the leaching of contaminants through soils, and hence, protect the groundwater where guidelines for contaminant levels are more stringent *(3)*. This technology has also been used to prevent off-site migration of contaminants in groundwaters *(4)*. Finally, within the petrochemical industry there is recognition of bioremediation strategies such as land farming, so gaining acceptance for phytoremediation should be easier than in other commercial enterprises.

Before detailing the potential clean-up options currently available it is necessary to understand something of the nature of petroleum hydrocarbons or total petroleum hydrocarbons (TPHs) as they are frequently referred to analytically. TPHs are a whole collection of organic compounds that include aromatics, PAH, alkanes, and others, all with different physicochemical properties. **Table 1** illustrates the carbon chain lengths of the alkanes associated with the different fuel and oil formulations and how these affect boiling point. Increasing carbon chain length will also reduce contaminant solubility and increase its partition to soil and hence reduce the availability of the contaminant to plants.

In the early years of the clean-up operations of polluted soil sites the focus was often on the total remediation of the contamination. Owing to the magnitude of most soil contamination problems this often led to a stagnation of the overall remediation process because of the high costs involved. Legislators have subsequently tried to avoid this by achieving a balance between environmental effectiveness and cost using the "suitable-for-use" cleanup criterion. Phytoremediation has a role in this process because of its low cost and positive environmental impact.

At present, phytoremediation of TPH is still not a proven technology. Rigorous science is now required to determine when it is an appropriate option so the technology is not rejected because of failures from inappropriate implementation. Further information is required on those species that tolerate/degrade TPH; how the system acts upon "aged" and recent contamination; and the influence of soil organic matter in the toxicity of TPH *(5)*. The chemical component that causes phytotoxicity still needs to be elucidated. If these questions can be addressed then phytoremediation may be applicable to higher concentrations than those currently being considered. The role of phytoremediation in ecological restoration also requires further examination—it may well be a major application. The best combination of phytoremediation with other technologies when addressing TPH contamination also needs to be investigated. With this knowledge the quality and repeatability of the clean up of TPH by phytoremediation will be improved and its adoption by industry for routine use may then be achieved. The methodology outlined here will be useful for progressing the science underpinning phytoremediation of TPHs and other organic contaminants.

1.2. Traditional Clean-up Options for Soil TPHs

1.2.1. Excavation

This involves the removal of the contaminated material from the site to landfill. It has become an increasingly difficult option to adopt because of the conflicts it creates with the sustainable goals of most brownfield-site remediation projects. Additionally, the increasing costs of disposal to landfill and the potential problems this creates with liability will see further declines in this strategy.

1.2.2. Air Sparging

During air sparging volatile organic compounds are removed from soil by mechanically drawing or venting air through the soil matrix. Volatiles contain much of the toxicity and this approach is most efficient for the lighter fractions of TPH, e.g., benzene, toluene, ethylbenzene, xylene, and petroleum. This process has the benefit that it is relatively cheap, however, it often leaves a residual fraction and is only really useful for removing the volatile organic compounds. The main advantage of air sparging is the speed of this process, which can take a few weeks to 6 mo.

1.2.3. Bioremediation

Bioremediation involves the use and/or enhancement of micro-organisms, which degrade the pollutants of concern. For biodegradation to be successful four conditions must be met. First, the contaminants have to act as the primary carbon source for the microbial population; this will be the electron donor. Second, an electron acceptor must be available so that energy can be extracted at environmentally significant rates; this can be oxygen, nitrate, sulfate, carbon dioxide, or organic carbon. Oxygen is usually selected for its high efficiency and can be provided by aerating formed piles *(6)*. Nitrate is also often added as a fertilizer to bioremediation systems. Third, macro- and micronutrients need to be available for the production of cellular material, usually in the ratio 100:10:1 for carbon, nitrogen, and phosphorus. Finally, conditions must not be inhibitory to the indigenous microflora, these should be soil moisture at 50–80% field capacity, pH 5.5–8.5, temp 15–45°C, and an absence of microbial toxicants.

Bioremediation is attractive because it can be effective on some of the more recalcitrant components of oils, however, it can also increase costs because the soil area being treated will often need to be manipulated to accommodate the treatment (e.g., formed piles), chemicals may need to be applied, and extra degrading microbes may also be added (bioaugmentation). The time-span required can be long and hence monitoring and management costs will increase. However, bioremediation can potentially be a relatively fast treatment and can take as little as 8 wk *(7)*.

1.2.4. Land Farming

In this system contaminated soils are spread and tilled regularly and left to degrade by biological processes, in many ways it is a low-input form of bioremediation. Plants are not a component of this system. Repeated applications of contaminated material to the area being farmed will maintain the pollutant-degrading microbial population. Fertilizers may also be added to improve bacterial populations. Tilling the land improves contact between the soil and the contaminated material and is used to increase effectiveness. Land farming is a low-cost option that is considered to be a slow process, for example the treatment of drill cuttings might require 2–4 yr *(8)*. The previously described processes can be used in combination and, for example, soil vapor extraction followed by bioremediation is considered to be a very effective strategy. They can also be used in combination with phytoremediation. **Table 2** provides an overview of the abilities and costs of the different strategies.

2. Materials
2.1. Choice of Plant

Grasses have been proposed for phytoremediation by many researchers because they have fibrous roots that penetrate a large soil volume and can quickly provide a good surface cover to suppress dust. Nedunuri et al. *(9)* found

Table 2
Summary of Remediation Technologies Available for Clean Up of the Petroleum Hydrocarbons[a]

Strategy	Application	Advantages	Limitations	Costs
Air sparging	Gasoline	Can remove some compounds resistant to biodegradation	VOCs only	Low: low capital cost, short treatment time
Bioremediation	Gasoline, fuel oils	Effective on some nonvolatile compounds	Possible lengthy clean up time	Moderate: because of capital outlay and management
Land farming	Gasoline, fuel oils, coal tar residues	Uses natural degradation processes	Some residuals remain	Moderate: because of initial capital outlay.
Phytoremediation	Gasoline (?), fuel oils (?)	Uses natural degradation process, ecologically benign, and soil stabalization	Efficacy not really known	Low: low capital cost

[a]Adapted from **ref. 26**.

ryegrasses and St. Augustine's grass to enhance TPH degradation. However, other researchers have found little treatment effect when using grass species *(10)*. Adam and Duncan *(11)* found cultivated crops the most tolerant of diesel-range organics with many grasses sensitive. Plant choice is further complicated in that there may be differences in tolerance within individual species *(12)*. Many of the recommended species in the Phytopet© database *(13)*, which contains details of species known to be tolerant or to degrade TPH, are also grasses. Overall, therefore, there is no consensus on those species to be recommended for phytoremediation of TPH. In the light of these findings, perhaps the best strategy is to follow the Research Technologies Development Forum (RTDF) approach (*see* **Note 1**), in which a combination of cool and warm season grasses and a legume, which are known to be adapted to the local environment, are chosen. Grasses are chosen for the reasons previously outlined and a legume is included to provide nitrogen, which can be beneficial for microbial action. If contaminated groundwater needs to be treated, then a high water-use plant for example cottonwood, willow, or poplar needs to be used, which will draw the contaminated plume upward or prevent its migration off site.

3. Methods
3.1. Assessing Phytotoxicity

Some plants are clearly tolerant of high levels of TPH, for example Lindau and Delaune *(14)* applied 2 L/m of South Carolina crude oil to *Sagittaria lancifolia* and found only short-term toxicity effects on plant growth and function. This increase in tolerance has also been reported for other wetland macrophytes *(15)* and short-term toxicity has been seen by a number of other researchers; Donnetti *(16)* reported it was completely gone after 16 wk, whereas others have found it significantly reduced after 2 wk *(11)*. Most studies have found that TPH effects on germination occur around 5000 ppm on sensitive species, and nearly all species are affected at 25,000 ppm, but some species can tolerate concentrations of up to 50,000 ppm *(16–19)*. Planting tolerant species at 30,000 ppm (3%) has been proposed *(20)* and this would seem an appropriate level. If the contamination has been subject to "aging" the tolerable contamination levels may be higher because of the loss of the phytotoxic fractions.

3.2. Implementing a Phytoremediation Strategy

The first decision that must be made is whether phytoremediation is appropriate for the site. For example, are the conditions conducive to plant growth? This is defined by the Interstate Technology Resource Center (ITRC) (*see* **Note 2**) as greater than 200 d with temperatures higher than freezing and annual rainfall of greater than 10 in. If the TPH levels are such that phytotoxicity is a problem then the removal of the volatile fraction using air sparging can be considered. This might reduce toxicity to a level where healthy plant growth can be achieved. To assess whether this is appropriate for a given soil contamination the benzene, toluene, ethylbenzene, and xylene compounds could be monitored — these are a good indicator of this volatile toxic fraction and are often routinely analyzed in the initial site investigation if TPH are known to be present. Active remediation measures may also be required to deal with nonaqueous phase liquids as these cannot be treated by phytoremediation alone *(21)*. If the TPH level is equal to or less than 3% of soil by weight then planting tolerant species can proceed either using an RTDF mix or high water use plants for polluted groundwaters.

Even with nitrogen fixation from the legume in the recommended RTDF seed mix, application of fertilizer is considered good practice to encourage the phytoremediation proecess. It has been found that the best reduction in TPH in vegetated treatments occurred when the fertilizer ratio was 100:2:0.2 (C:N:P) *(10)*. This is more than the amount required for plant growth and probably a ratio of 100:10:1 should be the target to enhance microbial action. An annual application of fertilizer may be required. In addition aeration is also advised, this is similar to the principle for bioremediation systems *(21)*. Initially the aeration could be provided by agricultural operations such as ploughing once the plants are established; the action of the roots should aerate the soils.

If, after the site has been put under a phytoremediation system, no decline in TPH is evident, then some additional active measures may be required to start the degradation process. Once the TPH levels have been reduced to less than 1%, then species can be planted to ecologically restore the habitat. At this point a risk assessment should be carried out to see if the levels of TPH are such that the site is no longer classed as contaminated and no longer poses a risk. A schematic of these decisions based on previous research *(20,22)* is provided in **Fig. 1**. (For helpful data sources and previous field trials *see* **Notes 3** and **4**, respectively.)

4. Notes

1. RTDF is a partnership of the US Environmental Protection Agency (EPA) to develop and improve remediation technologies.
2. The ITRC is a web-based open access tool.
3. There are a number of useful data sources for those wishing to implement phytoremediation of TPH. The best database for plant selection is operated by the University of Saskatchewan *(13)*. The PhytoPet database contains detailed information on a plant's phytoremediation potential or tolerance of hydrocarbons and the literature source this came from. At present the data are predominantly for grass species and are biased toward Canadian conditions, but the database is being updated continuously and can be accessed over the Internet. The ITRC provides a web-based open access tool for the application of phytoremediation to a range of contaminants *(22)*. This tool is in the form of a decision tree, which runs through a series of questions and provides recommendations as to whether phytoremediation is appropriate for the site, as well as guidance on which species and process management decisions to use. Some of these decisions have been incorporated into **Fig. 1**. Finally, there are a number of US EPA documents, which can be used for additional guidance and background to the phytoremediation of TPH *(23–25)*.
4. Although a lot of research has been carried out testing phytoremediation on TPH-contaminated soils, there are very few examples of field scale trials of phytoremediation of TPH. If it is to be adopted as a technology, more "proof of concept" at this scale is required. The largest range of field trails to date has been a state-wide survey in the United States and Canada by the RTDF. Thirteen sites were chosen with a range of source terms such as a closed refinery and a manufactured gas plant; and a range of climates from subarctic to warm temperate. The research is ongoing, but to date, the results have been mixed. In only two instances have there been clear phytoremediation effects. There are a number of potential reasons for this, but possibly the greatest is that the contamination at a number of the sites is extremely variable and may have obscured some findings, despite a good plot size (min 400 sq ft) and four replicates (P. Kulakow personal communication). Overall this seems to be a disappointing finding for phytoremediation protagonists.

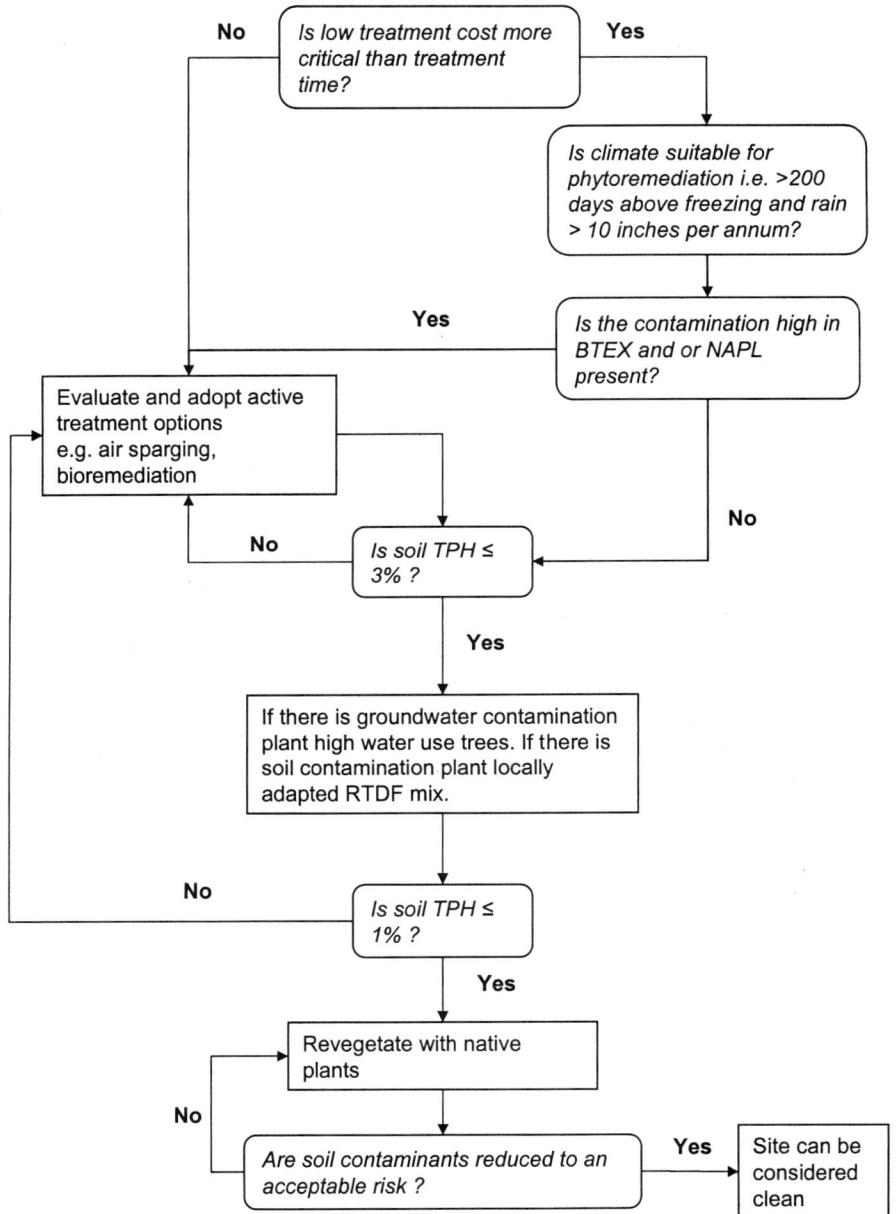

Fig. 1. Decision tree for the phytoremediation of total petroleum hydrocarbon-contaminated sites. (Adapted from **refs. 20** and **21**.)

References

1. World Petroleum Consumption. Energy: http://www.eia.doe.gov/emeu/aer/pdf/pages/sec11_21.pdf: Table 11.10. Last Accessed 05/01/04.
2. API (2003) *U.S. Oil and Natural Gas Industry's Environmental Expenditures 1992-2001*. American Petroleum Institute, Washington DC, pp. 1–17.
3. Susarla, S., Medina, V. F., and McCutcheon, S. C. (2002) Phytoremediation: An ecological solution to organic chemical contamination. *Ecol. Eng.* **18,** 647–658.
4. Hong, M. S., Farmayan, W. F., Dortch, I. J., Chiang, C. Y., McMillan S. K., and Schnoor, J. L. (2001) Phytoremediation of MTBE from a groundwater plume *Environ. Sci. Tech.* **35,** 1231–1239.
5. Robinson, S. L., Novak, J. T., Widdowson, M. A., Crosswell, S. B., and Fetterolf, G. J. (2003) Field and laboratory evaluation of the impact of tall fescue on polyaromatic hydrocarbon degradation in an aged creosote-contaminated surface soil. *J. Environ. Eng.-A* **129,** 232–240.
6. Dupont, R. R. (1993) Fundamentals of bioventing applied to fuel contaminated sites. *Environ. Prog.* **12,** 45–53.
7. Hildebrandt, W. W. and Wilson, S. B. (1991) On-site bioremediation systems reduce crude-oil contamination. *J. Petr. Technol* **43,** 18–22.
8. Zimmerman, P. K. and Robert, J. D. (1991) Oil-based drill cuttings treated by land-farming. *Oil Gas J.* **89,** 81–84.
9. Nedunuri, K. V., Govindaraju, R. S., Banks, M. K., Schwab, A. P., and Chens, Z. (2000) Evaluation of phytoremediation for field-scale degradation of total petroleum hydrocarbons. *J. Environ. Eng.-A* **126,** 483–490.
10. Hutchinson, S. L., Banks, M. K., and Schwab, A. P. (2001) Phytoremediation of aged petroleum sludge: Effect of inorganic fertilizer. *J. Environ. Qual.* **30,** 395–403.
11. Adam, G., and Duncan, H. J. (1999) Effect of diesel fuel on growth of selected plant species. *Environ. Geochem. Health* **21,** 353–357.
12. Wiltse, C. C., Rooney, W. L., Chen, Z., Schwab, A. P., and Banks, M. K. (1998) Greenhouse evaluation of agronomic and crude oil phytoremediation potential among alfalfa genotypes. *J. Environ. Qual.* **27,** 169–173.
13. PhytoPet© A Database of Plants that Play a Role in the Phytoremediation of Petroleum Hydrocarbons. http://www.phytopet.usask.ca/mainpg.php. Last accessed 05/01/04.
14. Lindau, C. W. and Delaune, R. D. (2000) Vegetative response of Sagittaria lancifolia to burning of applied crude oil. *Water, Air, Soil Pollut.* **121,** 161–172.
15. Pezeshki, S. R., Jugsujinda, A., and Delaune, R. D. (1998) Responses of selected US Gulf coast marsh macrophyte species to oiling and commercial cleaners. *Water, Air, Soil Pollut.* **107,** 185–195.
16. Donnetti, C. (2003) Plant microbe interactions for the clean up of contaminated land. MSc Thesis, Imperial College, London, UK.
17. Adam, G. and Duncan, H. (2002) Influence of diesel fuel on seed germination. *Environ. Pollut.* **120,** 363–370.

18. Harvey, B. S. (2002) Screening plant species for phytoremediation potential. MSc Thesis, Imperial College, London, UK.
19. Lin, Q. X., Mendelssohn, I. A., Carney, K., Bryner, N. P., and Walton, W. D. (2002) The dose-response relationship between No. 2 fuel oil and the growth of the salt marsh grass, *Spartina alterniflora. Mar. Pollut. Bull.* **44,** 897–902.
20. Brown, J. L. and Nadeau, R. J. (2002) Restoration of petroleum contaminated sites using phased bioremediation. *Biorem. J.* **6,** 315–319.
21. Dowty, R. A., Shaffer, G. P., Hester, M. W., Childers, G. W., Campo, F. M., and Greene, M. C. (2001) Phytoremediation of small-scale oil spills in fresh marsh environments: a mesocosm simulation. *Mar. Environ. Res.* **52,** 195–211.
22. ITRC. Phytoremediation Online Decision Tree Document. http://www.itrcweb.org/user/webphyto/envdept/phyto/wwwphyto/index.htm. Last accessed 05/01/06.
23. Environmental Protection Agency (1999) *Phytoremediation Resource Guide*, EPA 542-B-99-003.
24. Environmental Protection Agency (2000) *Introduction to Phytoremediation*, EPA/600/R-99/107.
25. Environmental Protection Agency (2001) *Brownfields Technology Primer: Selecting and Using Phytoremediation*: EPA 542-R-01-006.
26. Kujat, J. D. (1999) A comparison of popular remedial technologies for petroleum contaminated soils from leaking underground storage tanks. *E. Green J.* **11,** 1–13.

9

Uptake, Assimilation, and Novel Metabolism of Nitrogen Dioxide in Plants

Misa Takahashi, Toshiyuki Matsubara, Atsushi Sakamoto, and Hiromichi Morikawa

Summary

To understand the uptake and assimilation of nitrogen dioxide (NO_2) in various plants, quantification of both inorganic nitrogen such as nitrate, nitrite and ammonium ions, and organic nitrogen (or Kjeldahl nitrogen) is vital. Thus, we first describe the quantification of these ions by the capillary electrophoresis method. It is noteworthy that the nitrite ion concentrations in plant tissues are somewhat controversial, and that we have previously reported possible causes of nitrite ion contamination from experimental tools. Details of fumigation of plants with NO_2, and of nitrogen analysis of fumigated plant tissues are described. According to plant physiology textbooks, the total nitrogen taken up into the plant body should equal the sum of the inorganic nitrogen plus organic or Kjeldahl nitrogen. However, we have unexpectedly discovered that about one-third of the total nitrogen derived from NO_2 taken up in the leaves of *Arabidopsis thaliana* is converted to neither inorganic nor Kjeldahl nitrogen, but instead to an as yet unknown nitrogen. We hereafter designate this nitrogen unidentified nitrogen (UN). Some details for the determination of the UN also are described.

Key Words: Kjeldahl nitrogen; inorganic nitrogen; nitrogen dioxide; NO_2; unidentified nitrogen; genetic manipulation; *Arabidopsis thaliana*; tobacco.

1. Introduction

Nitrogen dioxide (NO_2) is a major air pollutant that causes acid rain and forms photo-oxidants such as ozone by photochemical reactions with hydroxyl radicals *(1)*. According to estimation in 1980, the total natural and anthropogenic emission of nitrogen oxides (NOx), which includes nitric oxide and NO_2, was 150 million tons per year with more than half of it emitted by natural sources. Road transport, the major anthropogenic source of NOx in many developed countries, accounted for up to 75% of the NOx in some metropolitan cities in 1984, and this value is still rising. In many developing countries, petro-fueled motor vehicles are the principal source of NOx *(2)*.

Plants are reported to assimilate NO_2 into organic compounds through the primary nitrate assimilation pathway *(3–5)*. Therefore, natural and artificial vegetation can potentially act as a major sink for atmospheric pollutants in terrestrial ecosystems. However, the atmospheric level of NOx is rising all over the world, which suggests that the capacity of naturally occurring sinks such as plants may be already becoming saturated. Innovative methods to decrease the atmospheric level of NOx, or to improve existing sinks, are therefore an issue of considerable urgency. We previously reported that, among 217 taxa, there is more than a 600-fold variation in the capability of plants to assimilate NO_2 *(6,7)*. NO_2 taken up by plant leaves through stomata is reportedly assimilated to organic nitrogen such as amino nitrogen.

In our attempt to study the mechanism of NO_2 metabolism in the plant-mediated decontamination of this major air pollutant *(7–17)* we unexpectedly discovered *(18)* that about one-third of the total nitrogen derived from NO_2 taken up in the leaves of *Arabidopsis thaliana* was converted to neither inorganic nor Kjeldahl nitrogen, but instead to an as yet unknown nitrogen. We hereafter designate this nitrogen unidentified nitrogen (UN).

2. Materials

2.1. Seedlings of Model Plants

Plants of *A. thaliana* (L.) Heynh. ecotype C24, *Cucurbita maxima* Duchesne ex. Lam cv. Hohka-aokawa-amaguri, *Erechtites hieracifolia* Raf., *Glycine max* (L.)Merr. cv. Okuhara-wase, *Hordeum vulgare* L. cv. Uzuakashinri, *Oryza sativa* cv. Akiroman, *Spinacia oleracea* L. cv. Taiheiyo, *Triticum aestivum* L. cv. Shirasagi-komugi grown using standard procedures *(8,18)*.

1. Seedlings grown in pots containing vermiculite and perlite (1:1 [v/v]) placed in a growth chamber (model ER-20-A; Nippon Medical and Chemical Instruments Co. Osaka, Japan) for 6 wk under continuous light (70–100 µmol photons/m/s) at 25.0 ± 0.3°C at a relative humidity of 70 ± 4% with irrigation at 4-d intervals with a half-strength solution of the inorganic salts of Murashige and Skoog's medium *(19)* that contains 19.7 mM nitrate and 10.3 mM ammonium salts. Prior to fumigation with NO_2 supply all plants with tap water every 3–7 d for 2 wk *(11)*.

2.2. Seedlings of Transgenic Arabidopsis and Tobacco

Seedlings of wild-type and transgenic *A. thaliana*, and wild-type tobacco (*Nicotiana tabacum* L. cv. Xanthi XHFD8) and a transgenic tobacco (*N. tabacum* clone 271) that contains the chimeric nitrite reductase (NiR) cDNA in an antisense orientation *(20)*, grown using standard procedures *(17)*.

1. *Arabidopsis* plants grown in pots containing vermiculite and perlite (1:1 [v/v]) placed in a growth chamber for 5–6 wk under continuous light (70–100 µmol photons/m^2/s) at 22.0 ± 0.3°C at a relative humidity of 70 ± 4% with irrigation at 4-d intervals with a half-strength MS salts *(19)*.
2. Tobacco seedlings from seeds sown in vitro on B-medium (pH 5.6), which contains 20 m*M* KNO$_3$ as the sole nitrogen source and 2 wk after sowing, transferred to B-medium containing 10 m*M* ammonium succinate as the sole nitrogen source and grown for a further 4 wk *(21)*.

2.3. Seedlings and Cuttings of Woody Plants

Eucalyptus viminalis Labill., *Pittosporum tobira* Ait. and *Hibiscus cannabinus* L. grown using standard procedures *(6,7)*.

1. Seedlings grown in a greenhouse under natural light for 40–60 d using a weekly supply of 0.1% (v/v) Hyponex (Hyponex Japan, Osaka) that contains 0.4 m*M* nitrate and 0.07 m*M* ammonium salts.
2. *Rhododendron mucronatum* G. Don. plants are propagated by cuttings, and cultured for about 6 mo in the greenhouse using a weekly supply of 0.1% (v/v) Hyponex *(7)*.
3. Additionally, in vitro cultured *Ficus thunbergii* Maxim. *(16)*, if desired.

2.4. Plants Collected From Local Roadsides

In our studies, 50 taxa of wild herbaceous plants and their seeds (42 genera, 15 families) were collected *(11)* from 18 stations along roadsides polluted with NO$_2$ (0.03–0.06 µmol/mol) (*see* **Note 7**).

1. Plants transfered to pots containing artificial soil (sand and leaf mold, 1:1 [v/v]) and grown for 1–2 wk in a greenhouse and supplied with tap water every 3 d.
2. Wild herbaceous plants grown from seed in pots containing vermiculite, and given tap water daily until germination, then for 1–2 mo in a greenhouse given 0.1% (v/v) Hyponex solution (N:P:K = 5:10:5) every 3 d.

2.5. Plants Purchased From Local Shops and Botanical Gardens

Sixty taxa of cultivated herbaceous plants (55 genera, 30 families) and 107 taxa of woody plants (74 genera, 45 families), provided by the Hiroshima Botanical Garden (Hiroshima, Japan) or purchased from commercial shops, were maintained in a greenhouse for 1–2 wk before use under the same conditions as the herbaceous seedlings *(11)*. The heights of individual plants varied from 30 to 50 cm.

2.6. Nitrogen Analyses

1. Water purified with a 7MQ2167 Milli-Q system.
2. 0.1% Solution of SDS.
3. Mortar and pestle made of agate (Sanshyo Corp., Tokyo, Japan).
4. Chloroform.
5. Capillary ion analyzer (CIA; CODE CIA; Millipore Corp., Milford, MA).

6. Running buffer A: 450 mM NaCl and 2.5% OFM Anion-BT (Millipore Corp.).
7. EA-MS analyzer consisting of an elemental analyzer (EA1108 CHNS/O; Fisons Instruments, Milan, Italy) directly connected to a mass spectrometer (Delta C; Thermo-Finnigan, Bremen, Germany) *(15)*.
8. Dowex 50W hydrogen form (Dow Chemical, Midland, MI, USA) for ion chromatography with Devarda's alloy *(22)*.

2.7. Fumigation

1. Indoor fumigation chamber (Nippon Medical and Chemical Instruments Co., model NC-1000-SC) installed inside a building with artificial light from fluorescent lamps.
2. Outdoor fumigation chamber (Nippon Medical and Chemical Instruments Co., model NC-1000-P1SC) in a confined greenhouse with natural light conditions.

3. Methods

3.1. Quantification of Nitrate, Nitrite, and Ammonium Ions by Capillary Electrophoresis

Nitrate, nitrite, and ammonium ions are key metabolites in the NO_2 assimilation in plants. However, only limited information about the levels of these ions in plant tissues has been published *(8)*. The presence of nitrite ions in plant tissues is controversial (*see* **ref. 8** and references therein).

3.1.1. Extraction of Nitrate and Nitrite Ions From Plant Tissues

1. Rinse freshly harvested leaves (100 mg fresh weight) with pure water, and then homogenize in 50 µL of a 0.1% solution of SDS with an agate mortar and pestle (*see* **Note 1**).
2. Add 200–500 µL of chloroform to the homogenate, and homogenize the tissue further.
3. Clarify each homogenate by centrifuging twice at 18,000g for 10 min.
4. Use the resulting aqueous layer to determine levels of NO_3^- and NO_2^- ions, and the chloroform layer for quantification of chlorophyll *(8)*.

3.1.2. Extraction of Ammonium Ions From Plant Tissues

1. Homogenize freshly harvested leaves (100 mg fresh weight) in 800 µL of pure water with the agate mortar and pestle.
2. Use an aliquot of the homogenate for quantification of chlorophyll. Clarify the remainder by centrifuging twice at 18,000g for 10 min, and determine the level of ammonium ions in the supernatant *(8)* (*see* **Note 2**).

3.1.3. Capillary Electrophoresis

1. Quantify the ions with the capillary ion analyzer. For nitrate and nitrite use running buffer A.

2. Fix the wavelength of the ultraviolet detector at 214 nm for the analysis of nitrate and nitrite ions and at 185 nm for that of ammonium ions *(8)*.

3.2. Fumigation Chamber

3.2.1. Fumigation Conditons

1. Fumigate plants with 4.0 ± 0.4 µL/L or with 0.1 ± 0.01 µL/L $^{15}NO_2$ (51.6 atom% ^{15}N) for 4–8 h in the light (70 µmol photons/m^2/s) or under natural light at 22–25°C, a relative humidity of 70 ± 4%, and atmospheric CO_2 concentration (340 ± 80 µL/L) in a fumigation chamber (*see* **Notes 3** and **4**).
2. Harvest the leaves, lyophilize, grind into powder, and store in a desiccator until use *(11)* (*see* **Notes 5** and **6**).

3.3. Nitrogen Analysis

Total, Kjeldahl and inorganic nitrogen content in each sample was obtained from the following equation:

$$\text{nitorgen content} = (B - 0.3663)/100 \times A \times 100/C \qquad (1)$$

where the values *A*, *B* and *C* correspond respectively to the amount of ^{15}N plus ^{14}N in the sample, the atomic percentage of ^{15}N [$^{15}N/(^{15}N + {}^{14}N)$] in the sample and the atomic percentage of ^{15}N in the NO_2 gas. An atomic percentage of 0.3663 corresponds to the natural abundance of ^{15}N *(23)*.

3.3.1. Total Nitrogen

1. To quantify the total nitrogen, place about 4 mg of the powdered leaves in tin containers for use in the EA-MS analyzer, and determine the value *A* and the amount of ^{15}N in each sample.
2. From A and ^{15}N, determine the value *B*.

3.3.2. Kjeldahl Nitrogen

1. To estimate the Kjeldahl nitrogen, digest 10–70 mg of the powdered leaves by the Kjeldahl method *(17)*, but with the catalytic reagents of sodium thiosulfate and copper sulfate omitted from the digestion mixture.
2. Obtain the value *A* or the amount of ammonia nitrogen by titration analysis of the distilled ammonia.
3. Concentrate ammonia in the distillates using the Conway diffusion method *(24)*, after which the value *B* for ammonia can be determined using the EA-MS analyzer (*see* **Note 5**).

3.3.3. Inorganic Nitrogen

The nitrogen of nitrate and nitrite in each sample was considered here to represent the inorganic nitrogen *(8)*.

1. Quantify the nitrate and nitrite concentrations by capillary electrophoresis *(8)* to give the value *A*.
2. Determine the amount of ^{15}N in each sample by first collecting them by ion-exchange chromatography and then reduction to ammonia by Devarda's alloy *(22,23)*.
3. Concentrate the ammonia using the Conway diffusion method *(24)* and analyze the concentrated ammonia using the EA-MS analyzer.
4. From the two values previously listed, determine the value *B* for inorganic nitrogen.

3.3.4. Unidentified Nitrogen

UN was calculated from the following equation:

$$UN = \text{total nitrogen} - (\text{Kjeldahl nitrogen} + \text{inorganic nitrogen}) \qquad (2)$$

4. Notes

1. Under the analytical conditions described, the limits of detection are 0.05 ppm for nitrate ion, 0.05 ppm for nitrite ion, and 0.2 ppm for ammonium ion. The background levels of these ions are below the limits of detection when the agate mortar and pestle are used. When a ceramic mortar and pestle are used, the background levels of nitrate and nitrite ions are about six times higher than the limit of detection, whereas that of ammonium ions remains less than the limit of detection.
2. To esimate UN (*see* **Note 5**), addition of catalytic reagents such as sodium thiosulfate and copper sulfate into the Kjeldahl digestion mix must be avoided because addition of such catalytic reagents enhances the conversion of some inorganic nitrogen, such as nitrate nitrogen and UN-bearing compounds, to ammonia.
3. Plants emit NO *(25–32)*. Thus, NO emitted from plants may interfere with the concentration of NOx in the fumigation chamber. Therefore, the concentration of NO_2, but not NOx must be monitored.
4. Using kenaf as a model plant, Takahashi et al. *(33)* compared uptake and assimilation under artificial light from fluoresencent lamps (70 µmol photons/m²/s) and under natural light. They found that the uptake and assimilation of NO_2 under both light conditions coincided with experimental errors. This implies that artificial fluorescent light provides enough light energy to kenaf leaves for the uptake and assimilation of NO_2. This behavior of plants deserves further investigation.
5. According to the Kjeldahl method, nitrogen in biological systems (including plants) is classified into two forms: inorganic nitrogen (excluding ammonia) that is almost exclusively nitrate (and nitrite), and organic nitrogen (and ammonia) that is stoichiometrically recoverable by the Kjeldahl method (and therefore called Kjeldahl nitrogen). The century-old Kjeldahl method is still the sole method available for the analysis of organic nitrogen in biological systems *(34)*. We unexpectedly discovered *(18)* that about one-third of the total nitrogen derived from NO_2 taken up in the leaves of *A. thaliana* was converted to neither inorganic nor Kjeldahl nitrogen, but instead to an as yet unknown nitrogen. This nitrogen was designated UN. We therefore addressed the formation of UN in various plants in

response to fumigation with NO_2, the formation of UN in tobacco fed with nitrate, and also in naturally fed plants. In all of the cases explored here, the formation of UN was always ascertained *(18)*.
6. The NO_2, after uptake into the plant leaf cells, reacts with water to yield nitric acid, nitrous acid, and NO, although the quantitative details remain unclear *(12–14,18)*. Therefore, those enzymes involved in the primary nitrate assimilation pathway such as nitrate reductase (NR), NiR, and glutamine synthetase (GS), which are, respectively, the first, second, and third enzyme in the primary nitrate metabolism, should play key roles in the metabolism of NO_2. We have been investigating the GS gene from *Arabidopsis (6)*, and the NiR gene from *Arabidopsis (35)* and tobacco *(36)*. Chimeric plasmids harboring tobacco NR cDNA, spinach NiR cDNA, and *Arabidopsis* GS cDNA were introduced into the root sections of *Arabidopsis (17)*. Each of the transgenes was under the control of cauliflower mosaic virus (CaMV) 35S promoter and nopaline synthase terminator. Three transgenic lines had significantly higher NO_2-derived NR (at most by 40% increase) than that of the control ($p < 0.01$). Neither the NR- nor GS-transformants showed a significant increase in NO_2-derived RN *(17)*. The flux control coefficient, which is a measure of the effect of change in a single enzyme activity on the flux *(37)* of NiR for NO_2 assimilation was about 0.4. The flux control coefficients of NR and GS were much smaller than this value (–0.01 and –0.1). More recently we have found that the overexpression of *S*-nitrosoglutathione reductase, formerly known as glutathione-dependent formaldehyde dehydrogenase *(38)*, increased both the uptake and assimilation of NO_2 by *A. thaliana*. (our unpublished results).
7. We investigated genetic transformation of evergreen shrubs such as *P. tobira (39)* and *Rhaphiolepis umbellate (38)*, both of which are used as roadside trees. They ranked 137th and 134th, respectively, in their NO_2-assimilation capability among 217 taxa *(11)*. Young hypocotyls were bombarded with a mixture of plasmid vector pANiR and pCH. pANiR bears the NiR cDNA from *A. thaliana* under the control of the cauliflower mosaic virus (CaMV) 35S promoter and the nopaline synthase terminator. pCH *(42)* bears the hygromycin-resistance gene under the control of the CaMV 35S promoter and nopaline synthase terminator. Transgenic shoots were obtained but no transgenic plantlets were obtained to date in *P. tobira (39)*. Transgenic plants of *Rhaphiolepis umbellate* were obtained, and more recently a number of transgenic plants of this woody species were produced by use of *Agrobacterium*-mediated transformation (unpublished results).

Acknowledgments

This work was supported in part by Grants-in-Aid for Scientific Research (nos. 13556002 and 16208033) from the Japan Society for the Promotion of Science, by the Research for the Future Program, Japan Society for the Promotion of Science (JSPS-RFTF96L00604) and by a Grant-in-Aid for Creative Scientific Research (no. 13GS0023) from the Japan Society for the Promotion of Science.

References

1. Wellburn, A. R., Barnes, J. D., Lucas, P. W., McLeod, A. R., and Mansfield, T. A. (1997) Controlled O_3 exposures and field observations of O_3 effects in the UK. In: *Forest Decline and Ozone,* (Sandermann, H., Wellburn, A. R., and Heath, R. L., eds.), Springer Verlag, Berlin, Germany, pp. 201–248.
2. Yunus, M., Singh, N., and Iqbal, M. (1996) Gobal status of air pollution: an overview. In: *Plant Response to Air Pollution,* (Yunus, M. and Iqbal, M., eds.), John Wiley and Sons Ltd., New York, NY, pp. 1–34.
3. Zeevaart, A. J. (1976) Some effects of fumigating plants for short periods with NO_2. *Environ. Pollut.* **11,** 97–108.
4. Yoneyama, T. and Sasakawa, H. (1979) Transformation of atmospheric NO_2 absorbed in spinach leaves. *Plant Cell Physiol.* **20,** 263–266.
5. Wellburn, A. R. (1994) Nitrogen oxides. In: *Air Pollution and Climate Change: The Biological Impact,* (Wellburn, A. R., ed.), Longman Scientific and Technical, Essex, England, 57–82.
6. Morikawa, H., Takahashi, M., and Irifune, K. (1998) Molecular mechanism of the metabolism of nitrogen dioxide as an alternative fertilizer in plants. In: *Stress Responses of Photosynthetic Organisms,* (Satoh, K. and Murata, N., eds.), Elsevier Science, Amsterdam, The Netherlands, pp. 227–237.
7. Takahashi, M., Kondo, K., and Morikawa, H. (2003) Assimilation of nitrogen dioxide in selected plant taxa. *Acta Biotechnol.* **23,** 241–247.
8. Kawamura, Y., Takahashi, M., Ariumura, G., et al. (1996) Determination of levels of NO_3^-, NO_2^-, and NH_4^+ ions in leaves of various plants by capillary electrophoresis. *Plant Cell Physiol.* **37,** 878–880
9. Kawamura, Y., Fukunaga, K., Umehara, A., Takahashi, M., and Morikawa, H. (2002) Selection of *Rhodedendron mucornatum* plants that have a high capacity for nitrogen dioxide uptake. *Acta Biotechnol.* **22,** 113–120.
10. Hakata, M., Takahashi, M., Zumft, G., Sakamoto, A., and Morikawa, H. (2003) Conversion of the nitrogen of nitrate and nitrogen dioxide to nitrous oxide in plants. *Acta Biotechnol.* **23,** 249–257.
11. Morikawa, H., Higaki, A., Nohno, M., et al. (1998) More than a 600-fold variation in nitrogen dioxide assimilation among 217 plant taxa. *Plant Cell Environ.* **21,** 180–190.
12. Morikawa, H. and Erkin, O. C. (2003) Basic processes in phytoremediation and some applications to air pollution control. *Chemosphere* **52,** 1553–1558.
13. Morikawa, H., Takahashi, M., Hakata, M., and Sakamoto, A. (2003) Screening and genetic manipulation of plants for decontamination of pollutants from the environments. *J. Biotechnol. Adv.* **22,** 9–15.
14. Morikawa, H., Takahashi, M., and Kawamura, Y. (2003) Air pollution clean up using pollutant-philic plants—metabolism of nitrogen dioxide and genetic manipulation of related genes. In: *Phytoremediation: Transformation and Control of Contaminants,* (McCutcheon, S. C. and Schnoor, J. L., eds.), John Wiley and Sons, Inc., New York, NY, pp. 765–786.

15. Goshima, N., Mukai, T., Suemori, M., Takahashi, M., Caboche, M., and Morikawa, H. (1999) Emission of nitrous oxide (N_2O) from transgenic tobacco expressing antisense nitrite reductase mRNA. *Plant J.* **19,** 75–80.
16. Takahashi, M., Kohama, S., Hakata, M., et al. (2001) Production of mutants that have high ability to assimilate nitrogen dioxide by the irradiation of ion beams in *Ficus stipulata, Ann. Rep. TIARA* **39,** 62–63.
17. Takahashi, M., Sasaki, Y., Ida, S., and Morikawa, H. (2001) Enrichment of nitrite reductase gene improves the ability of *Arabidopsis thaliana* plants to assimilate nitrogen dioxide. *Plant Physiol.* **126,** 731–741.
18. Morikawa, H., Takahashi, M., Sakamoto, A., et al. (2004) Formation of unidentified nitrogen in plants: an implication for a novel nitrogen metabolism. *Planta* **219,** 14–22.
19. Murashige, T. and Skoog, F. (1962) A revised medium for rapid growth and bioassays with tobacco tissue cultures. *Physiol Plant* **15,** 473–497
20. Vaucheret, H., Kronenberger, J., Lepingle, A., Vilaine, F., Boutin, J. P., and Caboche, M. (1992) Inhibition of tobacco nitrite reductase activity by expression of antisense RNA. *Plant J.* **2,** 559–569.
21. Bourgin, J. P., Chupeau, Y., and Missonier, C. (1979) Plant regeneration from mesophyll protoplasts of several Nicotiana species. *Physiol Plant.* **45,** 288–292.
22. Gatley, S. J. and Shea, C. (1991) Radiochemical and chemical quality-assurance methods for [^{13}N]-ammonia made from a small volume $H_2^{16}O$ target. *Appl. Rad. Isotop.* **42,** 793–796.
23. Mariotti, A. (1983) Atmospheric nitrogen is a reliable standard for natural ^{15}N abundance measurements. *Nature* **303,** 685–687.
24. Conway, E. J. and Byrne, A. (1933) An absorption apparatus for the micro-determination of certain volatile substances. I. The micro-determination of ammonia. *Biochem J.* **27,** 419–429.
25. Durner, J. and Klessig, D. F. (1999) Nitric oxide as a signal in plants. *Curr. Opin. Plant Biol.* **2,** 369–374.
26. Delledonne, M., Xia, Y., Dixon, R. A., and Lamb, C. (1998) Nitric oxide functions as a signal in plant disease resistance. *Nature* **394,** 585–588.
27. Tun, N. N., Holk, A., and Scherer, G. F. E. (2001) Rapid increase of NO release in plant cell cultures induced by cytokinin. *FEBS Lett.* **509,** 174–176.
28. Wendehenne, D., Pugin, A., Klessig, D. F., and Durner, J. (2001) Nitric oxide: comparative synthesis and signaling in animal and plant cells. *Trends Plant Sci.* **6,** 177–183.
29. Chandok, M. R., Ytterberg, A. J., van Wijk, K. J., and Klessig, D. F. (2003) The pathogen-inducible nitric oxide synthetase (iNOS) in plants is a variant of the P protein of the glycine decarboxylase complex. *Cell* **113,** 469–482.
30. Rockel, P., Strube, F., Rockel, A., Wildt, J., and Kaiser, W. M. (2002) Regulation of nitric oxide (NO) production by plant nitrate reductase *in vivo* and *in vitro*. *J. Exp. Bot.* **53,** 103–110.

31. Yamasaki, H., Sakihama, Y., and Takahashi, S. (1999) An alternative pathway for nitric oxide production in plants: new features of an old enzyme. *Trends Plant Sci.* **4**, 128–129.
32. Wildt, J., Kley, D., Rockel, A., Rokel, P., and Segschneider, H. J. (1997) Emission of NO from several higher plant species. *J. Geophys. Res.* **102**, 5919–5927.
33. Takahashi, M., Konaka, D., Sakamoto, A., and Morikawa, H. (2005) Nocturnal uptake and assimilation of nitrogen dioxide by C3 and CAM plants. *Z. Naturforsch.* **60c**, 279–284.
34. Bradstreet, R. B. (1965) The Kjeldahl digestion. In: *The Kjeldahl Method for Organic Nitrogen,* (Bradstreet, R. B., ed.), Academic Press, New York, NY, pp. 9–88.
35. Tanaka, T., Ida, S., Irifune, K., Oeda, K., and Morikawa H. (1994) Nucleotide sequence of a gene for nitrite reductase from Arabidopsis thaliana. *DNA Sequence* **5**, 57–61.
36. Kato, C., Takahashi, M., Sakamoto, A., and Morikawa, H. (2004) Differential expression of the nitrite reductase gene family in tobacco as revealed by quantitative competitive RT-PCR. *J. Exp. Bot.* **55**, 1761–1763.
37. Kacser, H. and Porteous, J. W. (1987) Control of metabolism: what do we have to measure? *Trends Biochem. Sci.* **12**, 5–14.
38. Sakamoto, A., Ueda, M., and Morikawa, H. (2002) *Arabidopsis* glutathione-dependent formaldehyde dehydrogenase is an *S*-nitrosoglutathione reductase. *FEBS Lett.* **515**, 20–24.
39. Kondo, K., Takahashi, M., and Morikawa, H. (2002) Regeneration and transformation of a roadside tree *Pittosporum tobira* A. *Plant Biotechnol.* **19**, 135–139.
40. Erkin, O. C., Takahashi, M., Sakamoto, A., and Morikawa, H. (2003) Development of regeneration and transformation systems for *Rhaphiolepis umbellata* L. plants using particle bombardment. *Plant Biotechnol.* **20**, 145–152.
41. Shigeto, J., Yoshihara, S., Adam, S. E. H., Sueyoshi, K., Sakamoto, A., Morikawa, H., and Takahashi, M. (2006) Genetic engineering of nitrite reductase gene improves uptake and assimilation of nitrogen dioxide by *Rhaphiolepis umbellate* (Thumb) Makino. *Plant Biotechnology* **23**, 111–116.
42. Goto, F., Toki, S., and Uchiyama, H. (1993) Inheritance of a co-transferred foreign gene in the progenies of transgenic rice plants. *Transgen. Res.* **2**, 300–305.

II

MANIPULATING CONTAMINANT AVAILABILITY AND DEVELOPING RESEARCH TOOLS

10

Testing the Manipulation of Soil Availability of Metals

Fernando Madrid Diaz and M. B. Kirkham

Summary

Manipulating heavy-metal availability with chelating agents is a way to accelerate natural phytoremediation of contaminated soils. Nevertheless, increasing metal availability also increases the risk of metal movement through the soil profile, and consequently the contamination of ground water, exacerbating the environmental problem. Knowledge of metal displacement is, therefore, necessary before attempting manipulations of metal availability at a contaminated site. Experiments done in columns packed with the contaminated soil, to compare effects with and without plants, with and without chelates (e.g., ethylenediaminetetraacetic acid), and with a monitored watering regime, are an easy and feasible way to study different parameters in the soil, plants, and water. Using this method, the utility of the technique for phytoremediation can be tested. Measurements can include, for example, metal content in the leachate, metal movement through the soil profile, metal availability and uptake by plants, and plant growth.

Key Words: Soil contamination; column experiment; assisted-phytoremediation; EDTA; barley; sewage sludge; soil profile; metal availability; leaching.

1. Introduction

Because of various human activities and natural processes, many ecosystems are changing, often resulting in the contamination of the environment. One of the consequences is soil contamination with chemical compounds related to human activities (e.g., industries, mining, traffic, and agriculture). Remediation of these areas should be a priority of countries and improving remedial techniques a duty of the scientific community. Remediation techniques may be grouped into two categories: (1) *ex situ* techniques, which require removal of contaminated soil for treatment, and (2) *in situ* methods, which remediate without excavation of contaminated soil *(1)*. This second group of techniques are preferred because of lower cost and reduced impact on the environment, although at present they are only applicable to sites with low or medium contamination levels.

It is important to distinguish between contamination with organic compounds and with metals. Organic pollutants may be mineralized into relatively nontoxic products such as carbon dioxide, nitrate, chlorine, and ammonia *(2)* and this process of decomposition may help during remediation of sites contaminated with this kind of pollutant. However, as metals cannot be degraded, alternative strategies have to be applied. Stabilization or extraction of metals is the ultimate aim of most soil-remediation techniques. Stabilization tries to control the contamination by transforming soil metals to less toxic forms while avoiding the migration of metals to groundwater or by leaching to other sites as dust. This can be achieved by adding organic or inorganic amendments to soil that form stable complexes with metals *(3,4)*.

Heavy metals may also be stabilized using plants (phytostabilization) that sequester the metals in roots, which prevents leaching from the site. For extraction of metals *in situ,* the use of plants is also possible. A selected crop is planted on the contaminated soil, it takes up metals during its growth, and at the end of the season or after a given period of time, aboveground tissues of the plants that contain a certain content of metals are harvested and kept in a safe place. This is the aim of phytoextraction. Both phytostabilization and phytoextraction are subtypes of phytoremediation, which is defined as the use of plants to remediate contaminated soils *(5)*. When possible, the second technique is preferred because the pollutants are actually removed from the soil. Nevertheless, because of the low availability of metal compounds, this could become an endless process. Furthermore, transport of metals from the roots to the shoots before harvesting is also necessary for successful phytoextraction, and this step is also a limiting factor of the technique.

During recent decades, metal hyperaccumulator plants have been discovered and developed using genetic engineering, and over 400 plant species have been identified *(5)* that might help to accelerate the phytoextraction process. Several problems appear when using these plants; (1) they tend to accumulate one specific metal, whereas many metal-contaminated sites have a high concentration of several metals, so different crops would have to be used; (2) plants take up the more available metal fraction, but less available fractions cannot be extracted; and (3) hyperaccumulators often have low biomass, which results in a low amount of metal extracted from the site. Therefore, increasing availability of metals is an alternative that has to be studied. Various organic compounds have been used to increase extractability, such as ethylenediaminetetraacetic acid (EDTA), *N*-hydroxyethyl-ethylenediamine-*N,N',N'*-triacetic acid (HEDTA), diethylenetrinitrilopentacetic acid (DTPA), and also more natural low molecular weight acids such as citric or malic acid. The most frequently used is EDTA, which has been reported as more effective than other synthetic chelators for several heavy metals *(6,7)*. It was first suggested in 1974 *(8)* that metal–EDTA complexes formed in the soil could increase the potential for plant uptake. These acids form a complex with the metal, preventing it from being retained by the soil. EDTA forms hexahedral metal ion complexes that enhance metal uptake by plants and translocation to shoots *(2,9)*.

The use of chelating agents to increase metal availability to plants, with the aim of extracting them from soil is called "assisted," "induced," or "enhanced" phytoextraction. However, increasing metal mobility in soil also increases the risk of pollutants leaching into groundwater *(10–12)*. This limits the usefulness of the technique to situations in which soil containment and hydrological control can be provided *(5)*. When planning the design of an experiment to study induced phytoextraction, not only metal-extraction capacity of plants has to be studied but also metal leaching. Column experiments have been widely used to study metal content in leachate *(13–15)*. A method to study trace and heavy-metal movement in a sewage-sludge amended soil, in the presence of a chelating agent (EDTA) and plants (*Hordeum vulgare*, barley) is described in this chapter.

2. Materials
2.1. Experimental Apparatus

1. Clear plastic tubes of 15 to 20 cm inner diameter, with lengths depending on the depth of soil profile to be studied, and a leachate collection system at their base. The simplest leachate collection system is to put the column directly on a pie pan, and, if possible, to maintain the column a few centimeters over the pan, let free leaching occur through small holes and collect the leachate daily in a container. Columns are wrapped with aluminium foil to prevent algae from growing along the sides of the column.
2. (EDTA)Na$_2$·2H$_2$O (purity grade >99%). EDTA is a tetraprotic acid, and is found commercially in many forms. To have good solubility and neutral pH in the solution to be added to soil, the disodium salt is recommended. A rate of 0.5 g/kg soil is added as a solution (0.1 M; 37.2 g/L). (EDTA)Na$_2$ solution is stable for several months in a dark container at room temperature.

2.2. Laboratory Analysis

1. Fuming hydrochloric acid (38%) and nitric acid (65%). Analytical-grade acids are used for soil analysis and super-pure reagent grade is recommended for plant and water analysis.
2. 0.05 M EDTA, pH 7.0 *(16)*. Weigh 73.1 g of analytical-grade EDTA (not to be confused with partially substituted sodium salt; *see* **Subheading 2.1.2.**). Add 4 L water, stir the solution, and add 65 mL 25% ammonia solution. Filter if the solution is not clear, adjust the pH to 7.0 with ammonia solution (if pH <7.0) or hydrochloric acid (if pH >7.0), and make up to 5 L with distilled water. This solution is stable for 1 mo at room temperature.

3. Methods
3.1. Experimental Protocol

1. For four treatments replicated three times, 12 columns are filled with dried soil to reach the desired height. The columns are watered at the top to homogenize soil packing (*see* **Note 1**). The leachate obtained during this period is collected and discarded. When water has been absorbed by the columns, liquid biosolids are added to the soil surface at the desired rate (*see* **Note 2**). The soil profile is maintained in the dark by wrapping the columns with aluminium foil, which can be unrolled daily to observe the growth of roots.
2. To create a nonsaturated zone and aerobic conditions at the soil surface, water is not added for 2 or 3 d. Then 20 barley (*H. vulgare* L.) are planted (*see* **Note 3**) in half the columns, randomly selected. Germination of the seeds is recorded and 14 d after planting (14 DAP), all the columns are thinned to the same number of plants (e.g., 10). Plant height and root penetration depth are measured daily (*see* **Note 4**). The watering regime during this period depends on the desired experimental conditions (*see* **Note 5**). It might be adapted to the regime of the contaminated site, or to a constant daily rate (e.g., 10 mm/d). The second option is preferable for modeling the metal movement though the soil profile when EDTA is added. During the above period of time leachate is discarded.
3. 1 wk after thinning (21 DAP; *see* **Note 6**), Na_2EDTA solution is added to half the columns with plants and to half the columns without plants, randomly selected. This results in four treatments replicated three times; no plants+no EDTA (P0A0), no plants+EDTA (P0A1), plant+no EDTA (P1A0), and plants+EDTA (P1A1). To get a homogenous addition, EDTA is added as a solution as per **Subheading 2.1.2.** (*see* **Note 7**), onto the surface of the soil in each column. After the addition of EDTA, leachate volume is measured and analyzed daily for trace metal content (*see* **Subheading 3.2.1.**). Plant growth observation is continued by measuring plant height and root-penetration depth. The time-course of the experiment depends on the final aim (*see* **Note 8**). If soil physical properties are known, columns may be watered until an amount equivalent to two pore volumes has been added *(13)*, i.e., when it is considered that the initial water in the column has been replaced.
4. The day after watering is stopped, the aerial plant shoots are harvested, weighed fresh and dry, and analyzed for metal content (*see* **Subheading 3.2.2.**). Roots in the first 5 cm of the soil profile are sampled, weighed, and analyzed for metals. Roots deeper in the profile will probably be too sparse to extract and analyze. To study metal content in the soil column, it is divided into 5 cm layers (*see* **Note 9**) and each analyzed separately. Aqua regia- and EDTA-extractable contents are analyzed in each layer (*see* **Subheadings 3.2.3.** and **3.2.4.**).

3.2. Laboratory Analytical Procedures

1. Leachate analysis is carried out by acidification at 2% with concentrated HNO_3 (super-pure reagent grade) and trace metals determination by inductively coupled plasma—optical emission spectrometry (ICP-OES) (*see* **Note 10**).

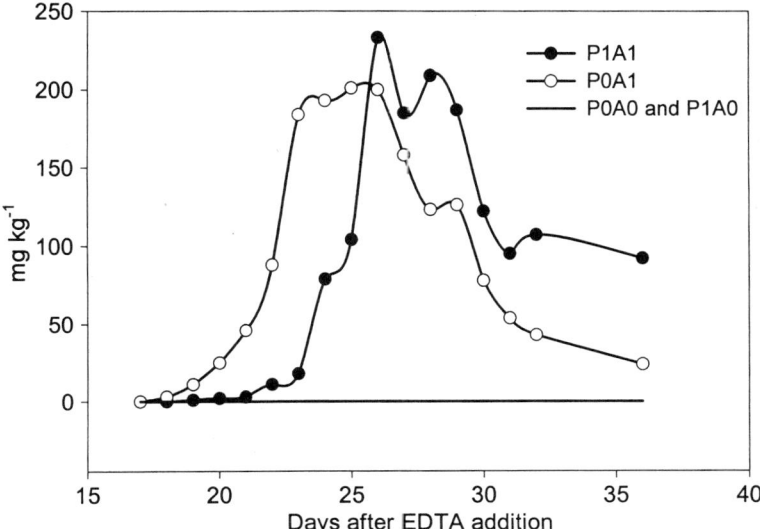

Fig. 1. Fe content in leachate. Notation of treatments as explained in **Subheading 3.1.**

2. Analysis of plant material is carried out by nitric acid digestion and trace metal determination by ICP-OES. After harvesting the shoots, plants are washed once with tap water and twice with distilled water, dried at 70°C for 48 h and ground. Microwave digestion (0.5 g) with nitric acid (4 mL), filtered to a final volume of 50 mL, is carried out and trace metals are determined by ICP-OES. The same procedure is applied for root analysis, although a more careful washing has to be carried out to be certain that no soil remains on the roots.
3. Aqua regia-extractable metal content of soil is determined by aqua-regia digestion of soil and trace metal determination by ICP-OES. Samples are dried at 40°C, sieved at 2 mm, and a subsample (5–10 g) ground to <60 µm. Microwave digestion (0.5 g) with aqua-regia (3 mL concentrated HCl:1 mL concentrated HNO_3), filtered to a final volume of 50 mL, is carried out, and trace metals determined by ICP-OES.
4. EDTA-extractable metal content of soil is determined by extraction with EDTA solution (0.05 M, pH 7.0) and trace-metal determination by ICP-OES. Weigh 2.5 g of sieved soil at 2 mm, extract with 25 mL of the EDTA solution (0.05 M, pH 7.0), and determine trace metals by ICP-OES. These solutions must be kept in a refrigerator and analyzed as soon as possible (less than 1 wk).

3.3. Interpretation of Results

Trace-metal content in the leachate may show a time-course similar to that in **Fig. 1**. In **Fig. 1** the effects of EDTA addition result in increased Fe content of leachate compared with treatments with no EDTA. There is also a delay in metal leaching in the presence of plants. Trace-metal content (aqua regia-extractable) can be studied using depth profiles as in **Fig 2**. Lower metal content in surface

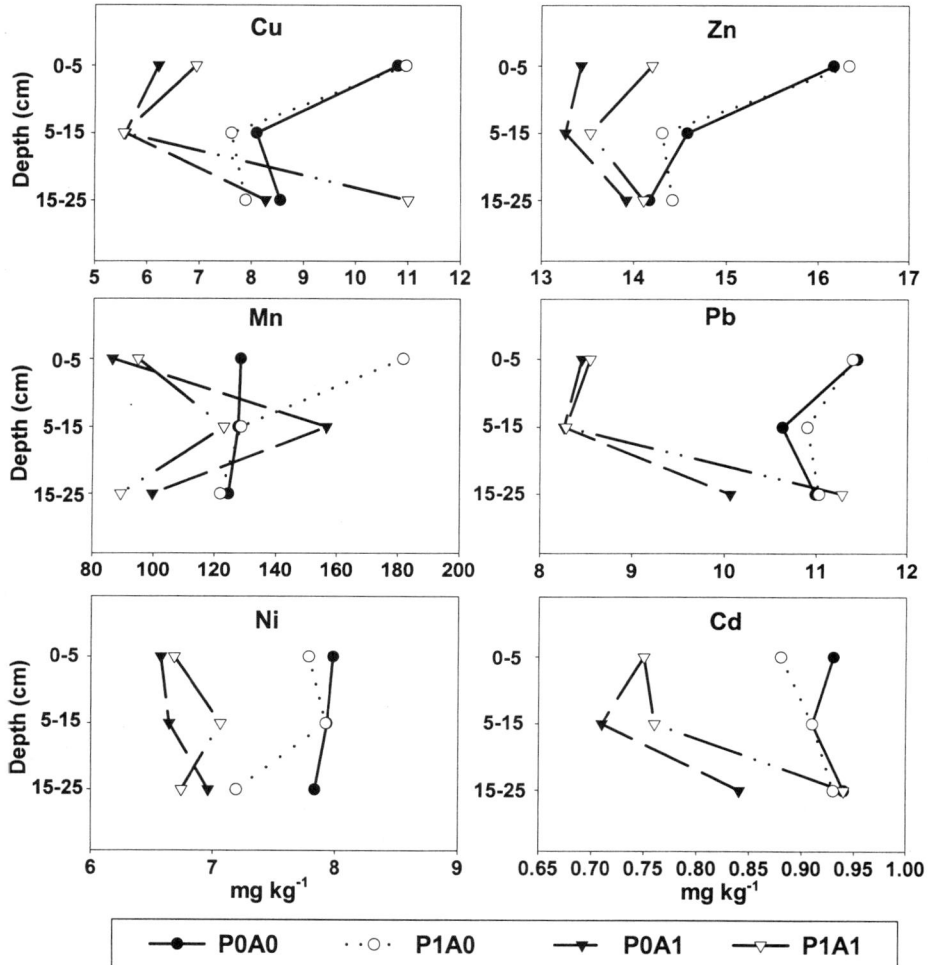

Fig. 2. Trace-metal content (aqua regia-extracted) throughout soil profile.

layers of columns treated with EDTA (as shown in **Fig. 2**) indicates metal leaching through the soil profile. The EDTA-extractable:aqua regia-extractable metal ratio indicates if metal availability is affected by the treatments (*see* **Note 11**). Plant metal content and biomass show if metal extraction is significantly increased by EDTA.

4. Notes

1. When the soil is still dry, add the water carefully to avoid holes that could disturb the surface. One liter H_2O per 3 kg of soil is enough to pack the column. When water is added, soil height usually decreases 1 or 2 cm. Using undisturbed soil is

a possibility, if a suitable system to fill up the column at the field site in this way is available.
2. High application rates to study metal movement are desirable, although this can affect plant growth. Availability of sludge-borne metals and problems related to its application to soils have recently been reviewed *(17)*. Clearly, if one wants to study a soil previously contaminated by any process, one can study heavy-metal movement in the soil profile without adding any polluted amendment.
3. Although barley cannot be considered a hyperaccumulator, several authors *(18)* have suggested that barley has, at least, equal phytoremediation potential to Indian mustard (*Brassica juncea*), a plant widely used in phytoremediation research. The highly developed root system of barley increases the contact surface between soil and plant tissues that may take up metals, and it also improves the chance to observe its growth through the clear plastic of the column. Furthermore, barley is a salt-tolerant crop, which helps to reduce the negative effects that the sodium added with the EDTA solution may cause in other crops.
4. Different methods may be followed for the measurements. One may select the maximum height, the mean height of all plants, and so on. Measurement of the three highest plants and the three longest roots per column, to obtain mean data for each parameter, is recommended.
5. If it is necessary to fertilize plants, a nutrient solution may be prepared to be added during the watering when needed. KH_2PO_4 (0.33 g/kg soil) and NH_4NO_3 (0.43 g/kg soil) are recommended *(19)*. How and when to fertilize plants used for chelate-induced phytoremediation are aspects that have not been studied.
6. The schedule may change, depending on the crop used, the experimental design and the final aim of the experiment. EDTA must be added when the vegetative growth rate is high, to take advantage of element uptake by plants. Another aspect related to the schedule of EDTA addition and final harvest is the plant biomass. This has to be enough to be analyzed, and at least 1 g dry weight per column will be needed.
7. Actually, the EDTA concentration of the added solution may vary (0.01–0.15 *M*) depending on the watering regime of the columns. If a high irrigation rate is being used, a dilute solution may be applied *(20)*. Furthermore, the rate of 0.5 g EDTA/kg soil may also be changed depending on the soil profile depth that one wants to study. A profile longer than 30 cm may need a very high EDTA concentration, if the watering regime is normal, but the plants may then be affected when the solution is added *(13)*. Instead, a rate based on the surface area has also been used (75 g EDTA/m^2 soil) *(21)*. Recently, a direct comparison between two strategies used for EDTA addition for assisted phytoremediation has been reported *(11)*, applying the chelate once a few days before harvest or gradually during the growth period. These authors concluded that split applications were generally more effective for phytoextraction than the same amount of EDTA added in a single addition.
8. If the aim of the assay is to study heavy-metal accumulation of plants after EDTA addition, short-term experiments have to be carried out, with a low watering rate, to maintain EDTA in the rooting zone. For a wider perspective, including

environmental aspects as explained in this paper, longer experiments and watering times are necessary.
9. If the soil profile is deeper than 30 cm, select a surface layer of 5 cm, but the rest of the column may be divided into layers of 10 cm.
10. The acidified leachate (0.5 mL HNO_3 per 25 mL of leachate) may be kept in a glass flask for several weeks in a refrigerator until analysis. To analyze trace-metal content, ICP-OES is recommended because of its low detection limits and simultaneous multielement analysis. If ICP-OES is not available, atomic absorption spectroscopy is also recommended for most trace metals (Fe, Cu, Mn, Zn, Ni, and Cd; higher detection limits for this technique are found for Pb and Cr).
11. Many other parameters may be studied if they are considered necessary, for example those related to plant physiology or nutrition.

References

1. Khan, A. G., Kuek, C., Chaudhry, T. M., Khoo, C. S., and Hayes, W. J. (2000) Role of plants, mycorrhizae and phytochelators in heavy metal contaminated land remediation. *Chemosphere* **41**, 197–207.
2. Meagher, B. R. (2000) Phytoremediation of toxic elemental and organic pollutants. *Curr. Opin. Plant Biol.* **3**, 153–162.
3. Chlopecka, A. and Adriano, D. C. (1997) Influence of zeolite, apatite and Fe-oxide on Cd and Pb uptake by crops. *Sci. Tot. Environ.* **207**, 195–206.
4. Lombi, E., Zhao, F., Zhang, G., et al. (2002) In situ fixation of metals in soils using bauxite residue: chemical assessment. *Environ. Pollut.* **118**, 435–443.
5. McGrath, S. P. and Zhao, F. (2003) Phytoextraction of metals and metalloids from contaminated soils. *Curr. Opin. Biotech.* **14**, 277–282.
6. Huang, J. W., Chen, J., Berti, W. R., and Cunningham, S. D. (1997) Phytoremediation of lead-contaminated soils: role of synthetic chelates in lead phytoextraction. *Environ. Sci. Technol.* **31**, 800–805.
7. Chen, H. and Cutright, T. (2001) EDTA and HDETA effects on Cd, Cr and Ni uptake by *Helianthus annuus*. *Chemosphere* **45**, 21–28.
8. Wallace, A., Mueller, R. T., Cha, J. W., and Alexander, G. V. (1974) Soil pH, excess lime and chelating agents on micronutrients in soybeans and bush beans. *Agron. J.* **66**, 698–700.
9. Piechalak, A., Tomaszewski, B., and Baralkiewicz, D. (2003) Enhancing phytoremediative ability of *Pisum sativum* by EDTA application. *Phytochem.* **64**, 1239–1251.
10. Lombi, E., Zhao, F. J., Dunham, S. J., and McGrath, S. P. (2001) Phytoremediation of heavy-metal contaminated soils: natural hyperaccumulation versus chemically enhanced phytoextraction. *J. Environ. Qual.* **20**, 1919–1926.
11. Wenzel, W. W., Unterbrunner, R., Sommer, P., and Sacco, P. (2003) Chelate-assissted phytoextraction using canola (*Brassica napus* L.) in outdoors pot and lysimeter experiments. *Plant Soil* **249**, 83–96.
12. Jiang, X. J., Luo, Y. M., Zhao, Q. G., Baker, A. J. M., Christie, P., and Wong, M. H. (2003) Soil Cd availability to Indian mustard and environmental risk following EDTA addition to Cd-contaminated soil. *Chemosphere* **50**, 813–818.

13. Madrid, F., Liphadzi, M. S., and Kirkham, M. B. (2003) Heavy metal displacement in chelate-irrigated soil during phytoremediation. *J. Hydrol.* **272,** 107–119.
14. Grcman, H., Velikonja-Bolta, S., Vodnik, D., Kos, B., and Lestan, D. (2001) EDTA enhanced heavy metal phytoextraction: metal accumulation, leaching and toxicity. *Plant Soil* **235,** 105–114.
15. Sun, B., Zhao, F. J., Lombi, E., and McGrath, S. P. (2001) Leaching of heavy metals from contaminated soils using EDTA. *Environ. Pollut.* **113,** 111–120.
16. Ure, A. M., Quevauvillier, P. H., Muntau, H. J., and Griepink, B. (1993) Speciation of heavy metals in soils and sediments. An account of the improvement and harmonisation of extraction techniques undertaken under the auspices of the BCR of the Commission of the European Communities. *Int. J. Environ. Anal. Chem.* **51,** 135–151.
17. Merrington, G., Oliver, I., Smernik, R. J., and McLaughlin, M. J. (2003) The influence of sewage sludge properties on sludge-borne metal availability. *Adv. Environ. Res.* **8,** 21–36.
18. Ebbs, S. D. and Kochain, L. V. (1998) Phytoextraction of zinc by oat (*Avena sativa*), barley (*Hordeum vulgare*) and Indian mustard (*Brassica juncea*). *Environ. Sci. Technol.* **32,** 802–806.
19. Wu, L. H., Luo, Y. M., Christie, P., and Wong, M. H. (2003) Effects of EDTA and low molecular weight organic acids on soil solution properties of a heavy metal polluted soil. *Chemosphere* **50,** 819–822.
20. Clothier, B. E., Green, S. R., Robinson, B. H., et al. Contaminants in the root zone: bioavailability, uptake and transport, and their implications for remediation. In: *Chemical Bioavailability in Terrestrial Environments,* (Naidu, R., ed.), CSIRO Publishing, Collingwood, Australia, pp. in press.
21. Madrid, F., Liphadzi, M. S., and Kirkham, M. B. EDTA-assissted phytostabilization by barley roots contaminated with heavy metals. In: *Chemical Bioavailability in Terrestrial Environments,* (Naidu, R., ed.), CSIRO Publishing, Collingwood, Australia, pp. in press.

11

Testing Amendments for Increasing Soil Availability of Radionuclides

Nicholas R. Watt

Summary

Phytoextraction has been shown to be potentially feasible for some radionuclides. It is, however, likely that soil amendments will be needed to make contaminant radionuclides sufficiently available for plant uptake over the long time-scales likely to be required for phytoextraction. A method is described here for investigating the effect of soil amendments on the long-term uptake of ^{137}Cs in a laboratory trial. The method described uses large containers filled with artificially contaminated soil, which are harvested at 8-wk intervals and then replanted. Consideration is also given to applying this method to other radionuclides.

Key Words: Phytoextraction; phytoremediation; soil amendment; radionuclide; ^{137}Cs.

1. Introduction

Phytoextraction of radionuclides has been shown to be at least theoretically feasible for a number of elements including isotopes of U *(1)*, Sr *(2,3)*, and Cs *(4–7)*. However, although plants have been shown to readily remove some radionuclides under certain conditions *(1,8)*, it would appear that availability of radionuclides for plant uptake may be the rate-limiting step for contaminant removal *(8)*. Two possible reasons for the limited availability of radionuclides for plant uptake are (1) the chemical species found in the soil is unsuitable for plant uptake *(9)* and (2) fixation of the radionuclide to the substrate as is the case, for example, for ^{137}Cs *(10)*. Some amendment of soil availability may, therefore, be necessary if phytoextraction is to become a practically useful remediation technology for radionuclides *(11,12)*. Future investigations of the potential for phytoextraction to remediate radioactively contaminated soils will therefore need to examine the effects of soil amendments on reducing remediation times.

Given the relatively slow removal of ions from soils by plants, time-scales for phytoextraction of radionuclides are likely to be measured in years. However, phytoextraction trials reported in the literature where soil amendments have been used have tended to utilize time-scales of less than 12 wk *(4–7)*, hence little is known about the potential performance of soil amendment-assisted phytoextraction over practical time-scales. Furthermore, the herbaceous plants that might be useful in phytoextraction systems will require repeated planting and harvesting, at least in the case of ^{137}Cs. Although the literature contains many examples of ^{137}Cs uptake over long time-scales following weapons testing and the Chernobyl accident, repeated planting and harvesting data are rarely reported.

The current limited knowledge of the effects of soil amendments on the long-term behavior of radionuclides in phytoextraction systems suggests that it would be inappropriate to establish trials of the use of soil amendments at field sites. First, there is a risk of mobilizing the radionuclide off-site and second, in the United Kingdom, for example, regulatory approval may be unlikely to be forthcoming for such trials. Furthermore, the complexities of interpreting results from field sites suggest that controlled laboratory experiments represent a sensible starting point for investigations into the longer term phytoextraction of radionuclides combined with the use of soil amendments.

Here, a method is described for investigating the effects of soil amendments on ^{137}Cs uptake by herbaceous plants grown under a repeated cropping regime. ^{137}Cs is of particular interest at radioactively contaminated sites in the United Kingdom because it is a high-yield fission product and has a relatively long half-life (~30 yr) *(8)*. Details are provided in the notes as to how the method might be adapted for other radionuclides.

2. Materials
2.1. Soil

Soil used in experiments can be either soil sampled from contaminated sites or, where this is not practically possible or experimentally desirable, clean soil with radionuclides added (*see* **Note 1**). Where clean soil is employed the radionuclide being investigated will have to be applied in aqueous form. The limits on the activities used in the experiment will vary depending on local laboratory rules. Sufficient activity should be applied to ensure that measurable quantities of the radionuclide are removed by the particular plant species being used (*see* **Note 2**), within limits that ensure that dose rates to laboratory workers are kept as low as reasonably practicable.

2.2. Soil Amendments

Soil amendments can be selected from a wide range of chemical compounds (*see* **Note 3**). As a starting point for investigations of ^{137}Cs phytoextraction, monovalent cations including NH_4^+, Cs^+, and K^+ can be used. Where an amendment can degrade over time consideration should be given to the most appropriate application regime.

2.3. Plant Species

Although any herbaceous plant species could potentially be used in this experiment, as a starting point a species that is known to possess reasonably high uptake should be selected to ensure that measurable quantities of the contaminant are taken up by the plant. For example, in the case of ^{137}Cs, *Beta vulgaris* (leaf beet) is a suitable species (*see* **Note 4**).

3. Methods
3.1. Experimental Setup

1. Four open-topped plastic containers (approximate dimensions 40 × 60 cm and equipped with drainage holes), are filled with a quantity of the soil to be investigated to a depth marginally greater than the rooting depth of the plant (*see* **Note 5**). The containers are placed in trays to prevent cross-contamination of the treatments and leaching of any radioactive solution.
2. Where clean soil is the starting point, each container is contaminated with sufficient ^{137}Cs in a deionized water solution to provide a soil ^{137}Cs-activity concentration of approx 4 Bq/g (*see* **Note 6**). The soil is then thoroughly mixed using a trowel, and samples are taken from throughout the soil and analyzed for ^{137}Cs content to ensure a heterogeneous ^{137}Cs distribution. Following addition of the ^{137}Cs, the soil is allowed to stand for 4 wk to allow for equilibration of the ^{137}Cs with the substrate. Where soil is used from a contaminated site, mixing and sampling is still carried out to ensure the distribution of activity is heterogeneous.
3. The soil amendments being investigated are then applied to the containers in sufficient quantify to be uniformly distributed throughout the soil (*see* **Note 7**).

3.2. Growing and Harvesting Plants

1. *B. vulgaris* var. Swiss chard "Fordhook Giant" seeds are broadcast into each container by hand. The containers are then covered with cling film to encourage germination and placed on a bench equipped with supplementary lighting (mean PAR = 200 µmol/m/s) for 16 h each day (*see* **Fig 1**). Following germination the cling film is removed and the containers are watered with deionized water to keep them in a sufficiently moist state for plant growth but not to a level such that the amendment or radionuclide solution is flushed from the soil.
2. The position of the containers is rotated at weekly intervals so that each container occupies each position on the bench for 2 wk during each harvest period.

Fig. 1. Experimental arrangement for a ^{137}Cs phytoextraction trial using four experimental treatments.

3. At 8-wk intervals the aerial plant parts are harvested, weighed, placed in envelopes, and dried for 48 h in an oven at 80°C. The dried plant material is then ground to a powder in a boiling tube, transferred to a 70-mm diameter plastic pot, and weighed. Samples are analyzed for ^{137}Cs content with the plastic pot in contact with the end-cap of a high-purity Ge detector for 3600 s (*see* **Note 8**).
4. Following harvesting, the soil in the containers is turned over using a hand trowel and watered with deionized water. Where degradable soil amendments are being used, these are replenished as appropriate. The containers are then replanted with *B. vulgaris* and the harvest and planting regime is repeated for as long as is required for the experiment.
5. Concurrently with the plant growth in the containers, four 12-cm diameter plastic plant pots are prepared that duplicate the experimental treatments in all respects except that they do not contain any ^{137}Cs. This allows the concentrations of the soil amendments to be determined in ^{137}Cs-free solutions (*see* **Note 9**). Samples are taken from each pot at intervals throughout the experiment and analyzed using standard techniques for the soil amendment in question, for example, for NH_4^+ and NO_3^-, conductivity detection can be used.

4. Notes

1. The type of soil used will depend on the purpose of the experiment. For work based on actual contaminated sites, soil from the site being investigated should be

used. This may not always be possible if activity levels, and hence dose rates, are too high to permit the soil to be moved into a laboratory or greenhouse environment. For general investigations concerning the effectiveness of amendments for a particular radionuclide, consideration should be given to using a soil which does not significantly reduce plant availability, for example, an organic soil for isotopes of Cs. For some radionuclides, aging processes are an important factor in their availability for plant uptake over time. In particular, where artificially contaminated soils are used, experimenters will need to consider over what time-period the radionuclide should be allowed to "age" after application and before planting takes place. Care will need to be taken when applying the results from experiments using artificially contaminated soils to field applications.

2. Depending on existing knowledge of uptake by a particular species for a given radionuclide in a given substrate, it may be useful to conduct pot experiments to ensure that measurable levels of the radionuclide can be removed.

3. Soil amendments could potentially be almost any chemical compound. Selection will depend in part on the mechanism of improving plant uptake of the radionuclide in question. For example, amendments could be targeted at desorbing the radionuclide from the soil (for example, monovalent cations, and ^{137}Cs), increasing the solubility of a radionuclide, or chelating the radionuclide (for example, EDTA and Pb). In addition, there are many agrochemicals that have been developed to affect soil processes, especially for some plant nutrients, and much is known about the behavior and use of these compounds in soils. Future research efforts should consider testing compounds from the wide range of agrochemicals available. Some agrochemicals do not have a direct effect on nutrient availability, rather they increase the persistence of other agrochemicals, for example, nitrification inhibitors can maintain a higher NH_4^+ concentration in the soil, which is of interest to ^{137}Cs phytoextraction.

4. The choice of plant species may be constrained by the ability of a particular species to grow in the soil being tested and an understanding of the uptake characteristics of the species considered for the trial. Approaches to aid species selection include considering species that are indigenous to the contaminated site, in the United Kingdom for example, using the Natural History Museum's Postcode Plant Database *(8)* and in other countries regional/national equivalents, or selecting species from higher taxonomic groups that are known to have higher uptake of the radionuclide being investigated. For example, in the case of ^{137}Cs, the Caryophyllidae clade has been shown to include species with high-uptake characteristics for ^{137}Cs including *B. vulgaris* *(13)*.

5. The size of, and number of, containers used will depend on the number of treatments being investigated and the space available, which can be limited for experiments involving radionuclides. The volume of soil in the container should be large enough to allow root growth to be uninhibited by the container over the duration of the harvesting period. For manual handling purposes, consideration should be given to the weight of the container when filled with wet soil.

6. A soil-activity concentration of approx 4 Bg/g ^{137}Cs should be sufficient for measurable quantities of ^{137}Cs to be taken up by plants over a time-scale of some months while keeping the radiation dose as low as reasonably practicable. Suitable adjustments will need to be made for other radionuclides. Low soil-activity concentrations allow the performance of phytoextraction to be assessed at contaminant concentrations that might be expected to be difficult to remediate and to establish whether a body of soil can be remediated to levels that are below regulatory concern for a particular radionuclide.
7. It may be useful in experiments designed to simulate field trials to consider the unit of application in agricultural terms especially where amendments are agrochemicals. For example, for trials involving ^{137}Cs, NH_4^+ can be applied in quantities equivalent to 300 kg/ha.
8. Where it is difficult to estimate the uptake of a radionuclide by plants from soils containing low radionuclide-activity concentrations, more sensitive analytical techniques may need to be used. For example, for ^{137}Cs uptake at soil-activity concentrations of approx 4 Bq/g, a high-purity Ge detector has been used for the analysis because it is a more sensitive technique than a NaI (Tl) detector *(14)*. In practice, the sensitivity of the detector will need to be considered in conjunction with the number of samples that can be processed in each batch. NaI (Tl) detectors found in some laboratories can be capable of processing many more samples in a batch than high-purity Ge detectors.
9. Measurement of soil-amendment concentrations over time are of interest and may be of particular importance where amendments degrade over time, as is the case with some nitrification inhibitors (for example, dicyandiamide), or where the amendment is itself taken up by the plant. This presents a practical difficulty where facilities are not available to perform chemical analysis of radioactive samples. Furthermore, these techniques can be expensive to source from external laboratories. Attempts should therefore be made to replicate the conditions found in each experimental treatment as closely as possible but without any radionuclides being present.

References

1. Huang, J. W., Blaylock, M. J., Kapulnik, Y., and Ensley, B. D. (1998) Phytoremediation of uranium-contaminated soils: Role of organic acids in triggering uranium hyperaccmulation in plants. *Environ. Sci. Technol.* **32,** 2004–2008.
2. Entry, J. A., Vance, N. C., Hamilton, M. A., Zabowski, D., Watrud, L. S., and Adriano, D. C. (1996) Phytoremediation of soil contaminated with low concentrations of radionuclides. *Water, Air Soil Pollut.* **88,** 167–176.
3. Fuhrmann, M., Lasat, M. M., Ebbs, S. D., Kochian, L. V., and Cornish, J. (2002) Uptake of cesium-137 and strontium-90 from contaminated soil by three plant species: Application to phytoremediation. *J. Environ. Qual.* **31,** 904–909.
4. Lasat, M. M., Norvell, W. A., and Kochian, L. V. (1997) Potential for phytoextraction of ^{137}Cs from a contaminated soil. *Plant Soil* **195,** 99–106.
5. Lasat, M. M., Furmann, M., Ebbs, S. D., Cornish, J. E., and Kochian, L. V. (1998) Phytoremediation of a radiocesium-contaminated soil: Evaluation of cesium-137 bioaccumulation in the shoots of three plant species. *J. Environ. Qual.* **27,** 165–169.

6. Dushenkov, S., Mikheev, A., Prokhnevsky, A., Ruchko, M., and Sorochinsky, B. (1999) Phytoremediation of radiocaesium-contaminated soil in the vicinity of Chernobyl, Ukraine. *Environ. Sci. Technol.* **33,** 469–475.
7. Willey, N., Hall, S., and Mudiganti, A. (2001) Assessing the potential of phytoremediation at a site in the U.K. contaminated with ^{137}Cs. *Int. J. Phytorem.* **3,** 321–333.
8. Watt, N. R. (2004) *Assessing the Potential of Phytoextraction to Remediate Land Contaminated with ^{137}Cs at Nuclear Power Station Sites.* PhD Thesis, University of the West of England, Bristol, UK.
9. Marschner, H. (1995) *Mineral Nutrition of Higher Plants, 2nd ed.* Academic Press, London, UK.
10. Squire H. M. and Middleton, L .J. (1966) Behaviour of ^{137}Cs in soils and pastures; a long term experiment. *Rad. Bot.* **6,** 413–423.
11. Cunningham, S. D. and Ow, D. W. (1996) Promises and prospects of phytoremediation. *Plant Phys.* **110,** 715–719.
12. Dushenkov, S., Kapulnik, Y., Blaylock, M., Sorochisky, B., Raskin, I., and Ensley, B. (1997) Phytoremediation: a novel approach to an old problem. *Global Environ. Biotech.* **1,** 563–571.
13. Broadley, M. R. and Willey, N. J. (1997) Differences in root uptake of radiocaesium by 30 plant taxa. *Environ. Pollut.* **97,** 11–15.
14. Gilmore, G. and Hemmingway, J. D. (1995) *Practical Gamma-Ray Spectrometry.* John Wiley and Sons, Chichester, New York, pp. 196–197.

12

Using Electrodics to Aid Mobilization of Lead in Soil

David J. Butcher and Jae-Min Lim

Summary

Lead is a significant contaminant in soil that poses a challenge to phytoremediation because of its low bioavailability induced by complexation with soil components. Conventional phytoremediation approaches employ chemical chelating agents, such as ethylenediaminetetraacetic acid, to increase this element's availability to plants. The addition of chemicals adds an increased cost and may induce other environmental impacts. An alternative approach involves the use of electrodic-remediation strategies, in which an electric field is introduced into the soil by electrodes to mobilize contaminants. In this work, greenhouse studies were conducted to evaluate electrodic phytoremediation for lead using soil collected from Barber Orchard, Haywood County, NC, a US Environmental Protection Agency Superfund site. Our instrumentation and methodology are described for optimization studies of the magnitude of applied electric potential, harvest time after application, and daily application time. Representative results are presented that demonstrate the effectiveness of this approach.

Key Words: Phytoremediation; phytoextraction; lead; electrodics; EDTA; inductively coupled plasma optical emission spectrometry; ICP-OES; *Brassica juncea*; Indian mustard.

1. Introduction

Chemical contaminants have been introduced into soil by a variety of anthropological activities, including chemical production, weapons production/use, energy production, and agriculture. The development of economically feasible remediation methods remains a subject of considerable research. Conventional methods of soil cleansing involve the replacement of polluted soil by clean soil. Although this approach has been widely employed and is extremely effective, it is an expensive approach and has severe ecological impacts *(1–6)*.

An alternative strategy for soil remediation is the subject of this volume, phytoremediation. Major advantages of phytoremediation methods are its much lower cost and the concentration of waste into a reduced volume *(5–7)*. In addition, these techniques are aesthetically pleasing and less disruptive than conventional remediation approaches.

1.1. Lead Phytoremediation

Lead is a significant, persistent contaminant in residential and industrial sites *(8)*. Although lead has a number of physiological effects, including anemia, its best known and perhaps most severe effect is the impact on the central nervous system, resulting in mental retardation, especially in children. Risk assessment studies have suggested a maximum soil concentration of 300 mg/kg *(9)*.

A major impediment to the remediation of lead in soil by phytoremediation is the low bioavailability of this element in soil. Huang et al. *(3)* estimated that approx 0.1% of total lead in soil is available for extraction. The low availability of lead is caused by complexation with a variety soil components, including carbonates, phosphates, oxides, and hydroxides. Consequently, it has been necessary to apply chelating agents, such as ethylenediaminetetraacetic acid (EDTA) to mobilize lead and enhance phytoextraction. Although a variety of chelates have been investigated, EDTA has been most widely used because of its relatively low cost, relatively low toxicity, and effectiveness at enhancing lead translocation from roots to shoots *(5,7,10)*.

1.2. Electrodic Phytoremediation for Lead

However, the addition of chelating agents increases the cost of remediation, and may introduce other environmental problems. Hence, alternatives to chemically enhanced phytoextraction have been investigated. Electrodic- and electrokinetic-remediation techniques are innovative technologies that have been developed to remove heavy metals from saturated and unsaturated soil, sludges and sediments, and groundwater. The general principles of electrochemistry, which apply to the situation of electrodes in the soil–water system, are embodied in the study of electrodics *(11–13)*. This technology applies a high-voltage and low-level direct current to the polluted soil by electrodes placed in the ground to remove inorganic and organic contaminants from the soil.

1.3. Barber Orchard

Barber Orchard, Haywood County, NC, is a US Environmental Protection Agency (EPA) Superfund site located along US Highway 23/74 in Haywood County *(5,7,14–17)*. The 438-acre property was used for the commercial production of apples from 1903 to 1988. During the first half of the twentieth century, one of the approved pesticides for orchards, lead arsenate, was employed for the control of coddling moths at this location. Application of pesticides throughout the orchard was performed using an underground galvanized copper pipeline system. In the late 1980s and early 1990s, the property was subdivided into residential housing. Extensive lead soil contamination was discovered at some of the home sites in the late 1990s. The area was consequently declared a Superfund site by the US EPA in late 1999.

Here, we describe the application of electrodic phytoremediation for lead employing *Brassica juncea* (Indian mustard). This species has been commonly used for the phytoextraction of lead and other elements because of its high biomass, rapid growth, and ability to hyperaccumulate metals *(5,7,14,18–20)*. The contaminated soil employed for the greenhouse experiments described here was obtained from Barber Orchard. The goal of this study was to investigate the combination of a chemical chelating agent (EDTA) with electrodic phytoremediation for lead. Pertinent parameters were optimized, including operating voltage/current, EDTA concentration, and the application time of electric potential. Considerable potential has been demonstrated for the remediation of lead from contaminated soils.

2. Materials

2.1. Soil and Plants

1. For all experiments, lead-contaminated soil (approx 50 kg) from Barber Orchard screened with 1.0-cm sieves to remove roots and other debris, and homogenized manually (*see* **Note 1**).
2. 12 wk old *B. juncea* plants, grown from seeds washed, rinsed, and germinated on filter paper for 2–3 d, then transferred to 1.2 kg lead-contaminated homogenized soil, watered daily, and fertilized weekly in the Western Carolina University Greenhouse.

2.2. Harvesting and Digesting

1. Deionized water.
2. Ball Mill (Spex 8000).
3. Concentrated HNO_3.
4. 30% H_2O_2.
5. Inductively coupled plasma optical emission spectrometer (ICP-OES) (Perkin-Elmer, Optima, cat. no. 4100DV, Wellesley, MA).
6. National Institute of Standards and Technology (NIST) standard reference material (SRM 2710 Montana Soil, Highly Elevated Metals).

2.3. Electrodic Apparatus

1. DC power supply (400 V, 1 A), an electronic ammeter, and copper sheet electrodes (12 × 3 cm).

3. Methods
3.1. Plant Harvesting/Homogenization

1. Following phytoremediation experiments, *B. juncea* is harvested by cutting the stem 1 cm above the soil surface.
2. Plant tissue is washed and rinsed with deionized water.
3. Plant tissue is dried at 60°C and homogenized using a ball mill.

3.2. Soil/Plant Digestion and Analysis

1. Homogenized and triplicate 0.2 g samples of the soil or plant material were digested for 2 h on a heating block (lab manufactured) with 5 mL of concentrated HNO_3.
2. The samples were allowed to cool; 1 mL of 30% H_2O_2 was added and the samples were digested for an additional hour.
3. The digested samples were diluted to 100 mL with deionized water.
4. Lead concentrations were determined by inductively coupled plasma optical-emission spectrometry (*see* **Note 2**).
5. NIST standard reference material is analyzed (in triplicate) for every digestion set for quality control purposes. The mean lead concentrations are consistently within ±10% of the certified levels in the SRM.

3.3. Apparatus for Electrodic Phytoremediation

1. *B. juncea* were placed in 10 cm (bottom diameter) pots with the electrodes spaced 8 cm apart and around the plant. The electric potential applied varied depending upon the experiment performed (*see* **Subheading 3.4, 3.5, 3.6, 3.7.**).

3.4. Optimization of Electric Potential and EDTA Concentration

1. *B. juncea* is cultivated as described in **Subheading 2.**
2. Plants are treated with EDTA at concentration values between 0 and 5 mmol/kg with electric potentials between 0 and 40 V. Three plants are employed for each EDTA concentration/electric potential data-point.
3. Electric potential is applied for 1 h each day for 9 d.
4. Shoots and roots are harvested after 9 d and prepared for analysis as described in **Subheadings 3.1. and 3.2.**
5. This procedure is performed three times (*see* **Note 3**).

3.5. Optimization of Harvest Time After Application of EDTA and Electric Potential

1. *B. juncea* are cultivated as described in **Subheading 2.**
2. EDTA is applied to the plants and electric potential is applied for 1 h each day.
3. Plants (triplicates) are then harvested 3, 6, 9, 12, and 15 d after the beginning of the experiment.
4. The harvested plants are prepared for analysis as described in **Subheadings 3.1. and 3.2.** (*see* **Note 4**).

Mobilization of Lead in Soil

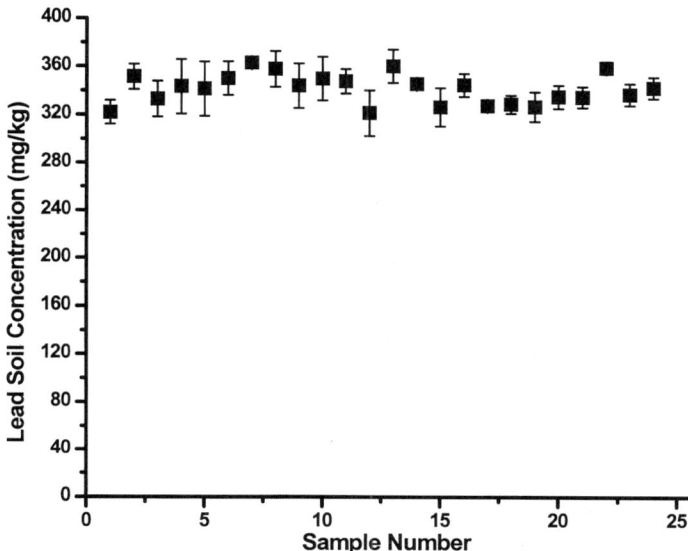

Fig. 1. Evaluation of homogeneity of lead concentrations in soil employed in phytoremediation experiments. Data-points represent the mean ± standard deviation of three replicates.

3.6. Optimization of Application Time of Potential Per Day

1. *B. juncea* are cultivated as described in **Subheading 2**.
2. EDTA (2 mmol/kg) is applied to the plants.
3. Plants are exposed to 10 V of DC potential for 9 d for 0 min (control), 30 min, or 60 min.
4. The harvested plants are prepared for analysis as described in **Subheadings 3.1.** and **3.2.** (*see* **Note 5**).

3.7. Optimized Conditions for Electrodic Phytoremediation

1. The combination of electrodics with EDTA has been shown to be the most effective approach.
2. Use 5 mmol EDTA/kg with 40 V applied potential.
3. Harvest plants after approx 9 d of EDTA/potential application.
4. Apply potential for 1 h per day.

4. Notes

1. **Figure 1** shows representative data for lead. The average concentration of lead was determined to be 341 ± 12 mg/kg.
2. **Table 1** shows the operating conditions employed for ICP-OES.
3. The use of electrodic phytoremediation in combination with EDTA was shown to increase lead accumulation compared with EDTA alone. **Figure 2** shows the observed enhancement obtained with 40 V compared with 0 V (no electrodic enhancement).

Table 1
Operating Conditions for the Perkin-Elmer Optima 4100 DV ICP-OES

Radio frequency	40 MHz
Radio frequency power	1.3 kW
Argon flow rates	
Nebulizer gas	0.8 L/min
Auxiliary gas	0.2 L/min
Plasma gas	15 L/min
Lead emission lines	220.353 nm
	217.000 nm
	283.306 nm

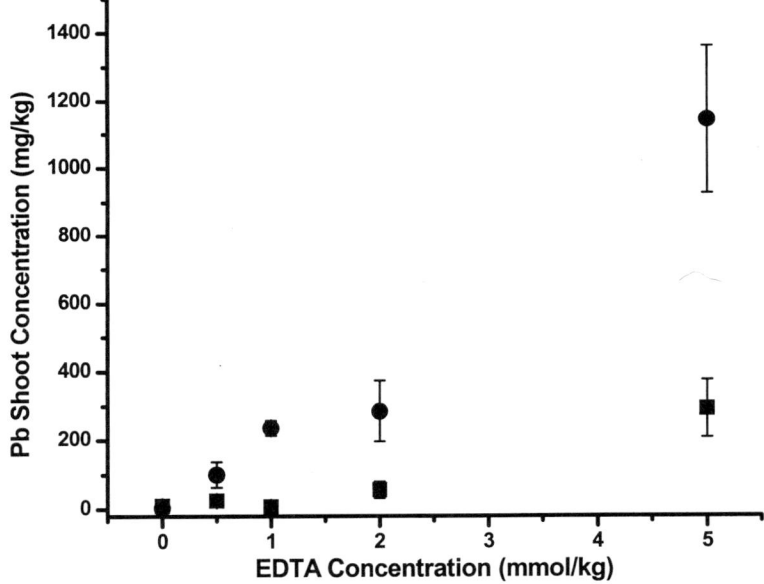

Fig. 2. The effect of operating voltage and EDTA concentration on lead accumulation in *Brassica juncea*: (■) 0 V electric potential and (●) 40 V electric potential. Datapoints represent the mean ± standard deviation of three replicates from three plants.

4. Using EDTA in the absence of electric potential, the maximum accumulation of lead occurred 12 d from the beginning of the experiment (**Fig. 3**). However, the combination of EDTA and electric potential induced a maximum accumulation after only 9 d. Almost no phytoremediation was achieved with electric potential in the absence of EDTA (data not shown).
5. The amount of lead remediated increased with daily application time up to 60 min, as shown in **Fig. 4**. Application times more than 60 min significantly affected the biomass (and the phytoremediation) achieved.

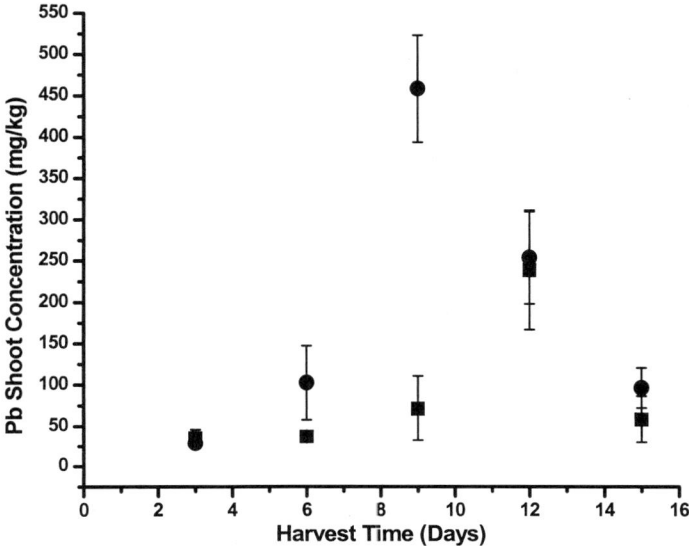

Fig. 3. Effect of EDTA and electric potential (1 h/d) on shoot lead accumulation in *Brassica juncea*: (■) 2 mmol/kg EDTA, 0 V electric potential, and (●) 2 mmol/kg EDTA, 20 V electric potential. Application of EDTA and electric potential began at time equals zero. Data-points represent the mean ± standard deviation of three replicates from three plants.

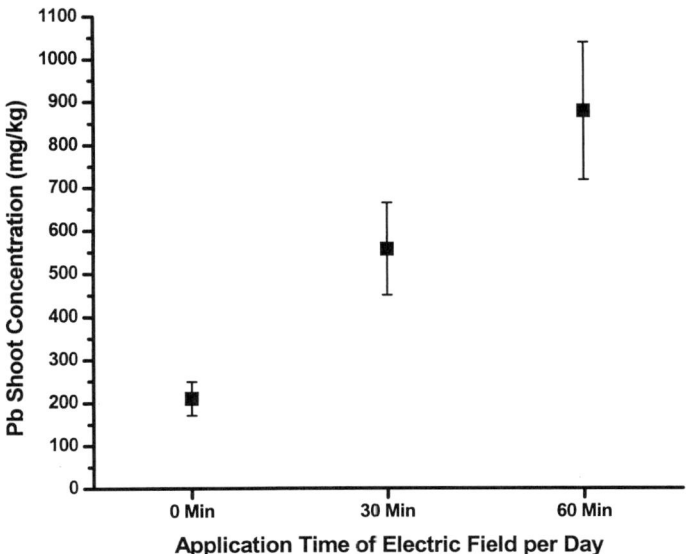

Fig. 4. Effect of daily application time on lead accumulation in *Brassica juncea*. 10 V of DC current were applied with 2 mmol/kg EDTA. Data-points represent the mean ± standard deviation of three replicates from three plants.

Acknowledgments

This research was supported by Mr. and Mrs. Blanton Whitmire. The ICP-OES instrument employed in this research was partially funded by an award from the National Science Foundation Division of Undergraduate Education (NSF Award 995022 CCLI-A&I).

References

1. Van der Lelie, D., Schwitzguebel, J., Glass, D. J., Vangronsveld, J., and Baker, A. (2001) Assessing phytoremediation's progress in the United States and Europe. *Environ. Sci. Technol.* **31,** 182A–186A.
2. Blaylock, M. J., Salt, D. E., Dushenkov, S., et al. (1997) Enhanced accumulation of Pb in Indian mustard by soil-applied chelating agents. *Environ. Sci. Technol.* **31,** 860–865.
3. Huang, J. W., Chen, J., Berti, W. R., and Cunningham, S. D. (1997) Phytoremediation of lead-contaminated soils: role of synthetic chelates in lead phytoextraction. *Environ. Sci. Technol.* **31,** 800–805.
4. Watanabe, M. E. (1997) Phytoremediation on the brink of commercialization. *Environ. Sci. Technol.* **31,** 182A–186A.
5. Salido, A. L., Hasty, K. L., Lim, J.-M., and Butcher, D. J. (2003) Phytoremediation of arsenic and lead in contaminated soil using Chinese brake ferns *(Pteris vittata)* and Indian mustard *(Brassica juncea)*. Int. J. Phytorem. **5,** 89–103.
6. Raskin, I. and Ensley, B. D. (eds.) (2000) *Phytoremediation of Toxic Metals: Using Plants to Clean up the Environment.* John Wiley, New York, NY.
7. Lim, J.-M., Salido, A. L., and Butcher, D. J. (2004) Phytoremediation of lead using Indian mustard *(Brassica juncea)* with EDTA and Electrodics. *Microchem. J.* **76,** 3–9.
8. Elless, M. P. and Blaylock, M. J. (2000) Amendment optimization to enhance lead extractability from contaminated soils for Phytoremediation. Int. J. Phytorem. **2,** 75–89.
9. Dudka, S. and Miller, W. P. (1999) Permissible concentrations of arsenic and lead in soils based on risk assessment. *Water Air Soil Pollut.* **113,** 127–132.
10. Wu, J., Hsu, F. C., and Cunningham, S. D. (1999) Chelate-assisted Pb phytoextraction: Pb availability, uptake, and translocation constraints. *Environ. Sci. Technol.* **33,** 1898–1904.
11. Yong, R. N. (2001) *Geoenvironmental Engineering: Contaminated Soils, Pollutant Fate and Mitigation.* CRC Press, Boca Raton, FL.
12. Ko, S.-O., Schlautman, M. A., and Carraway, E. R. (2000) Cyclodextrin-enhanced electrokinectic removal of Phenanthrene from a model clay soil. *Environ. Sci. Technol.* **34,** 1535–1541.
13. Maini, G., Sharman, A. K., Sunderland, G., Knowles, C. J., and Jackman, S. A. (2000) An integrated method incorporating sulfur-oxidizing bacteria and electro-kinetics to enhance removal of copper from contaminated soil. *Environ. Sci. Technol.* **34,** 1081–1087.

14. Lim, J.-M. (2003) *Electrodic and EDTA-Enahnced Phytoremediation of Lead using Indian mustard (Brassica juncea)*, MS thesis, Western Carolina University Cullowhee, NC.
15. Porter, K. M. (2003) *Speciation of lead and arsenic in contaminated soil*, MS thesis, Western Carolina University Cullowhee, NC.
16. Embrick, L. (2004) *Determination of lead and arsenic concentrations in samples from barber orchard using ICP, along with determination of arsenic uptake in the brake fern (Pterris vittata) from Barber Orchard soil*, MS thesis, Western Carolina University Cullowhee, NC.
17. Kondrad, S. L. (2002) *Use of lead isotopes to assess sources and mobility of lead arsenate from Barber's Orchard, Haywood County, North Carolina*, MS thesis, Western Carolina University Cullowhee, NC.
18. Ebbs, S. D. and Kochian, L. V. (1998) Phytoextraction of zinc by oat (*Avena sativa*), barley (*Hordeum vulgare*), and Indian mustard (*Brassica juncea*). *Environ. Sci. Technol.* **32,** 802–806.
19. Sarret, G., Vangronsveld, J., Manceau, A., et al. (2001) Accumulation forms of Zn and Pb in *Phaseolus vulgaris* in the presence and absence of EDTA. *Environ. Sci. Technol.* **35,** 2854–2859.
20. Whiting, S. N., Leake, J. R., McGrath, S. P., and Baker, A. J. M. (2001) Hyperaccumulation of Zn by *Thlaspi caerulescens* can ameliorate Zn toxicity in the rhizosphere of cocropped *Thlaspi arvense. Environ. Sci. Technol.* **35,** 3237–3241.

13

Stable Isotope Methods for Estimating the Labile Metal Content of Soils

Andrew J. Midwood

Summary

Estimation of the labile or available metal content of soils relies extensively on the use of chemical-extraction techniques. However, where interest is focused on only one or two elements that have more than one isotope, an alternative approach is to use isotope-dilution analysis and measure E- or L-values. The E-value is an isotopic estimate of the labile metal content of the soil. It can be established by suspending a soil in water, adding an isotope tracer, and then measuring the isotope content of the solution phase after equilibration. The labile metal content of a soil may also be measured by growing plants in soil to which an isotope of the metal of interest has been added. The isotopic signature within the plant tissue provides an estimate of the labile metal content of the soil and is referred to as the L-value. The L-value is a biological estimate of the labile metal pool. The use of isotopes to determine E- and L-values is described here, with an emphasis on the use of stable isotopes.

Key Words: Stable isotopes; heavy metals; L-value; E-value; phytoremediation; soils; Cd; Zn.

1. Introduction

The labile metal content of soils is important because it represents the pool that is available to plants and potentially mobile. This can have important ramifications in, for example, soil polluted with heavy metals. It is important to be able to measure the amount of labile metal available to growing plants or potentially entering a water course. Traditionally, the metal content of soil has been assessed by the use of various chemical extractants, often used in sequential extraction schemes. Such schemes use a number of extractants of increasing severity to extract metal from a soil. Typically, the first two or three extractants yield the labile or bioavailable metal content of the soil, whereas later extractants provide estimates of increasingly refractory forms of the metal(s) of interest. A wide range of sequential extraction schemes exist, as illustrated by Kersten and Förstner (1), who identified 25 different schemes in use between 1973 and 1993.

Isotope dilution offers an alternative to chemical extraction and is particularly appropriate where the focus of attention is restricted to only one or two elements. Isotope dilution has been used in the past primarily to assess P availability to plants by utilizing the radioisotope ^{32}P (*see* **ref. 2** and references therein). However, over recent years the scope of application has expanded to include Ni (3), Cd (4–6), Cd and Ni (7,8), and Se (9). These studies all used radioactive isotopes as tracers, the analysis of which requires not only specific activity to be measured but also the metal content of each sample. However, the work of Ahnstrom and Parker (10) on Cd and Ayoub et al. (11) on Cd and Zn are notable exceptions. These authors used stable isotopes and were able to directly quantify the labile metal content of soils by characterization of the added spike material.

Stable isotope-dilution techniques allow E- or L-values to be quantified. Briefly, the E-value or isotopically exchangeable metal content of a soil is measured by adding an isotope tracer to a soil suspension, where the liquid phase is either water or a neutral salt solution such as 0.1 M Ca(NO$_3$)$_2$. The extent of dilution of the added tracer as measured in the water provides an estimate of the labile pool size. As discussed by Hamon et al. (2) when making E-value measurements it is important to recognize the distinction between the total available metal pool and that which is exchangeable. The available metal pool relates to the metal present in soil solution plus that associated with the solid phase and in rapid equilibrium with the soil solution. In contrast, the exchangeable pool relates only to that associated with the solid-phase component of the soil.

An L-value experiment involves the addition of an isotope to a soil in which a plant is subsequently grown, usually from seed. By measuring the isotopic composition of the metal in the plant tissue, the size of the labile pool available to the plant in the soil may be estimated. Effectively, an L-value is an estimation of the labile pool as measured in an E-value experiment, but is sampled using a plant.

2. Materials

Stable isotopes can be purchased from a number of suppliers across the world. A good list of current suppliers can be found on the pages of the discussion group Isogeochem (www.eeb.cornell.edu/isogeochem/). The cost of stable isotopes varies enormously and depends on the enrichment and form of the isotope. The form should be considered when planning experiments. A water-soluble salt, for example, may be more convenient for plant and soil studies than a pure metallic form, which may have to be acid solubilized before use. The cost of isotopes and budget available can have an overriding influence on the experimental design and scope of a study, and should be considered at the outset before any experimental work (*see* **Notes 1** and **2**).

3. Methods
3.1. Stable Isotope Spike Solution and Analysis

1. For any isotope-dilution experiment it is necessary to characterize a working solution of the isotope spike in terms of its abundance and concentration. The isotope abundance may be provided by the supplier and can be checked in the laboratory using mass spectrometric analysis. This involves measuring all the isotopes in a single analysis, the feasibility of which will depend on the type of mass spectrometer available (*see* **Subheading 3.2.**). The concentration of the working solution is generally determined by conducting a reverse isotope-dilution experiment. This involves adding a known quantity of spike material to a solution in which the concentration of the element of interest is precisely known. By applying the principles of the isotope dilution (*see* **Eq. 1**) the concentration of the working solution can be accurately determined.

2. The analysis of stable metal isotopes is achieved using thermal ionization mass spectrometry (TIMS) or inductively coupled plasma-mass spectrometry (ICP-MS (**Figs. 1** and **3**)). TIMS involves isolation of the metal to be analyzed from the sample matrix, which in E- and L-value experiments, for example, can be an aqueous solution or plant tissue. Isolation of the metal is achieved using specific ion chromatography techniques (**Fig. 2**). However, plant material must first be digested using mineral acids.

3. Dried and ground plant material is digested with 50% (v/v) HNO_3 in a beaker covered with a watch glass. The beaker is heated on a hot plate to gently reflux the acid for 30 min and any residue removed by filtration. Alternatively, a microwave digestion system can be used in which the sample is contained in a Teflon bomb with concentrated HNO_3 and digested using a combination of heat and pressure. Such sealed systems are particularly appropriate for volatile metals.

4. After digestion, ion chromatography is used to purify the metal of interest (typically up to 1 µg), which is then deposited in a concentrated spot onto a fine wire filament ready for introduction into the mass spectrometer. Within the mass spectrometer the filament is heated using a current of several amps causing the deposited metal to vaporize and ionize. The ions generated are then separated according to the mass-to-charge ratio. For an overview of TIMS analysis, *see* **ref. *12*** (*see* **Notes 3** and **4**).

5. In a tracer study the presence of an isotope spike such as ^{114}Cd is measured as an isotope ratio relative to another Cd isotope, for example this may be ^{111}Cd. It is this ratio, $^{114}Cd/^{111}Cd$, which is determined by the mass spectrometer and used in

Fig. 1. Thermal ionization mass spectrometer.

all subsequent calculations. Because stable isotopes occur naturally it is important to be able to measure a change in the natural ratio with confidence. For example, the addition of a small amount of spike ^{114}Cd to a soil containing Cd will result in a decrease in the overall Cd isotope ratio (^{114}Cd/^{111}Cd) from the natural level. The magnitude of this change will depend on the amount of Cd in the soil. A small amount of spike added to a large pool will produce a very slight change in the ratio and may not be detectable unless the isotope analyses are very precise. So it is important to be aware of the precision of isotope ratio analyses when planning any experiment. This can be assessed by the use of quality control data from the mass spectrometer. For example, repeated measurements of a Cd standard should ideally provide a mean ^{114}Cd/^{111}Cd of 2.240 ± a standard deviation. As a guide the mean ratio ± 5 times the standard deviation will provide an indication of the minimum ratio change, which can be attributed with confidence to the presence of an isotope spike. It is critical that sufficient spike is used to ensure the enrichment level remains detectable throughout the experiment.

6. E- or L-values based on stable isotope ratio measurements may be calculated using this well-established equation:

$$E\text{- or } L\text{-value} = \frac{w'c'A_R(RY'-X')}{wA_R'(X-RY)} \qquad (1)$$

where w is the weight of sample solution or plant tissue digested, A_R is the relative atomic mass of the element being determined, X and Y are the natural isotopic

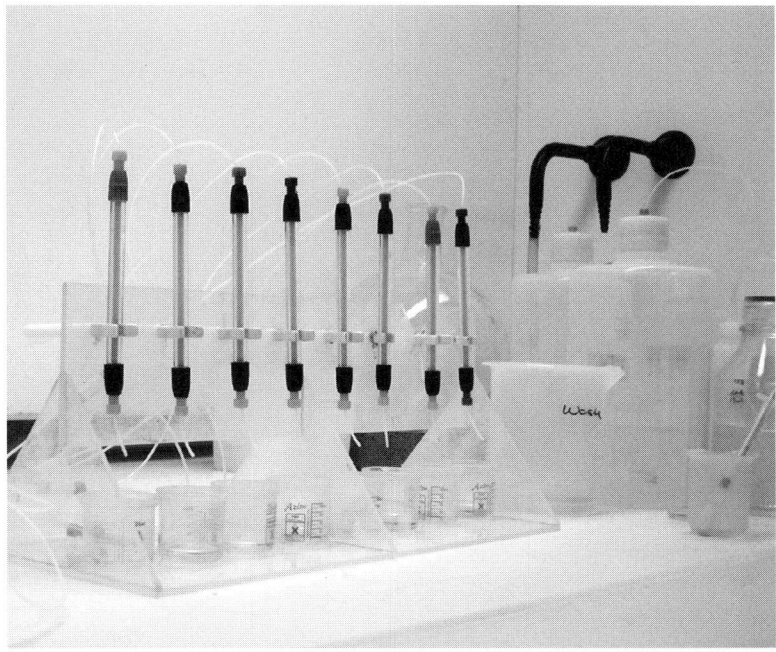

Fig. 2. Ion chromatography columns used to purify samples prior to analysis on a thermal ionization mass spectrometer.

abundances of the isotopes (atom %) of the element being analyzed, w' is the weight of spike solution used, c' the concentration of spike solution as determined by reverse isotope dilution, and X' and Y' are the isotope abundances (atom%) of the spike solution. One of these isotopes will be the spike isotope and will have an abundance significantly greater than the natural isotope level and in many cases close to 100%. A_R is the relative atomic mass of the spike material and is calculated from the spike abundance measurements (*see* **Note 5**).

7. For comparative purposes the labile metal content of a soil is often expressed as a percentage of the total metal content of the soil, calculated as:

$$\% \, Labile \, Metal = \frac{E\text{-} or \, L\text{-} Value \, (\mu g / g)}{Total \, Metal \, Content \, (\mu g / g)} \times 100 \tag{2}$$

3.2. E-Value Experiments

1. E-value estimates are made by placing a weighed amount of soil into a bottle with water or salt solution. The bottle should be sufficiently large to allow for mixing and maintenance of a suspension—typically achieved using an end-over-end shaker. Quantitative soil analyses are usually based on soil which has been air-dried (25°C) or oven-dried (105°C) and sieved to 2 mm.

Table 1
Conditions for E-Value Experiments

Dispersant	Volume (mL)	Soil (g)	Initial dispersal time	Metal/ isotope	Refs
Water	200	1.25–5	18 h	^{114}Cd & ^{67}Zn	*11*
	200	10	18 h	^{32}P	*14*
	132	13.2	18 h	^{32}P	*15*
	99	10	Overnight	^{32}P	*16*
	99	10	Overnight	^{63}Ni	*3*
0.01 M CaCl$_2$	50	2.5	24 h	^{109}Cd	*4*
0.1 M Ca(NO$_3$)$_2$	25	2–5	5 d	^{109}Cd	*8*
	25	5	5 d	^{109}Cd	*6*
	25	5	5 d	^{109}Cd	*5*

2. A summary of soil:water or salt solution ratios and initial mixing times that have been used in the past are shown in **Table 1**. Young et al. *(8)* found that E-value measurements were insensitive to soil:0.1 M Ca(NO$_3$)$_2$ ratios in the range 1:50 to 1:12.5 (*see* **Notes 6** and **7**).
3. After an appropriate dispersion time, isotope spike is added to the soil suspension. To determine the E-value an equilibration must be reached between the added isotope and the labile metal content of the soil. The time allowed for this to occur in studies using ^{32}P has often been just 1 min, E-values based on this approach may have a subscript 1 (e.g., E_1-value) *(14–19)*. To measure E-values of Cd in soil, equilibration times between 24 h *(20)* and 7 d *(4)* have been used. A number of studies have sampled over time periods of hours to days and monitored the extent of isotope exchange *(8)*. In a recent study by Ayoub et al. *(11)* it was found that Zn and Cd stable isotope tracers equilibrated in a soil:water suspension after approx 70 h. Equilibration times for heavy metals in general are considerably longer than, for example, P and should be at least 24 h.
4. At the end of the selected equilibration time an aliquot of the suspension is sampled, centrifuged, and filtered through a 0.2-μm filter. This isolates the solution from the solid phase and prevents any further isotopic exchange. The isotope ratio of the solution is then used to calculate the E-value (*see* **Eq. 1**).

3.3. L-Value Experiment

1. The general protocol again involves air-dried soil sieved to 2 mm. The soil is spread out on a plastic tray or waxed paper and a characterized isotope solution (in terms of concentration and isotopic abundance), ideally in an aqueous form, is sprayed onto the soil as a fine mist. The soil is then mixed thoroughly and placed in plastic pots ready to receive the plants. As with the E-value experiments, when dealing with expensive stable-isotope labels the amount of tracer to use should be tailored to the experiment. Using too little tracer will result in an

inability to detect a change in the natural abundance level in the plant. Without a measurable change no estimate of the labile metal pool is possible. Too much isotope is simply unnecessary, and something most tightly constrained experimental budgets should avoid.
2. Typically, at the end of the experiment the plant material is harvested, dried, milled, and digested using strong mineral acids, as described earlier (*see* **Subheading 3.1.**). The isotope ratio of the element of interest is then measured. By knowing the amount of isotope present in each pot, the size of the labile metal pool in the soil accessed by the plant can be estimated from the isotope ratio of the plant tissues (*see* **Eq. 1**) (*see* **Note 8**).

4. Notes

1. A chemical extraction procedure will yield an extract which can be analyzed for any number of elements. Isotope-dilution techniques by marked contrast provide detailed information only about those metals for which an isotope spike has been used. As such, it is a technique normally applied to only one or two elements at a time. It is not intended that isotope-dilution techniques should be seen as a replacement for chemical-extraction procedures. Both techniques provide different information. Indeed circumstances may well arise where the combination of both techniques will be a powerful approach to the characterization of, for example, a polluted soil. Clearly, the multielement approach allows an inventory of the metal content to be assessed before a specific isotope technique is used to focus attention on one particularly metal.
2. A key consideration to work with stable isotopes is cost. The isotopes are expensive and individual experiments may consume appreciable quantities so that care needs to be exercised when planning each study. In addition to the cost of the isotopes spikes, analysis costs can also be significant. For most laboratories an instrument such as an ICP-MS is a sizeable capital investment with a high annual running cost. As a result, the majority of E- and L-value experiments in the past have been conducted using radioisotopes, developed from earlier ^{32}P experiments (P having only one stable isotope). Work with radioisotopes however, has safety implications relating to the laboratory organization and sample disposal. Isotope techniques stable or otherwise, offer a unique way of studying the behavior of metals in soils, which can provide information extremely difficult to obtain by any other means.
3. Although offering unrivaled levels of precision and accuracy, TIMS (**Figs. 1** and **2**) has the disadvantage of requiring specialized sample pretreatment that is time consuming, and also the time taken for each analysis can run into several hours per sample. These constraints mean that TIMS is an expensive analysis option and is probably not the most appropriate for this type of work. Isotope analysis using ICP-MS (**Fig. 3**) in comparison is rapid and does not necessarily require isolation of the metal of interest from the sample matrix. Liquid samples, either aqueous solutions or sample digests, can be introduced directly into an ICP-MS system. The sample is drawn into a nebulizer where a fine aerosol is generated and sprayed

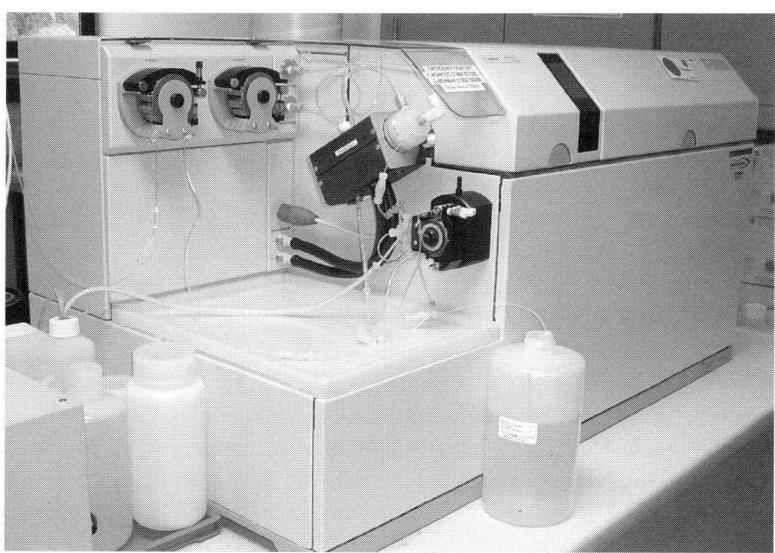

Fig. 3. Inductively coupled plasma-mass spectrometer.

into an argon gas-generated plasma, which dries and ionizes the elemental content of the aerosol. The ions are then accelerated into the mass spectrometer for isotopic analysis. The precision of isotope ratio analyses by an ICP-MS is typically between 0.1 and 0.05% Relative Standard Deviation (RSD), in comparison a TIMS can attain a precision of 0.002% RSD *(13)*. However, the level of precision achievable with an ICP-MS may be acceptable in tracer studies where relatively large changes in the measured isotope ratio may be observed with respect to the analysis precision.

4. Isotope analysis by ICP-MS is complicated by the presence of various spectral interferences, which are compounds or elements entering the mass spectrometer that have the same nominal mass-to-charge ratio as the analyte of interest. The interferences may be caused by isobaric, polyatomic, oxide, and doubly charged ions. Isobaric interferences arise through the ionization of isotopes of other elements that have the same nominal mass, and can be corrected for mathematically using predefined equations. Polyatomic interferences are small molecular ions formed from a number of precursors such as the sample matrix, the reagents used for preparation, the plasma gas, and entrained atmospheric gases. These interferences together with the presence of doubly charged or oxide ions are dependent on the operation of the instrument and sampling handling. Such interferences can make characterizing the isotopic abundance of a spike isotope used in an E- or L-value experiment difficult. For example, Cd has six stable isotopes and it is unlikely all of these masses will be free from spectral interference. It is not impossible to make such measurements using ICP-MS, but does require careful application of correction equations. Use of ion chromatography to isolate the metal of interest prior to analysis can overcome some polyatomic interferences. It is arguably more

straightforward to conduct these measurements using TIMS because not only is the element isolated prior to analysis, but differences in elemental behavior on the heated wire filament related to first ionization potentials reduce isobaric interferences. There are other considerations when analyzing stable isotopes on either TIMS or ICP-MS, such as mass discrimination or fractionation, which are out with the scope of this discussion but can be addressed in a number of ways.

5. It should be noted that not all elements have a predictable natural stable-isotope abundance, for example, Pb has a variable isotope content so it may be necessary to measure the isotope abundance of Pb in an unspike soil or plant to establish the parameters to be used in the previously mentioned equation.

6. Experiments to determine the availability of P to plants using the radioisotope ^{32}P have in the past been based on a soil-to-water ratio of 1:100, 1 g of soil to 100 g water. Being a major plant nutrient, P is typically present at a relatively high concentration. Such an approach may not be appropriate or possible for all elements. For example, heavy metals may be present at much lower concentrations, which are nevertheless environmentally important. Ideally E-value experiments should be planned to provide an optimal metal concentration for isotope ratio analysis. If the concentration of the metal in the suspension is too low, accurate isotope ratio analysis will not be possible.

7. The amount of soil to use for an E-value experiment depends on the labile metal content of the soil. Although this is the parameter to be measured, it is extremely useful if some estimate can be made of the likely labile metal content of the soil. The total metal content of a soil is often determined in E and L-value experiments with the labile content expressed as a percentage of this value (*see* **Eq. 2**). The total metal content represents the maximum labile content. If the soil to be analyzed has a high-metal content, then the amount needed for an experiment will be low. Use of too much soil runs the risk of not being able to detect the added isotope or requiring so much isotope that the experiment becomes too expensive to perform. If the soil has a low-metal content, the amount of soil should be sufficient so that the metal content in the suspension remains detectable and at a level that does not compromise the quality of the isotope analysis.

8. An L-value experiment is a definitive measure of the metal available to the plant. By growing the plant from seed (so that any metal contribution is negligible) in an isotopically labeled soil, a direct measure of the labile metal pool accessed by the plant is possible. The choice of plant species for such an experiment depends on the questions to be answered, and may relate to, for example, the efficiency of hyperaccumulating plant species or susceptibility of food plants to a particular metal pollutant. The absolute amount of metal accumulated by a plant is not important when estimating the labile metal content of the soil. This is fortuitous because it is known that, for example, Cu uptake in plants eventually reaches a plateau at a level, which does not correlate with the available pool. A strength of isotope-labeling techniques is that from the isotope signature of the metal that is absorbed, the size of the total available pool can be estimated regardless of the absolute metal uptake by the plant.

References

1. Kersten, M. and Förstner, U. (1995) Speciation of trace metals in sediments and combustion waste, in *Chemical Speciation in the Environment,* (Ure, A. M. and Davidson, C. M., eds.), Blackie Academic Press, Glasgow, Scotland, UK. pp. 243–275.
2. Hamon, R. E., Bertrand, I., and McLaughlin, M. J. (2002) Use and abuse of isotopic exchange data in soil chemistry. *Austr. J. Soil Res.* **40,** 1371–1381.
3. Echevarria, G., Klein, S., Fardeau, J. C., and Morel, J. L. (1997) Mesure de la fraction assimilable des elements en traces du sol par la methode des cinetiques d'echange isotopique: cas du nickel. *Geochim.* **324,** 221–227.
4. Smolders, E., Brans, K., Foldi, A., and Merckx, R. (1999) Cadmium fixation in soils measured by isotopic dilution. *Soil Sci. Soc. Am. J* **63,** 78–85.
5. Hutchinson, J. J., Young, S. D., McGrath, S. P., West, H. M., Black, C. R., and Baker, A. J. M. (2000) Determining uptake of 'non-labile' soil cadmium by *Thlaspi caerulescens* using isotopic dilution techniques. *New Phytol.* **146,** 453–460.
6. Stanhope, K. G., Young, S. D., Hutchinson, J. J., and Kamath, R. (2000) Use of isotopic dilution techniques to assess the mobilization of nonlabile Cd by chelating agents in phytoremediation. *Environ. Sci. Technol.* **34,** 4123–4127.
7. Hamon, R., Wundke, J., McLaughlin, M., and Naidu, R. (1997) Availability of zinc and cadmium to different plant species. *Austr. J. Soil Res.* **35,** 1267–1277.
8. Young, S. D., Tye, A., Carstensen, A., Resende, L., and Crout, N. (2000) Methods for determining labile cadmium and zinc in soil. *Eur. J. Soil Sci.* **51,** 129–136.
9. Goodson, C. C., Parker, D. R., Amrhen, C., and Zhang, Y. (2003) Soil selenium uptake and root system development in plant taxa differing in Se-accumulating capability. *New Phytol.* **159,** 391–401.
10. Ahnstrom, Z. A. S. and Parker, D. R. (2001) Cadmium reactivity in metal-contaminated soils using a coupled stable isotope dilution-sequential extraction procedure. *Environ. Sci. Technol.* **35,** 121–126.
11. Ayoub, A. S., McGaw, B. A., Shand, C. A., and Midwood, A. J. (2003) Phytoavailability of Cd and Zn in soil estimated by stable isotope exchange and chemical extraction. *Plant Soil* **252,** 291–300.
12. Heumann, K. G. (1988) Isotope dilution mass spectrometry. In: *Inorganic Mass Spectrometry,* (Adams, F., Gijbels, R., and Van Grieken, R., eds.), John Wiley and Son, New York, NY, pp. 301–376.
13. Becker, J. S. and Dietze, H. J. (2000) Precise and accurate isotope ratio measurements by ICP-MS. *Fres. J. Anal. Chem.* **368,** 23–30.
14. Morel, C., Blaskiewitz, J., and Fardeau, J. C. (1995) Phosphorus supply to plants by soils with variable phosphorus exchange. *Soil Sci.* **160,** 423–430.
15. Tran, T. S., Fardeau, J. C., and Giroux, M. (1988) Effects of soil properties on plant-available phosphorus determined by the isotopic dilution phosphorus-32 method. *Soil Sci. Soc. Am. J.* **52,** 1383–1390.
16. Fardeau, J. C., Morel, C., and Boniface, R. (1991) Cinetiques de transfert des ions phosphate du sol vers la solution du sol: parametres caracteristiques. *Agron.* **11,** 787–797.

17. Fardeau, J. C. and Guiraud, G. (1974) Utilisation des filtres millipores en chimie du sol: apllication à la caratérisation des cinétiques de dilution isotopique des ions dans le sol. *Annal. Agron.* **25,** 113–117.
18. Fardeau, J. C. and Jappé, J. (1976) Nouvelle méthode de détermination du phosphore assimilable per les plantes: extrapolation des cinétiques de dilution isotopique. *C. R. Acad. Sci. Paris* **282D,** 1137–1140.
19. Fardeau, J. C. and Jappé, J. (1978) Analyse par dilution isotopique de la fertilité et de la fertilisation phophorique de quelques sols du Quebec. *Canad. J. Soil Sci.* **58,** 251–258.
20. Nakhone, L. N. and Young, S. D. (1993) The significance of (radio-) labile cadmium pools in soil. *Environ. Pollut.* **82,** 73–77.

14

In Vitro Hairy Root Cultures as a Tool for Phytoremediation Research

Cecilia G. Flocco and Ana M. Giulietti

Summary

Plant model systems are needed in which to conduct basic laboratory studies prior to field applications of phytoremediation. In vitro plant cultures are a useful tool for such research purposes. This chapter focuses on the generation of hairy root cultures and their use as a laboratory root model for the in vitro study of the removal of aromatic compounds. A protocol for the generation of hairy root cultures of *Armoracia lapathifolia* L. by infection with *Agrobacterium rhizogenes* (a soil-borne bacterium that induces the differentiation of roots at the infection spots) is described. A second protocol describes the application of hairy root cultures in assays for studying the removal of organic compounds, exemplified with phenol, a model organic contaminant. *A. lapathifolia* roots contain high levels of peroxidases (E.C. 1.11.1.7), enzymes that are known to be involved in the detoxification of phenols and other aromatic compounds. Briefly, the cultures are exposed to different concentrations of the contaminant under study and the remaining amounts of it, and some physiological parameters, are monitored along defined time intervals. This experimental procedure permits the estimation of the capability of a plant species to remove a contaminant and also the main variables that may affect the remediation process. This information is essential for assessing the feasibility of a remediation process prior to its field application. This can be done at relatively low analytical expense and in short periods of time with the use of in vitro plant models. Hence, the hairy root culture constitutes a valuable tool for phytoremediation research and development.

Key Words: Phytoremediation; bioremediation; in vitro culture; hairy roots; transformed roots; plant model system; *Armoracia lapathifolia*; *Agrobacterium*; organic pollutants; aromatic compounds; phenol.

1. Introduction

Phytoremediation takes advantage of the ability of plants and associated micro-organisms to remove, contain, or render harmless inorganic as well as

From: *Methods in Biotechnology, vol. 23: Phytoremediation: Methods and Reviews*
Edited by: N. Willey © Humana Press Inc., Totowa, NJ

organic contaminants *(1)*. In recent decades, this area of biotechnology experienced increasing interest, mainly in Europe and the United States, but also in developing countries, where the first field applications are being successfully implemented. These facts are making it necessary to gain insight into the mechanisms underlying the decontamination process. Prior to the application of a phytoremediation protocol to field conditions, it is necessary to conduct laboratory-scale studies, for which plant model systems are required. These model systems permit the control and reproducibility in experimental conditions necessary for conducting basic phytoremediation research.

Phytoremediation studies carried out with whole plants provide useful information related to the removal capabilities of the plant species under study and also about the phytotoxic effects of the contaminant. These studies include plants grown in soil or in hydroponics. However, when the aim is to study the specific interaction of a contaminant with the roots (the main plant organ involved in uptake), the use of alternative cultures can be advantageous. In vitro cultures of roots are of particular interest for studying the interaction of contaminants with this plant organ. This type of isolated organ culture permits the characterization of the uptake capability of the roots while avoiding the interference of translocation to other plant tissues. The drawback of root cultures is their low growth rate, which can impair experimental procedures. But the in vitro culture of transformed roots ("hairy roots") can overcome this drawback. The transformed roots are adventitious roots that are obtained by infection with the soil pathogen *Agrobacterium rhizogenes,* a Gram-negative bacterium that belongs to the Rhizobiaceace family. This bacterium is attracted to wounded sites of the plant and subsequently induces the formation of adventitious roots in a wide range of plant species. The development of these roots is the result of natural genetic engineering in which a specific region of bacterial DNA contained in the *Ri* (*root inducing*) plasmid of *A. rhizogenes* is transferred from the bacterial cell to the plant cell. This fragment of transferred DNA (T-DNA) is integrated into the plant genome and expressed *(2–4)*. The T-DNA carries the information for the synthesis of enzymes related to production of plant-growth regulators and/or for enzymes that regulate the sensitivity of plant cells to these compounds. **Figure 1** summarizes the main events involved in the generation of hairy roots. A detailed account of the molecular events of the transformation process is described by Sheng and Citovsky *(5)*. The metabolic changes produced by the expression of these bacterial genes are responsible for the development of adventitious roots and also for their main characteristics, namely fast growth, a high degree of lateral branching, the profusion of root hairs (because of this the transformed roots are called "hairy roots"), and the absence of geotropism *(6)*.

The transformed roots can be cultured under axenic conditions and are easily established and propagated in the laboratory. As mentioned before, they are

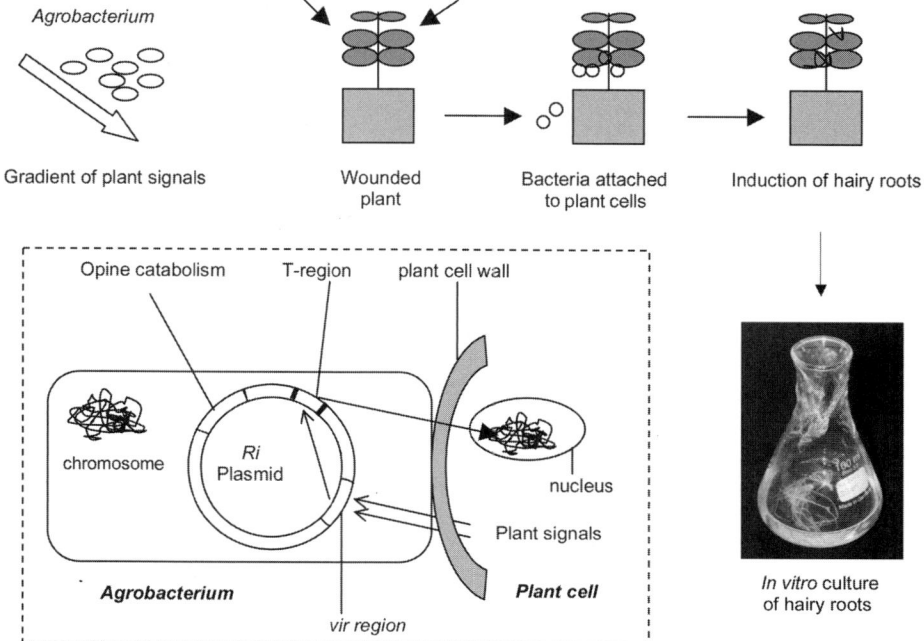

Fig. 1. Simplified scheme of the generation of hairy roots by infection with *Agrobacterium rhizogenes*. In the dashed box, the main molecular events involved in the transformation process are represented.

characterized by fast growth (compared to nontransformed ones), exhibit genetic and biochemical stability, and have a pattern of metabolites similar to that of the original plant *(2,7)*. Furthermore, the hairy root culture allows propagation and indefinite utilization of the tissue of the same plant, avoiding the variability that may exist among individuals of the same species *(8)*. The hairy root cultures have proven to be successful in vitro systems for studying the phytoremediation of a wide range of organic contaminants as well as inorganic ones. Some examples are presented in **Table 1**.

This chapter describes the protocols for the establishment and culture of hairy roots and also for their use as a plant model system in phytoremediation research, exemplified by the study of the removal of a model organic contaminant from aqueous solutions. *Armoracia lapathifolia* L. is the species selected to establish hairy root cultures for investigating the phytoremediation of phenol, a model and a frequently occurring organic contaminant. *A. lapathifolia* plants are cultured for their value as a condiment and for their high root content of peroxidases (E.C. 1.11.1.7) *(9)*. These and other oxidoreductive enzymes, such as catechol oxidases (E.C. 1.10.3.1) and laccases (E.C. 1.10.3.2), can decrease the

Table 1
Examples of the Use of Hairy Root Cultures for Phytoremediation Research

Hairy root culture	Compound under study	Reference
Beta vulgaris	Cadmium	*(19)*
Nicotiana tabacum		
Solanum nigrum	Cadmium	*(20)*
Thlaspi caerulescens	Cadmium	*(21)*
		(22)
Rubia tinctorum	Copper	*(23)*
Brassica juncea	Uranium	*(24)*
Chenopodium amaranticolor		
Alyssum spp.	Nickel	*(25)*
S. nigrum	Polychlorinated biphenyls	*(26)*
Catharanthus roseus	Trinitrotoluene	*(27)*
Armoracia lapathifolia	Phenol	*(28)*
A. lapathifolia	Phenol	*(29)*
Daucus carota	Phenol	*(29)*
		(30)
Brassica napus	2,4-dichlorophenol	*(30)*
		(31)

bioavailability of phenolic compounds by catalyzing their polymerization with subsequent precipitation into a solution, or by coupling them to structural components in plants or to humus in soils. The result of any of these processes is a diminished availability of the parent compound, which is generally (but not always) associated to a reduction of its toxic action *(10–12)*. In summary, hairy root cultures constitute a useful tool for studying plant uptake and transformation of pollutants. In fact, this root model system permits the evaluation of these variables, as well as physiological parameters, in short periods of time and at lower analytical expense, which is essential for developing and improving plant-based remediation strategies.

2. Materials

2.1. Establishment and In Vitro Culture of Hairy Roots of Armoracia lapathifolia

1. A plant of *A. lapathifolia* with green and healthy leaves.
2. A 48 h culture of *A. rhizogenes* strain LBA 9402 in YMB medium.
3. 4% Solution of sodium hypochlorite with Triton X-100 0.1% (v/v).
4. Sterile distilled water.
5. YMB medium: 0.5 g/L K_2HPO_4, 0.2 g/L $MgSO_4·7H_2O$, 0.1 g/L NaCl, 0.4 g/L yeast extract, and 10 g/L mannitol; pH 7.0.

6. Liquid MS medium *(13)* containing. Macronutrients: 1900 mg/L KNO_3, 1650 mg/L NH_4NO_3, 170 mg/L KH_2PO_4, 440 mg/L $CaCl_2·2H_2O$, and 370 mg/L $MgSO_4·7H_2O$. Micronutrients: 27.8 mg/L $FeSO_4·7H_2O$, 22.3 mg/L $MnSO_4·4H_2O$, 8.6 mg/L $ZnSO_4·7H_2O$, 6.2 mg/L H_3BO_3, 0.83 mg/L KI, 0.025 mg/L $CuSO_4·5H_2O$, 0.25 mg/L $Na_2MoO_4·2H_2O$, 0.025 mg/L $CoCl_2·6H_2O$, and 37.3 mg/L Na_2EDTA.
7. 15–20 Conical flasks of 250 mL.
8. Sterile Petri dishes.
9. 15–20 Plates with 20 mL of MS medium (with 3% [w/v] sucrose and solidified with 0.8% agar).
10. Sterile hypodermic syringe with needle.
11. Razor blade, forceps.
12. Filter-sterilized ampicillin (sodium salt).

2.2. In Vitro Assay of Phenol Removal by Hairy Root Cultures of A. lapathifolia

1. 30-d-old hairy root cultures of *A. lapathifolia* grown in MS liquid medium.
2. 25-mL Conical flasks (for incubations of phenol–plant tissues).
3. Sterile aqueous solutions of phenol (25, 50, 100 mg/L).
4. 30% Hydrogen peroxide.
5. Catalase (Sigma, cat. no. C3515).
6. 6 *M* Ammonium hydroxide.
7. 2% (w/v) 4-Aminoantipirine (aqueous solution).
8. 8% (w/v) Potassium ferricyanide (aqueous solution).
9. Guaiacol (2-methoxy phenol).
10. 100 m*M* Potassium phosphate buffer, pH 6.0.
11. 100 m*M* Potassium phosphate buffer, pH 7.4.
12. Glass test tubes (of approx 3 mL, for peroxidase assay)
13. Glass vials with Teflon screw cap of 7–10 mL (for taking samples of the incubation mixtures containing phenol).

3. Methods

3.1. Establishment and In Vitro Culture of Hairy Roots of A. lapathifolia

The main experimental steps described in this protocol are illustrated in **Fig. 2**.

3.1.1. Plant Material

1. Grow roots of *A. lapathifolia* (which can be bought at the vegetable market) in pots with commercial soil in the greenhouse or under room temperature (approx 20°C) and natural light. Once the leaves are developed, separate those to be used as explants (starting material) for the production of hairy roots.
2. Surface sterilize the leaves by a 5-min immersion in 4% solution of sodium hypochlorite containing 0.1% (v/v) of Triton X-100, using a conical flask under agitation (approx 80 rpm).

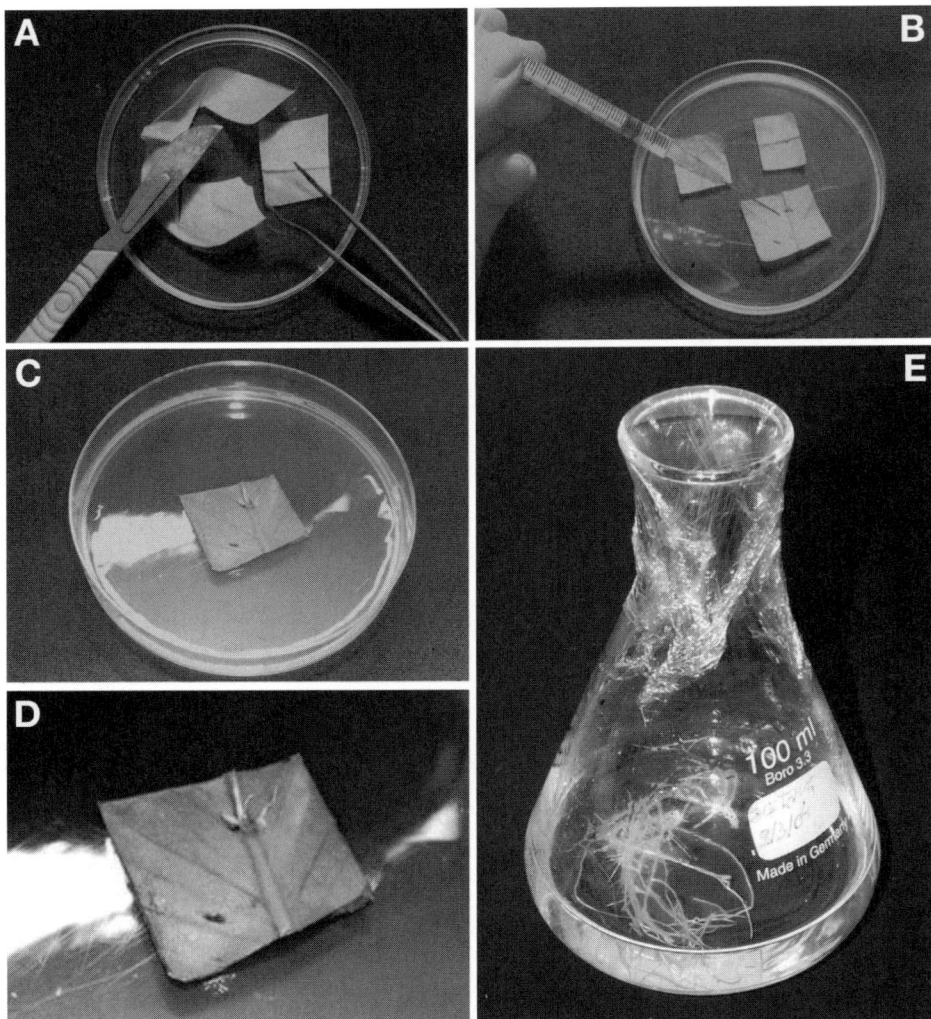

Fig. 2. Experimental steps for the transformation of *Armoracia lapathifolia* with *Agrobacterium rhizogenes*. **(A)** Surface-sterilized leaves are cut into square sections to obtain explants for infection. **(B)** The main vein and secondary ones are wounded by using a sterile needle and subsequently a 2-d-old *A. rhizogenes* culture is inoculated into the wounded spots. **(C,D)** After 2–3 wk of the infection, hairy roots appear at the inoculated spots. Explants containing hairy roots are cultured in liquid MS containing ampicillin for eliminating the agrobacteria. **(E)** Axenic hairy root cultures are established in liquid media by cultivating root tips.

In Vitro Hairy Root Cultures

3. Under laminar flow, rinse three times with sterile distilled water. Save for subsequent infection with *A. rhizogenes*.

3.1.2. Preparation of Inoculum and Infection of Plants

1. Inoculate a conical flask containing approx 20 mL of YMB medium with *A. rhizogenes* strain LBA 9402. Grow under agitation at 200 rpm, 28°C for 48 h.
2. Cut sterilized leaves of *A. lapathifolia* into sections of approx 2–3 cm^2 under laminar flow using sterile forceps and a razor blade.
3. Using a sterile syringe and needle make some incisions on major veins of the explant and inoculate 1–2 drops of a 48 h culture of *A. rhizogenes* into the wounded tissues.
4. Transfer the infected explants into Petri dishes containing 20 mL of MS medium, with the addition of 3.0% (w/v) of sucrose and solidified with 0.8% (w/v) agar (*see* **Note 1**).
5. Incubate the infected explants at 24 ± 2°C (low light or darkness). After approx 2–3 wk, the roots will differentiate around the inoculation spots.

3.1.3. Generation and Maintenance of In Vitro Cultures of Hairy Roots

1. Under laminar flow, excise the *A. lapathifolia* hairy roots that appear at the inoculation sites, leaving a small portion of the surrounding parent tissue. Transfer into 250 mL conical flasks with 50 mL of a hormone-free MS liquid medium (3.0% [w/v] of sucrose) containing 0.05% ampicillin. Culture the tissues at 24 ± 2°C in an orbital shaker at 80 rev/min, under a 16 h photoperiod (given by cool white fluorescent lamps at a light intensity of approx 1.8 W/m^2).
2. After 2–3 d, select single root tips and transfer each one to fresh culture medium (with 0.05% ampicillin) and incubate as described to establish clonal lines of hairy roots.
3. Every 2 wk cut the tips of the established clonal lines of hairy roots and transfer those into 250 mL conical flasks with 50 mL of the same liquid culture medium. Subculture the root tips every 2 wk gradually reducing the concentration of antibiotic until axenic cultures are obtained.
4. Maintain the hairy root cultures by monthly subculture of root tips in liquid MS medium without antibiotic, under the culture conditions as described in **Subheading 3.1.3.1**.
5. The transformation can be confirmed by the PCR (*see* **Note 2**).

3.2. In Vitro Assay of Phenol Removal by Hairy Root Cultures of A. lapathifolia

3.2.1. Phenol Removal Assay

1. Grow hairy root cultures as previously described and plan to start the removal assay when the cultures are 30 d old (*see* **Note 3**).
2. In the laminar flow hood, weigh portions of 0.4 g of 30-d-old *A. lapathifolia* hairy roots and distribute each of them to 25 mL conical flasks containing 10 mL of aqueous solutions of phenol (25, 50, and 100 mg/L). Prepare three replicate flasks for each concentration and for each incubation period to be analyzed.

3. Start the removal assay by the addition of hydrogen peroxide (final concentration 1.5 mM), which is necessary for the peroxidase-catalyzed oxidation of phenolic compounds (*see* **Note 4**).
4. Repeat the procedures described in **steps 2** and **3** using hairy root cultures, which were previously boiled for 30 min. This set of flasks represents the control of adsorption to root biomass.
5. Repeat the procedure described in **step 2** and omit the addition of hydrogen peroxide. This set of flasks represents the control for mechanisms involved in the removal of phenol other than peroxidases.
6. Incubate all the sets of flasks on a rotary shaker at 70 rpm, at 24 ± 2°C.
7. Take samples of the aqueous phenol solutions and of the root biomass at regular time intervals (for example after 30 min, 1, 5, 24, and 48 h after incubation). Sacrifice three replicate flasks of each set of flasks for each sampling time.
8. When taking samples, stop the reactions by addition of catalase (1.0 U/mL) to the flasks that contain hydrogen peroxide (*see* **Note 5**).
9. Separate the roots and thoroughly wash them with distilled water. Remove excess water with paper tissue and record the fresh weight. Store at –18°C for further analysis (for example, protein content or enzyme activity such as that described in **Subheading 3.2.3.**).
10. Take a 5 mL aliquot of the remaining solution and save at 4°C for analysis of the remaining amounts of phenol.

3.2.2. Quantification of the Remaining Amounts of Phenol by a Colorimetric Assay

1. React a 5 mL aliquot of the phenolic solution to be analyzed with 0.025 mL 6 M ammonium hydroxide, 0.025 mL of a 2% aqueous solution of 4-aminoantipirine, and 0.05 mL potassium ferricyanide (8% [w/v]). Mix by vortexing.
2. After 5 min add 2.5 mL of chloroform and extract the dye formed.
3. Measure the absorbance of the chloroformic extract at 510 nm.
4. Convert the absorbance data to phenol concentrations using a calibration curve obtained with phenol standards of known concentrations.
5. Alternatively, the remaining amounts of phenol can be measured by gas chromatography (*see* **Note 6**).

3.2.3. Enzyme Extraction and Peroxidase-Activity Assay

1. Take approx 100 mg of each hairy root samples saved for analysis and homogenize the plant material in a mortar with 3 mL of 100 mM of phosphate buffer, pH 6.0.
2. Centrifuge the homogenate at 8000g for 5 min and save the supernatant.
3. Prepare the reaction mixture in a test tube with 3 mL of guaiacol reagent (0.35% guaiacol in 100 mM potassium phosphate buffer [pH 7.4] and 10 µL of the supernatant obtained as described in the previous step). Add 10 µL of 30% hydrogen peroxide, invert the tube for mixing three times, and immediately read the absorbance at 430 nm. Continue reading it for 2 min every 30 s. The

described kinetic method corresponds to that of Tijssen et al. *(14)*, with minor modifications.
4. Express the results of peroxidase activity as indicated in **Note 7**.

3.3. Data Analysis (see Note 8)

1. Perform analyses of variance (ANOVA) and a Tukey's HSD (Tukey's honestly significant difference) at the level of 95% to test for mean differences of the dependent variables, under different experimental conditions.
2. Fit appropriate models to describe the kinetics of the contaminant removal (*see* **Note 9**).

4. Notes

1. It is important to ensure that the infected side of the tissues does not contact the agar or plate surfaces to prevent the growth of bacteria over the explant and agar surface. When possible, larger explants harboring a stem portion can be placed vertically in closed jars by introducing the stem into the agar medium, so the leaves with infected spots do not contact either the jar or the agar.
2. Adventitious roots can be generated in the absence of infection with agrobacteria. However, these will not have all the characteristics of transformed ones. The transformed nature of the roots can be established by analyzing the presence of bacterial genes in the plant tissues. For this purpose, it is essential to ensure that no agrobacteria are present because they would mask the result. Briefly, a segment of 700 bp is amplified by PCR using specific primers for a bacterial gene (*rol b*) using as a template DNA extracted from root tissues. The products are visualized by agarose gel electrophoresis with ethidium bromide staining. The transformed root should yield the desired PCR product, as well as a control with agrobacterial DNA as template *(15)*.
3. Previous growth kinetics studies carried out in our laboratory with *A. lapathifolia* hairy root cultures showed that the highest levels of peroxidases are produced at the end of the exponential phase of growth *(16)*. Activities of other oxidoreductive enzymes, which can also be involved in the polymerization of phenolic compounds (such as catechol oxidases and laccases), were not detected in these cultures. Hence, *A. lapathifolia* tissues are a good model for studying the implications of peroxidases in the removal process without interference of those enzymes of very similar catalytic activity.
4. Hydrogen peroxide is a cosubstrate for the peroxidase-catalyzed oxidations. The optimal concentration of this reagent for the in vitro removal assays was established in previous experiments *(16)*.
5. Catalase is an enzyme that rapidly converts hydrogen peroxide to water. Its addition is important not only for stopping any hydrogen peroxide-dependent reaction, but also for preventing possible interferences of this reagent with other type of analyses that might be also accomplished (for example, acute toxicity analyses).
6. The colorimetric quantification of phenol described in **Subheading 3.2.2.** is a simple method that can be accomplished in a laboratory with basic analytical instruments,

such as a spectrophotometer. If available, the quantification of the remaining amounts of phenol in the reaction solution can be determined by liquid–liquid extraction followed by gas chromatography. The procedures are as follows: aqueous samples of 5 mL are acidified to pH <2.0 with sulfuric acid and sequentially extracted in a separation funnel by the addition of three portions of 2 mL of dichloromethane (liquid–liquid extraction, EPA method 3510C, adapted for 5 mL samples). The extracts are concentrated using nitrogen blow down to reduce the volume by a factor of 10. All prepared samples are identified and quantified with gas chromatography. Injections are performed into a glass insert packed with silanized glass wool. Detection is performed with a flame ionization detector and operated with nitrogen make-up of 25 mL/min, a hydrogen flow of 30 mL/min, and an air flow of 300 mL/min. Typically, the GC oven is operated with the following temperature program: the column temperature is held at 60°C for 2 min and then ramped at 20°C/min to 150°C where it is held for 5 min. The injector temperature is 280°C. Separation is accomplished with a fused-silica capillary column (cross-linked 5% PH ME siloxane) with a film thickness of 0.25 μm, 30 mm × 0.25 mm I.D. (Hewlett Packard, Palo Alto, CA). The carrier gas velocity (He) is set at 31 cm/s. Linear-regression analysis of peak heights vs compound concentrations are performed with standard solutions. p-Cresol is used as internal standard.

7. One enzymatic unit (U) was arbitrarily defined as the amount of enzyme that produces a change in A_{470} of 0.021/min at 25°C under the assay conditions. The results can be expressed in U/mL = $(V_f/V_o \times a \times l) \times \Delta Abs/min$, where: V_f is the final volume in the reaction mixture (3 mL), V_o is the volume of peroxidase extract used in the reaction mixture (0.01 mL), a is an extinction coefficient of the oxidation of guaiacol at 470 nm (6.4/μM/cm), l is light path (1 cm), and ΔAbs/min is the increase in absorbance per minute. The results can be normalized to U/g of plant tissue as follows: U/mL × V_e/ FW_e, where V_e is the volume of buffer used for preparing the enzymatic extract (3 mL) and FW_e is fresh weight (g) of the tissues used for preparing the enzymatic extract.

8. The parameters described are the basic ones for the initial assessment of the phytoremediation capability. However, the detoxification effect is not always guaranteed by the removal of the parent compound because toxic intermediates can be formed during the process. Hence, a complete assessment of the phytoremediation potential requires the analysis of the possible reaction intermediates and also of the toxicity of the resulting reaction mixtures. When polymeric products are expected, methods such as gas chromatography or mass spectrometry with fast atom bombardment ionization might not be appropriate for their analysis mainly because of the high molecular weight of the compounds. Mass spectrometry with electrospray ionization can overcome this drawback and analyze compounds of high molecular weight. For evaluating the toxicity of the samples along the removal assay, the Microtox® assay can be used. This test evaluates the acute toxicity by exposing a suspension of luminescent bacteria (*Vibrio fischeri*) to the sample under study and measuring the reduction in luminescence. The degree of luminescence reduction is indicative of the toxicity of the sample *(17)*. When

studying compounds that are known or suspected carcinogens or that may undergo metabolic activation, it is appropriate to include mutagenicity assays. The Ames assay *(18)* is a plate assay that employs mutant *Salmonella typhymurium* strains, which can reveal the presence of mutagenic compounds. Although the lack of positive results of this assay is not a definitive proof of the absence of mutagenic compounds, it can be used as a screening procedure.

9. In removal assays carried out in our laboratory with similar initial concentrations of phenol, it was observed that the amount of phenol removed was greater than 70% after 3 h of incubation of the hairy roots in the presence of hydrogen peroxide. In the absence of the oxidant and cosubstrate for the peroxidase-catalyzed oxidation, the removal values were 55–30% *(16)*.

Acknowledgments

The authors are thankful to Secretaria de Ciencia y Técnica de la Universidad de Buenos Aires (UBACyT) and Consejo Nacional de Investigaciones Científicas y Técnicas (CONICET), for supporting this work.

References

1. Cunningham, S. D., Anderson, T. A., Schwab, A. P., and Hsu, F. C. (1996) Phytoremediation of soils contaminated with organic pollutants. *Adv. Agron.* **56,** 55–114.
2. Tepfer, M. (1984) Transformation of several species of higher plants by *Agrobacterium rhizogenes*, sexual transmission of the transformed genotype and phenotype. *Cell* **37,** 959–967.
3. Zambryski, P. (1992) Chronicles from the *Agrobacterium*-plant cell DNA transfer story. *Ann. Rev. Plant Physiol. Plant Mol. Biol.* **43,** 465–490.
4. Toivonen, L. (1993) Utilization of hairy root cultures for production of secondary metabolites. *Biotechnol. Prog.* **9,** 12–20.
5. Sheng, J. and Citovsky, V. (1996) Agrobacterium-plant cell DNA transport; have virulence proteins, will travel. *Plant Cell* **8,** 1699–1710.
6. Hamill, J. D., Parr, A. J., Rhodes, M. J. C., Robins, R. J., and Walton, N. J. (1987) New routes to plant secondary products. *Bio-Technol.* **5,** 800–804.
7. Flocco, C. G., Alvarez, M. A., and Giulietti, A. M. (1998) Peroxidase production in vitro by *A. lapathifolia* lapathifolia (horseradish)-transformed root cultures: effect of elicitation on level and profile of isoenzymes. *Biotechnol. Appl. Biochem.* **28,** 33–38.
8. Pollard, A. J. and Baker, A. J. M. (1996) Quantitative genetics of zinc hyperaccumulation in *Thlapsi caerulescens*. *New Phytol.* **132,** 113–118.
9. Krell, H. W. (1991) Peroxidase, an important enzyme for diagnostic test kits. In: *Biochemical, Molecular and Physiological Aspects of Plant Peroxidases*, (Lobarzewski, J., Greeping, H., Penel, C., and Gaspar, T., eds.), University of Geneva, Switzerland, pp. 469–478.
10. Dec, J. and Bollag, J.-M. (1994) Use of plant material for the decontamination of water polluted with phenols. *Biotechnol. Bioeng.* **44,** 1132–1139.

11. Dec, J., Shuttleworth, K. L., and Bollag, J.-M. (1990) Microbial release of 2,4- dichlorophenol bound to humic acid or incorporated during humification. *J. Env. Qual.* **19,** 546–551.
12. Sun, W. Q. and Payne, G. F. (1996) Tyrosinase-containing chitosan gels, a combined catalyst and sorbent for selective phenol removal. *Biotechnol. Bioeng.* **51,** 79–86.
13. Murashige, T. and Skoog, F. (1962) A revised medium for rapid growth and bio-assays with tobacco tissue culture. *Physiol. Plant.* **15,** 473–497.
14. Tijssen, T. (1985) Enzymes for immunoassays. In: *Practice and Theory of Enzyme Immunoassays,* (Burdon, R. H. and van Knippenberg, P. H., eds.). Elsevier, Amsterdam, pp. 173–220.
15. Hamill, J. D., Rounsley, S., Spencer, A., Todd, G., and Rhodes, M. J. C. (1991) The use of polymerase chain reaction in plant transformation studies. *Plant Cell Rep.* **10,** 221–224.
16. Flocco, C. G. (2002) Phytoremediation, use of plant model systems for the study of the remediation of organic compounds. *Doctoral Thesis.* Universidad de Buenos Aires, Buenos Aires, Argentina.
17. Aitken, M. D., Massey, J. I., Chen, T., and Heck, P. E. (1994) Characterization of reaction products from the enzyme catalyzed oxidation of phenolic pollutants. *Wat. Res.* **28,** 1879–1889.
18. Ames, B. N., McCann, J., and Yamasaki, E. (1975) Methods for detecting carcinogens and mutagens with the Salmonella/mammalian microsome mutagenicity test. *Mutat. Res.* **31,** 347–364.
19. Metzger L., Fouchault, I., Glad, C., Prost, R., and Tepfer, D. (1992). Estimation of cadmium availability using transformed roots. *Plant Soil* **143,** 249–257.
20. Macek, T., Kotbra, P., Suchova, M., Skacel, F., Demnerova, K., and Ruml, T. (1994) Accumulation of cadmium by hairy-root cultures of *Solanum nigrum. Biotechnol. Lett.* **16,** 621–624.
21. Nedelkoska, T. J. and Doran, P. M. (2000) Hyperaccumulation of cadmium by hairy roots of *Thlapsi caerulescens. Biotechnol. Bioeng.* **67,** 607–615.
22. Boominathan, R. and Doran, P. M. (2003) Cadmium tolerance and antioxidative defenses in hairy roots of the cadmium hyperaccumulator, *Thlaspi caerulescens. Biotechnol. Bioeng.* **83,** 158–167.
23. Maitani, T., Kubota, H., Sato, K., Takeda, M., and Yoshikira, K. (1996) Induction of phytochelatin (class III metallothionein) and incorporation of copper in transformed hairy roots of *Rubia tinctorum* exposed to cadmium. *J. Plant Physiol.* **147,** 743–748.
24. Eapen, S., Suseelan, K. N., Tivarekar, S., Kotwal, S. A., and Mitra, R. (2003) Potential for rhizofiltration of uranium using hairy root cultures of *Brassica juncea* and *Chenopodium amaranticolor. Environ. Res.* **91,** 127–133.
25. Nedelkoska, T. J. and Doran, P. M. (2001) Hyperaccumulation of nickel by hairy roots of alyssum species: comparison with whole regenerated plants. *Biotechnol. Prog.* **17,** 752–759.
26. Macková, M., Macek, T., Kučerová, P., Burkhard, J., Pazlarová, J., and Demnerová, K. (1997) Degradation of polychlorinated biphenyls by hairy root culture of *Solanum nigrum. Biotechnol. Lett.* **8,** 787–790.

27. Hughes, J. B., Shanks, J., Vanderford, M., Lauritzen, J., and Bahdra, R. (1997) Transformation of TNT by aquatic plants and plant tissue cultures. *Env. Sci. Technol.* **31,** 266–271.
28. Flocco, C. G. and Giulietti, A. M. (1998) Removal of phenol by *Armoracia lapathifolia* hairy roots. *Int. J. Biodet. Biodeg.* **42,** 248–249.
29. Flocco, C. G., Gulietti, A. M., Araujo, B. S., Charlwood, B. V., and Pletsch, M. (1999) Removal of phenolic compounds by hairy root cultures. In: *Proceedings of the 9th European Congress on Biotechnology,* (Hofman, M., ed.), Branche Belge de la Société de Chimie Industrielle, Belgium, ECB9/2781, pp. 1–6.
30. Santos, D. Charlwood, B. V., and Pletsch, M. (2002) Tolerance and metabolism of phenol and chloroderivatives by hairy root cultures of *Daucus carota* L. *Environ. Pollut.* **117,** 329–335.
31. Agostini, E., Coniglio, M. S., Milrad, S. R., Tigier, H. A., and Giulietti, A. M. (2003) Phytoremediation of 2,4-dichlorophenol by Brassica napus hairy root cultures. *Biotechnol. Appl. Biochem.* **37,** 139–144.

15

Sectored Planters for Phytoremediation Studies

Chung-Shih Tang

Summary

Field practice of phytoremediation involves complicated and variable conditions. There is a need for methods that can transpose the contaminated site to a controlled greenhouse environment so that screening of phytoremediators and evaluation of efficacy can be carried out objectively. The "sectored planter" method is designed to compare and evaluate preselected phytoremediators and offers validation to a field experiment. The method described here recapitulates our recent work at a managed coastal phytoremediation site on Oahu, Hawaii where deep coastal soil contamination of petroleum hydrocarbon occurred. The contaminants are of relatively low concentrations and unevenly distributed, so field evaluation of efficacy would be impractical. Because the contaminants were accumulated at the third soil layer beneath a top sandy loam and a middle silt layer, a trisector planter was designed to simulate the field conditions, including soil profiles and field management of the three selected tree species. Known quantities of six diesel-fuel components were used to spike soils in the third or bottom section of the planter. The reduction of concentrations of these compounds was determined and results on the three tree species and the no-plant control were compared after 200 d growth in the greenhouse. The experience obtained in this experiment suggests that the "sectored planter" method offers a wide range of applications in phytoremediation studies. These possibilities are discussed.

Key Words: Phytoremediation; deep-soil contamination; coastal-soil contamination; trisector planter; sectored planters; PHC; PAH; rhizosphere; soil micro-organisms.

1. Introduction
1.1. Background

Regardless of many successful examples, the efficiency of phytoremediation can vary. In reviewing the literature, it has become clear that the propriety of phytoremediation must be determined on a case-by-case basis. Location, plant and microbial community, soil characteristics, and method of management and type of contaminants are just a few factors that determine the phytoremediation efficacy. The sectored planter method is designed to grow remediation plants under simulated field conditions of soil and microbial profile, depth of contaminants, and field management practice. Concentrations of contaminants may be enriched to facilitate quantitation, soil samples may be removed from the windows on the planter for analysis, and the bottom sector detached at the end point and replaced with new samples for additional study using the same plant. It is therefore a method with considerable flexibility.

The particular example described here is a trisector planter *(1)* because three soil layers are involved. In other applications, the number of sectors and dimensions of the planter can vary according to needs. Three Hawaiian tree species were evaluated in the greenhouse using the sectored planter method. They were among the trees cultivated at a contaminated site for the reduction of deep coastal soil contamination by petroleum hydrocarbons (PHCs). In this chapter, emphasis will be placed on the construction and maintenance of the sectored planter and its potential in meeting the needs in phytoremediation studies. Methods of instrumental analysis and enumeration of soil micro-organisms are only briefly described because analytical methods will differ from application to application.

1.2. Characteristics of the Field Demonstration Site

At the John Rogers Tank Farm (JRTF) of Hickam Air Force Base, Honolulu, Hawaii, a demonstration site for agriculture-based phytoremediation (ABRP) was established by CH2M HILL, Inc. in 2000 to decontaminate PHCs embedded in the deep-soil layer in a tropical coastal environment *(2)*. The PHCs at the field site are diesel and gasoline-range PHCs. However, because the contaminants are detected only at relatively low concentrations, and are unevenly distributed, it would be impractical to evaluate the efficacy by analyzing and comparing the field contamination levels before and after the field demonstration. The trisector planter method describes greenhouse and laboratory experiments to help validate the field practice of ABRP *(3)*. There are several characteristics of the site that require special consideration: (1) the coastal soils are of high salinity; (2) the soil profile of the JRTF site from surface down is, in general, 90–120 cm of coral sand and gravel, followed by ca. 30–60 cm of silt, then at the bottom, a layer of sandy loam; (3) low levels of PAH contaminants were detected in the top portion of the bottom sandy loam at a depth of approx 160 cm; (4) the contaminants are unevenly distributed; and (5) relatively low microbial population was found in the (contaminated) bottom sandy loam.

1.3. Objectives of the Trisector Planter

The trisector planter is designed to simulate the field conditions at JRTF. Soils collected from the site were used for planting to ensure that the initial microbial populations were similar to those of the field. Three Hawaiian woody plants were chosen for comparative studies based on results from a preliminary screening experiment for tolerance of salt and diesel fuel in coastal soils *(4)*. To solve the problem of low initial soil concentrations of PHC contaminants, six representative diesel constituents were used to spike the sandy loam layer to known concentrations, so their levels can be measured at the end of the experiment, and thus, the effectiveness of the three phytoremediator plants can be evaluated and compared.

2. Materials

1. Plants: three tropical woody plants, milo (*Thespesia populnea*), kou (*Cordia subcordata*), and false sandalwood (*Myoporum sandwicense*), were chosen based on resistance to high-salinity treatment (2% NaCl) and two diesel-fuel levels (5 and 10 g no. 2 diesel fuel/kg soil) in prescreening trials using soils from the site under greenhouse conditions *(4)*. Young plants relatively uniform in size (approx 1 yr old) were purchased from local nurseries for growing in the trisector planter.
2. Soils: the texture of the soils obtained from the JRTF site consists of a top layer of coral sand with various sizes of coral gravels, a middle layer of silt, and a bottom layer of sandy loam. The top and bottom layers were passed through a 2 mm sieve prior to use. After sieving, the refined top-layer soil was classified as sandy loam according to its particle-size distribution *(5)*. The middle-silt layer was left unsieved. The top sandy loam, middle silt, and bottom sandy loam were stored separately under 4°C until use.
3. The planter: inert materials were used for the construction of the planter to minimize contamination. The trisector planter (**Fig. 1**) was constructed from Plexiglas pipes, Teflon and glass tubings, and perforated Teflon plates (approx 1 mm thickness).
4. Chemicals: the six standard PHCs for spiking the soils were pyrene (four aromatic rings), phenanthrene (three aromatic rings) and pristine (branched alkane, C19), *n*-alkanes hexadecane (C16), eicosane (C20), and docosane (C22) (Aldrich, Milwaukee, WI). All solvents used in diluting the PHCs, soil extraction, and instrumental analysis were of gas chromatography Grade (GC).

3. Methods

The example described here is a trisector planter with the top two sections simulating the upper two soil layers in the field and the bottom-section soil, where contamination occurred, spiked with the six PHCs to known concentrations. The height of each sector is proportional to the depth of the corresponding soil layer in the JRTF site. Effects of three factors, i.e., plant presence, plant species, and passive aeration of the bottom-section soil, on the degradation of the spiked PHCs are examined at the end of incubation (200 d). Four planters with fittings for passive aeration of the bottom section and four without fittings (**Fig. 1**) are planted with each of the three plants (milo, kou, and

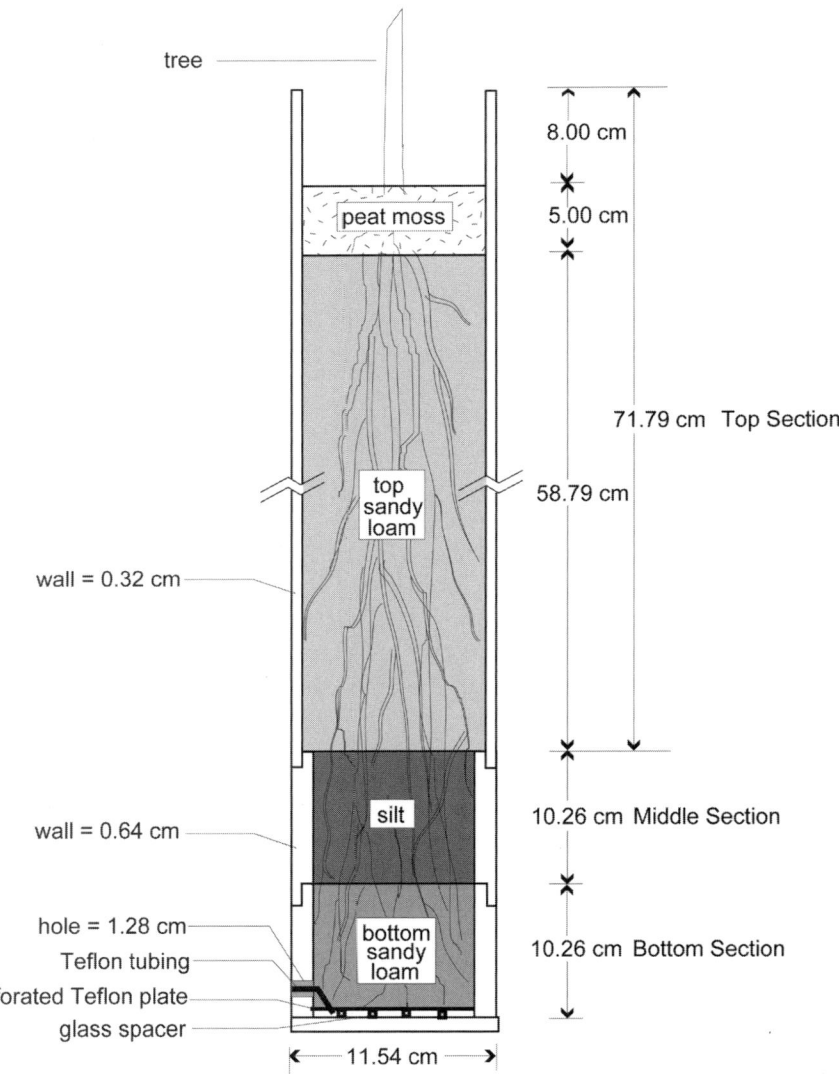

Fig. 1. An example of the sectored planter: the trisector planter with a tree and a soil profile similar to that of the field demonstration site. When the roots reached the bottom of the middle section, the bottom section with sandy loam spiked with diesel components of known concentrations was attached. After incubation under the rhizospheric influence of the test plant for 200 d, the bottom-section soil was collected for residual diesel chemical analysis and bacterial enumeration. (© 2004 from Evaluation of Agriculture-Based Phytoremediation in Pacific Island Ecosystems Using Trisector Planters by Tang et al. Reproduced by permission of Taylor & Francis, Inc., http://www.taylorandfrancis.com.)

false sandalwood), or left unplanted (no-plant control). The 32 planters are distributed in the greenhouse in a completely randomized design.

3.1. Construction of the Trisector Planter

The trisector planter (**Fig. 1**) consists of three sections staggered snugly together by counter-bore fittings. The bottom section is glued onto a Plexiglass plate. A perforated Teflon plate supported by spacers made of short sections of 0.64-cm glass tubing (**Fig. 1**) provided large contact surface of soil to air. A hole bored at the lower edge with a Teflon tubing insert at the bottom section reaching below the perforated plate serves as an air vent for passive aeration. For no-air experiments, the bottom hole is plugged. Additional holes (not shown in **Fig. 1**) of appropriate size may be bored on any section of the planter for specific purposes, such as taking soil samples before the end point for analysis, or inserting probes for continuous monitoring of various physical factors. These holes can be conveniently plugged using rubber stoppers wrapped with Teflon tape (*see* **Notes 1** and **2**).

3.2. Soils in the Trisector Planter

To simulate the field-soil profile, **Fig. 1** shows the structure, dimensions, and soil types in the trisector planter. The top section (71.12 cm height, 0.32 cm wall thickness) of the planter is filled with the top sandy loam and above that, a layer of peat moss is added to simulate the organic mulch used in the field. The middle section is filled with the silt, and the bottom section with the bottom sandy loam. The layers are not compacted. The planters are wrapped in aluminum foil during tree growth to prevent light penetration, except when observations of root growth were made.

3.3. Spiking With Selected PHC Standards

The sandy loam in the bottom section of all trisector planters is spiked with a mixture of the six selected constituents of diesel fuel. The hydrocarbons are first dissolved in a 1:1 mixture of acetone and methylene chloride (*6*) and added to the air-dried soil manually and then mechanically mixed for 2 h. The soil was stored in a bucket overnight at room temperature to allow solvents to evaporate. Concentrations of the PHCs are: 200 mg/kg soil each of pyrene, phenanthrene, and pristane, and 500 mg/kg each of hexadecan, eicosane, and docosane. A preliminary experiment indicated that spiking the soil with these PHCs in solvents had no adverse effect on the microbial populations.

3.4. Plant Growth Conditions

A young plant is transplanted to the top section of a planter, which is rested on a Petri dish. The middle section is attached when roots reached down to the bottom of the top section; the time to reach this stage varies dependent on the rate of root extension of each species. Finally, the bottom section is attached when the roots reached the bottom of the middle section.

The temperature in the greenhouse ranges from 36°C at midday to 19°C during the night. Mist is applied for 2 min five times during the daytime to control the greenhouse temperature. Initially, each planter is manually watered at approx 1 L/d, dependent on needs. Peters Professional liquid nutrients (30N-10P-10K, Scotts-Sierra Horticultural Products Company, Marysville, OH) are incorporated into the irrigation water. After the plants were fully established (2 mo), 61 g/planter of slow-release Nutricote fertilizer (13N-13P-13K; Nutricote, Inc., HI) is top-dressed. Automated-drip irrigation is used three times a day for 1 min, at approx 1.5 L/d. The properties of the JRTF site are maintained, i.e., water saturation in the bottom section without any effluent water.

3.5. PHC Analysis

3.5.1. Soil Sampling and Schedule

The root extension in the planter is monitored through a window on the aluminum foil wrapping. Because of their different growth rate, root extension reaching the bottom of a given section in each planter would take different lengths of time; therefore, the bottom section with the spiked compounds should be attached at different times accordingly. However, a compromise to accommodate the analytical schedule for residual PHCs is necessary, so harvesting is arbitrarily grouped into three increments based on the root growth rate. Thus, the fast-growing milo plants were in the first increment and most of the slow-growing kou plants are in the third increment.

At 120 d after attachment of the bottom section to milo, the progress of the PHC depletion is determined by a gas chromatograph equipped with flame ionization detectors (GC-FID) analysis *(6)*. One gram (dry weight) of the spiked soil is scooped out from a hole on the bottom section (1.5 cm, not shown in **Fig. 1**) for the estimation of residual PHCs. Results from this explorative analysis indicated that it would require 6 mo to observe significant differences. Harvesting dates of plant roots, therefore, were set in three increments 200 d after attaching the contaminated bottom section.

Prior to harvesting, irrigation is held off for 48 h to reduce moisture in the bottom section, so the soil had the proper consistency for handling. The planter with the tree was placed in a horizontal position on the floor and the bottom section detached. The tree with two remaining sections is replaced on the bench for the next experiment, if needed. Roots and the associated thin layer of soil (<1 mm) are removed from the bulk soil in the bottom section for microbial enumerations *(7)*.

3.5.2. PHC Extraction and Analysis

For PHC concentration determination, the freshly spiked soil and the harvested soil in each bottom section is passed through a 2 mm sieve and a portion is stored in short wide-mouth glass jars (125 mL) with Teflon-lined caps at 4°C for shipping. Samples are sent by express mail to the Applied Sciences Lab., CH2M HILL, Corvallis, OR, for quantitative GC–mass spectrometry analysis by standard methods. Data are statistically treated using ANOVA two-way analysis (*see* **Notes 3** and **4**).

3.5.3. Microbial Analysis

Microbial enumerations are made on selected treatments to assess the degree of proliferation of phenanthrene-degrading bacteria, hexadecane-degrading micro-organisms, and total heterotrophic bacteria in response to plant roots in the presence of contaminants *(1,7)* (*see* **Notes 5–7**).

4. Notes

1. All three tree species had healthy and relatively uniform growth, indicating that the planter provided adequate environment for relatively long-term growth of these trees. Passive aeration did not show obvious effects on the growth of trees.
2. Milo root extension was the fastest among the three testing plants. False sandalwood roots were plentiful and healthy but performed the worst in hydrocarbons (HC) degradation. Kou showed the slowest root growth among the three.
3. Effects of each plant species on the reduction of individual PHC concentrations are assessed based on quantitative GC and two-way ANOVA analysis. Only the results with significant reductions in comparison to those in the no-plant treatments are listed and were as follows: (1) phenanthrene was the most biodegradable compound among the six spiked hydrocarbons. It was reduced by approx one-third of the initial (day 0) concentration even in the no-plant treatment *(8)*. In the presence of plants, phenanthrene was reduced to levels less than the GC-detection limit. (2) Milo significantly reduced hexadecane by 33%, and pyrene by 42% *(9)* compared to the concentrations of these compounds in the no-plant treatment. (3) Kou reduced hexadecane by 38%, eicosane by 24%, docosane by 19%, and pyrene by 55%. (4) False sandalwood reduced the level of pyrene by 37%. (5) Pristane is a highly branched, saturated PHC and there was no plant effect on pristine. (6) The benefit of aeration was small. We do not recommend this practice because field installment of deep-soil aeration is costly.

    When comparing their levels to those on day 0, no significant reduction in hexadecane, eicosane, and docosane, but significant losses of pristane, phenanthrene, and pyrene by 13, 31, and 22.5%, respectively, occurred in the unplanted soil. Fertilization and irrigation are routine measures for bioremediation of PHCs in soil *(10)*. Thus, biodegradation of diesel constituents would be expected in unplanted soil, although for some compounds, such as phenanthrene, the rate of degradation is further enhanced by plants *(11)*.

4. An important observation in the bottom section was that in some plant treatments, especially those of false sandalwood for which the roots distributed in soil only sparsely, pyrene degradation occurred at rates significantly faster than those of the no-plant treatments. This result suggests that irrigation effectively carried organic root exudates from the top sections to the bottom section, enhancing the activities of microbial hydrocarbon degraders in this "expanded rhizosphere" *(1)*. Accordingly, ABRP in the field will effectively reduce contaminants in soils to the depth where root exudates can reach, rather than being limited to the rhizosphere in a conventional sense, which is less than a few millimeters from the rhizoplane *(12)*.
5. The original microbial populations in the bottom sandy loam were very small but increased drastically in the rhizosphere at the end of the experiment. There was also a selective stimulation of the hydrocarbon-degrading micro-organisms in the rhizosphere. A similar selective effect was exerted on phenanthrene-degrading micro-organisms by the rhizosphere of slender oats (*Avena barbata* Pott ex Link) grown in phenanthrene-spiked soil *(11)*.
6. By using the trisector planter to grow selected tropical coastal plants under a controlled environment, we have obtained greenhouse and laboratory data supporting the practice of agriculture-based phytoremediation. Our data suggest translocation of organic root exudates by the downward irrigation water may accelerate the breakdown of PHCs in relatively deep soil layers in the Pacific island ecosystems.
7. There are many potential uses of the sectored planter in phytoremediation studies. The sectored planter method previously described is based on a specific application of the sectored method, but it is expected that a similar approach can be used for other needs in wide ranges of phytoremediation studies including: (1) final-stage screening of potential phytoremediators for deeper soil contamination (i.e., other than surface contamination). The type of contaminants are not limited to PHCs, others such as pesticides, industrial organic chemicals, and heavy metals are equally applicable. (2) The contaminated bottom sectors can be detached from the growing plants for physical, chemical, and microbiological analyses, and fresh sectors loaded for a repeated study. This saves much time compared to growing new plants. Also, the same established perennial plant can be used for the study of different target compounds. (3) Sampling holes on each section could provide the opportunity for kinetic study of the contaminants or the microbial populations (*see* **Subheading 3.1.**). (4) With excessive irrigation, effluent water may be continuously collected for the study of root exudates. (5) By comparing the effects under different environmental factors such as nutrients, water, temperature, lighting, and microbial inoculation, the sectored planter can be used to optimize field management parameters for a selected phytoremediator in ABRP.

Acknowledgments

The author wishes to thank W. H. Sun, F. M. Robert, M. Toma, and R. K. Jones for their contributions in establishing the trisector planter method. **Figure 1** has been adopted from the original paper with the kind permission of Taylor and Francis Inc, Philadelphia, PA, on behalf of the New Phytologist Trust.

References

1. Tang, C. S., Sun, W. H., Toma, M., Robert, F. M., and Jones, R. K. (2004) Evaluation of agiculture-based phytoremediation in Pacific Island ecosystems using trisector planters. *Int. J. Phytorem.* **6,** 17–33.
2. CH2M HILL (2001) *Hickam Air Force Base Phytoremediation Annual Report.* June 2000. Prepared for US Army Corps of Engineers.
3. CH2M HILL (2003) *Agriculturally Based Bioremediation of Petroleum-Contaminated Soil and Shallow Groundwater in Pacific Island Ecosystems.* Final Report. Prepared for US Army Corps of Engineers.
4. Sun, W. H., Lo, J. B., Robert, F. M., Ray, C., and Tang, C. S. (2004) Phytoremediation of petroleum hydrocarbons in tropical coastal soils. I. Selection of promising woody plants. *Environ. Sci. and Pollut. Res.* **11,** 260–266.
5. American Society for Testing and Materials (1999) *ASTM Book of Standards, Sec. 4, Vol. 04.08*. West Conshohoken, PA.
6. Banks, M. K., Govindaraju, R. S., Schwab, A. P., and Kulakow, P. (2000) Part I: field demonstration. In: *Phytoremediation of Hydrocarbon-Contaminated Soil*, (Fiorenza, S., Oubre, C. L., and Ward, C. H, eds.), Lewis Publishers, Boca Raton, FL, pp. 3–88.
7. Jones, R. K., Sun, W. H., Tang, C. S., and Robert, F. M. (2004) Phytoremediation of petroleum hydrocarbons in tropical coastal soils II. Microbial response to plant roots and contanminant. *Environ. Sci. and Pollut. Res.* **11,** 260–266.
8. Liste, H. and Alexander, M. (1999) Rapid screening of plants promoting phenanthrene degradation. *J. Environ. Qual.* **28,** 1376–1377.
9. Liste, H. and Alexander, M. (2000) Plant-promoted pyrene degradation in soil. *Chemosphere* **40,** 7–10.
10. Atlas, R. M. and Bartha, R. (1998) *Microbial Ecology, Fundamentals and Applications*, 4th ed., Addison Wesley Longman, Inc., New York, New York.
11. Miya, R. K. and Firestone, M. K. (2000) Phenanthrene-degrader community dynamics in rhizosphere soil from a common annual grass. *J. Environ. Qual.* **29,** 584–592.
12. Hiltner, L. (1904) Uber neuere erfahrungen und problem auf dem gebiet der bodenbakteriodologie und unter besonderer berucksichtigung der grundungung und brache. *Arb. Btsch. Landwirt. Ges.* **98,** 59–78.

16

Phytoremediation With Living Aquatic Plants
Development and Modeling of Experimental Observations

Steven P. K. Sternberg

Summary

This chapter provides a summary of the mathematical analysis and experimental design of laboratory measurements of the bioremoval potential for living aquatic plants. This process is called phytoremediation, bioremoval, biosorption, or bioaccumulation. The mathematical models are based on the concept of the conservation of mass and include descriptive equations, including adsorption of the metal onto living and growing biomass. The models describe the concentration of metal in solution as a function of time. An example case from previously published data is included to demonstrate the use of the models. The results from the mathematical models can be used to scale up a process, or to answer questions of how long to run an experiment, how much biomass material is required, what the expected level of removal is, and to help set benchmarks to determine how well a process is working. In addition to presenting model equations, a summary of experimental considerations, such as statistical design, choice of variables, and result quantification has been included. The information provided allows good experimental data to be collected such that a maximum amount of information is obtained with the minimum amount of effort.

Key Words: Phytoremediation; bioremoval; biosorption; bioaccumulation; phytoremoval; experimental model; statistical design; mass balance.

1. Introduction

Bioremoval is a process that can remove soluble heavy metal contaminants from aqueous solutions by an adsorption process *(1–11)*. This process is also called phytoremediation, bioremoval, biosorption, bioaccumulation, or phytoremoval. The adsorbant is biomass, often obtained from an aquatic plant, which may be living or dead *(12–14)*. The biomass is capable of accumulating the metal at concentrations many times greater than that of the solution *(15–21)*.

From: *Methods in Biotechnology, vol. 23: Phytoremediation: Methods and Reviews*
Edited by: N. Willey © Humana Press Inc., Totowa, NJ

The metal is removed from the system by a simple filtration of the biomass. This process has the advantages of being simple, with a low cost to build and to maintain *(22,23)*. It may not require changes in water quality, such as pH, temperature, or nutrient loadings *(24,25)*. For these reasons, this technology is currently being studied for potential use in waste water pretreatment from industrial facilities *(26–28)*. Some current shortcomings include the need for long retention times (2–48 h), and the existence of an equilibrium concentration beyond which no removal occurs, even when additional biomass is added. The need for additional research focuses on identification of biomass sources for particular metals, examinations of bioremoval potentials for multiple metal contaminants, and the use of multiple sources of biomass.

This chapter provides a summary of the mathematical analysis and experimental design for laboratory measurements of the bioremoval potential of living aquatic plants. The intended audience is an interested researcher starting a heavy metal bioremoval project. I combined ideas from many engineering and science fields to develop a procedure for the design of experiments and the presentation of collected information. This work by no means provides the final word on how to conduct such experiments or their interpretation; rather it seeks to provide a common starting point.

The main issues this chapter discusses are the development of a general system for modeling a bioremoval reactor, the effective design of experiments using statistics, and the identification of typical dependent and independent variables. Also, there are several decisions to make in determining how to design a system for making measurements of the bioremoval potential: use of living or dead plants, type of reactor (batch or flow), and water pretreatment. This discussion will center on using living aquatic plants in a batch reactor to remove a metal ion from water.

1.1. Modeling

An important aspect of present-day engineering is the ability to adequately model the behavior of a system before committing to large-scale investment. Developing a good model requires a basic theoretical understanding of the system, and experimental observation and measurement of the system or parts of the system. (*See* List of Symbols on p. 203.)

This chapter attempts to develop a general model based on standard engineering practice. The model describes the concentration of a toxic metal in an aqueous solution that is being treated with plant biomass for the purpose of removing the metal. In this chapter, I will model the bioremoval of the metal ions as an adsorption phenomenon. The main modeling tool, however, will be the species mass balance:

$$\text{Mass}_{in} - \text{Mass}_{out} + \text{Generation/Removal Rate} = \text{Accumulation} \quad (1)$$

which can be written mathematically as

$$M_{ads}^{in} - M_{ads}^{out} + \int_V r_{ads} dV = \frac{dM_{ads}}{dt} \qquad (2)$$

where the ads subscript refers to the metal being adsorbed. The integral in the rate term allows generation to vary in space throughout the reactor. The last term uses a derivative to represent the change in mass in the solution over time. Most batch reactors are well mixed and so the rate term can be considered constant throughout the reactor. This equation will be used to model the change in metal in the water over time, the amount of biomass, and the amount of metal adsorbed to the biomass.

1.1.1. Batch Reactor Models

The system of the metal solution and biomass can be modeled as a chemical reactor. Time is the key design variable of interest for a laboratory reactor (consisting of a beaker filled with the waste water into which the biomass is introduced). This type of reactor is called a batch reactor.

When time is an important consideration, we need to include a study of the reaction/adsorption kinetics. Kinetics is the study of how fast a reaction (or adsorption) occurs under given conditions (temperature, competing reactions, and catalysts). The equilibrium conditions can be obtained from these models by assuming a large value for the time variable (where the rate of adsorption is zero).

We start by defining the adsorption rate, an intensive variable that depends on temperature and concentration, which describes the mass of metal adsorbed per unit time. An example equation,

$$-r_{ads} = k_1 C_{ads} \qquad (3)$$

describes the adsorption rate as a linear function of a rate constant (which may depend on temperature) and the concentration of the adsorbed species. We will explore more interesting forms of this function in the section on adsorption (*see* **Subheading 1.1.2**).

A batch reactor typically consists of a holding vessel into which all components of the reaction or adsorption process are added. It then is allowed to react for a certain period of time, after which it is drained, cleaned, and the process begun again. It works well for slow reactions, low volumes, or new/novel systems. It is time consuming and requires thorough cleaning after each run. It is the typical reactor chosen for laboratory studies. The main consideration for design purposes is the time needed for the reaction.

The mass balance (**Eq. 2**) description for a batch reactor includes no mass in or mass out terms (there is only a single instantaneous input at time = 0, which is modeled mathematically by using an initial condition). The generation/removal term describes the rate of metal ion adsorption onto the biomass and it is assumed to be the same everywhere. The accumulation term describes the metal ion concentration in solution. Excluding zero terms and assuming that the reactor has a constant volume (which allows simple conversion from mass to concentration, C = M/V):

$$r_{ads} = \frac{dC_{ads}}{dt} \quad (4)$$

where r_{ads} has units of mass of metal per volume of solution per unit time. Realizing that the rate of adsorption is dependent on the quantity of biomass, we can modify this equation by relating the rate of adsorption to the rate of adsorption per unit biomass:

$$r_{ads} = m_x\, r_{bio} = \frac{dC_{ads}}{dt} \quad (5)$$

where m_x is the amount of biomass per unit volume of solution, and r_{bio} has units of mass of metal adsorbed per unit biomass per unit time. These equations, combined with an equation for the adsorption rate, provide the model equation for the batch reactor experiments.

We are not considering the other types of reactors like the continuous stirred tank reactor or plug flow reactors. These are flow-based reactors as opposed to the time-based batch reactor, *see* **ref. 29** for more details on any of these reactor types.

1.1.2. Adsorption Models

The next consideration in building our system model equations is to describe various adsorption models. These identify the adsorption rate term (r_{bio}) used in the reactor model equations. This chapter assumes all the removed metal is adsorbed by the plant biomass. The term could be expanded to include adsorption by sediments, additional organisms (each with its own behavior), or the reactor vessel itself.

The models should all describe the potential of the biomass to adsorb a given metal. We would expect this to be fastest when the biomass is first exposed to the metal, then to slow, and to eventually approach zero as equilibrium is reached. How fast the transition occurs can be modeled using adjustable parameters. The adjustable parameters are obtained experimentally. I will develop two example models, each having three adjustable parameters. One of these three parameters is used to describe the equilibrium concentration in the solution, found after long exposure times of the biomass to the metal solution. Note that we expect the rate to depend on the difference between the actual and equilibrium concentration, this may be described as a driving potential for the adsorption. This concentration potential is described with the term $(C_{ads} - C_{eq})$, where C_{ads} represents the adsorbed metal concentration and C_{eq} represents the equilibrium concentration for a particular metal and type of biomass.

Next, two adsorption rate models are proposed. Each has the characteristics of starting large in value and becoming zero as the concentration in the solution approaches the equilibrium value. This idea is based on the concept of mass transfer, and it is from this area of study that these models are found *(30)*.

Model one has two adjustable parameters

$$r_{bio} = k_1 (C_{abs} - C_{eq})^n \tag{6}$$

k_1 may be a temperature dependent term, if necessary, and n is the rate power. When n has the value of one, this is called a first order model.

Model two also has two adjustable parameters

$$r_{bio} = \left[\frac{k_{2A}(C_{ads} - C_{eq})}{1 + k_{2B}(C_{ads} - C_{eq})} \right] \tag{7}$$

Both of the parameters, k_{2A} and k_{2B}, may be temperature dependent. When the concentration term is large the rate is a constant, and when it is small the rate term is also small. **Figure 1** shows the general shapes of these functions. For multiple species of metal adsorbing on common sites this equation can be generalized for metal species, i:

$$r_{bio} = \frac{K_1 C_i}{1 + K_a C_c + K_b C_b} \tag{8}$$

where C_i is concentration of the species of interest, C_a is the concentration of species A, and C_b is the concentration of species B. The temperature dependence of the constants is usually described with an equation of the following type

$$k_1(T) = A \exp\left(\frac{-E_a}{RT}\right) \tag{9}$$

Where A is some pre-exponential frequency factor, E is an activation energy, R is the ideal gas constant, and T is the absolute temperature.

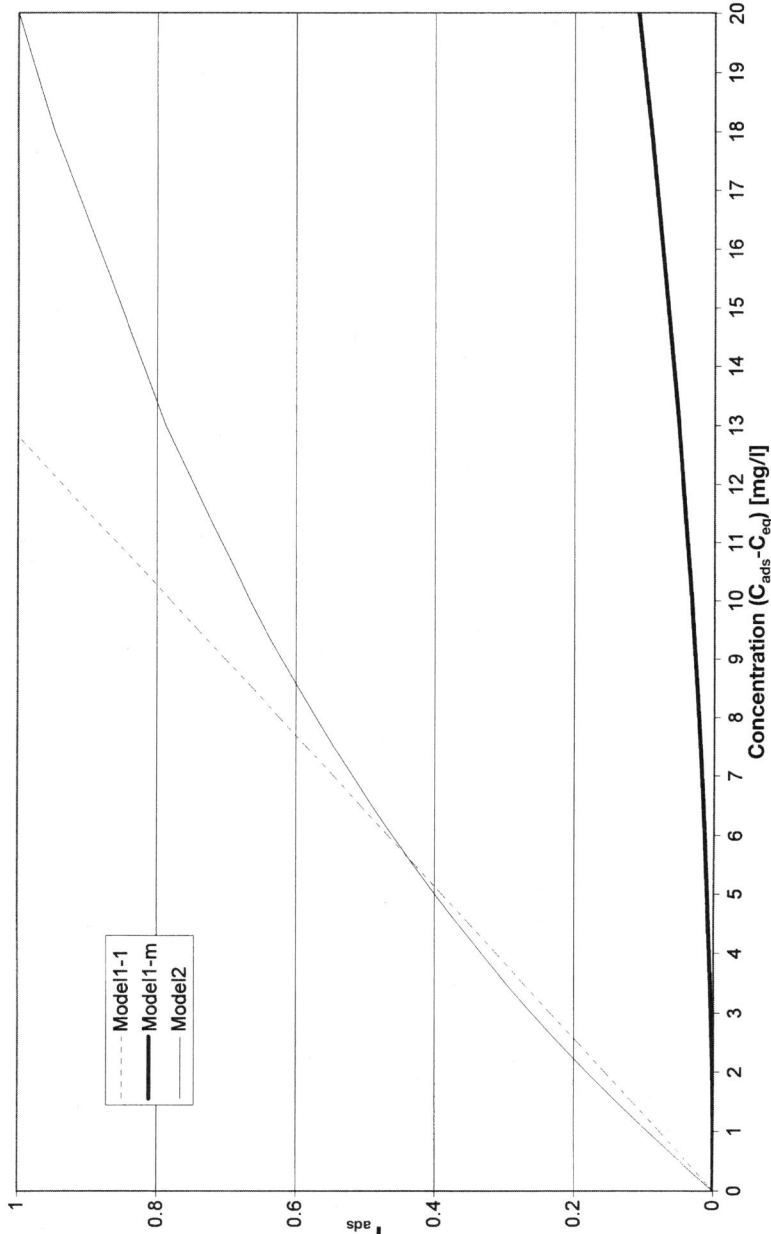

Fig. 1. Rate of adsorption models.

Phytoremediation With Living Aquatic Plants

These adjustable models can describe many simple types of relation between the adsorption rate and the concentration difference. The shapes all monotonically decrease as the concentration difference becomes smaller, and include linear, concave-up, and concave-down forms. A comparison of the different models, for a dataset as described in the example data section, is shown in **Fig. 1**.

1.1.3. Biomass Growth Rate

Using living biomass, such as *Lemna (31,32)* or a macro alga *(33)* can provide an advantage during the bioremoval process—it can increase in mass, thereby increasing the total amount of metal that could be adsorbed. However, this additional biomass does not change the equilibrium amount of metal in the solution *(34)*. The growth of the biomass may be modeled using the mass balance (**Eq. 2**). Again, for a batch reactor there will be no in or out terms, other than the initial amount present at the time zero.

$$\frac{dm_x}{dt} = m_x r_x \quad (10)$$

where m_x represents the amount of biomass per unit volume and r_x is the biomass growth rate with units 1/time. The simplest model describes the growth rate as a constant, k_{bio}. This is acceptable for short time periods and single species batch reactors. More interesting and realistic models have been described elsewhere *(35)*. For the initial condition $m_x = M_0$ at time zero, this equation becomes

$$m_x = M_0 \exp(k_{bio} t) \quad (11)$$

1.1.4. System Model

Now we can combine our reactor models with the adsorption rate equations and biomass growth equation to derive a relationship between the reactor design variable (time for batch, volume for flow) and the concentration of metal in solution.

Combining the mass balance (**Eq. 5**), the adsorption rate laws (**Eqs. 6 and 7**), and the biomass rate model (**Eq. 10**) yields the following solutions:
Model 1
Model 1–1 case one: with $n = 1$

$$C_{ads} = (C_I - C_{eq}) \exp\left[\frac{M_0 k_1}{k_{bio}} \left(\exp(k_{bio} t) - 1\right)\right] + C_{eq} \quad (12)$$

Model 1–n case two: with $n > 1$

$$C_{ads} = \left[(1-n) \left[\frac{M_0 k_1}{k_{bio}} \left(\exp(k_{bio} t) - 1 \right) \right] + (C_I - C_{eq})^{(1-n)} \right]^{\frac{1}{1-n}} + C_{eq} \quad (13)$$

where C_I is the initial concentration. The parameter k_{bio} can be found by measuring the amount of biomass. It may not be possible to measure the mass during the experiment, so only the initial and final values will be known. **Equation 11** only requires two mass measurements to be known to determine a value of k_{bio}. Case one only requires one parameter to be fit with experimental data, k_1. The values of k_1 and n can be found using nonlinear best-fit statistics between the model and the measured values of concentration over time. The best model parameters will minimize the sum of squares difference (SSD).

$$SSD = \sum_i \left[C_{model} - C_{data} \right]^2 \quad (14)$$

Model 2:

$$C_{ads} = (C_I - C_{eq}) \exp \left[k_{2B}(C_I - C_{ads}) - \left[\frac{M_0 k_{2A}}{k_{bio}} \left(\exp(k_{bio} t) - 1 \right) \right] \right] + C_{eq} \quad (15)$$

Again, the parameters M_0 and C_{eq} can be obtained directly from the experimental measurements, k_{bio} from biomass measurements and **Eq. 11**, and k_{2A} and k_{2B} can be found using nonlinear best-fit statistics by minimizing the SSD. Comparison between these three models for a set of batch reactor data is included at the example data section.

When using dead biomass or biomass that does not grow ($k_{bio} = 0$), the previously listed equations may be modified by making the following substitution:

$$\left[\frac{M_0}{k_{bio}} \left(\exp(k_{bio} t) - 1 \right) \right] \equiv t \quad (16)$$

1.1.5. Metal Adsorbed on Biomass

The next part of the system to model is the interaction between the contaminant metal and the biomass. Specifically we need to know how much metal is adsorbed onto a unit mass of the biomass. This calculation is relatively straightforward. From the mass balance we know (assume) that any metal ion mass that leaves the water must be adsorbed onto the biomass:

$$M_{bio} = (C_I - C_{ads}) * Volume \quad (17)$$

where C_I is the initial concentration and *Volume* represents the total liquid volume of the batch reactor. We can calculate the concentration of metal in the biomass by dividing this value by the amount of biomass:

$$q_{biomass} = M_{bio}/M_{biomass} \quad (18)$$

Finally, this allows the calculation of the concentration factor

$$CF = q_{biomass}\rho/C_I \quad (19)$$

where ρ is the solution density and CF has units of mass of solution per unit mass of biomass. It can be described as "1 g of biomass contains as much metal as CF grams of solution." The best biomass treatment systems will maximize this number.

1.2. Example Data

The previously mentioned equations are used to model a set of experimental data *(28)* in which the macroalga, *Cladophora parriaudii*, was used to remove cadmium from synthetic waste water in a batch reactor. The information was obtained by measuring concentration and biomass over time in a batch reactor in five simultaneous trials with identical initial conditions. The model parameters that are determined from these initial conditions used the average value from all five trials. The two parameters that are statistically fit were determined by minimizing the SSD over all five trials, with the reported SSD being the average of the five sets. **Table 1** summarizes the information used and the final values of the parameters for the two models, including both cases for model 1. The experimental data and the three model equations are shown in **Fig. 2**. All three models appear to fit the data well, though a comparison of the SSD values for each shows that, of the three models, model 1–*n* provides the best fit.

2. Materials

The three primary materials used in these experiments are the water source, the form of the metal contaminant, and the type of biomass (*see* **Notes 1–3**). Any experiment must choose these carefully, with full appreciation of how the final results will be used.

Table 1
Experimental Data and Model Parameters

Model	1–1	1–n	2
CI	4.4	4.4	4.4
Ceq	0.4	0.4	0.4
Parameter 1	M = 1	m = 1.75	k_2A = 0.1
Parameter 2	k_1 = 0.078	k_1 = 0.05	k_2B = 0.05
Initial biomass	1.03 g	1.03 g	1.03 g
Kbio	0.00118	0.00118	0.00118
Sum of squares	0.67	0.37	0.76
Bioremoval percent	91	91	91
CF	393	393	393

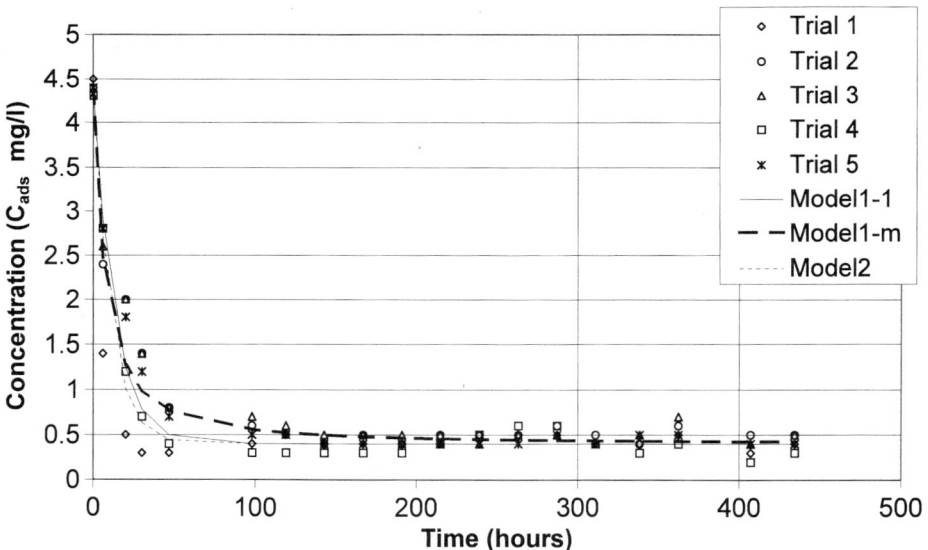

Fig. 2. Model comparisons.

2.1. Experimental Trials

1. 1 L Plastic beakers. Polyethylene or polypropylene has been found to not interact with the water, metals, or plants used in these experiments.
2. Plastic stir bar. Polyethylene or polypropylene.
3. Time recorder.
4. Micropipets with disposable tips. Volume range should include 100 μL to 1 mL. These are used to add metal solution, to collect samples for metal analysis, and to dilute samples to the operating range of the metal analysis equipment. These will also be used for the metal analyses.

2.2. Sample Metal Analysis

For atomic adsorption spectrophotometer, either flame or graphite furnace mode. Refer to the manufacturer for operation procedure. Details of methods, preparation of standards, and discussion of matrix modifiers will, in general, be different for each metal *(36)* (*see* **Note 4**).

2.3. Biomass Metal Analysis

1. Filtration equipment. Use weighed, dried ashless filter papers to collect biomass.
2. Porcelain crucibles with lids, approximate volume of 50 mL.
3. Hot plate with chemically resistant top.
4. Drying oven, temperatures of 105°C for 48 h.
5. Muffle furnace, capable of temperatures to 500°C for 2 h.
6. Concentrated nitric acid, metal-analysis grade.
7. Concentrated hydrochloric acid, metal-analysis grade.
8. Small glass (Pyrex) beaker to mix 25 mL Aqua Regia solution.

3. Methods

This section discusses the procedure used to determine the bioremoval potential and concentration factor, and good experimental technique for developing the information needed to understand the bioremoval processes. The first section provides a list of steps for performing the experiments. The second section is a set of general guidelines for the statistical design of experiments. The third section provides some ideas and suggestions for developing good experimental data. An understanding of the system model equations helps determine which variables to measure.

3.1. Experimental Trials

The experimental trials can be commenced once the experimental design has been finalized, and the type of water, form of metal, and type of biomass have been chosen. The experiments are relatively straightforward:

For each trial:

1. Fill a 1L container with 700 mL source water (*see* **Note 1**).
2. Add metal source to container (*see* **Note 2**), and allow to equilibrate, usually 2 h. Sample this solution to obtain the time zero data.
3. Add the appropriate amount of weighed wet biomass (*see* **Note 3**).
4. Sample the solution at the required time intervals (*see* **Note 5**). Liquid samples may be stored for short times (days) by slight acidification with dilution water (2 wt percent solution of nitric acid).
5. Measure the concentration of metal in water as per standard methods *(36)*. Atomic adsorption with flame or graphite furnace will allow very low concentrations (milligrams per liter to micrograms per liter) to be measured.
6. After completion of the solution sampling, collect the biomass for metal analysis.

3.1.1. Biomass Metal Analysis

1. Filter the biomass from the solution at the conclusion of the experimental trial.
2. Record wet weight.
3. Dry in drying oven. This will take 24–48 h, until a consistent weight is obtained.
4. Record dry weight.
5. Place in ashing crucible, and incinerate at 500°C in muffle furnace for 2 h, until all organic mater is removed and a consistent weight is obtained.
6. Record ash weight.
7. Digest ash in Aqua Regia (1:3 mixture of nitric acid and hydrochloric acid).
8. Dissolve in Aqua Regia; will at first be a dark mixture.
9. Heat mixture over a chemically resistant hot plate to evaporate water; never allow mixture to completely dry.
10. Add more Aqua Regia and heat.
11. Repeat until solution is clear.
12. Resuspend solution in dilute hydrochloric acid.
13. Measure concentration of this solution to obtain the mass of metal adsorbed by the biomass, using the same technique as trial samples.

3.2. Statistical Design of Experiments

The most effective experiments are those that result from a conscious and well-thought out plan to determine a specific objective. The commonly accepted tool for creating such experiments is the statistical design of experiments *(37–39)*. The statistical designs will maximize the amount of information obtained from any experiment while minimizing the amount of effort needed to complete the project.

Generally accepted requirements for a good experimental design require that the project first has a clearly defined objective (thesis). This includes an understanding of what may and what may not be learned from the experimental data, as well as what size of an effect can be expected to not be overlooked (relative to the experimental error). Second, the planned experiments must estimate the precision of the results. Without this estimation, there is no way to determine if the observed effects are real or just the byproduct of random errors. This estimation is obtained by replication of each and every experimental trial (*see* **Note 6**). Enough replications must be performed to prove that effects that are large enough to have practical significance will be statistically significant. Third, each variable should have a control case, to ensure that the observed effect is actually correlated with the design variable. With the objective in mind, the experiments may be designed by considering the independent variables (experimentally controlled and varied); the dependent variables (measured responses); the types and frequency of measurements; the relationship between the measured variables and the objective; cost of the experiment; precision of the measurements; and what prior knowledge exists. These will be discussed in more detail in the second section of this topic.

Two types of precision are discussed, as they are all too frequently confused. First is the precision of a measurement. This is determined by performing multiple analysis on a given sample from an experimental trial. It is assumed that a given sample is homogeneous and that each portion of a sample will yield similar results. This is important in demonstrating the value of the analysis technique. The second type of precision involves replication of each trial of an experiment. These may be done in different physical containers simultaneously or at different times. This precision demonstrates the variability in the overall process, and it represents the quantification of the importance of each independent variable. If the measurement precision is poor, then it may be difficult to determine whether or not the experimental precision is significant. Excellent measurement precision will not improve the experimental precision if the independent variables do not significantly affect the dependent variables. An ideal experiment uses a measurement that is very precise to clearly establish the importance of each independent variable on each dependent variable.

Controls are special trials added to an experiment to verify certain hypothesis used in developing the experimental design. The use of a control should establish, within the experimental precision limits, that the considered independent variable does cause the observed response in the dependent variable (*see* **Note 7**). An example may illustrate the point: to prove that biomass is indeed removing metal in an experimental trial, a control experiment might include everything in the trial, except the biomass. The expected response would be that no removal is observed during the experiment in the control. Actual measurement should confirm this. If it is not true, then something else must be causing the removal (such as adsorption onto the walls of the experiment container).

An experiment can be designed once the objectives have been discovered; the independent and dependent values are identified, and the variable levels of the independent variables are chosen. There are many types of experimental design *(39)*. I will discuss one very simple type called a factorial design. The number of experimental trials is determined by multiplying the number of levels of each independent variable together. For example, if there are three variables, two of which will be examined at two levels and one at three levels, then the number of trials is $2 \times 2 \times 3$, or 12 trials. Additionally, each trial must be replicated in at least duplicate to provide the information needed to determine the experimental precision. Triplicates are often used to help improve the estimate of the experimental precision. Triplication of the trials in this example would lead to 36 total experimental trials. Additional trials for the controls also need to be included and replicated. It is important to consider the total number of trials when choosing the number of variables and levels. In general, only two levels of a variable should be considered unless it is strongly suspect that the relationship between the variables is very nonlinear (perhaps going through a maximum or minimum within the range of values). Analysis of the results from a factorial experiment is straightforward, though tedious. Many statistical software packages will perform this analysis, such as SAS institute Inc. JMP *(40)*, or Minitab *(41)*. This method has a very large advantage over one at a time variable testing, as it can identify the interactions between variables with the least amount of experimental work.

3.3. Bioremoval Experiments

A typical objective for bioremoval experiments may be to determine the kinetic and equilibrium model parameters for the adsorption of a metal by some type of biomass. The important dependent variables include the equilibrium concentration of metal in solution, the bioremoval potential, and the concentration factor of metal adsorbed on the biomass. Additional dependent variables to consider are the growth rate (or lack thereof) of the biomass, and time required to reach equilibrium. Determination of these quantities requires measurement of the concentration of metal in solution at several times (*see* **Note 3**), the amount of biomass, and the amount of metal in the biomass at the beginning and end of the experiment. The standard procedures for the measurements of metal concentrations can be found in **ref. 36**. It is especially important to properly prepare all standards (*see* **Note 8**).

The experimental or independent variables may include the amount and type of biomass, the amount and type of metal contaminants, and water quality (especially pH and temperature). These variables may be actively controlled or just monitored during the experiments. Only the actively controlled variables will be considered when designing the experiment. For each of the controlled variables, the experimenter chooses two or more levels at which to investigate it. The levels should be chosen far enough apart so that the measured response can be expected to show a significant difference if it exists. The levels may be found by considering the measurement precision. All of these variables must also be measured, usually at the beginning of the experiment.

A simple batch experiment for examining the bioremoval of a single metal using a single plant will include three types of control trials, each replicated for determination of precision. The first control uses a trial without any metal or plant. This control may show if additional sources of metal or plant are present or it may detect contamination from poor laboratory technique. The second control uses a trial with just the biomass. This control will allow determination of reduction in growth rates because of the metal. The third control uses a trial with just metal. This control will expose any additional removal sources, such as adsorption to the container walls, or a chemical or physical precipitation not associated with the biomass. Each control helps determine if the assumptions used to build the model are adequate. Finally, the experiment trials will include both metal and plant perhaps at two levels each. The results from the controls will allow the experimenter to determine the significance of the controlled variables (*see* **Note 9**).

4. Notes

1. Water: typical sources of water include lab water (distilled/deionized), synthetic waste water, and field water. Each has advantages and disadvantages. Deionized water will provide the results with the best precision and reproducibility; however, it is a poor choice for growing biomass in and will probably be the least useful choice in providing information for a particular field site or engineering bioremoval project. Synthetic waste water is made in the lab and will contain many of the additional impurities that field site waste water would contain, though in carefully controlled and measured quantities. Results using this type of water will more closely match the field site, but it may contain too many variables to allow consideration of each one. Field water is a sample taken directly from the source to be treated. It is an excellent choice for developing site-specific data. It will suffer from unknown materials and its composition may vary almost randomly over time. There is no one best choice and the selection should be based on how the results will be used.
2. Metal: source and type of metal contaminant include specific metal compounds such as soluble salts or organometallic complexes. Simple salts provide easy-to-reproduce results, which do not fully model the extreme variability in water chemistry of field waste water. More complex sources will be more realistic, but may be difficult to fully quantify in a reproducible manner. There are many possible combinations. Any choice must be based on consideration of how the results will be used.
3. Biomass: source and type of biomass used is probably the most important, and least understood, variable in this work. Many different plants have the ability to adsorb metals from aqueous solution and the adsorption can occur on living or dead biomass. The overall process is very similar to ion exchange using beads or resin, except that there is no regeneration step. Instead, the plants either are replaced or simply continue to grow in the reactor. The metal is removed by harvesting an appropriate amount of the biomass over time. A few concerns when choosing a biomass source include: use of dead or living biomass, bioremoval potential, concentration factor, availability of supply, ease of removal from the solution (harvesting), growth rates and ease of growth, nutrient, light requirements, synergistic/antagonistic effects with multiple metals or multiple plants, and interaction with any sediments. The biomass can be grown separately from the waste stream and the harvested material can be placed in the waste-removal stream, with no concern for keeping it healthy and alive, or it can be grown directly in the waste stream, if the waste-stream conditions are not too harsh. Living biomass will replace itself over time if provided the proper conditions for growth. Dead biomass can be used in waste streams too harsh for the biomass to live in. It is noted that there may be no need to kill, dry, or cure the biomass before using its removal potential. Particle size

of the biomass does not appear to be an important variable *(5)*. Sediment interaction may be especially important for emergent plants—those with roots in the sediment—where they may help remove metal from sediments, but may cause problems when harvested. Synergistic/antagonistic effects relate to how the biomass interacts with other biomass and how it interacts with multiple species of metals in the system. Sometimes multiple component systems can do more together than the individual parts, sometimes less. These effects are difficult to predict without experimental testing.

4. The preparation of standards of low concentration metals may be significantly altered by adsorption onto the glass walls of labware. All standard solutions should be acidified, as discussed in standard methods *(40)*. A typical solution for standards is 2% by weight nitric acid.

5. The growth and care of the plant biomass may be the most difficult part of this research. All plants need light, nutrients, air, and space. Water quality may be especially important for some species. If harvesting from a local source, measure and test as many water quality parameters as possible. If possible, find someone who has experience with the particular class of plants. In terms of ease of care, the *Lemna* species are very robust and capable of handling some neglect. Macroalga can also be easy to care for, though they need constant attention and may quickly die when neglected for even short times (days). Microalga are very easy to grow but are extremely difficult to harvest (filter) from a solution. They also tend to clog any pumps, airlines, or other water treatment equipment. They are extremely unpopular with waste water-treatment personnel.

6. There is no one best plant and several species may be used either together or individually. True removal occurs when the plant is harvested from the solution. The metal will have been concentrated by 100- to 10^6-fold. The plant mass may need to be treated as a hazardous waste, though it may have much less volume than a chemical precipitation sludge, especially if the biomass volume can be further reduced by composting or air-drying. For single metal species removal, some consideration may be given to recovery of the metal *(42,43)*.

7. Have the greatest concentration measurement frequency early in the run, as it is at the early times that the most significant changes in concentration occur. I recommend samples at times of 0, 2, 4, 8, 12, and 24 h, and then once per day afterwards. It is only necessary to run until you determine what the equilibrium concentration is. This may occur between 48 and 168 h.

8. Always provide replication of all measurements and all experiments. Unreplicated data is worthless.

9. Use the control trials to ensure that there are no additional sources or sinks for the metals or biomass. These surprises can cost significant time when found late in the experimental trials.

References

1. Muramoto, S. and Oki, Y. (1993) Removal of some heavy metals from polluted water by water hyacinth. *Bull. Envir. Contam. Toxic* **30,** 170–177.

2. Axtell, N. A., Sternberg, S. P. K., and Claussen, K. (2003) Lead and nickel removal using microspora and *Lemna minor*. *Bioresource Technol.* **89,** 41–48.
3. Sobhan, R. (1997) *Cadmium Removal Using Living Aquatic Plants*. MS thesis, University of North Dakota, Grand Forks, ND.
4. Haq, N. (1998) *In-situ Bioremediation of Aqueous Lead and Cadmium Using Plants*. MS thesis, University of North Dakota, Grand Forks, ND.
5. Dorn, R. (1998) *Cadmium Removal Using* Chladophora *in a Flow Reactor*. MS thesis, University of North Dakota, Grand Forks, ND.
6. Gardea-Torresdey, J. L., Gonzalez, J. H., Tiemann, K. J., and Rodriguez, O. (1998) Biosorption of cadmium, chromium, lead, and zinc by biomass of *Medicago sativa* (Alfalfa). *J. Haz. Mat.* **57,** 29–39.
7. Kratachvil, D. and Volesky, B. (1998) Advances in the biosorption of heavy metals. *Trends Biotechnol.* **16,** 291–300.
8. Adou, C. (1999) *Bioremediation of Zinc and Nickel Using Living Aquatic Plants*. MS thesis, University of North Dakota, Grand Forks, ND.
9. Roditi, X., Hudson, A., Fisher, X., Nicholas, S., Sanudo-Wilhelmy, and Sergio A. (2000) Field testing a metal bioaccumulation model for zebra mussels. *Environ. Sci. Technol.* **34,** 2817–2825.
10. Omar, H. H. (2002) Bioremoval of zinc ions by *Scenedesmus obliquus* and *Scenedesmus quadricauda* and its effect on growth and metabolism. *Internat. Biodet. Biodeg.* **50,** 95–100.
11. Dursun, A. Y., Uslu, G., Tepe, O., Cuci, Y., and Ekiz, H. I. (2003) A comparative investigation on the bioaccumulation of heavy metal ions by growing *Rhizopus arrhizus* and *Aspergillus niger*. *Biochem. Eng. J.* **15,** 87–92.
12. Kim, I.-S., Kang, K.-H, Johnson-Green, P., and Lee, E.-J. (2003) Investigation of heavy metal accumulation in *Polygonum thunbergii* for phytoextraction. *Environ. Pollut.* **126,** 235–243.
13. Gupta, M. and Chandra, P. (1998) Bioaccumulation and toxicity of mercury in rooted-submerged macrophyte *Vallisneria spiralis*. *Environ. Pollut.* **103,** 327–332.
14. Theegala, C. S., Robertson, C., Carriere, P. E., and Suleiman, A. A. (2001) Phytoremediation potential and toxicity of barium to three freshwater microalgae: *Scenedesmus subspicatus, Selenastrum capricorntum,* and *Nannochloropsis* sp. *Prac. Per. Haz. Tox. Radioac. Waste Manag.* **5,** 194–202.
15. Foster, P. (1976) Concentrations and concentration factors of heavy metals in brown algae. *Environ. Pollut.* **10,** 45–53.
16. Wang, H. K. and Wood, J. M. (1984) Bioaccumulation of nickel by algae. *Environ. Sci. Technol.* **18,** 106–109.
17. Maine, M. A., Sune, N.L., and Lagger, S.C. (2004) Chromium bioaccumulation: comparison of the capacity of two floating aquatic macrophytes. *Water Res.* **38,** 1494–1501.
18. Konhauser, K. O. and Fyfe, W. S. (1991) Biogeochemical cycling of metals on freshwater algae from Manaus and Carajas, Brazil. *Biorecovery* 595–608.
19. Roy, D., Greenlaw, P. N., and Shane, B. S. (1992) Adsorption of heavy metals by green algae. *J. Environ. Sci. Health,* **A28,** 37–50.

20. Chen, C. Y. and Folt, C. L. (2000) Bioaccumulation and diminution of arsenic and lead in freshwater food web. *Environ. Sci. Technol.* **34,** 3878–3884.
21. Saeed, N. and Muhammed, I. (2003) Bioremoval of cadmium from aqueous solution by black gram husk (*Cicer arietinum*). *Water Res.* **37,** 3472–3480.
22. United States Environmental Protection Agency (1990) *Removal and recovery of metal ions from groundwater. Superfund innovative technology evaluation.* EPA/540/S5-90/005 and EPA/540/F-92/003.
23. Ngo, V. (1995) Lemna solves algae problems in Ashland Chemical Polishing Ponds, Lemna USA, Inc. *Treater's Digest,* **3**.
24. Dirilgen, N. (1998) Effects of pH and chelator EDTA on Cr toxicity and accumulation in *Lemna minor*. *Chemosphere* **37,** 771–783.
25. Remoudaki, E., Hatzikioseyian, A., Kousi, P., and Tsezos, M. (2003) The mechanism of metals precipitation by biologically generated alkalinity in biofilm reactors. *Water Res.* **37,** 3843–3854.
26. Aksu, Z., Ozer, D., Ozer, A., Kutsal, T., and Caglar, A. (1998) Investigation of the column performance of cadmium (II) biosorption by *Cladophora crispate* flocs in a packed bed. *Separ. Sci. Technol.* **33,** 667–682.
27. Sobhan, R. and Sternberg, S. P. K. (1999) Cadmium removal using *Cladophora*. *J. Environ. Sci. Health* **A34,** 53–72.
28. Sternberg, S. P. K. and Dorn, R. (2002) Cadmium removal using *Cladophora* in batch, semi-batch and flow reactors. *Bioresource Technol.* **81,** 249–255.
29. Fogler, H. S. (1999) *Elements of Chemical Reaction Engineering, 3rd ed.* Prentice Hall, Upper Saddle River, NJ.
30. Incropera, F. P. and Dewitt, D. P. (1996) *Fundamentals of Heat and Mass Transfer, 4th ed.* Wiley, New York, NY.
31. Kuhn, D. J. (1969) The duckweed. *The American Biology Teacher* **31,** 328–329.
32. Hillman, W. S. and Culley, D. D., Jr. (1978) The uses of duckweed. *American Scientist* **66,** 442–451.
33. Dodds, W. K. and Gudder, D. A. (1992) The ecology of *Cladophora*. *J. Phycol.* **28,** 415–427.
34. Rahmani, N. and Sternberg, S. P. K. (1999) Bioremoval of lead using *Lemna minor*. *Bioresource Technol.* **70,** 225–230.
35. Murray, J. D. (1989) *Mathematical Biology*. Springer-Verlag, New York, NY.
36. Eaton, A. D., Clesceri, L. S., and Greenberg, A. E. (1995) *Standard Methods for the Examination of Water and Wastewater,* American Public Health Association, Washington, DC.
37. Keppel, G. (1982) *Design and Analysis: A Researcher's Handbook*. Prentice-Hall, Englewood Cliffs, NJ.
38. Lawson, J. and Erjavec, J. (1998) *Modern Statistics for Engineering and Quality Improvement*. University of North Dakota, Grand Forks, ND.
39. Berger, P. D. and Maurer, R. E. (2002) *Experimental Design*. Duxbury, Thomas Learning, Inc., Belmont CA.
40. Sall, J., Lehman, A., and Creighton, L. (2000) *JMP Start Statistics, 2000 2nd ed.* Sas Institute Inc., Brooks Cole.

41. Ryan B. F. and Joiner B. L. (2000) *MINITAB; Handbook 4th ed.* Brooks Cole.
42. Tsezos, M. (1984) Recovery of uranium from biological adsorbents: desorption equilibrium. *Biotechnol. Bioeng.* **26,** 973–981.
43. Sakaguchi, T., Nakajima, A., Honma, S., Aoyama, M., and Kasai, A. (1996) Recovery and removal of uranium by hardwood barks. *Resource Environ. Biotechnol.* **1,** 129–143.

Appendix: List of Symbols

A	Pre-exponential frequency factor
C_{ads}	Concentration of metal in solution to be adsorbed
C_{eq}	Equilibrium concentration of metal in solution to be adsorbed
C_I	Initial concentration
CF	Concentration factor
E_a	Activation energy
k_1	Model 1 parameter
k_{2A}	Model 2 parameter
k_{2B}	Model 2 parameter
k_{bio}	Biomass growth rate parameter
M_{ads}	Mass of metal in solution to be adsorbed
M_x	Mass of
M_0	Initial amount of biomass
M_{bio}	Mass of metal adsorbed onto biomass
$M_{biomass}$	Mass of biomass
n	Model 1 parameter
$q_{biomass}$	Concentration of metal on biomass
r_{ads}	Rate of adsorption
r_{bio}	Rate of adsorption per unit biomass
r_x	Rate of biomass growth
R	Ideal gas constant
SSD	Sum of squares difference
t	Time
T	Absolute temperature
V	Volume
ρ	Solution density

17

Near-Infrared Reflectance Spectroscopy
Methodology and Potential for Predicting Trace Elements in Plants

Rafael Font, Mercedes del Río-Celestino, and Antonio de Haro-Bailón

Summary

Near-infrared spectroscopy (NIRS) has been applied for decades to the analysis of agri-food products, and in recent years its use has been extended to the determination of mineral species and trace elements in organic and inorganic matrices. The near-infrared region (NIR) spectrum contains physical and chemical information of the product being analyzed. The spectral information has its origin in the different vibrational modes of the molecules caused by their interaction with the electromagnetic radiation absorbed at wavelengths between 750 and 2500 nm. The use of chemometrics allows the relevant information contained in the NIR spectra to be extracted to develop calibration models that permit the prediction of the composition of unknown samples. The technique is rapid and, in contrast to the standard techniques of analysis, can be performed at a low analytical cost and without using chemicals. In addition, those error sources related with laboratory analysis are avoided. The control of those sources of error specific to the NIR analysis leads to equations of high accuracy and precision. The application of NIRS to the determination of arsenic, lead, copper, and zinc in wild and cultivated plant species has revealed its potential in the screening of these elements for phytoremediation purposes.

Key Words: Arsenic; chemometrics; near-infrared spectroscopy; NIRS; phytoremediation; trace elements.

1. Introduction

Although Herschel discovered light in the near-infrared region (NIR) as early as 1800, acceptance of the NIR region of the electromagnetic spectrum as a valuable tool can be attributed to two researchers. First, Karl Norris in the 1960s who worked on instruments that could record NIR spectra and applied multivariate treatment of the spectra to determine major plant components *(1)*.

Second, Phil Williams in the 1970s, who recognized the potential of the technology to segregate wheat grain according to protein content *(2)*.

Since then, the development of equipment featuring improved electronic and optical components, as well as the advent of computers capable of effectively processing the information contained in NIR spectra, has facilitated the expansion of this technique in an increasing number of fields *(3)*. In recent years, the use of near-infrared spectroscopy (NIRS) for the assessment of mineral composition and trace elements in plant and animal tissues, has opened new horizons to NIR spectroscopists dealing with the application of this technique to the agricultural, medical, food safety and environmental fields. Since 1998, the Department of Agronomy and Plant Breeding (DAPB) at the Institute for Sustainable Agriculture (IAS, CSIC, Córdoba, Spain), has been applying NIRS in the framework of different international, national, and regional research projects concerning phytoremediation.

In this chapter, we describe some aspects of the chemical principles of NIR absorption and the protocol we use to complete the whole analytical procedure for NIRS analysis, with comments related to trace metal analysis. Major procedures involved in the protocol include sampling and sample presentation to the NIR instrument, selection of representative groups of samples as calibration and validation sets, development and validation of the calibration equation, and additional considerations concerning the control of humidity and temperature in the laboratory, which represent two major sources of error in NIR analysis (**Fig. 1**).

When molecules are irradiated with an external source of energy they acquire the potential for energy changes. The electromagnetic spectrum consists of energy vibrations ranging from wavelengths several meters in length to less than 10^{-2} nm. The visible region ranges from about 300 to 750 nm, the NIR region between 750 and 2600 nm, and the mid-infrared region lies from 2600 to 25000 nm *(4)*. NIR spectra result from light absorption by organic molecules in the electromagnetic segment previously mentioned. All the absorption bands are the results of overtones or combinations of overtones originating in the mid-infrared region. The majority of the overtone bands are a consequence of the stretching modes of the R-H groups (O-H, C-H, N-H, S-H, and so on) *(5)*. The vibration of molecules responds to the harmonic oscillator model, by which the energy of different, equally spaced levels is determined by the equation

$$E_{vib} = \left(\upsilon + \frac{1}{2}\right)\frac{h}{2\pi}\sqrt{\frac{k}{\mu}} \qquad (1)$$

where is υ the vibrational quantum number, h is the Planck constant, k the force constant, and μ the reduced mass of the bonding atoms. Most fundamental

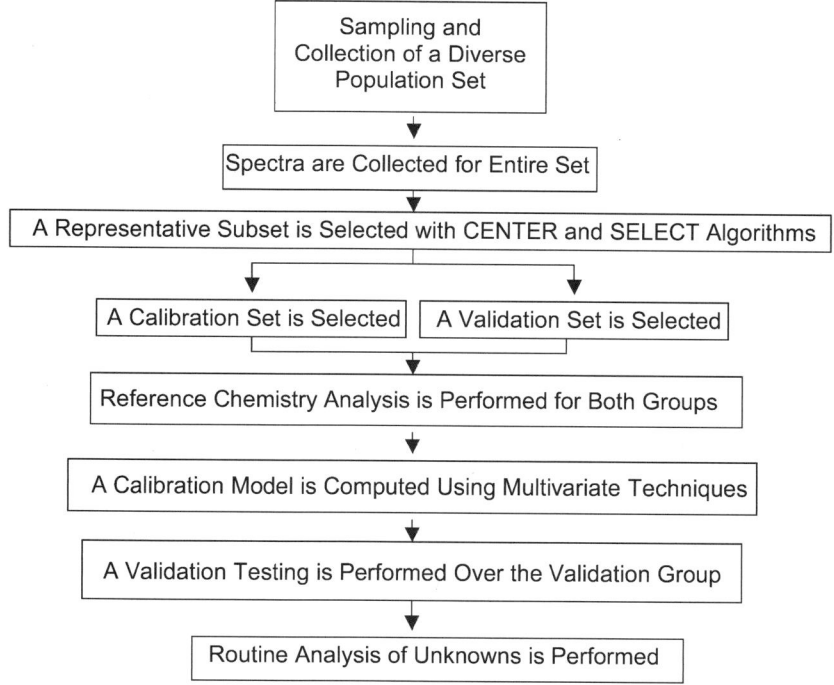

Fig. 1. Near-infrared spectroscopy global procedure.

molecular vibrations occur at frequencies outside the NIR region. The fact that any overtones are seen is evidence that the R-H vibration is not strictly harmonic. Thus, the anharmonicity increases with the amplitude of atomic oscillation. The anharmonicity can result in transitions between vibrational energy states where $\Delta\upsilon = \pm2, \pm3$, and so on. These transitions between noncontiguous vibrational states yield absorption bands known as *overtones*. As the number (n) of atoms in an absorbing molecule increases, the number of fundamental vibrations increases at the rate of $3n–6$ *(4)*. Furthermore, the fundamentals may interact with overtones, and overtones with each other, to produce combinations bands. NIR combination bands take place between 1900 and 2500 nm. The intensity of the absorption of the NIR bands is related to the anharmonicity. On the other hand, hydrogen is the lightest atom and thus, exhibits the largest vibrations and anharmonicity. These facts make most bands in the NIR region correspond to bonds containing H in combination with other atoms such as C, N, or S.

No single technique can solve all analytical problems, but NIRS has many advantages over chemical and other instrumental methods *(6)*. However, NIRS has a number of limitations.

1.1. Advantages

1. The low absorptivities of absorption bands are compatible with moderately concentrated samples and longer path lengths than those used in the mid-infrared region.
2. It is a nondestructive technique because sample preparation is avoided.
3. The previously mentioned factors result in NIRS analysis being very simple to perform so that there are few operator-induced sources of error.
4. Measurement and result delivery is fast.
5. It is an environmentally friendly analytical technique, as no chemicals are used during the process.
6. Many components of the material being analyzed can be determined simultaneously from a unique spectrum, and not only chemical but also physical parameters can be determined on the sample.
7. The accuracy of an NIR analysis is comparable to that of the chemistry reference method, and its precision is usually high because the avoidance of sample treatment.

1.2. Disadvantages

1. The complexity of the NIR spectrum requires chemometric techniques to extract the relevant information to the component being measured.
2. In constructing the calibration models the whole physical and chemical variability predicted to be present in the population must be added to the calibration set of samples. This implies continuous addition of new samples to the original set to encompass all variations.
3. No specific methodology for transferring calibrations between instruments has gained widespread acceptance in recent years.

The analysis of mineral elements in plants by NIRS was first documented by Shenk et al. in forages *(7,8)*. In recent years, different authors have explored the potential of this technique for predicting the mineral composition of agricultural products *(9–12)*, as well as for monitoring trace elements in plants concerning environmental studies *(13,14)*. Trace elements and minerals exist in the plant as organic complexes, chelates with other minerals, salts, and in ionic forms. Elements are present in relatively low concentrations in the plant tissues, and thus, the NIR absorptions owing to the presence of these elements are difficult to detect directly *(15)*. In addition, the two main types of molecular motion are caused by vibrational and rotational energy transitions *(4)*, which are the basis for NIRS spectral bands. For this reason mineral species are not expected to absorb in the NIR region, thus inorganic salts are transparent to NIRS radiation. However, NIRS can determine some cation concentrations because of their association with organic or hydrated inorganic molecules *(16)*. In addition, the inorganic forms of As, as well as other trace elements are responsible for large disturbances in the plant physiology when they enter the cell, affecting the metabolism of protein *(17)*, lipids *(18)*, starch *(19)*, or photosynthetic pigments *(20)*. These disturbances can be detected and measured by NIRS.

2. Materials

The development of NIRS has followed advances in the enabling technologies of optics, electronics, computer hardware and software, and, especially, chemometrics. Optical technology has advanced since 1945 from photographic recording of spectra from a spot galvanometer through holographic diffraction gratings and interference filters to diode array, acousto-optic tuneable monochromators and fiber optics *(6)* (*see* **Note 1**).

3. Methods
3.1. Sampling, Samples, and Sample Presentation

More than 30 factors affecting the accuracy and precision of an NIR analysis are attributable to sampling, samples, or sample presentation. Among them are those related to the type of material (*see* **Note 2**), size of the sample (*see* **Note 3**), storage, composition and physical texture (*see* **Note 4**), grinding, instrument of analysis, cell type, and so on. These factors can lead to the incorporation of unreal and inconsistent results if they are ignored. Management of most of these factors have been discussed by Williams *(21)*.

3.2. NIRS Analysis: Control of Humidity and Temperature in the Laboratory

The moisture status of a sample submitted to an NIR analysis is important at the time of the scanning because the scatter coefficient is proportional to the refractive index of the particles of the surrounding medium. Thus, variations in moisture status are accompanied by a shift in the spectra along the log 1/reflectance axis *(22)*. In addition to moisture, temperature of a sample at the time of analysis is recognized as another source of error if it differs widely from other samples in the calibration *(23)* (*see* **Note 5**).

3.3. Selecting Representative Sets of Samples as Calibration and Validation Sets

3.3.1. The Use of the Mahalanobis Distance in the Selection of Representative Sets of Samples

With proper calibrations, NIRS is an accurate, rapid, and cost-effective analytical method. Often, a large number of spectra are available, but obtaining chemistry reference values for all samples would be very expensive. Thus, calibration costs and the labor input necessary for developing the predictive mathematical models can be substantially reduced if an appropriate method of selection of representative samples is used. Protocols for calibration procedures include the use of algorithms. The algorithms called CENTER and SELECT provided by the software WINISI II (Infrasoft International, LLC, Port Matilda, PA) are an excellent tool for this purpose *(24,25)* (*see* **Note 6**).

3.3.2. Splitting Samples for Calibration and Validation: "Spectral" or "Chemical" Features

As has been previously stated, the successful performance of a calibration equation relies heavily in the proper selection of the samples. In the same way, the samples used for validating the equation must be similar to those used during the calibration process. Usually, spectra are sorted from lowest to highest reference values and divided into calibration and validation sets. Sample sets may be assembled in ratios of calibration:validation spectra of 3:1, 2:1, or 1:1 *(26)* (*see* **Note 7**).

Another strategy is to determine first the distance of Mahalanobis (H) from each spectrum to the mean spectrum of the population. Shenk and Westerhaus *(25)* proposed that only those samples whose H value was lower than three should be considered, rejecting extreme values. Once the boundary of the spectral population is established, spectra can be sorted for calibration and validation (*see* **Note 8**).

3.4. Developing and Validating NIR Equations

3.4.1. The Use of Derivatives and Other Algorithms in Developing the Calibration Model

Transformation of the original spectra to their first or second derivative prior to calibration is usually used by NIR researchers because this supposes an approach to solve those problems associated to overlapping peaks and base-line correction *(27)*. Spectral derivative reduces variance caused by particle size and, as a result, differences between spectra are more likely to be because of chemical rather than physical differences *(21)*. Most interesting is the second derivative spectrum of the log (1/reflectance) (*see* **Note 9**). This shows peaks and valleys corresponding to the points of maximum curvature in the original spectrum, each valley having a correspondence to each peak in the original. Second derivative transformation produces a clear separation between peaks, which overlap substantially in the original *(6)*. Additional algorithms for baseline correction as a result of differences in particle size or path length variation among samples are the standard normal variate transformation and detrending *(28)*.

3.4.2. Regressing Spectra Against Chemistry Reference Values

Derived or nonderived spectra are regressed against the chemistry reference values for the analyte studied to develop the final predictive model calibration equation. Alternatives to wavelength selection methods, as multiple regression or step-up regression, which uses selected wavelengths to develop the calibration model, are the full-spectrum methods, mainly principal component regression and partial least squares (PLS) regression (*see* **Note 10**). Both methods construct factors from the original spectral data to reduce the quantity of spectral data to avoid over-fitting problems without discarding any useful information. Further discussion about full-spectrum methods can be found in Martens and Naes *(29)* and Osborne et al. *(6)*. The calibration equation thus obtained has to be validated before it is applied on unknown samples.

3.4.3. External Validation vs Cross-Validation

An external validation implies the use of samples not used in the calibration process. This approach can be carried out if a sufficient number of samples exist, thus allowing the splitting of the whole set of samples into calibration and validation sets. In using this validation strategy, both groups must be similar in chemical composition and physical features (*see* **Note 7**). But often, the number of samples is low, and then an alternative to the method is to validate the equation using cross-validation.

Cross-validation is an internal validation method that, like the external validation approach, seeks to validate the calibration model on independent test data, but it does not waste data for testing only, as occurs in external validation. This procedure is useful because all available chemical analyses for all individuals can be used to determine the calibration model without the need to maintain separate validation and calibration sets. The method is carried out by splitting the calibration set into M segments and then calibrating M times, each time testing about a $(1/M)$ part of the calibration set *(29)* (*see* **Note 11**).

3.4.4. Some Statistics and Their Significance in NIR Spectroscopy for Evaluating the Performance of an Equation

Different statistical tests can be performed to establish the accuracy and precision of an NIR equation, which have been widely discussed by different authors *(29,30)*. Statistics of common use in validating an NIR equation are the coefficient of determination (r^2) *(31)* (**Eq. 2**) and the standard deviation (S.D.) to standard error of performance (SEP) ratio, which is known as RPD *(26)* (**Eq. 3**).

$$r^2 = \left[\sum_{i=1}^{n} (\hat{y} - \bar{y})^2 \right] \left[\sum_{i=1}^{n} (y_i - \bar{y})^2 \right]^{-1} \qquad (2)$$

where $\hat{y}$ = NIR measured value; $\bar{y}$ = mean "y" value for all samples; y_i = lab reference value for the ith sample.

$$RPD = S.D. \left\langle \left[\sum_{i=1}^{n} (y_i - \hat{y}_i)^2 \right] (N-1)^{-1} \right]^{1/2} \right\rangle^{-1} \qquad (3)$$

where y_i = lab reference value for the ith sample; $\hat{y}$ = NIR measured value; N= number of samples, S.D. = standard deviation.

The r^2 statistic is the total explained variation. This statistic allows us to determine the amount of variation in the data that is adequately modeled by the calibration equation as a total fraction of 1.0. *(30)* (*see* **Note 12**).

The statistic RPD, is used by NIR analysts as a guide to the usefulness of a calibration. It includes the SEP in its formula, which is defined as the standard deviation of the residuals because of differences between the reference chemistry values and the NIR predicted values for samples outside the calibration set. The variability of the NIR determinations is indicated by the magnitude of deviations from the SEP *(32)*. If the SEP value is similar to the S.D. of the reference values, the instrument is not predicting the standard values any better than the S.D. of the original data. Different RPD values have been assigned in literature depending on the parameter being analyzed to determine the predictive potential of an equation (*see* **Note 13**).

4. Notes

1. To collect near-infrared spectra we use at DAPB an NIRS spectrometer model 6500 (Foss-NIRSystems, Inc., Silver Spring, MD) equipped with a transport module. The monochromator 6500 consists of a tungsten bulb and a rapid-scanning holographic grating with detectors positioned for transmission or reflectance measurements. To produce a reflectance spectrum, a ceramic standard is placed in the radiant beam and the diffusely reflected energy is measured at each wavelength. The actual absorbance of the ceramic is very consistent across wavelengths.
2. NIR calibration is sometimes performed with different materials to get a more robust equation at the time that prediction is simplified, which is known as multiproduct calibration *(33)*. Most times, this procedure yields lower prediction ability equations than single-product calibrations do, although the opposite has been demonstrated under determined circumstances *(34)*. In our experience with determination of trace metals in plants, single-product calibrations always exhibit higher accuracy than multiproduct calibrations.
3. Usually older plants show thicker cell walls than younger. This fact can increase the difficulty of grinding the plant tissues to a similar particle size than that of the younger individuals. If the plant sample is not ground to a similar particle size than the rest of the samples in the calibration, then it introduces an additional noise in the analysis. Software applications allow the use of derivatives and/or other algorithms *(28)* to reduce the base-line offset produced by scattering effects, although this correction is not complete.
4. Plants are stored in a refrigerator at –18°C until NIR analysis. For As determination, samples are freeze-dried and ground. For Pb, Zn, and Cu analysis, samples are oven-dried at 70°C and ground before NIR analysis.
5. To avoid variations in moisture and temperature in plant samples during NIR analysis at the DAPB, the monochromator is located at a humidity–temperature-

controlled environment. These parameters are maintained constant at 60–64% and 24°C throughout, respectively.

6. The algorithm CENTER establishes population boundaries with a standardized H distance, which can be defined. The criterion of establishing the population boundary at an $H = 3$ *(25)* for most agricultural applications performed over seed samples is valid, in our opinion, for plant shoot. The use of the previous algorithm together with SELECT, which allows the elimination of neighbor samples (an $H < 0.6$ is widely accepted), provides an acceptable number of samples for calibration.

7. In performing an external validation, we have adopted the general criterion of a ratio of calibration:validation of 2:1, as it has been efficient through different materials and components.

8. Different authors *(14,15)* have demonstrated that major plant components (i.e., fiber, chlorophyll, and so on) correlate well with some mineral species and trace elements in plant tissues, and that NIRS uses such correlations to develop the calibration equations for these components, as it can be seen in the loading plots for the first three terms of the equation for As in *Amaranthus blitoides* (**Fig. 2**), where absorption by chlorophyll at 674 nm, as well as lipids, protein, and cell wall components participated actively in developing the equation. Thus, when samples for calibration and validation groups are selected on the basis of their spectral features (H distance), usually both groups show the whole physical and chemical diversity of the population for the trace element studied.

9. We usually test different derivatives on the original 1/reflectance values. In determining As in *A. blitoides* *(14)* the first derivative transformation yielded slightly higher prediction ability than the second derivative. However, the second derivative gave the highest accuracy in the determination of Pb, Cu, and Zn in *Brassica juncea* plants *(13)*.

10. Through experience we have corroborated the higher performance of the PLS regression as opposed to wavelength-selection methods, concerning trace element determination in plant tissues. In particular, modified PLS regression is often more stable and accurate than the standard PLS algorithm because it standardizes the NIR residuals before the next factor of the equation is calculated.

11. In cross-validation, the number of groups used in validating the equation is dependent on the number of samples in the calibration file. The optimum number of validation groups is selected automatically by the application as a function of the number of samples, although this option can be selected by the user.

12. In our work *(14)*, the values shown by r^2 of the best prediction ability equation for As in the cross-validation indicated good quantitative information *(31)*, meaning that 70% of the As chemical variability in the plants were explained by the model in the cross-validation. In the external validation (**Fig. 3**), the r^2 obtained was characteristic of equations that can be used for a good separation of the samples in the validation set into high, medium, and low As contents. Similar r^2 values to those for As have been obtained in a previous study performed on *B. juncea* plants *(13)*. In that work, equations for Pb and Cu showed r^2 values in cross-validation of 0.65 and 0.60, respectively. Clark et al. *(35)* reported the prediction of sulfur in alfalfa (*Medicago sativa* L.) and tall fescue

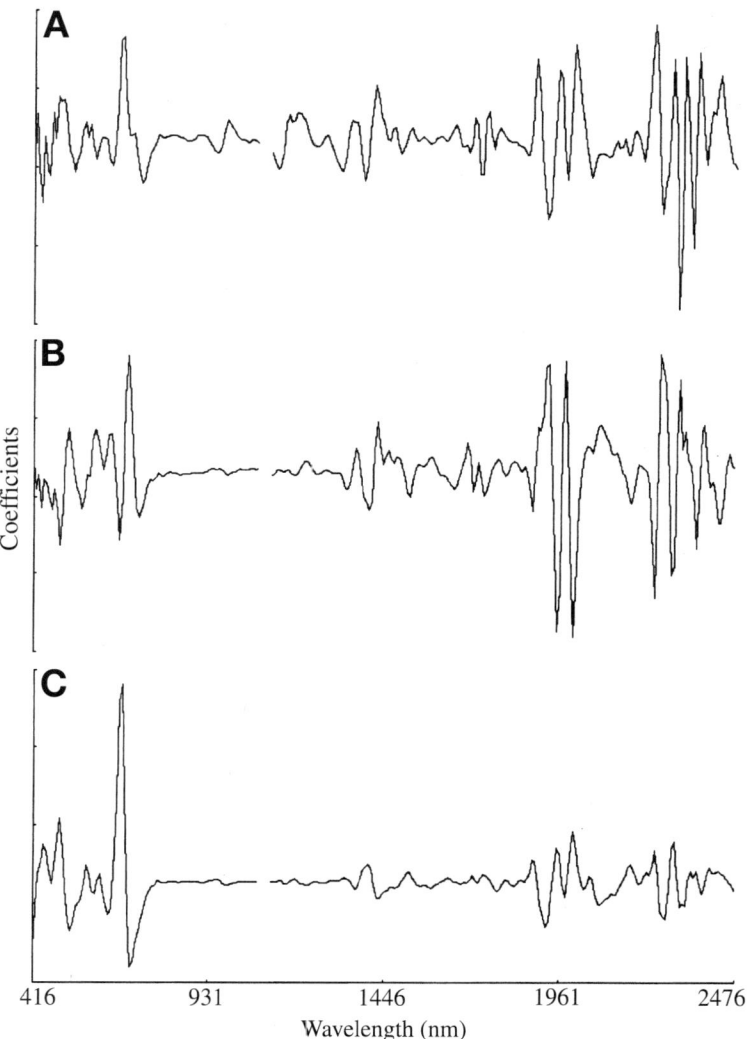

Fig. 2. Modified partial least squares loading spectra for As in the samples of *Amaranthus blitoides* for the second derivative (2, 5, 5, 2; SNV+DT) transformations. **A**, **B**, and **C** correspond to the first, second, and third loadings, respectively.

(*Festuca arundinacea* Schreb.) with similar results to those obtained for As in *A. blitoides*, but lower than those obtained by the same authors in predictions of Al and Li in tall fescue ($r^2 = 0.80$). Shenk and Westerhaus *(15)* also reported predictions of different minerals, obtaining r^2 that ranged from 0.35 (magnesium) to 0.92 (phosphorus).
13. The RPDs obtained by us in the prediction of trace metals in *A. blitoides* and *B. juncea* tissues are of the same order as those previously reported for trace

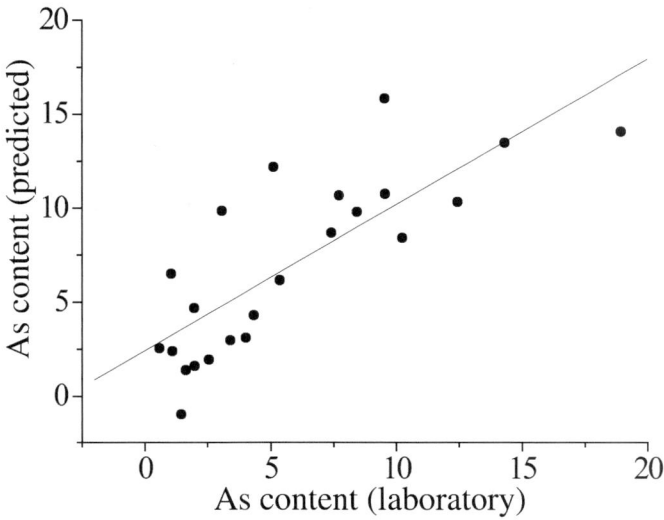

Fig. 3. External validation scatter plot for As (laboratory) vs predicted by near-infrared spectroscopy (milligram/kilogram dry weight) in *Amaranthus blitoides* plants.

elements and minerals in different plant species *(13,14,16,35)*, varying between 1.34 for Zn in *B. juncea*, and 1.72 (Pb in *B. juncea*) and 1.74 (As in *A. blitoides*). In accordance with the RPD values shown by these equations, and considering the limits for RPD recommended by Dunn et al. *(36)*, these equations are acceptable for trace metal prediction in these species.

References

1. Norris, K. H. (1996) History of NIR. *J. Near Infrared Spectrosc.* **4,** 31–37.
2. Williams, P. C. (1995) Near infrared technology in Canada. *NIR News* **6,** 12–13.
3. Blanco, M. and Villarroya, I. (2002) NIR spectroscopy: a rapid-response analytical tool. *Trends Anal. Chem.* **21,** 240–250.
4. Murray, I. and Williams, P. C. (1987) Chemical principles of near infra-red technology. In: *Near-Infrared Technology in the Agricultural and Food Industries,* (Williams, P. C. and Norris, K., eds.), American Association of Cereal Chemists Inc., St. Paul, MN, pp. 17–34.
5. Ciurczak, E. (1992) Principles of near-infrared spectroscopy. In: *Handbook of Near-Infrared Analysis,* (Burns, D. A. and Ciurczak, E. W, eds.), Dekker Inc., New York, pp. 7–11.
6. Osborne, B. G., Fearn, T., and Hindle, P. H. (1993) *Practical NIR Spectroscopy with Applications in Food and Beverage Analysis.* Longman Scientific and Technical, Essex, England.
7. Shenk, J. S., Westerhaus, M. O., and Hoover, M. R. (1979) Analysis of forages by infrared reflectance. *J. Dairy Sci.* **62,** 807–812.

8. Shenk, J. S., Landa, I., Hoover, M. R., and Westerhaus, M. O. (1981) Description and evaluation of a near infrared reflectance spectroscopy computer for forage and grain analysis. *Crop. Sci.* **21,** 355–358.
9. García-Ciudad, A., García-Criado, B., and Ponton San Emeterio, C. (1985) Determination of fluoride in plant samples by a potentiometric method and near-infrared reflectance spectroscopy. *Comm. Soil Sci. Plant Anal.* **16,** 1107–1122.
10. McClure, W. F., Crowell, B., Stanfield, D. L., Mohapatra, S., Morimoto, S., and Batten, G. (2002) Near infrared technology for precision environmental measurements: part 1. Determination of nitrogen in green-and dry-grass tissue. *J. Near Infrared Spectrosc.* **10,** 177–185.
11. Morón, A. and Cozzolino, D. (2002) Determination of macro-elements in alfalfa and white clover by near-infrared reflectance spectroscopy. *J. Agr. Sci.* **139,** 413–423.
12. Sauvage, L., Frank, D., Stearne, J., and Millikan, M. B. (2002) Trace metal studies of selected white wines: an alternative approach. *Anal. Chim. Acta* **458,** 223–230.
13. Font, R., Del Río, M., and De Haro A. (2002) Use of near infrared spectroscopy to evaluate heavy metal content in *Brassica juncea* plants cultivated on the polluted soils of the Guadiamar river area. *Fresen. Environ. Bull.* **11,** 777–781.
14. Font, R., Del Río, M., Vélez, D., Montoro, R., and De Haro A. (2004) Use of near-infrared spectroscopy for determining the total arsenic content in prostrate amaranth. *Sci. Total Environ.* **327,** 93–104.
15. Shenk, J. S., Workman, J. J., Jr., and Westerhaus, M. O. (1992) Application of NIR spectroscopy to agricultural products. In: *Handbook of Near-Infrared Analysis,* (Burns, D. A. and Ciurczak, E. W, eds.), Dekker Inc., New York, pp. 383–431.
16. Clark, D. H., Mayland, H. F., and Lamb, R. C. (1987) Mineral analysis of forages with near infrared reflectance spectroscopy. *Agron. J.* **79,** 485–490.
17. Grill, E., Winnacker, E. L., and Zenk, M. H. (1987) Phytochelatins, a class of heavy-metal-binding peptides from plants, are functionally analogous to metallothioneins. *Proc. Natl. Acad. Sci.* **84,** 439–443.
18. Hartley-Whitaker, J., Ainsworth, G., and Meharg, A. A. (2001) Copper- and arsenate-induced oxidative stress in *Holcus lanatus* L. clones with different sensitivity. *Plant Cell Environ.* **24,** 713–722.
19. Simola, L. K. (1977) The effect of lead, cadmium, arsenate and fluoride ions on the growth and fine structure of *Sphagnum nemoreum* in aseptic culture. *Can. J. Bot.* **90,** 375–405.
20. Meeta-Jain, R. P. and Gadre, R. P. (1997) Effect of As on chlorophyll and protein contents and enzymatic activities in greening maize leaves. *Water Air Soil Pollut.* **93,** 109–115.
21. Williams, P. C. (1992) Samples, sample preparation, and sample selection. In: *Handbook of Near-Infrared Analysis,* (Burns, D. A. and Ciurczak, E. W, eds.), Dekker Inc., New York, pp. 281–315.
22. Williams, P. C. and Norris, K. H. (1983) Effect of mutual interactions on the estimation of protein and moisture in wheat. *Cereal Chem.* **60,** 202–207.
23. Williams, P. C., Norris, K. H., and Zarowski, W. S. (1982) Influence of temperature on estimation of protein and moisture in wheat. *Cereal Chem.* **59,** 473–477.

24. Shenk, J. S. and Westerhaus, M. (1991) Population definition, sample selection, and calibration procedures for near infrared reflectance spectroscopy. *Crop Sci.* **31**, 469–474.
25. Shenk, J. S. and Westerhaus, M. (1991) Population structuring of near infrared spectra and modified partial least squares regression. *Crop Sci.* **31**, 1548–1555.
26. Williams, P. C. and Sobering, D. C. (1996) How do we do it: a brief summary of the methods we use in developing near infrared calibrations. In: *Near Infrared Spectroscopy: The Future Waves,* (Davies, A. M. C. and Williams, P. C., eds.), Nir Publications, Chichester, UK, pp. 185–188.
27. Giese, A. T. and French, C. S. (1955) The analysis of overlapping spectral absorption bands by derivative spectrophotometry. *Appl. Spectrophot.* **9**, 78–96.
28. Barnes, R. J., Dhanoa, M. S., and Lister, S. J. (1989) Standard normal variate transformation and de-trending of near-infrared diffuse reflectance spectra. *Appl. Spectrosc.* **43**, 772–777.
29. Martens, H. and Naes, T. (1989) *Multivariate Calibration.* John Wiley and Sons, New York, NY.
30. Workman, J. J., Jr. (1992) Nir spectroscopy calibration basics. In: *Handbook of Near-Infrared Analysis,* (Burns, D. A. and Ciurczak, E. W., eds.), Dekker Inc., New York, NY, pp. 247–280.
31. Shenk, J. S. and Westerhaus, M. O. (1996) Calibration the ISI way. In: *Near Infrared Spectroscopy: The Future Waves,* (Davies, A. M. C. and Williams, P. C., eds.), Nir Publications, Chichester, UK, pp. 198–202.
32. Williams, P. C. (1987) Variables affecting near-infrared reflectance spectroscopic analysis. In: *Near-Infrared Technology in the Agricultural and Food Industries,* (Williams, P. C. and Norris, K., eds.), American Association of Cereal Chemists, Inc., St. Paul, MN, pp. 143–167.
33. Shenk, J. S. and Westerhaus, M. O. (1993) Near infrared reflectance analysis with single- and multiproduct calibrations. *Crop Sci.* **33**, 582–584.
34. Font, R., Del Río, M., Fernández, J. M., and De Haro, A. (2003). Acid detergent fiber analysis in oilseed *Brassicas* by near-infrared spectroscopy. *J. Agr. Food Chem.* **51**, 2917–2922.
35. Clark, D. H., Cary, E. E., and Mayland, H. F. (1989) Analysis of trace elements in forages by near infrared reflectance spectroscopy. *Agron. J.* **81**, 91–95.
36. Dunn, B. W., Beecher, H. G., Batten, G. D., and Ciavarella, S. (2002) The potential of near infrared reflectance spectroscopy for soil analysis - a case study from the Riverine Plain of south-eastern Australia. *Aust. J. Exp. Agr.* **42**, 607–611.

III

CURRENT RESEARCH TOPICS IN PHYTOREMEDIATION

18

Using Hydroponic Bioreactors to Assess Phytoremediation Potential of Perchlorate

Valentine Nzengung

Summary

Studies on plant-mediated decontamination of polluted waters commonly rely on hydroponic bioreactor systems for the rapid screening of different plant species and the investigation of the phytoremediation mechanisms. This chapter describes the use of ebb-and-flow and sealed root-zone hydroponic bioreactors (some mounted with dissolved oxygen, pH, and E_H probes) to investigate, under aerobic and anaerobic rhizosphere conditions, respectively, the phytoremediation potential of the new and emergent contaminant perchlorate. The experimental data show slower kinetics and higher fraction uptake and phytodegradation of perchlorate by plants grown under predominantly aerobic root-zone conditions. Meanwhile, for plants grown in sealed (predominantly anaerobic) root-zone hydroponic systems, the bulk solution chemistry is influenced by the plant photosynthesis and respiration cycles, and both phytodegradation and rhizodegradation are the predominant phytoprocesses. In ongoing research, various approaches are being developed to minimize the undesired uptake and slow phytodegradation process and enhance the favorable and rapid rhizodegradation process. Additionally, planted soil bioreactor experiments that simulate phytoremediation of perchlorate under natural field conditions are described.

Key Words: Ebb-and-flow; rhizosphere; rhizodegradation; phytodegradation; phytoremediation; perchlorate; hydroponic bioreactors; nitrate.

1. Introduction

Basic research on the development of phytoremediation technologies has utilized the centuries old practice of growing plants in nutrient solutions (water-containing fertilizers), with or without the use of a solid medium to provide mechanical support, called hydroponics (hydro = water, ponos = labor, i.e., working water). Different types of planted bioreactors are used in laboratory and bench-scale tests to screen plants suitable for phytoremediation and to investigate the mechanisms by which plants remove contaminants from air, soil, and water. The mechanisms of phytoremediation identified in the published literature include rhizodegradation, rhizostabilization, phytodegradation, phytovolatilization, and phytoaccumulation *(1)*. The choice of reactor is determined by the physicochemical properties of the contaminant of concern (COC), plant physiology, and the phytoprocesses being studied. For volatile compounds, the challenge is to design a bioreactor to grow the test plants under less stressful conditions while minimizing losses of the probe compound by volatilization. For highly soluble, nonvolatile compounds such as perchlorate, nitrate, sulfate, phosphate, and so on, the design may be fairly simple, except when there is a need to measure the gases produced in the root zone.

The hydroponic technique offers a number of advantages in plant screening tests: (1) bioavailability of the COC is greatly increased; (2) there is better control of the environment (i.e., root zone, nutrient feeding, light, temperature, humidity, and air composition); and (3) a simpler root-zone environment is created to study plant-mediated processes without the complexity introduced by soil minerals. The choice of nutrient solution used in hydroponic growing of plants during phytoremediation is important to the success of the technology because the reaction of plants to good or poor nutrition is extremely fast *(2)*. Additionally, high concentrations of competing terminal electron acceptors in the root-zone solution may inhibit phytoreduction of some oxidized COCs. For example, high concentrations of oxygen and/or nitrate nitrogen in nutrient solution inhibit the reduction of perchlorate in plant rhizosphere *(3,4)*. The mortality rate is higher in hydroponic systems than in soil bioreactors, as the plant response to unfavorable growth conditions is rapid—nutrient and pH changes being the most common unfavorable growth conditions observed. The use of probes for regular measurement is advisable.

This chapter describes the use of different designs of hydroponic bioreactors for perchlorate phytoremediation screening tests and for the evaluation of the influences of aerobic and anaerobic conditions created in the bulk root-zone environment. Aerobic conditions were created and maintained in an ebb-and-flow bioreactor, whereas anaerobic conditions were created in sealed root-zone bioreactors. Some of the sealed bioreactors were equipped with dissolved oxygen (DO), pH, and E_H probes to monitor changes in the solution chemistry during phytoremediation. Planted soil bioreactor experiments that simulate phytoremediation performance under natural environmental conditions are also described.

2. Ebb-and-Flow Bioreactor Experiments

The ebb-and-flow bioreactor provides the flexibility of varying the residence time and composition of the water undergoing treatment in the rhizosphere.

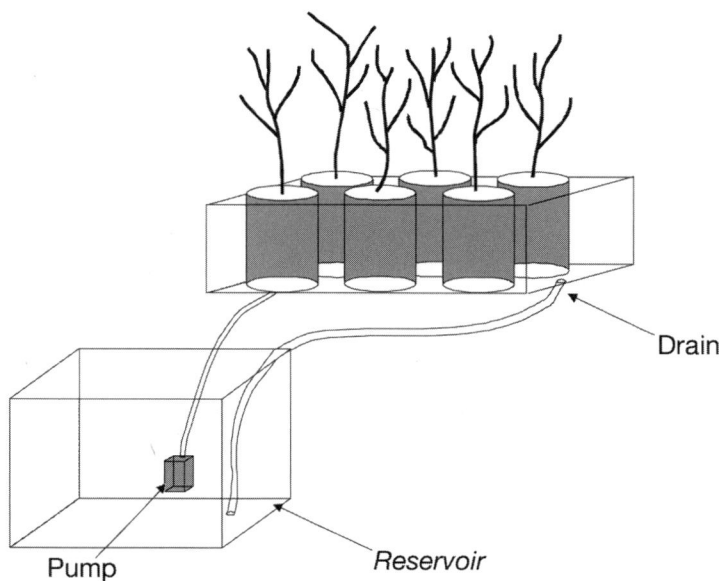

Fig. 1. Schematic of ebb-and-flow hydroponic bioreactor used to grow six trees under predominantly aerobic conditions. Each willow tree was supported in individual pots by porous baked-clay balls. (© 2004 from **ref. 5**. Reproduced by permission of Taylor & Francis Group, LLC., http://www.taylorandfrancis.com.)

Also, the design can be easily adapted to hold emergent and submergent aquatic plants. The most important use of the ebb-and-flow system is for the measurement of contaminant uptake by plants under predominantly aerobic conditions (DO>3.0 mg/L and pH 5.0–6.5). The components of an ebb-and-flow bioreactor include an upper hydroponic tray that holds the plants and a lower reservoir that holds the bulk of the contaminated water (**Fig. 1**). The size of the upper holding tray can be varied to hold as many wetland or terrestrial plants as desired. In an ebb-and-flow bioreactor, the contaminated water mixed with plant nutrients is continuously circulated from the reservoir tank through the rhizosphere to ensure that the roots are kept aerated. An air pump is used to continuously aerate the reservoir and maintain aerobic conditions throughout the study.

For tree studies, each tree is supported in individual pots by porous baked clay balls. The trees are established for about 6 mo in the system prior to dosing or exposure to the desired contaminants. The nutrient concentration in solution is monitored weekly and replenished as needed. The pH of the nutrient solution is monitored and buffered within a pH range of 5.0 to 6.5 to optimize bioavailability of plant nutrients. Samples are taken for analysis from the root zone in the upper tray and from the reservoir. The plant leaves are harvested and analyzed to quantify the fraction of the COC taken-up and phytoaccumulated. As for all bioreactors, persistent plant infestation by spider mites needs to be checked and controlled immediately.

Nzengung et al. *(5)*, Nzengung and McCutcheon *(6)*, and O'Niell et al. *(7)* used an ebb-and-flow bioreactor system to show that perchlorate was mainly removed from aqueous solution by uptake and phytodegradation under aerobic rhizosphere conditions. The contribution of rhizodegradation to perchlorate removal from aerobic ebb-and-flow bioreactors is insignificant, resulting in the persistence of perchlorate in solution. This result occurs because, unlike rhizodegradation, uptake and phytodegradation is a slower process for removal of perchlorate from water. The absence of rhizodegradation and predominance of uptake and phytodegradation in natural systems contributes to phytoaccumulation of high concentrations of perchlorate (up to 1000 mg/kg on a dry weight basis) *(5)*. A similar high fraction uptake of perchlorate into plant tissues has been observed in laboratory tests and field sites in the presence of a competing terminal electron acceptor such as nitrate. The high fraction uptake of perchlorate by plants grown in hydroponic bioreactors under aerobic conditions and high nitrate concentration in solution can be easily reversed. In experiments and/or environments where natural or artificial carbon and electron sources, (e.g., dead roots, acetate, molasses, lactate, pyruvate, and so on) are provided, these competing terminal electron acceptors (O_2 and NO_3^-) are quickly used up and biostimulation of perchlorate degradation is observed. Monitoring of the perchlorate concentration in young, old, and senescenced leaves has provided evidence of temporal phytoaccumulation, but no strong evidence of hyperaccumulation of perchlorate by the plants tested *(5)*. The latter authors observed higher biomass production in the ebb-and-flow hydroponic system than in sealed root-zone bioreactors.

3. Sealed Root-Zone Bioreactors

The sealed root-zone bioreactor consists of a flask with a screw-top opening and one or two side ports. Erlenmeyer flasks outfitted with side sampling (bottom port) sealed with a screw valve with septum and feeding ports (top port) and a screw-cap opening (**Fig. 2**) is the most common design used in hydroponic experiments *(8,9)*. Prior to use in experiments, the tree cuttings are dipped in a rooting hormone (Green Light® RooTone® rooting hormone with fungicide powder) and placed in aerated water and allowed to preroot for seven or more days. The root zone may be completely sealed off from the atmosphere by wrapping the tree cutting at the level of the top of the flask with Parafilm laboratory film and sealed into the cut-out screw caps with Elmer's® ProBond clear silicone rubber sealer. In some experiments, it is especially important to completely seal off the root zone and growth solution from the atmosphere to

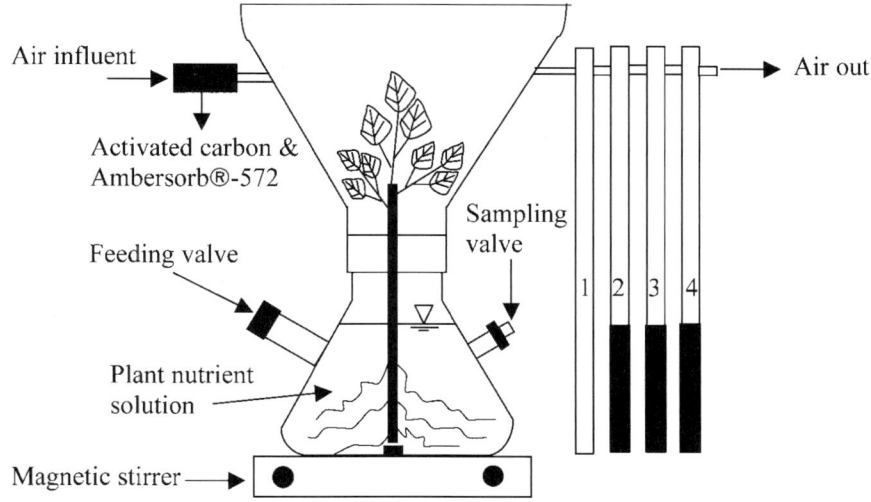

Fig. 2. Schematic of sealed root-zone bioreactor. The top feeding port is omitted in this diagram.

prevent losses of volatile compounds dissolved in solution. The flasks are filled with a diluted solution of the desired growth media and dosed with the contaminant. A calibrated reservoir unit holding diluted nutrient solution or water is connected to the feeding port of the planted bioreactor using nonreactive feeding tubes. Throughout the experiment, any water lost by evapotranspiration is continuously replenished from the reservoir. The volume of water evapotranspired by the tree is recorded daily. Each planted bioreactor is wrapped in aluminum foil to keep out light from the tree roots and prevent algae growth in the nutrient-rich growth solution. Samples for analysis are taken via the sampling port near the bottom of the reactor, at desired time intervals, using a syringe with a 10 cm long needle.

Kinetic and metabolite data on perchlorate removal from sealed root-zone bioreactors provides evidence of the mechanistic changes during phytoremediation of perchlorate. The published data shows an initial slow removal of perchlorate from solution mainly by uptake followed by a more rapid rate of removal once biostimulation of rhizodegradation is achieved. Unlike ebb-and-flow bioreactors, where slow removal of perchlorate from the rhizosphere by uptake into the plant under aerobic conditions predominates, both processes play a role in sealed root-zone bioreactors. The lag time that precedes rhizodegradation of perchlorate varies with the tree species. Plants with a higher fraction root mass tend to have a shorter lag time and faster rhizodegradation kinetics. This often accounts for differences in performance within species. Also, willow trees tend to perform better than cottonwoods in hydroponic reactors

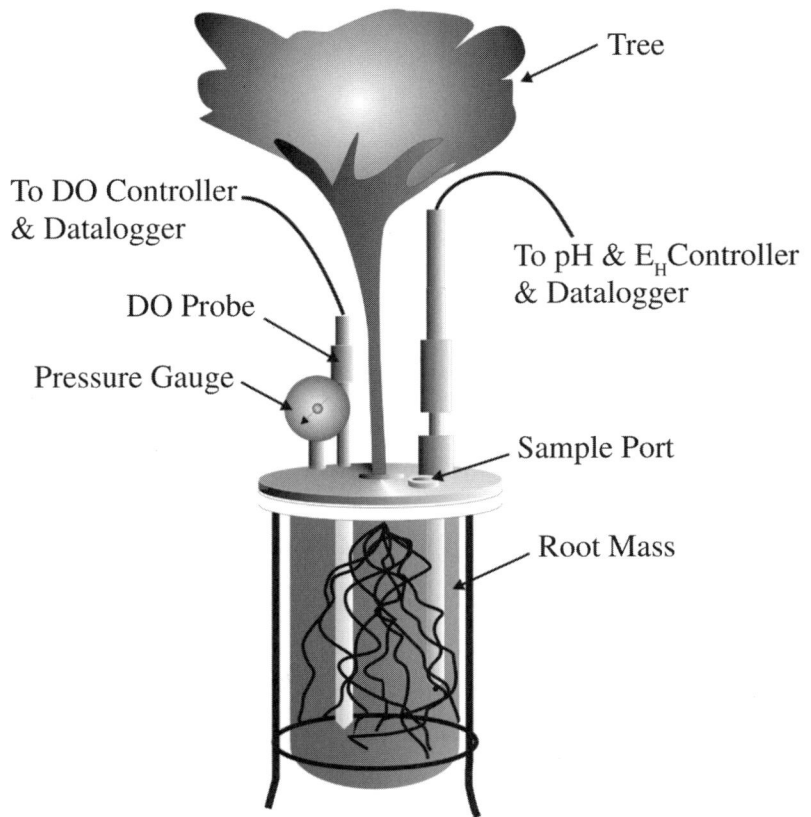

Fig. 3. Schematic of sealed root-zone bioreactor with pH, ORP, and dissolved oxygen probes and digital recorders used to monitor changes in root-zone conditions. The whole bioreactor was sealed with aluminium foil to keep out light and prevent algae growth in the nutrient-rich growth solution. An aluminium cover on an inspection window is used for occasional inspection of the root zone. (© 2004 from **ref. 5**. Reproduced by permission of Taylor & Francis Group, LLC., http://www.taylorandfrancis.com.)

for this reason. Nzengung et al. *(5)* used radiolabeled (^{36}Cl labeled) perchlorate to quantify the contribution of phytodegradation and rhizodegradation processes, respectively, during phytoremediation of perchlorate in sealed root-zone bioreactors. The authors showed that the type and concentration of nitrogen source in the growth media determine the predominant phytoprocesses in sealed root-zone bioreactors, as explained next.

4. Sealed Root-Zone Bioreactors With Probes

A 4 L nonjacketed fermentation vessel marketed by Cole Parmer® Instrument Company (Chicago, IL) (**Fig. 3**) and mounted with multiple probes on a steel-head plate provides the capability to continuously monitor changes in root-zone solution chemistry during phytoremediation studies. The probes include Cole Parmer autoclavable fermentation electrodes attached to Chemcadet® pH/mV and Cole Parmer DO controllers used for pH/E_H and DO measurements. The data is recorded using a CR21X Mircologger by Campbell Scientific, Inc.® (Logan, UT). A prerooted tree cutting is inserted through the center port on the head plate. The rhizosphere is sealed by wrapping the stem of each cutting with Parafilm laboratory film and applying an outer layer of silicone seal (Aquarium Sealant 100% silicone, DAP Inc., Baltimore, MD). The interface of the septum and the tree cuttings is sealed with parafilm to minimize losses of water vapor and gases from the reactors by volatilization.

The bulk root-zone solution is slowly mixed with a magnetic stirrer during the experiment to avoid stratification and to get representative samples for analysis. Nutrient solution is replenished continuously by a valve-controlled feeding system so that no headspace develops in the reactor. The feeding system consists of a 1 L calibrated flask connected to the reactor by a nonreactive feeding tube. The capability of measuring evapotranspiration on a daily basis exists. Prior to dosing the planted bioreactor, the E_H, DO, and pH changes in the bulk solution are recorded until the system stabilizes. At least 2 d are needed for the bulk solution chemistry to stabilize after dosing before recording of the redox, DO, and pH changes at fixed intervals is resumed. A long steel needle (15 cm) is inserted into sampling ports to collect liquid samples for analysis. Each reactor is completely wrapped with aluminum foil to minimize algal growth on the inner walls of the bioreactor, roots, and submerged stem. Algal growth in the nutrient-rich media is always a concern when growing plants hydroponically.

Unlike soil bioreactors, hydroponic bioreactors allow for accurate measurement of the total plant biomass at the beginning and termination of the study. At the end of the experiment, the plants are removed from solution, weighed, and sectioned into roots, upper and lower stems, branches, and leaves. To obtain the dry weight, each plant fraction is dried prior to grinding, extraction, cleanup, and analysis of COCs and metabolites taken up into the respective plant tissues *(10)*.

Using the sealed root-zone bioreactor with probes previously described, Nzengung et al. *(5)* have documented for the first time the profile of E_H, DO, and pH changes in planted soil-less bioreactors during phytoremediation of perchlorate. The data obtained by these authors show diurnal (light and dark) changes in E_H, DO, and pH in bulk solution of hydroponically grown willow trees. The pattern was disrupted mostly on cloudy days when sunshine was diminished. In the bulk solution, the E_H increased from a minimum of –150 mV at sunrise (varies seasonally) to a maximum of 300 mV at about midday, and remained at this maximum for hours before decreasing to the daily minimum after sunset. A steady increase in the mean-E_H value was observed after dosing with perchlorate and as the perchlorate concentration in the rhizosphere decreased (5). The daily maximum E_H amplitude decreased from 450 mV before dosing to 250 mV after dosing with perchlorate. Meanwhile, the daily minimum E_H values increased as perchlorate was degraded to chloride and oxygen. Specifically, the E_H range stayed between –50 and 50 mV during active phytoremediation of perchlorate, then increased to an E_H range of between 100 and 250 mV after perchlorate was completely removed from the bioreactor.

DO data followed the same diurnal trend as the E_H with the maximum concentrations of DO (1–1.5 mg/L) observed during afternoon hours when photosynthesis was optimal. At night, the oxygen was completely consumed through bacteria and/or plant respiration regardless of whether the willow tree was dosed with perchlorate or not. After the bioreactors were dosed with perchlorate, the same DO diurnal trend was observed but on a smaller scale, with the highest DO only approaching 0.1 mg/L. The latter observation could mean that dosing the planted bioreactors with perchlorate apparently triggered a shut down of the pressurized ventilation system that supplies the rhizosphere with oxygen (11). The DO probe was calibrated routinely using an anoxic potassium metabisulfite solution, whereas the pH/mV probe was calibrated with freshly prepared ZoBell's solution following United States Geological Survey (USGS) protocol (12).

The pH of the bulk solution showed small daily variations. In the same studies where Nzengung and coworkers (5) measured changes in E_H and DO, the pH varied between 5.9 and 6.05 prior to dosing the reactors. An abrupt pH increase of 0.8 units (from 5.8 to 6.6) occurred immediately after dosing. Thereafter, the pH decreased from 6.6 to a steady pH reading of about 5.8 during phytoremediation of perchlorate. Small daily cycles (amplitude of 0.1 pH unit) were recorded during phytoremediation experiments with daily maxima occurring in the early afternoon and minima occurring at night. As expected, the daily pH cycles followed the photosynthesis and respiration cycle of plants. At night, during respiration, carbon dioxide was released into the rhizosphere resulting in the formation of carbonic acid and an accompanying decrease in pH. With the onset of light at dawn, photosynthesis replaced respiration and oxygen was released into the rhizosphere by plants, which decreased the proton activity and increased the pH. The sudden increase in pH immediately after the reactors were dosed may have resulted from the introduction of oxygen with fresh nutrient solution to the bioreactors. Subsequently, the pH decreased as the plant exudates (sugars and other dissolved organic compounds) were metabolized by the root-zone microorganisms to CO_2. Following the complete removal of perchlorate from solution, the pH was steady at 5.8, suggesting that the bulk solution was buffered by carbonic acid and bicarbonate formed from dissolved CO_2.

5. Nutrient Solution

Nitrate is a competing terminal electron acceptor, which, when present at high concentrations, tends to inhibit rhizodegradation of perchlorate *(3)*. Therefore, the choice of nutrient solution is very important in phytoremediation of perchlorate in hydroponic or soil-less systems. Use of dilute (25–30%) Hoagland's nutrient stock solution type 1 (calcium nitrate 0.095%, potassium nitrate 0.061%, magnesium sulfate 0.049%, ammonium phosphate 0.012%, ferric tartrate 0.0005%, water 99.7825%) introduces high nitrate-N concentration in solution *(2)*. All of the nitrogen in type 1 Hoagland's solution is in the nitrate (NO_3^-) form. Using nutrients with an ammonium/urea nitrogen source minimizes the nitrate–nitrogen effect. Stern's Miracle-Gro® (6.8% ammoniacal nitrogen and 8.2% urea nitrogen), available widely from local plant nurseries, was diluted with deionized water to make the desired strength of nutrient solution (0.5 g/L). Pure Miracle-Gro contains 15% N, 30% P_2O_5, and 15% K_2O.

The influence of the type of nitrogen source in the nutrient solution has been determined in sealed root-zone bioreactor experiments dosed with ^{36}Cl-labeled perchlorate *(5)*. In the presence of nitrate N-source (25% Hoagland's solution), the removal of 76 ± 14% (mean ± 95% CI) of the initial $^{36}Cl^-$ perchlorate in solution is attributed to rhizodegradation. Meanwhile, in the presence of ammonium/urea N-source, the contribution of rhizodegradation increases to 96.1 ± 4.5% (mean ± 95% CI). In these experiments, the ^{36}Cl activity and perchlorate measured in the plant tissues accounts for 27 and 4% of the initial perchlorate added to the planted bioreactors with nitrate and urea N-sources, respectively.

6. Planted Soil Bioreactors

Soil bioreactor studies better simulate the performance of phytoremediation under natural conditions and account for the influences of soil minerals on microbial and plant activities. Evidence from multiple experiments conducted from 1999 to 2004 indicates that the rate of perchlorate removal from planted soil bioreactors is faster than in anaerobic hydroponic bioreactors by a factor of two. For soils rich in organic carbon, it is observed that the rates of perchlorate biodegradation in planted and unplanted (control) soil bioreactors are similar *(12)*. Thus, plants need not be present for perchlorate degradation to occur in carbon- and nutrient-rich anaerobic environments. The plants used in the soil studies have included terrestrial plants (*Salix nigra* [black willow], *Populus deltoides* Bartr. ex Marsh. [cottonwood], and *Pinus taeda* L. [loblolly pine]) and multiple wetland species (*Typha latifolia* L. [cattail], *Spirodela polyrhiza* [L.] Schleid [duck weed], and *Myriophyllum aquaticum* [Vell.] Verdc. [parrot feather]). The rate of plant growth in hydroponic bioreactors under greenhouse conditions is slower than in soil bioreactors. This may account in part for differences in biomass production and concentration of perchlorate in the plant tissues of soil-less and soil bioreactors. The redox

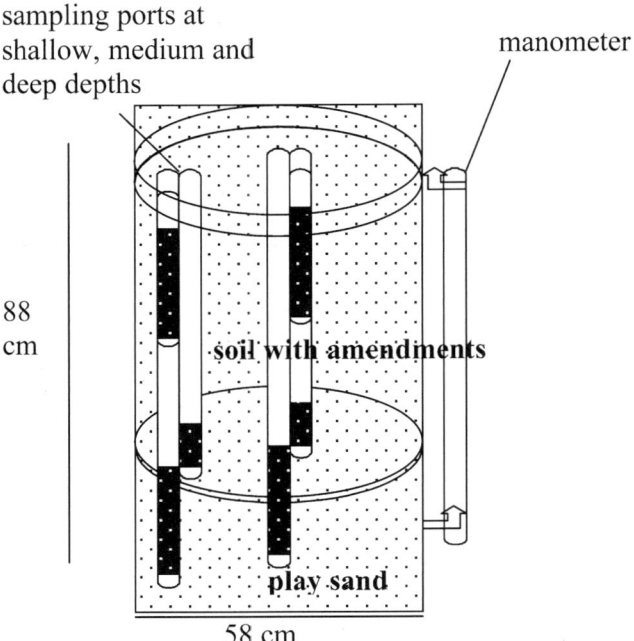

Fig. 4. Schematic of soil bioreactor showing a cluster of sampling ports and a manometer. The sand layer is completely saturated and the manometer is used to monitor the depth of water in the reactor. The sampling ports are installed at multiple depths, but only sampling ports installed at saturated depths are monitored.

changes observed in hydroponic systems do not apply and are difficult to monitor accurately in the more complex soil system where soil mineralogy has a greater influence on pore water chemistry than the plant.

The soil bioreactors are constructed from 5 gal (approx 19 L) plastic buckets or 55 gal (approx 193 L) metal drums (**Fig. 4**), respectively. Each reactor is mounted with a manometer on the side of the reactor to monitor the level of water in the bioreactor during phytoremediation. The manometer consists of a 0.5 by 0.5 in. (1.27 × 1.27 cm) hose barb to male internal pipe thread elbow inserted into a 0.5 in. (approx 1.27 cm) diameter hole cut at about 3 cm above the bottom of the reactor. The elbow is held in place by a faucet-coupling nut (0.5 in./1.27 cm IPS). The area around the hole is sealed with Plumber's Goop™ adhesive and sealant. A length of $^{1}/_{2}$ by $^{5}/_{8}$ in. (1.27 × 1.59 cm) vinyl tubing is attached to the hose barb and anchored up the side of the reactor. The inner male internal pipe thread elbow is covered by a small piece of WeedBlock® landscape fabric to prevent soil, sand, clay, and plant roots from entering the tubing. Each reactor is checked for leaks and to ensure that the manometer accurately represents the level of the water in the planted soil bioreactor.

The smaller (20 L or less) soil bioreactors were outfitted with one sampling port, whereas the larger reactors had two clusters of three sampling ports to monitor three depths (shallow, medium, and deep) at the edge and center of the reactor, respectively. The sampling ports were made from three different lengths of 0.5 in. (1.27 cm) Bristol pipe PVC. The longest pipes were cut to the full height of the reactor and drilled with randomly patterned 0.25 in. (0.64 cm) holes in the lowest 30 cm of the piping. The medium length set of sampling ports were 27.5 cm shorter than the longest set. The lowest, 5 cm, were drilled with 0.25 in. (0.64 cm) holes. The pipes for the third set of ports were 30 cm long and were drilled with 0.25 in. (0.64 cm) holes throughout the full length of the piping. The lower end of each of the pipes was covered with Parafilm laboratory film and 3M® duct tape to ensure that any water entering the sampling port came in from the depth at which the holes were drilled in the pipe. WeedBlock landscape fabric was secured over the drilled portion of each piece of PVC piping with Parafilm laboratory film. The top of each pipe was coded with colored tape to indicate its length. The pipes were bound together with Parafilm laboratory film in sets of three, with each set containing one pipe of each length.

Play sand (Garden Basics™ play sand) was added to the bottom of each reactor to a depth of 30 cm. The sand (0.64 cm) was smoothed to create a flat surface, and the reactor was filled within 4–5 cm from the rim with the desired soil type. The water level in each reactor was maintained at the top of the sand layer with the capillary fringe extending a few centimeters into the soil layer. Once the planted saplings were established, the reactors were dosed with a known concentration of perchlorate solution and monitored under natural conditions. The pore water (water within the saturated soil) in each reactor was sampled at the three depths and analyzed for perchlorate and water-quality parameters. Perchlorate and other anions were analyzed by ion chromatography methods and metals by ICP/MS.

Acknowledgments

Modified figures have been adapted from an original paper with the kind permission of Taylor and Francis publishers.

References

1. McCutcheon, S. C. and Schnoor, J. L. (eds.) (2003) Overview of phytoremediation and control of waste. In: *Phytoremediation: Transformation and Control of Contaminants*, John Wiley and Sons, Hoboken, NJ, pp. 1–52.
2. Jones, J. B. (1997) *Hydroponics: A Practical Guide for the Soilless Grower*. St. Lucie Press, Bocan Raton, FL.
3. Nzengung, V. A., Wang, C., and Harvey, G. (1999) Plant-mediated transformation of perchlorate into chloride. *Enivron. Sci. Technol.* **33**, 1470–1478.

4. Nzengung, V. A. and Wang, C. (2000) Influences on phytoremediation of perchlorate-contaminated water. In: *Perchlorate in the Environment,* (Urbansky, E. T., ed.), Kluwer Academic/Plenum Publishers, New York, pp. 219–229.
5. Nzengung, V. A., Penning, H., and O'Niell, W. (2004) Mechanistic changes during phytoremediation of perchlorate under different root-zone conditions. *Int. J. Phytorem.* **6,** 63–83.
6. Nzengung, V. A. and McCutcheon, S. C. (2003) Phytoremediation of perchlorate. In: *Phytoremediation: Transformation and Control of Contaminants,* (McCutcheon, S. C. and Schnoor, J. L. eds.), John Wiley and Sons, Hoboken, NJ, pp. 863–885.
7. O'Niell, W., Nzengung, V. A., and Adesida, A. (2000) Treatment of perchlorate-contaminated water in microbial mat, algae, and ebb-and-flow bioreactors. In: *The Second International Conference on Remediation of Chlorinated and Recalcitrant Compounds,* (Wickramanayake. G. B., Gavaskar, A. R., Gibbs, J. T., and Means, J. L. eds.), Battelle Press, Monterey CA.
8. Burken, J. G. and Schnoor, J. L. (1997) Uptake and metabolism of atrazine by poplar trees. *Environ. Sci Technol.* **31,** 1399–1406.
9. Burken, J. G. and Schnoor, J. L. (1998) Predictive relationships for uptake of organic contaminants by hybrid poplar trees. *Environ. Sci. Technol.* **32,** 3379–3385.
10. Van Aken, B. and Schnoor, J. L. (2002) Evidence of perchlorate (ClO_4^-) reduction in plant tissues (poplar tree) using radio-labeled perchlorate ($^{36}ClO_4^-$). *Environ. Sci. Technol.* **36,** 2783–2788.
11. Grosse, W., Jovy, K., and Tiebel, H. (1996) Influence of plants on redox potential and methane production in water-saturated soil. *Hydrobiologia* **340,** 93.
12. Dondero, A. C. (2001) *Phytoremediation of perchlorate under greenhouse and natural conditions.* PhD thesis, Department of Geology, University of Georgia, Athens, GA.

19

Using Plant Phylogeny to Predict Detoxification of Triazine Herbicides

Sylvie Marcacci and Jean-Paul Schwitzguébel

Summary

A plant useful for phytoremediation has first to grow in the presence of the target pollutant without being harmed. The plant must not only be resistant to the pollutant, but must also be able to remove it from the environment and transform it into nontoxic metabolites or end products. Differences in the ability of various plant species to accumulate and metabolize particular pollutants do exist, indicating that in choosing the most appropriate species in the development of any phytoremediation process, natural biodiversity should be better explored and exploited. Plant taxonomy and phytochemistry can help in the exploitation of biochemical specificities of plants that produce natural chemicals with structures similar to xenobiotic compounds. In the case of atrazine, however, numerous results obtained for agronomical purposes are extremely useful in choosing the most appropriate families or genera for phytoremediation.

Key Words: Atrazine; triazines; herbicides; detoxification; phytotransformation; biodiversity.

1. Interception of Atrazine in Runoff

Heavy environmental contamination by herbicides may arise from industrial point sources such as accidental spillage during production, wastewater from pesticide production plants, leakage of old stockpiles, storage and transport, or leachates from former dumping sites and municipal waste *(1)*. In contrast, sources of pollution arising from agricultural use of herbicides are considered to be diffuse because the compounds are distributed over large areas. Pre-emergence herbicides such as atrazine that are applied to the soil, as well as some of their degradation products, reach surface water and groundwater through leaching and runoff *(2)*. Most of the transport of atrazine in runoff occurs during the first rain or irrigation events after application. Most of the annual load in streams thus occurs over a relatively short period *(3)*; once atrazine reaches the surface water system, it is transported to the ocean without substantial loss, or ends up in long-term storage in lakes, reservoirs, and alluvial aquifers, where atrazine shows minimal loss by volatilization, sorption, or transformation.

Improving the quality of surface water has led to an emphasis on the best management practices for controlling agricultural nonpoint-source pollution *(4)*. One practice, which has received widespread interest, is the use of wetland vegetation or of natural, or artificial, vegetative filter strips to remove chemicals from the flow prior to their entry into stream, lake, or sea. Although atrazine has been applied in the field for over 30 yr, no enhanced degradation to complete mineralization by bacteria leading to adaptive soils has yet been reported on large scale, thus indicating the difficulty of rapid microbial breakdown in the field *(5)* and showing the usefulness of vegetative strips or constructed wetlands.

2. Plant Species Useful for Atrazine Phytoremediation

Publications on the metabolism of herbicides in plant species useful for phytoremediation are scarce compared with publications about plant metabolism for agronomic purposes. For example, the efficiency of a natural filter of bluegrass (*Poa annua*) and fescue (*Festuca* sp.) strips located immediately downslope from a standard erosion plot of 9% slope has been investigated *(4)*. Trapping efficiency of atrazine by a 4.5-m wide strip was 93%, i.e., of the same magnitude as dissolved phosphorus, nitrate, ammonium, and sediments, emphasising the relevance of grass filters as buffer strips, but the mechanism underlying atrazine disappearance was not studied. However, a possible explanation of atrazine disappearance in buffer strips has been reported *(6)*: *P. annua* and *Festuca sp.*, both belonging to the subfamily Festucoideae, take up atrazine, but without subsequent transformation.

Decontamination of water polluted with 6 ppm atrazine by several marsh plants, common club-rush (*Schoenoplectus lacustris*), bulrush (*Typha latifolia*), yellow iris (*Iris pseudacorus*), common reed (*Phragmites australis*) was observed and the disappearance of atrazine from water is a result of the action of rhizosphere micro-organisms *(7)*. The action of plants themselves was not explored, but was not excluded. Other authors also evaluated semiaquatic herbaceous perennial plants for their use in herbicide phytoremediation, such as canna (*Canna generalis*), pickerel (*Pontaderia cordata*), and iris (*Iris x Charjoys Jan*), and concluded that these taxa were not optimal for phytoremediation because the plants exposed to herbicides showed significantly reduced biomass *(8)*. In contrast, it seems that vetiver can tolerate 2 ppm atrazine without adverse effect *(9)*.

Hybrid-poplar buffer strips were first initiated and planted in rows along portions of streams at the end of the 1980s *(10)*. Poplar could take up atrazine in the transpiration stream, showing that poplar tree buffer strips are also effective in removing atrazine from agricultural percolation and runoff water. The only extensive study of plant metabolism of atrazine for a phytoremediation purpose is in poplar (*Populus deltoides x nigra*), which can take up atrazine and metabolize it mainly into dealkylates and to a lesser extend, into polar ammeline *(11,12)*.

3. Atrazine Metabolism in Agricultural and Weed Species

In contrast to scarce information on atrazine metabolism for phytoremediation, plant metabolism of pesticides is well documented for annual agricultural crops and grasses, to understand its role in herbicide selectivity *(13,14)* and to predict effective control of weeds *(6)*. Such fundamental knowledge is useful to discover new selective herbicides, as well as to obtain plants resistant to herbicides. Many plants are insensitive to atrazine. Some of them have acquired chloroplastic resistance, which is not useful for phytoremediation. Others tolerate atrazine by degradation of the active moiety through N-dealkylation, chemical hydroxylation, and/or conjugation. Selectivity of the herbicide atrazine is based on its tolerance by plants. For example, maize and sorghum-tolerate atrazine by chemical and enzymatic transformation of the herbicide.

3.1. Degradation of Atrazine by N-Dealkylation

The N-dealkylation is an enzymatic reaction leading to deethyl atrazine (DEA), deisopropyl atrazine (DIA), and didealkyl atrazine (DDA). This pathway is the major metabolization of atrazine found in *Asperillus fumigatus (15)*, pea (*Pisum sativum*) *(16)*, *Spartina alterniflora (17)*, and potato (*Solanum tuberosum*) *(18)*. But dealkylation was also found in maize, as a minor metabolic process *(19)*. A study on isolated chloroplasts from oat showed that the phytotoxic activitiy was reduced by 23 times for monoalkylates, as compared with atrazine *(20,21)*.

Species studied so far undergo dealkylation to different extents. In pea, dealkylation is fairly important, and confers intermediate resistance to atrazine *(16)*. The other species with slight dealkylation are sensitive to atrazine such as *A. fumigatus (15)*, *S. alterniflora (17)*, potato (*S. tuberosum*) *(18)*, soybean (*Glycine max*), and wheat (*Triticum sp.*) *(22)*. Monuron, chlorotoluron, prosulfuron, metolachlor, and alachlor herbicides are dealkylated by cytochrome p450 enzymes. Plant p450s are membrane bound to the endoplasmic reticulum *(23)*. They are powerful oxidizing catalysts, which activate molecular oxygen and typically insert one oxygen atom (as a hydroxyl group) into lipophilic substrates. But oxidation of heteroatoms like N was also found to give dealkylated products *(24)*. p450 involvement of in vitro plant microsomal preparations can be shown by its inhibition by CO *(25,26)*. The cytochrome p450 inhibitor 1-aminobenzotriazole, used in combination with simazine in *Lolium rigidum*, causes a greater reduction in dry mass of resistant plants than simazine applied

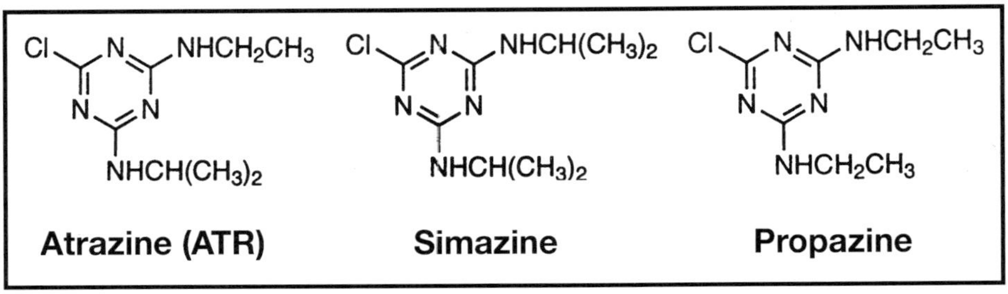

```
Cl     N     NHCH₂CH₃        Cl     N     NHCH(CH₃)₂       Cl     N     NHCH₂CH₃
   \\ //  \\ /                    \\ //  \\ /                   \\ //  \\ /
    N                               N                             N
   / \\                             / \\                           / \\
  N   N                            N   N                          N   N
   \\ /                             \\ /                           \\ /
    NHCH(CH₃)₂                      NHCH(CH₃)₂                     NHCH₂CH₃

  Atrazine (ATR)                    Simazine                       Propazine
```

Fig. 1. Triazine analogs : atrazine, simazine, and propazine.

alone *(27)*. This suggests involvement of oxidative enzymes in the mechanism of dealkylation of simazine. Triazines are analogs of atrazine and share exactly the same dealkylates, therefore conclusions for simazine could be extended to atrazine (**Fig. 1**).

3.2. Degradation of Atrazine by Hydroxylation

Chemical transformation of atrazine into hydroxyatrazine has been well studied in maize *(28,29)*. Hydroxylation was described not only on atrazine, but also on dealkylates DEA, DIA, and DDA *(30)*. The resultant metabolites are hydroxyatrazine, hydroxydeethyl atrazine, hydroxydeisopropyl atrazine, and hydroxydidealkyl atrazine (**Fig. 2**). The replacement of the chlorine atom by a hydroxy group results in nonphytotoxic metabolites *(28)*, explaining mainly maize tolerance to atrazine. Chemical pathways leading to the formation of the inactive hydroxyatrazine is the major form of metabolization inside the roots and, during the first week, inside the leaves of maize *(19)*. The formation of a glutathione-atrazine conjugate, owing to the activity of a glutathione-*S*-transferase (GST), although existing, is small, and anyway, is only fully effective after a 1 wk culture for maize plantlets.

Natural benzoxazinones were discovered 40 yr ago, when resistance against pathogenic fungi was investigated. They play a major role in the defense of cereals against insects *(31,32)*, fungi and bacteria *(31)*, chelation of Fe^{3+} *(33,34)* in allelopathic effects *(35,36)*, and in the detoxification of herbicides *(31,37,38)*. The family of benzoxazinones is divided into several classes, namely the cyclic hydroxamic acids, lactams, methyl derivatives, and benzoxazolinones. Benzoxazinones are predominantly found in the family Poaceae including genera *Aegilops, Arundo, Chusquea, Coix, Elymus, Secale* (rye), *Tripsacum, Triticale, Triticum* (wheat), and *Zea mays* (maize). They are not present in *Avena* (oat), *Hordeum* (barley), or *Oriza* (rice) *(31)*. But they were also identified in Acanthaceae, Ranunculaceae, and Scrophulariaceae. Interestingly, sorghum (Poaceae) is subject to contradictory information: some authors *(37,39)* found no benzoxazinone in sorghum, whereas others detected benzoxazinones in this

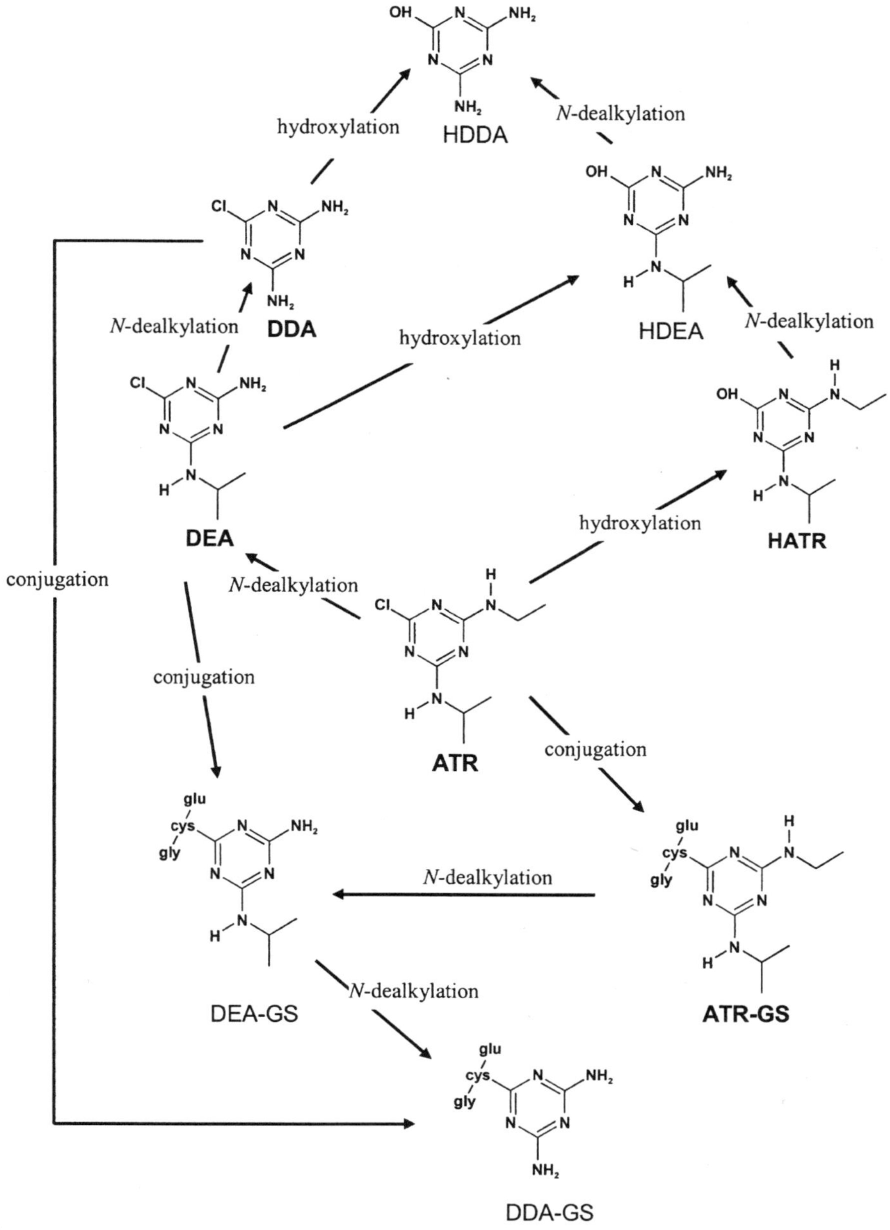

Fig. 2. Enzymatic pathways (dealkylation and conjugation) and chemical transformation (hydroxylation) of atrazine in maize and sorghum. (Molecules investigated in vetiver are represented with bold legend *[66]*. Monodealkylate DIA undergoes the same pathways as deethyl atrazine [not shown].)

plant *(31)*. It seems that plants lacking benzoxazinones fail to produce hydroxylates of atrazine. On the other hand, wheat contains benzoxazinones and soybean does not, and both plants are sensitive to atrazine *(37)*.

3.3. Degradation of Atrazine by Conjugation

The conjugation reaction means that the chlorine atom of the triazinic ring of atrazine is replaced by a substance produced by the plant, the tripeptide glutathione. GSTs are enzymes that act on hydrophobic, electrophilic, and cytotoxic substrates. GSTs have been found in virtually all living organisms. Cytotoxic substrates include xenobiotics, and they have been well studied with regard to herbicide detoxification. In sorghum, conjugated atrazine is not an end product, but is subject to further transformation. Four other successive conjugates have been identified, namely γ-glutamylcysteine, L-cysteine, *N*-acetyl-L-cysteine, and lanthionine conjugates *(40–42)*. Conjugation of atrazine was reported to occur in many species, in cultivated plants such as maize *(22,30,43,44)* and sorghum *(40–42)*, but also in weeds, such as *Digitaria sanguinalis (45)*, *Echinochloa crus-galli (45)*, *Panicum miliaceum (45,46)*, *Setaria* sp. *(45,47–49)*, and *Abutilon theophrasti (50,51)*. Very interestingly, if grouped taxonomically, atrazine-tolerant species are members of the subfamily Panicoideae, whereas sensitive species, such as quarkgrass, oats, and barley are in the Festucoideae *(6)*. To sustain this observation, more than 40 grass species belonging to Festucoideae and Panicoideae subfamilies have been studied; high yield of conjugation was observed in tolerant Panicoideae grasses such as *Setaria* sp. and *Sorghum* sp. compared with low or nil conjugation in sensitive Festucoideae like *Avena* sp., *Bromus* sp., and *Hordeum* sp. Maize tolerance tightly relies on GST's action when leaf treatment is applied, but when atrazine is applied to roots, hydroxylation seems to confer protection *(19,22,29,44)*. In sorghum, tolerance was found to be related to a high capacity of conjugation *(30,39–42)*.

4. Environmental and Health Hazards of Atrazine Metabolites

Plants act globally as detoxifiers, thanks to active p450s, benzoxazinones, and GSTs, and participate in the degradation of xenobiotics in the environment. Unlike animals, where most transformation products of xenobiotics are excreted, plant tissues store them in conjugated soluble form, or as insoluble bound residues. Pesticides that are particularly hazardous to all organisms are those that contain electrophilic sites, i.e., compounds that have centers of low electron density and can accept an electron pair to form a covalent bond. These chemicals can exert toxic effects by covalent binding to nucleophilic sites on cellular molecules. Electrophilic xenobiotics are particularly harmful because they can be cytotoxic or genotoxic *(23)*. Triazines possess the atom chlorine, which is an electrophile. The removal of chlorine could therefore be already considered as a beneficial transformation, as can hydroxylation and conjugation. Already in 1972, the dechlorination of atrazine was considered as a clear case of detoxification, because a lethal dose (LD_{50}) of glutathione conjugates is greater than 1000 mg/kg in the rat compared with 294 mg/kg for the herbicide *(52)*. Although mineralization (complete degradation of pesticides) is the desired endpoint in remediation, usually a few transformation reactions are sufficient to drastically change their biological activity *(53)*. Mineralization of atrazine is unlikely to occur, as total degradation of pesticides in plants has not so far been described.

Recently the US Environmental Protection Agency issued a report where atrazine, simazine, propazine, and their common metabolites, DEA, DIA, and DDA, could be grouped by a common mechanism of toxicity for disruption of the hypothalamic–pituary–gonadal axis in rats *(54)*. The exact mechanism of this endocrine disruptor action is currently being actively studied, and it seems that atrazine and DDA act on aromatase activity responsible for estrogen synthesis *(55)*. As dealkylates were found to exhibit endocrinal effects, dealkylation is not of high benefit for health. Nevertheless, phytotoxicity is reduced and this is positive for the environment. Conjugation and hydroxylation are the most interesting transformations, as these metabolites were not described to be acutely toxic, neither exhibiting endocrine effects. Based on the present knowledge of toxicity of atrazine metabolites, transformations of atrazine can be ranked in decreasing order of interest for atrazine remediation: total mineralization > conjugation = hydroxylation > dealkylation.

5. Resistance and Metabolism

A candidate species for phytoremediation has first to be screened for its resistance to pre-emergence herbicides to ensure its suitability either to the site to be decontaminated or to intercept runoff of triazine. Plant candidates must be resistant to atrazine, but if chloroplastic resistance is observed, there is no ecological benefit as no transformation of the target compound occurs. Such cases of resistance have been described in biotypes of *Senecio vulgaris (56)*, *Chenopodium album (57)*, *Brassica campestris*, *Solanum nigrum*, *P. annua*, *Setaria viridis*, and *Phalaris paradoxa (58)*. In total, 76 biotypes have been found to be resistant to atrazine thanks to chloroplastic resistance *(59)*.

Maize undergoes the three pathways of atrazine metabolization: hydroxylation, dealkylation, and conjugation *(22)*, and sorghum performs slight dealkylation and high conjugation *(40)* (**Fig. 2**). Tolerance of both crop plants and the weeds *Setaria adherens* and *S. verticillata* is inherent because of high metabolization *(48)*. Tolerance of crop plants to atrazine can be best explained by a high intensity of one metabolic pathway, as in sorghum, or by the addition of several metabolic pathways, as in maize (**Table 1**). In contrast, the absence of a

**Table 1
Metabolization Pathways of Atrazine in Selected Plant Species**

Plant		Dealkylation	Hydroxylation	Conjugation	Response to atrazine	Reference
Corn[a]	shoot	+	++++	+	tolerant	38,19
	roots	+	++++	+		
Corn[b]	shoot	+	+++	++	tolerant	
	roots	+	+++	+		
Sorghum	shoot	++	–	++++	tolerant	22,20
	roots	++	–	+		
Pea	shoot	++++	–	+	intermediate	
	roots	++++	–	+		
Soybean		++	–	+	sensitive	
Wheat		+	+	+	sensitive	

[a]From germination until 1-wk-old plant from Cherifi et al. *(19)*.
[b]1-mo-old plant *(19)*.

metabolic pathway explains the sensitivity of species like wheat and soybean, whereas dealkylation alone confers intermediate tolerance to atrazine in pea.

6. Vetiver as a Candidate for Atrazine Phytoremediation

It seems that dicotyledonous species are generally sensitive to atrazine, or resistant to atrazine thanks to chloroplastic resistance without atrazine transformation, or, as is the case for poplar trees, dealkylation is the major metabolic pathway of atrazine, with dealkylates still exhibiting endocrine effects on mammals. In contrast, the most interesting transformations of atrazine in terms of detoxification (hydroxylation and conjugation) are found in monocotyledons protected against competition from weeds: maize and sorghum via benzoxazinones and GSTs, respectively. In other words, monocotyledons are the most interesting plant species to be evaluated for atrazine remediation. Moreover, conjugation detoxification was observed in the subfamily Panicoideae, with the very well-studied case of sorghum. Therefore, it was believed that the exploration of a candidate for phytoremediation of atrazine should belong at least to the family Poaceae and if possible to the subfamily Panicoideae.

The monocotyledon vetiver belongs to the family Poaceae, subfamily Panicoideae, tribe Andropogonae, and subtribe Sorghinae, and the genus includes 10 species. The genus is related to the genera *Sorghum* and *Chrysopogon*. DNA fingerprinting revealed that *Vetiveria* and *Chrysopogon* cannot be distinguished and led to a merger of the genera *(60)*. Previously, vetiver was classified as *Vetiveria zizanioides*. However, it has recently been reclassified, and it should now be known botanically as *Chrysopogon zizanioides,* as recommended by the 2003 Catalogue of New World Grasses *(61)*.

Vetiver is a perennial tropical grass, also known as khus-khus *(62)*. The generic name Vetiveria comes from the Tamil word "vetiver" meaning "root that is dug up." It is native of India, but the exact location of origin is not precisely known. Vetiver is by nature a hydrophyte, but often thrives under xerophytic conditions: it grows particularly well on river-banks and in rich marshy soil. It can withstand periods of flood, as well as extreme drought, survives at temperatures of between –9 and 45°C, is fire resistant, and is able to grow in any type of soil regardless of fertility, salinity, or pH. Vetiver is a tall, fast growing, perennial grass with densely packed stiff, tough stems, which form a dense hedge when planted closely in rows. It grows large, densely tufted with a compact rhizome producing clumps up to 3 m high *(62,63)*.

The distribution of vetiver is pantropical, and some boundary strips are found in vetiver's native region of India *(64)*. It was introduced recently in Southern regions of Europe, such as Italy, Portugal, and Spain. Nonseeding vetiver plants are used in many countries for soil erosion control and many other applications: vetiver grass was first introduced for soil conservation and land stabilization in Fiji in the early 1950s *(63,65)*. Recognizing the potential in combating land degradation, the World Bank has promoted in the mid-1980s the vetiver grass system, which is now used worldwide as a low-cost, low-technology, and effective means of soil and water conservation and land stabilization in developing countries. The US Board of Science and Technology for International Development mentioned successful vetiver applications for stabilization of slopes, terraces, and channel banks in numerous tropical and subtropical countries: Australia, Bolivia, Brazil, China, Costa Rica, Ecuador, El Salvador, Guatemala, Honduras, India, Indonesia, Madagascar, Malawi, Malaysia, Mexico, Nepal, Nicaragua, Nigeria, Philippines, Sri Lanka, South Africa, Thailand, Zambia, and Zimbabwe *(64)*. Vetiver plantation for soil erosion control is mainly performed linearly, along fields, terraces, canals, streams, or rivers where the erosive force of water is at its greatest, lakeshores, artificial embankment, and little canals for irrigation or water drainage. It even can be planted across the river itself to slow down the flow of water (http://www.vetiver.org/).

Our experiments showed that vetiver is resistant to 20 ppm atrazine for at least 6 wk, even with a maximum bioavailability created by the use of a hydroponic system (**Fig. 3**). Atrazine resistance might be explained by plant metabolism, dilution of active ingredient into plant biomass, chloroplastic resistance, and sequestration of atrazine before it reaches its target site in leaves. It was found that vetiver thylakoids are sensitive to atrazine, excluding, therefore, chloroplastic resistance. Known plant metabolism of atrazine relies on (1) hydroxylation mediated by benzoxazinones, (2) conjugation catalyzed by GST,

Fig. 3. Vetiver grown under hydroponic conditions.

and (3) dealkylation probably mediated by cytochromes p450. Therefore, these metabolic pathways have been explored in vetiver to understand its resistance to atrazine and to evaluate benefits or risks of phytoremediation *(66)*.

Atrazine metabolism takes place in vetiver (**Table 2**). Small amounts of dealkylated products are found in roots and leaves, and conjugated atrazine is detected mainly in leaves confirming in vitro tests of GST activity. No benzoxazinones are detected in plant extracts, in agreement with the absence of hydroxyatrazine in vetiver organs. Altogether, these metabolic studies suggest that hydroxylation is not a metabolic pathway in vetiver; the plant behaves more like a related species, sorghum, where conjugation clearly dominates on dealkylation. Under transpiring conditions, conjugation in leaves is important, but under nontranspiring conditions, it appears that atrazine and its metabolites can be trapped in roots according to the partition–diffusion law. Over-concentration of atrazine is observed in oil from roots grown in soil, suggesting that during plant aging partition may play a significant role in retaining atrazine from agricultural runoff.

Vetiver-resistance mechanisms necessary to establish phytoremediation in soil or water contaminated with atrazine have been found. Major metabolism of atrazine in vetiver grown in a hydroponics system is conjugation mainly in leaves, a transformation known to be positive for the environment *(66)*. Phylogenetically, vetiver is close to sorghum, a plant described previously to

Table 2
Atrazine Metabolism in Vetiver: Detection of Metabolites *(66)*

Dealkylation	Hydroxylation		Conjugation	
In vivo dealkylates + in organs	In vivo hydroxylates in organs	−	In vivo conjugates in organs	++++
	Hydroxylates in entire plants	−	Conjugates in entire plant	++++
	In vitro Benzoxazinones extraction and in vitro test of hydroxylation		In vitro GST extraction and in vitro of conjugation	++++
	Benzoxazinones detection	−		
+	−		++++	

tolerate atrazine thanks to high conjugation capacity. It seems that, as in other Panicoideae plants, vetiver follows the same interesting detoxification pathway: conjugation.

7. Phytoremediation of Other Pre-emergence Herbicides

Studies with other pesticides would be relevant to see if vetiver's use as a tool against pesticide runoff could be extended. Moreover, atrazine is also used in combination with many other herbicides, such as alachlor, metolachlor, cyanazine, simazine, amitrole + simazine, or diuron + simazine *(67,68)*. In most pesticide-contaminated agrochemical facilities, atrazine is found in combination with other widely used agricultural chemicals *(69)*. Therefore, remediation strategies must cope with a multiple-contaminated environment.

The herbicides used most in the United States were in 1996, atrazine, metolachlor, and alachlor *(70)*. In the midwest of the United States, atrazine and metolachlor are frequently present in groundwater *(71)*. Atrazine and alachlor are also frequently detected in groundwater and rivers of many countries *(72,73)*. There is critical environmental concern about alachlor and one of its metabolites (2[(2′6′-diethylphenyl)(methoxymethyl)-amino]2-oxoethane-sulphonate) in the environment because it leaches much more rapidly through the soil than does the parent compound and makes an important contribution to the total organic contaminant load of groundwater in the central United States.

All these herbicides are used for pre-emergence treatment, but the persistence of herbicides is linked to their mode of action: only herbicides of post-treatment can be transitory *(74)*. In contrast, herbicides of pre-emergence must have an agronomical persistence of several weeks for an efficient action; they often need weeks to exert their phytotoxicity, and to kill weeds whose germination does not occur at the same time. Herbicides are generally retained in the soil because of their adsorption on superficial soil horizons, but most of the time, washing of herbicides occurs with the first rain following application. Besides leaching in the deep soil, there is also a risk of washing of soil particles on sloppy nude soils. The potential danger of groundwater contamination has been assessed and it appears that alachlor, atrazine, and simazine application should be avoided in sandy soils, and used only in nonirrigating crops *(75)*. In other words, pre-emergence herbicide triazines (ametryn, desmetryn, dimethymetryn, terbutryn, atrazine, propazine, cyprazine, simazine, cyanazine) and chloroacetanilides (alachlor, acetochlor, metolachlor, pretilachlor) are massively used but also detected in groundwater and surface water, except cyanazine, which is commonly found in surface water, but rarely in groundwater *(70)*. Methylthio-s-triazines (ametryn, desmetryn, dimethametryn, terbutryn) are not readily metabolized into water-soluble metabolites in excised sugarcane leaves and it was shown that methylthio-s-triazines are not substrates for GSTs isolated from corn *(41)*.

In contrast, plant species that have been shown to readily transport triazines acropetally from roots to leaves include corn, cucumber, spruce, black walnut, yellow poplar, poplar clones, radish seedlings, and barley *(76)*. In most species, plant metabolism of other triazines is similar to atrazine *(40,44)*. Many authors detected GST activities on triazines and chloroacetanilides *(45,49)*, and a positive correlation was found between plant tolerance to chloroacetanilide and triazine herbicides, best explained by conjugation to glutathione mediated by GSTs or not *(77)*. Moreover, in addition to common detoxification of triazines and chloroacetanilides in plants, atrazine and metolachlor mineralization is greater in rhizosphere collected from *Kochia scoparia* and *Brassica napus* than in bulk soil *(68)*. Vetiver has been shown to carry out conjugation of atrazine, and therefore it is believed that it is also capable of taking up and conjugating other triazines as well as chloroacetonilides, and not only atrazine in agriculture runoff.

References

1. Biziuk, M., Przyjazny, A., Czerwinski, J., and Wiergowski, M. (1996) Occurrence and determination of pesticides in natural and treated waters. *J. Chromat.* **754,** 103–123.
2. Coleman, J. O., Frova, C., Schröder, P., and Tissut, M. (2002) Exploiting plant metabolism for the phytoremediation of persistant herbicides. *Environ. Sci. Pollut. Res.* **9,** 18–28.
3. Capel, P. D. and Larson, S. J. (2001) Effect of scale on the behavior of atrazine in surface water. *Environ. Sci. Technol.* **35,** 648–657.

4. Barfield, B., Blevins, R., Fogle, A., et al. (1998) Water quality impacts of natural filter strips. *Am. Soc. Agric. Eng.* **41,** 371–381.
5. Wenk, M., Bourgeois, M., Allen, J., and Stucki, G. (1997) Effects of atrazine-mineralizing microorganisms on weed growth in atrazine-treated soils. *J. Agric. Food Chem.* **45,** 4474–4480.
6. Jensen, K., Stephenson, G., and Hunt, L. (1977) Detoxification of atrazine in three Gramineae subfamilies. *Weed Sci.* **25,** 212–220.
7. McKinlay, R. and Kasperek, K. (1998) Observations on decontamination of herbicide-polluted water by marsh plant systems. *Water Res.* **33,** 505–511.
8. Fernandez, T. R., Whitwell, T., Riley, M. B., and Bernard, C. R. (1999) Evaluating semi-aquatic herbaceous perennials for use in herbicide phytoremediation. *J. Amer. Soc. Hort. Sci* **124,** 539.
9. Cull, R., Hunter, H., Hunter, M., and Truong, P. (2000) Application of vetiver grass technology in off-site pollution control II. Tolerance to herbicides under selected wetland conditions. *Second International Vetiver Conference* pp. 404–408. January 18–22, 2000, Phetchaburi, Thailand.
10. Nair, D. R., Burken, J. G., Licht, L. A., and Schnoor, J. L. (1993) Mineralization and uptake of triazine pesticide in soil-plant systems. *J. Environ. Eng.* **119,** 842–854.
11. Burken, J. G. and Schnoor, J. L. (1996) Phytoremediation: plant uptake of atrazine and role of root exudates. *J. Environ. Eng.* **122,** 958–963.
12. Burken, J. G. and Schnoor, J. L. (1997) Uptake and metabolism of atrazine by poplar trees. *Environ. Sci. Technol.* **31,** 1399–1406.
13. Cole, D. and Edwards, R. (2000) Secondary metabolism of agrochemicals in plants. In: *Metabolism of Agrochemicals in Plants*, (Roberts, T., ed.), John Wiley and Sons, Chichester, UK, pp. 107–154.
14. Cole, D., Edwards, R., and Owen, W. (1987) The role of metabolism in herbicide selectivity. In: *Progress in Pesticide Biochemistry and Toxicology,* (Hudson, D., and Robert, T. eds.), John Wiley, Chichester, UK, pp. 57–104.
15. Kearney, P., Kaufman, D., and Sheets, T. (1965) Metabolites of simazine by *Asperillus fumigatus. J. Agric. Food Chem.* **13,** 369–372.
16. Shimabukuro, R., Kadunce, R., and Frear, D. (1966) Dealkylation of atrazine in mature pea plants. *J. Agric. Food Chem.* **14,** 392–395.
17. Pillai, C., Weete, J., and Davis, D. (1977) Metabolism of atrazine by *Spartina alterniflora*. 1-chloroform-soluble metabolites. *J. Agric. Food Chem.* **25,** 852–855.
18. Edwards, R. and Owen, W. (1989) The comparative metabolism of the s-triazine herbicide atrazine and terbutryne in suspension cultures of potato and wheat. *Pest. Biochem. Physiol.* **34,** 246–254.
19. Cherifi, M., Raveton, M., Picciocchi, A., Ravanel, P., and Tissut, M. (2001) Atrazine metabolism in corn seedlings. *Plant Physiol. Biochem.* **39,** 665–672.
20. Shimabukuro, R. and Swanson, H. (1969) Atrazine metabolism, selectivity and mode of action. *J. Agric. Food Chem.* **17,** 199–205.
21. Shimabukuro, R., Walsh, W., Lamoureux, G., and Stafford, L. (1973) Atrazine metabolism in sorghum: chloroform-soluble intermediates in the N-dealkylation and glutathione conjugation pathways. *J. Agric. Food Chem.* **21,** 1031–1036.

22. Shimabukuro, R., Swanson, H., and Walsh, W. (1970) Glutathione conjugation: atrazine detoxication mechanism in corn. *Plant Physiol.* **46**, 103–107.
23. Coleman, J. O., Blake-Kalff, M. M., and Davies, E. T. (1997) Detoxification of xenobiotics by plants: chemical modification and vacuolar compartmentation. *Trends Plant Sci.* **2**, 144–151.
24. Halkier, B. A. (1996) Catalytic reactivites and structure/function relationships of cytochrome P450 enzymes. *Phytochem.* **43**, 1–21.
25. Bolwell, P. G., Bozak, K., and Zimerlin, A. (1994) Plant cytochrome P450. *Phytochem.* **37**, 1491–1506.
26. Werck-Reichhart, D. (1995) Herbicide metabolism and selectivity: role of cytochrome P450. *Brighton Crop Protection Conference, Weeds.* **2**, 813–822.
27. Burnet, M. W., Loveys, B. R., Holtum, J. A., and Powles, S. B. (1993) Increased detoxification is a mechanism of simazine resistance in *Lolium rigidum. Pest. Biochem. Physiol.* **46**, 207–218.
28. Castelfranco, P., Foy, C., and Deutsch, D. (1961) Non enzymatic detoxification of 2-chloro-4,6-bis(ethylamino)-s-triazine (simazine) by extracts of *Zea mays. Weeds* **9**, 580–591.
29. Raveton, M., Ravanel, P., Serre, A.-M., Nurit, F., and Tissut, M. (1997) Kinetics of uptake and metabolism of atrazine in model plant system. *Pest. Sci.* **49**, 157–163.
30. Shimabukuro, R. (1968) Atrazine metabolism in resistant corn and sorghum. *Plant Physiol.* **43**, 1925–1930.
31. Niemeyer, H. M. (1988) Hydroxamic acids (4-hydroxy-1,4-benzoxazin-3-ones), defence chemicals in the *Graminae. Phytochem.* **27**, 3349–3358.
32. Niemeyer, H. M., Pesel, E., Francke, S., and Francke, W. (1989) Ingestion of the benzoxazinone DIMBOA from wheat plants by aphids. *Phytochem.* **28**, 2307–2310.
33. Pethö, M. (1992) Occurence and physiological role of benzoxazinones and their derivatives. III: possible role of 7-methoxy-benzoxazinone in the iron uptake of maize. *Acta Agron. Hung.* **41**, 57–64.
34. Pethö, M. (1992) Occurence and physiological role of benzoxazinones and their derivatives. IV: isolation of hydroxamic acids from wheat and rye root secretion. *Acta Agron. Hung.* **41**, 167–175.
35. Barnes, J. P. and Putnam, A. R. (1987) Role of benzoxazinones in allelopathy by rye (*Secale cereale* L.). *J. Chem. Ecol.* **13**, 889–906.
36. Nair, M., Whitenack, C. J., and Putnam, A. R. (1990) 2,2-oxo-1,1'-azobenzene. A microbially transformed allelochemical from 2,3-benzoxazolinone: I. *J. Chem. Ecol.* **16**, 353–364.
37. Hamilton, R. H. (1964) Tolerance of several grass species to 2-chloro-triazine herbicides in relation to degradation and content of benzoxazinone derivatives. *J. Agric. Food Chem.* **12**, 14–17.
38. Raveton, M., Ravanel, P., Kaouadji, M., Bastide, J., and Tissut, M. (1997) The chemical transformation of atrazine in corn seedlings. *Pest. Biochem.* **58**, 199–208.
39. Shimabukuro, R. (1967) Atrazine metabolism and herbicidal selectivity. *Plant Physiol.* **42**, 1269–1276.

40. Lamoureux, G. L., Shimabukuro, R. H., Swanson, H., and Frear, D. (1970) Metabolism of 2-chloro-4-ethylamino-6-isopropylamino-s-triazine (atrazine) in excised sorghum leaf section. *J. Agric. Food Chem.* **18**, 81–86.
41. Lamoureux, G. L., Stafford, L. E., and Shimabukuro, R. H. (1972) Conjugation of 2-chloro-4, 6-bis(alkylamino)-s-triazines in higher plants. *J. Agric. Food Chem.* **20**, 1004–1010.
42. Lamoureux, G. L., Stucki, G., Shimabukuro, R. H., and Zaylskie, R. G. (1973) Atrazine metabolism in sorghum: catabolism of the glutathione conjugate of atrazine. *J. Agric. Food Chem.* **21**, 1020–1030.
43. Dixon, D., Cole, D. J., and Edwards, R. (1997) Characterisation of multiple glutathione transferases containing the GST I subunit with activities toward herbice substrates in maize (*Zea mays*). *Pest. Sci.* **50**, 72–82.
44. Shimabukuro, R., Frear, D., Swanson, H., and Walsh, W. (1971) Glutathione conjugation: an enzymatic basis for atrazine resistance. *Plant Physiol.* **47**, 10–14.
45. Hatton, P. J., Dixon, D., Cole, D. J., and Edwards, R. (1996) Glutathione transferase activities and herbicide selectivity in maize and associated weed species. *Pest. Sci.* **46**, 267–275.
46. De Prado, R., Romera, E., and Menédez, J. (1995) Atrazine detoxification in *Panicum dichotomiflorum* and target site *Polygonum lapathifolium*. *Pest. Biochem. Physiol.* **52**, 1–11.
47. De Prado, R., Lopez-Martinez, N., and Gonzalez-Gutierrez, J. (2000) Identification of two mechanims of atrazine resistance in *Setaria faberi* and *Setaria viridis* biotypes. *Pest. Biochem. Physiol.* **67**, 114–124.
48. Giménez-Espinosa, R., Romera, E., Tena, M., and De Prado, R. (1996) Fate of atrazine in treated and pristine accessions of three *Setaria* species. *Pest. Biochem. Physiol.* **56**, 196–207.
49. Wang, R.-L. and Dekker, J. (1995) Weedy adaptation in *Setaria* spp. *Pest. Biochem. Physiol.* **51**, 99–116.
50. Gray, J. A., Balke, N. E., and Stoltenberg, D. E. (1996) Increased glutathione conjugation of atrazine confers resistance in a Wisconsin velvetleaf (*Abutilon theophrasti*) biotype. *Pest. Biochem. Physiol.* **55**, 157–171.
51. Plaisance, K. and Gronwald, J. (1999) Enhanced catalytic constant for glutathiones-s-transferase (atrazine) activity in an atrazine-resistant *Abutilon theophrasti* biotype. *Pest. Biochem. Physiol.* **63**, 34–49.
52. Crayford, J. and Hutson, D. (1972) The metabolism of the herbicide, 2-chloro-4-(ethylamino)-6-(1-cyano-1-methylethylamino)-s-triazine in the rat. *Pest. Biochem. Physiol.* **2**, 295–307.
53. Chaudhry, Q., Schröder, P., Werck-Reichhart, D., Grajek, W., and Mareck, R. (2002) Prospects and limitations of phytoremediation for the removal of persistant pesticides in the environment. *Environ. Sci. Pollut. Res.* **9**, 4–17.
54. US Enviromental Protection Agency (2002) *The Grouping of a Series of Triazine Pesticides Based on a Common Mechanism of Toxicity*. US EPA Office of Pesticide Programs: Health Effects Division.

55. Oh, S. M., Shim, S. H., and Chung, K. H. (2003) Antioestrogenic action of atrazine and its major metabolites *in vitro*. *J. Health Sci.* **49,** 65–71.
56. Ryan, G. (1970) Resistance of common groundsel to simazine and atrazine. *Weed Sci.* **18,** 614–618.
57. Souza Machado, V., Bandeen, J., Stephenson, G., and Jensen, K. (1977) Differential atrazine interference with the Hill reaction of isolated chloroplasts from *Chenopodium album* biotypes. *Weed Res.* **17,** 407–413.
58. Scalla, R. (1990) Obtention de plantes résistantes aux herbicides. In: *Les Herbicides, Mode d'Action et Principes d'Utilisation,* INRA Editions, Paris, France, P 450.
59. Le Baron, H. and Gressel, J. (1982) *Herbicide Resistance in Plants.* Wiley J and Sons Inc., New York, NY.
60. Adams, R., Zhong, M., Turuspekov, Y., Dafforn, M., and Veldkamp, J. (1998) DNA fingerprintings reveals clonal nature of *Vetiveria zizanioides* (L.) Nash, Gramineae and sources of potential new germplasm. *Molecular Ecol.* **7,** 813–818.
61. Zuloaga, F., Morrone, O., Davidse, G., et al. (2003) Catalogue of New World Grasses (*Poaceae*): III. Subfamilies Panicoideae, Aristidoideae, Arundinoideae, and Danthonioideae. *Contr. U.S. Natl. Herb.* **46,** 1–662.
62. Bertea, C. M. and Camusso, W. (2002) Anatomy, biochemistry and physiology. In: *Vetiveria,* (Maffei, M., ed.), Taylor and Francis, London and New York, pp. 19–43.
63. Leupin, R. E. (2001) *Vetiveria zizanioides: an approach to obtain essential oil variants via tissue cell culture.* PhD thesis, ETH, Zürich, Switzerland.
64. National Research Council (1993) *Vetiver Grass. A Thin Line Against Erosion.* National Academy Press, Washington, DC.
65. Dalton, P., Smith, R., and Truong, P. (1996) Vetiver grass hedges for erosion control on a cropped flood plain: hedge hydraulic. *Agric. Water Manag.* **31,** 91–104.
66. Marcacci, S. (2004) *A phytoremediation approach to remove pesticides (atrazine and lindane) from contaminated environment.* PhD thesis, EPFL, Lausanne, Switzerland.
67. Anhalt, J. C., Arthur, E. L., Todd, A. A., and Coats, J. R. (2000) Degradation of atrazine, metolachlor, and pendimethalin in pesticide-contaminated soils: effects of aged residues on soil respiration and plant survival. *J. Environ. Sci. Health B* **B35,** 417–438.
68. Arthur, E. L., Perkovich, B. S., Anderson, T. A., and Coats, J. R. (1999) Degradation of an atrazine and metolachlor herbicide mixture in pesticide-contaminated soils from two agrochemical dealerships in Iowa. *Water, Air, Soil Pollut.* **119,** 75–90.
69. Grigg, B. C., Bishoff, M., and Turco, R. F. (1997) Cocontaminant effects on degradation of triazine herbicides by a mixed microbial culture. *J. Agric. Food Chem.* **45,** 995–1000.
70. Thurman, E. and Meyer, M. (1996) Herbicide metabolites in surface water and groundwater: introduction and overview. In: *Herbicide Metabolites in Surface Water and Groundwater,* (Meyer, M. and Thurman, E., eds.), American Chemical Society Washington, DC, pp. 1–15.
71. Keller, K. E. and Weber, J. B. (1995) Mobility and dissipation of ^{14}C-labeled atrazine, metolachlor, and primisulfuron in undisturbed field lysimeters of a coastal plain. *J. Agric. Food Chem.* **43,** 1076–1086.

72. Radosevich, M., Traina, S. J., and Tuovinen, O. H. (1996) Biodegradation of atrazine in surface soils and subsurface sediments collected from an agricultural research farm. *Biodegrad.* **7,** 137–149.
73. Zargorc-Koncan, J. (1996) Effects of atrazine and alachlor on self-purification processes in receiving streams. *Water Sci. Technol.* **33,** 181–187.
74. Tissut, M., Arnaud, L., Nurit, F., and Ravanel, P. (1991) Présence des herbicides dans les eaux. Relations avec leur mode d'action. *Eau, Agric. Environ.* **30,** 157–162.
75. Businelli, M., Marini, M., Businelli, D., and Ggliotti, G. (2000) Transport to ground-water of six commonly used herbicides: a prediction for two Italian scenarios. *Pest Manag. Sci.* **56,** 181–188.
76. Wilson, C. P., Whitwell, T., and Klaine, S. J. (1999) Phytotoxicity, uptake, and distribution of ^{14}C simazine in *Canna hybrida* "Yellow King Humbert". *Environ. Toxicol. Chem.* **18,** 1462–1468.
77. Jablonkai, I. and Hatzios, K. (1993) *In vitro* conjugation of chloroacetanilide herbicides and atrazine with thiols and contribution of nonenzymatic conjugation to their glutathione-mediated metabolism in corn. *J. Agric. Food Chem.* **41,** 1736–1742.

20

Exploiting Plant Metabolism for the Phytoremediation of Organic Xenobiotics

Peter Schröder

Summary

Phytoremediation of organic pollutants has become a topic of great interest in many countries because of the increasing number of recorded spill sites. When applying plant remediation techniques to unknown pollutant mixtures, information on the uptake rates as well as on the final fate of the compounds is generally lacking. A range of compounds is easily taken up by plants, while others may stay motionless and recalcitrant in the soil or sediment. Uptake is a necessary prerequisite for close contact between the pollutant and the detoxifying enzymes, which are localized in the cytosol of living plant cells. The presence and activity of these enzymes is crucial for potential metabolism and further degradation of the chemicals under consideration. Conjugation to biomolecules is regarded as a beneficial detoxification reaction. The present chapter lists several prerequisites for pollutant uptake and summarizes information on conjugating detoxification reactions. The final fate of compounds and the contribution of rhizobacteria are critically discussed and perspectives for the development of this promising technology are given.

Key Words: Glycosyl conjugation; glutathione conjugation; xenobiotic metabolism; plant uptake.

1. Introduction

Today, we have to face the situation that hundreds of thousands of sites across Europe are characterized as polluted with organic chemicals because of industrial processes, spills, and accidents, or resulting from improper use of chemicals. Among them are chemicals of remarkable stability and recalcitrance in soils and water bodies. In Germany, more than 70,000 polluted sites have been identified [1], and the US superfund sites are also numerous [2]. Public awareness of environmental contamination is steadily increasing. Several authors have reviewed the situation in Europe and other continents, and it has become clear that under most conditions mixed pollution situations are found. Attempts to remove the unwanted chemicals from these sites include semi-industrial processes and incineration, but also more environmentally friendly methodologies such as bioremediation and phytoremediation. This is especially true for cases where pollution of water, soils, or sediments is not so severe that immediate removal of the whole environmental compartment (e.g., the soil, sediment, or water) is necessary, but also applies in areas without direct pressure of reuse where alternative technologies might be utilized to achieve control over pollution. Plant-based pollution treatment thus offers options for chemical removal or stabilization in environmental media and has the advantage that landscape and soil will not be destroyed. However, it has to be kept in mind that excessive pollution might kill the plants because plants and rhizobacteria have only limited capabilities to detoxify, for example, chlorinated xenobiotics with multiple ring structures such as polyaromatic hydrocarbons (PAH,) polychlorinated biphenyls (PCB,) dioxins, or related compounds.

In many cases the decision about which technique to be used is made by the land owner, and only in situations where publicly accessible sites are found to be polluted would local authorities be directly involved in the process. In both cases, however, at present the choice of plants seems to be more or less arbitrary, i.e., based on the availability of plants or the overall costs of the action. To make phytoremediation successful, it will be necessary to define a set of parameters for the measures to be taken that will describe uptake, transport, translocation, metabolism, and fate of the compounds of interest. The present chapter attempts to give some hints on the expert knowledge required for such an action.

2. Uptake and Transport

The first consideration to be taken into account, if phytoremediation is to be applied, is not the costs or the duration of the process, but the chances that the plants of choice would be able to access the pollutants in the soil with their root system *(3–5)*. This would mean that, depending on the location and depth of pollution, different plants would have to be chosen. Deep rooting species will be able to access pollutant plumes that have already moved to deeper zones of the soil horizon, whereas surface pollution might be easily controlled with shallow rooting plants. In cases where the chemicals have been aged in the soil, or have been complexed with minerals or organic matter, the rooting pattern of the plant under consideration is decisive for success. It is necessary to choose plants with a dense root system for such a purpose.

Once in direct contact with the root surface, the transfer from soil or water to the plant has to be mediated. This happens spontaneously and is diffusion driven for compounds with a lipophilicity close to that of the respective plant root. Root uptake and transport of organic xenobiotics has been reviewed by a number of authors *(6–9)*. Uptake of lipophilic and amphiphilic compounds from the soil has been intensively studied in the context of pesticide application. Data are available for the determination of the so-called root concentration factor (RCF). The RCF describes the potential of a given xenobiotic to accumulate in the plant root, and makes no differentiation between surface accumulation and uptake into the root tissue. However, as the RCF is heavily dependent on the log K_{ow}, i.e., the lipophilicity of the compound under consideration, it seems to be governed by the absorptive properties of the root epidermis.

Briggs et al. *(6)* reported the following relationship for barley:

$$\text{Log (RCF} - 0.82) = 0.77 \log K_{ow} - 1.52$$

Those compounds exhibiting a low K_{ow} (i.e., <1) will only sparsely penetrate the lipid-containing root epidermis, whereas those compounds of log K_{ow}>3 become increasingly retained by the lipid in the root epidermis and the mucilage surrounding the root as a result of their increasing hydrophobicity *(10)*. In this case, uptake might be minute, if detectable at all.

Uptake into the hydraulic system of the plant, and thus the path into stem and leaves, may be quantified by calculating the transpiration stream concentration factor. Here, compounds of intermediate solubility, weak acids, and amphiphilic substances are predisposed to transport. Compounds with a log $K_{ow} \approx 2$ are transported solely in the transpiration stream through xylem, whereas those with a log $K_{ow} \approx 1$ are both phloem and xylem mobile, although it is probable that only metabolites enter the phloem. For compounds with log K_{ow} between one and three, metabolism may occur in the leaf and stem tissue and there may be released into the atmosphere through leaf tissue, and additional "bound residue" can be created in the plants *(11)*.

Plants with a huge root system will be advantageous for the uptake of organics from the soil. Depending on the type and location of pollution in the soil, it might be interesting to distinguish between plants with a shallow or a deeply penetrating root system. Root growth and penetration into the soil can to some extent be influenced by fertilizer treatment. It is known, for example, that plants watered and fertilized well do not develop extensive root systems, whereas plants from nitrate-limited soils at lower water potential will develop more prominent root systems. Unfortunately for the process under consideration, organic pollution frequently coincides with surplus nutrients in the soil. To exploit a plant's potential for root growth, one option would be to remove the nutrients in advance of the remediation process. Underestimated in many cases, the role of rhizospheric bacteria and mycorrhizal fungi might be decisive for the solubilization and uptake of pollutants into the plant. Vice versa, the plant root system acts as a shuttle for the spread of rhizobacteria in the soil and provides the microbes with root exudates as nutrients. It has also been shown that, under the influence of certain plant root exudates, rhizobacteria can produce biosurfactants and thus facilitate the solubilization of pollutants from the soil *(12)*.

To date, the processes mediating xenobiotic loading into the xylem and entry into leaf tissue have not been well investigated but are thought to be analogous to herbicide movement in the plant. It is assumed, but not known, that metabolism of organic xenobiotics in plants is confined to root and leaf tissues, and is only scarcely taking place during transport in the plant's vasculature.

3. Detoxification Mechanisms of Plants

Both plant roots and leaves have been shown to possess elaborate detoxification mechanisms for organic xenobiotics, predominantly herbicides. It has been demonstrated that herbicide tolerance in numerous crops, as well as resistance in weeds, is caused by the action of these enzyme systems. Besides agrochemicals, few foreign compounds have been investigated, but the detoxification mechanisms have been explored with halogenated organic model compounds like chloroanilins and chlorobenzenes. It has generally been accepted that the enzyme systems responsible, although not physiologically connected, form a metabolic cascade for the detoxification, breakdown, and final storage of organic xenobiotic. It has been suggested that this network of reactions is analogous to a "green liver" *(13)*.

The detoxification cascade was first described by Shimabukuro *(14)* who subdivided xenobiotic metabolism into three distinct phases: (1) activation of the xenobiotic, (2) detoxification, and (3) excretion, in analogy to human hepatic metabolism. The cascade (**Fig. 1**) comprises activation reactions catalyzed by esterases, p450 mono-oxygenases, and peroxidases (POX), then true detoxification reactions in phase II performed by glutathione and glucosyltransferases, rendering the compound under consideration less toxic because of conjugation, and then a set of further reactions that include cleavage, rearrangement, secondary conjugation, and the like. Recently, this last phase has been proposed to be subdivided into two independent phases, one confined to transport and storage in the vacuole, and a second one taking final reactions, e.g., cell-wall binding or excretion, into account *(15)*.

Metabolic activation of xenobiotics is in most cases catalyzed by p450 mono-oxygenases or POX. These enzymes are localized in membrane fractions of plant cells (p450 mono-oxygenases), in the apoplast and in the cytosol (POX). Xenobiotic conjugation in plants was investigated in depth for pesticides, and several isoforms of glutathione-*S*-transferases (GST), glucosyltransferases (GT), malonyltransferases, and an array of processing enzymes have been identified in crops *(16–18)*. The initial phase of chemical activation is followed by such conjugation reactions; sugars, amino acids, or glutathione

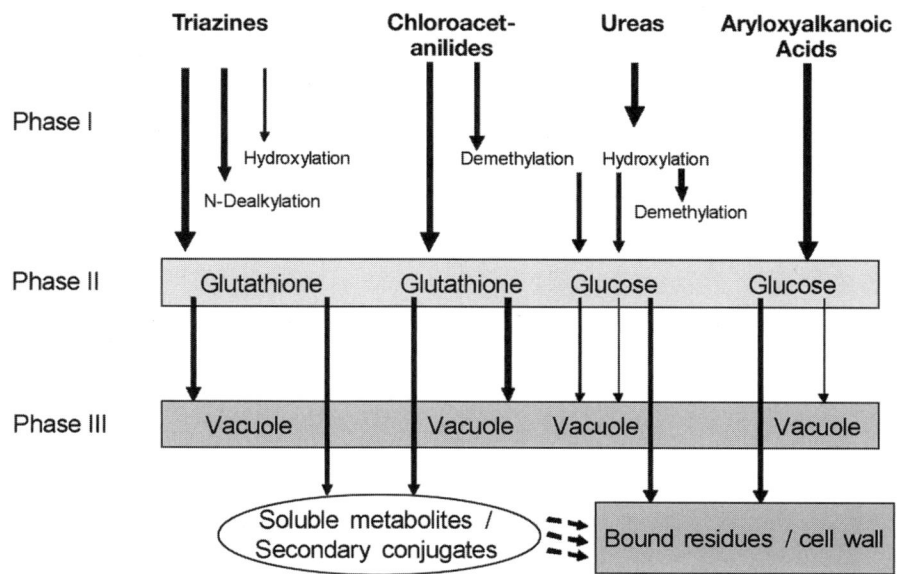

Fig. 1. Fate of herbicides in plants, according to the three-phase model first described by Shimabukuro et al. *(14)*, with modifications from Coleman et al. *(20)*.

may be transferred to the activated xenobiotic depending on the structure of the molecule and its active sites. OH⁻, NH₂⁻, SH⁻, and COOH⁻ functional groups on a molecule usually trigger glycosyl transfer (**Fig. 2**) mediated by GTs (E.C. 2.4.1.x *[19]*), whereas the presence of conjugated double bonds, halogen- or nitro-functional groups lead to glutathione conjugation catalyzed by GSTs (E.C.2.5.1.18 *[20]*). Amino acid conjugation has infrequently been described and seems to be a side reaction rather than a main detoxification step. Schröder and Collins have recently pointed out that, in the context of phytoremediation, it is crucial to know which type of primary conjugation occurs *(21)* because this will determine the final fate of the compound *(19)*.

The action of electrophilic xenobiotics in living tissue seems to depend on their nucleophilic cellular counterparts. There is a preference for reactions between xenobiotics and biomolecular partners, which may be explained by the concept of hard and soft nucleophiles/electrophiles *(20)*. Any reaction with hard electrophiles requires additional enzymatic support, which may be provided by GST isoenzymes. In any case, detoxification totally depends on the availability of glutathione. The homeostasis of glutathione inside the plant is maintained by a complex regulation process including synthesis, degradation, and long-range transport *(22)*.

Numerous herbicides are conjugated to sugars via *O*-glucosyl-transfer or *N*-glucosyl-transfer. These reactions are catalyzed by different enzyme families *(23)*.

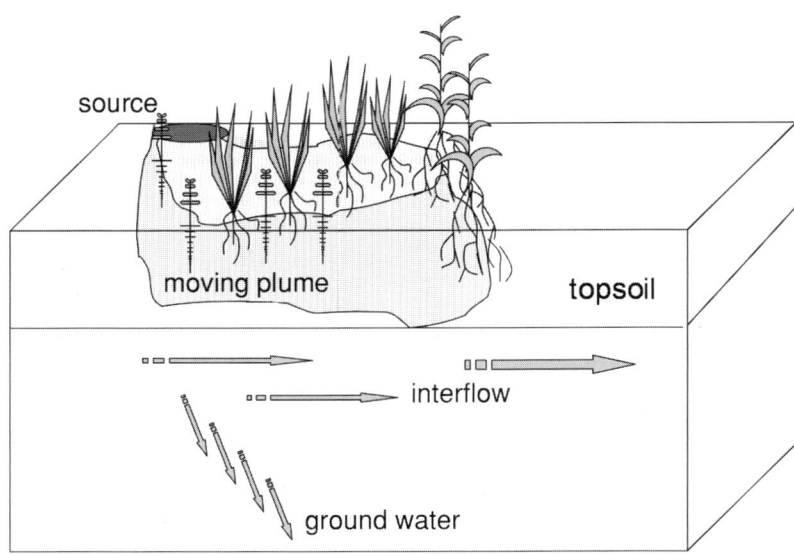

Fig. 2. Phytoremediation on a plume of soil pollutants. The advantage of mixed planting becomes obvious when rooting depth, rhizodiversity, and oxygen transport to the plume are considered.

The nonidentity of these enzymes has been demonstrated several times, although overlapping activities have been found in some cases. It is significant that conjugation may occur directly on the parent compound in many cases, but that sometimes activation may be needed in advance to provide the xenobiotic with the respective activated sites (see **Fig. 1**, Phase I). As p450 mono-oxygenases act on many of the mentioned compounds, hydroxylation will favor O-glucoside formation (**Fig. 1**). Conjugation may occur either at OH-groups of the molecule to form O-glucosides or at carboxy-functions to form acylglucosides *(13)*. For N-glucosyltransfer, coupling to NH_2-groups of the molecule is crucial. In summary, goals for the metabolic control of the remediation process will be have to take into account activating or detoxifying enzymes, but also the availability of the conjugation partners, e.g., UDP-glucose, amino acids, or glutathione and its analogs.

4. Species Specificity

Astonishingly, all plants possess the required enzyme classes *(24)*, however, specific activities of single enzymes for the metabolism of distinct chemicals might be not available in each plant species *(25,26)*. A prominent example for this is the selectivity of herbicides. Sulfonylureas and triazines are only detoxified by some species, because they possess the correct isoforms of p450 or GSTs capable of attacking the respective pesticide. Atrazine, for example, is

Exploiting Plant Metabolism

only detoxified by maize and sorghum species, whereas other plants and grasses are not equipped with the responsible isoenzymes and are killed on exposure to the compound.

Historically, our present state of knowledge of foreign-compound metabolism is focused on crops and a few ornamental plants. Only scattered reports exist for plants that are interesting for phytoremediation. The European Cooperation in the field of Scientific and Technical Research (COST) action 837 (lbewww.epfl.ch/COST837) has aimed to recognize and explore wild plants and crops for phytoremediation. A first candidate list of more than 50 species has been reduced, by practical studies as well as by surveying existing remediation sites, to a few species that have proven to be good candidates for several reasons. However, conjugative, i.e., detoxifying, metabolism in outstanding candidates *Arundo donax, Brassica juncea, Phragmites* sp., *Typha* sp., *Plantago majus, Populus* sp., *Salix* sp., to name but a few, have not been investigated in any depth. This situation is awkward because there are already numerous existing field sites that seem to be very successful in the removal of xenobiotics from soil and water. Knowing about the mechanisms involved, the efficiency of these systems could probably be improved when methods to increase metabolism rates are applied. One path could be to add inducers of herbicide resistance to the plants; another could be to enhance xenobiotic uptake and transport to tissues with high degradative activity by modern molecular methods. Each of these attempts would increase the chances for a commercial and publicly accepted use of phytoremediation and help to clean the environment.

In the past, most remediation sites were only planted with single plant species. This had the advantage that the fields were easily cultivated and maintained. Examples of the removal of pollutants by grass species, dicots, and trees have been published. However, it has long been known from other fields of applied botany, that mixed stands develop unique properties enhancing the performance of the whole system. To exploit plant metabolism for the decontamination of a site or a water body, it might be of great interest to have a pollution plume planted with a certain species that combines low transpiration and high-exudate production for bacterial stimulation, followed and secured by a belt of deep-rooting species with high transpiration and uptake, and a mixed canopy of perennial plants for protection of the site. These plantations might be adapted according to site-specific characteristics (**Fig. 2**).

The choice of plants might be assisted by expert systems, computer-aided pollution distribution calculations, and by novel decision tools based on the chemical ecology of plants. The latter approach takes into account that the normal physiology of plants includes the synthesis of complex secondary compounds that might resemble structures of organic pollutants. Enzymes in the metabolic pathway of these natural compounds are good candidates for the detoxification of structurally similar xenobiotics. Such an approach has been followed by Schwitzguebel et al. *(27)* for the phytoremediation of sulfonated anthraquinones. Ongoing discussions indicate (C. Collins, personal communication) that modern molecular tools in plant taxonomy will aid the search for promising species with the desired metabolic traits.

5. Increased Pollution Degradation by Enhanced Rhizosphere Activity

Co-metabolic processes in the root zone are responsible for most of the microbial pollutant degradation processes found. There is a large and so far unexploited potential to increase microbial metabolism in the root zone, be it by fertilizer application or plant choice. Under certain conditions, soil bacteria are also able to produce biosurfactants that will enhance solubilization and removal of organic pollutants from the soil. Thus, the selection of the proper plant might be of higher impact on the rhizosphere than any other process, as plants through their exudation patterns are able to influence the rhizosphere composition and turnover rates. For example, PAH degradation in an unrooted soil was shown to increase after fertilization, but degradation was highest when artificial root exudates (i.e., carboxy acids and sugars) were added *(12)*. It has been shown that such an effect will also occur under real-life conditions, and that microbial abundance is strongly increased in vegetated soils *(28)*. As a bacterial remediation process will usually include the action of various organism classes, one of the most important steps toward reliable performance of phytoremediation would be the secure utilization of the rhizobacterial biodiversity and its steering by plant exudation. However, we are far from understanding the exact biochemical background of these reactions, especially because many of the most interesting bacterial communities seem to be still unculturable *(29)*.

6. Induction of Detoxification Enzymes in Selected Species

In the context of increasing herbicide tolerance, the inducibility of both activating and conjugating enzymes has been studied in some detail in crop plants. p450 mono-oxygenases have been shown to be chemically induced by a number of organic xenobiotics, including PAH. The mechanism of induction seems to be analogous to animal p450 induction *(30,31)*, however no specific aromatic hydrocarbon receptors (AH-receptors) have so far been detected in plants. Nevertheless, plant p450s are induced up to 20-fold in plants, thus increasing the capacity for pesticide detoxification significantly *(32,33)*. Similarly, other phase I enzymes, e.g., the peroxidases, are easily induced by stressors of either chemical or biological origin.

It has been shown that the glucosyltransferase activities are hardly inducible in plants and might thus present a class of housekeeping enzymes *(34)*. This is contrary to GSTs, the inducibility of which has been demonstrated frequently

Exploiting Plant Metabolism 259

(16,17,34,35). Using this option, care should be taken to have a look at enzymes further downstream in metabolism of xenobiotics *(36)*. It is important to note in this context that the individual enzymes responsible for the respective reactions might well be under developmental control and that the conjugation of single xenobiotics cannot be expected to proceed throughout the plant's life and in every plant part.

7. Molecular Tools for Over-Expression of Detoxification Enzymes

The way out of this dilemma might be the use of molecular techniques to improve the plant's performance in phytoremediation. Different targets for these techniques need to be identified. Among them are the classical detoxification enzymes, but also enzymes further downstream in the transport or metabolism of xenobiotic compounds. Besides ABC transporters *(20)*, vacuolar-processing enzymes might be welcome targets for such an approach *(37)*. Some evidence has been found that coinduction is possible, for example with the herbicide antidote (safener) cloquintocet-mexyl *(38)*, but there are also data indicating that these enzyme systems can be stimulated separately. Several groups have provided evidence for the possible overexpression of detoxification enzymes in plants *(32,34,39)*, and also the enhanced biosynthesis of required metabolites, e.g., glutathione, has been demonstrated in transgenic plants.

However, as has been clearly pointed out in discussions of the COST Action 837, public opinion is strongly against the use of transgenic plants in phytoremediation. As long as the public feeling is against the use of these techniques, it would be wise to apply classical breeding techniques to avoid corruption of the acceptance of the methodology. Most of the plants used in phytoremediation these days possess good chances for propagation because they are perennial plants currently exploited in wetland management. *Typha*, *Arundo*, and *Phragmites* are usually propagated vegetatively, but also techniques for micropropagation from callus have been described *(40)*. The latter method is advantageous as it allows for the rapid complementation of the detoxification pathway.

8. Changing Views: Stabilization of Xenobiotics in Soil

Phytoremediation will change the nature of pollutants to an extent that makes them available for further metabolism. Although this means in the first instance that pollutants will become bioavailable to the plant in charge of the removal job, the enhanced water solubility of metabolites will also facilitate their escape into the surrounding media. This solubilization effect is a major constraint to heavy metal removal, but it might also play a role for organic xenobiotics *(41)*. We are presently lacking research on the binding of metabolized pollutants to the organic fraction of the soil or to soil minerals. The formation of these bound residues with the soil is, however, by no means to be regarded as critical, as it might be a step toward stabilization of the pollutant plume and prevent it from leaching into ground water or neighboring systems. Once covalently bound to a certain soil fraction, other measures like soil washing or treatment with microbes might become an additional option and a later step of the remediation process *(42)*.

9. Chances and Options: Designing Rhizospheres

With a view to the processes described, it will be of some significance to the whole field of biological pollutant removal to develop sound concepts and expert systems before explicit measures be taken. Especially the option of inoculating soils with potent rhizobacteria and using plants that exhibit specific traits concerning their detoxification capacity or concerning the exudate pattern, will be one of the most important tasks for the near future. On the other hand, the aforementioned stabilization of the pollutant in a certain depth and region of the soil should also be considered as an option in the case of multiple pollution with compounds of high recalcitrance. By this means it might be possible to either speed up the desired remediation process and/or to obtain the desired end products and land use options. In any case, it will be of paramount importance to assess bioavailability of pollutants and the potential toxicity of the released products. With a view to the sustainability of this emerging green technology, this is worth following and establishing in as many cases as possible.

References

1. Franzius, V. (1994) Aktuelle Entwicklungen zur Altlastenproblematik in der Bundesrepublik Deutschland. *Umwelt Technologie Aktuell* **6,** 443–449.
2. Bouwer, E., Durant, N., Wilson, L., Zhang, W., and Cunningham, A. (1994) Degradation of xeno-biotic compounds in situ: capabilities and limits. *FEMS Microb. Rev.* **15,** 307–317.
3. Schnoor, J. L., Licht, L. A., McCutcheon, S. C., Wolfe, N. L., and Carreira, L. H. (1995) Phytoremediation of organic and nutrient contaminants. *Environ. Sci. Technol.* **29,** 318–323.
4. Simonich, S. L. and Hites, R. A. (1995) Organic pollutant accumulation in vegetation. *Environ. Sci. Technol.* **29,** 2905–2914
5. Newman, L., Strand, S., Choe, N., et al. (1997) Uptake and biotransformation of trichloro-ethylene by hybrid poplars. *Environ. Sci. Technol.* **31,** 1062–1067.
6. Briggs, G. G. and Bromilow, R. H. (1983) Relationships between lipophilicity and the distribution of non-ionized chemicals in barley shoots following uptake by the roots. *Pestic. Sci.* **14,** 492–500.
7. Briggs, G. G., Bromilow, R. H., and Evans, A. A. (1982) Relationships between lipophilicity and root uptake and translocation of non-ionized chemicals by barley. *Pestic. Sci.* **13,** 495–504.
8. Behrendt, H. and Brüggemann, R. (1993) Modeling the fate of organic-chemicals in the soil-plant environment: model study of root uptake of pesticides. *Chemosphere* **27,** 2325–2332.

9. Rigitano, R. L. O. and Briggs, G. G. (1986) Phloem translocation of xenobiotics in plants - a physicochemical approach. *Pestic. Sci.* **17,** 62–63.
10. Sicbaldi, F., Sacchi, G. A., Trevisan, M., and Del-Re, A. A. M. (1997) Root uptake and xylem translocation of pesticides from different chemical classes. *Pestic. Sci.* **50,** 111–119.
11. Langebartels, C. and Harms, H. (1986) Plant cell suspension cultures as test systems for an ecotoxicological evaluation of chemicals. *Angew. Bot.* **60,** 113–123.
12. Joner, E., Corgie, S., Amellal, N., and Leyval, C. (2002) Nutritional constraints to PAH degradation in a rhizosphere model. *Soil Biol. Biochem.* **34,** 859–864.
13. Sandermann, H., Haas, M., Messner, B., Pflugmacher, S. Schröder, P., and Wetzel, A. (1997) The role of glucosyl and malonyl conjugation in herbicide selectivity, in *Regulation of Enzymatic Systems Detoxifying Xenobiotics in Plants,* (Hatzios, K. K., ed.), NATO ASI Series, Vol. 37, Kluwer, Kluwer Acad. Publ., Dordrecht, The Netherlands. pp. 211–231.
14. Shimabukuro, R. H., Walsh, W. C., and Hoerauf, R. A. (1979) Metabolism and selectivity of Diclofop-methyl in wild oat and wheat. *J. Agric. Food Chem.* **27,** 615–623.
15. Theodoulou, F. L. (2000) Plant ABC transporters. *Biochim. Biophys. Acta* **1465,** 79–103.
16. Lamoureux, G. L. and Rusness, D. G. (1989) The role of glutathione and glutathione S-transferases in pesticide metabolism; selectivity and mode of action in plants and insects. In: *Glutathione: Chemical Biochemical and Medical Aspects, Vol IIIB, Ser: Enzyme and Cofactors,* (Dolphin, D., Poulson, R., and Avramovic, O., eds.), Wiley and Sons, New York, pp. 153–196.
17. Schröder, P. (1997) Fate of glutathione S-conjugates in plants: cleavage of the glutathione moiety. In: *Regulation of Enzymatic Systems Detoxifying Xenobiotics in Plants,* (Hatzios, K. K., ed.), NATO ASI Series Vol. 37, Kluwer, Kluwer Acad. Publ., Dordrecht, The Netherlands. pp. 233–244.
18. Schröder, P. (2001) The role of glutathione and glutathione S-transferases in the adaptation of plants to xenobiotics. In: *Significance of Glutathione in Plant Adaptation to the Environment, Handbook Series of Plant Ecophysiology,* (Grill, D., Tausz, M., and DeKok, L. J., eds.), Kluwer, Kluwer Acad. Publ., Dordrecht, The Netherlands. pp. 157–182.
19. Frear, D. S. (1976) Pesticide conjugates: glycosides. In: *Bound and Conjugate Pesticide Residues,* (Kaufman, D. D., Still, G. G., Paulson, G. D., and Bandal, S. K., eds.), ACS Symposium 29, Washington DC, American Chemical Society, pp. 35–54.
20. Coleman, J. O. D., Randall, R. A., and Blake-Kalff, M. M. A. (1997) Detoxification of xenobiotics by plants: chemical modification and vacuolar compatimentation. *Trends Plant Sci.* **2,** 144–151.
21. Schröder, P. and Collins, C. J. (2002) Conjugating enzymes involved in xenobiotic metabolism of organic xenobiotics in plants. *Int. J. Phytorem.* **4,** 1–15.
22. Noctor, G., Gomez, L., Vanacker, H., and Foyer, C. H. (2002) Interactions between biosynthesis, compartmentation and transport in the control of glutathione homeostasis and signalling. *J. Exp. Bot.* **53,** 1283–1304.

23. Messner B., Thulke O., and Schäffner, A. R. (2003) Arabidopsis glucosyltransferases with activities toward both endogenous and xenobiotic substrates. *Planta* **217,** 138–146.
24. Kreuz, K., Tommasini, R., and Martinoia, E. (1996) Old enzymes for a new job. Herbicide detoxification in plants. *Plant Physiol.* **111,** 349–353.
25. Pflugmacher, S. and Sandermann, H. (1998) Taxonomic distribution of plant glucosyltransferases acting on xenobiotics. *Phytochemistry* **49,** 507–511.
26. Pflugmacher, S., Sandermann, H., and Schröder, P. (2000) Taxonomic distribution of plant glutathione S-transferases acting on xenobiotics. *Phytochemistry* **54,** 267–273.
27. Schwitzguebel, J. P., Aubert, S., Grosse, W., and Laturnus, F. (2002) Sulphonated aromatic pollutants. Limits of microbial degradability and potential of phytoremediation. *ESPR* **9,** 62–72.
28. Reynolds, C. M., Wolf, D. C., Gentry, T. J., et al. (1999) Plant enhancement of indigenous soil micro-organisms: a low cost treatment of contaminated soils. *Polar Record* **35,** 33–40.
29. Hoagland, R. E., and Williams, R. D. (1985) The influence of secondary plant compounds on the associations of soil microorganisms and plant roots. In: *The Chemistry of Allelopathy.* (Thompson, AcEd.) ACS Publications, Washington, DC, pp. 301–325.
30. Durst, F. and O'Keefe, D. P. (1995) Plant cytochromes P450: an overview. *Drug Metabol Drug Interact.* **12,** 171–187.
31. Morant, M., Bak, S., Moller, B. L., and Werck-Reichhart, D. (2003) Plant cytochromes P450: tools for pharmacology, plant protection and phytoremediation. *Curr. Opin. Biotechnol.* **14,** 151–162.
32. Werck-Reichhart, D. (1995) Herbicide metabolism and selectivity: role of cytochrome P450. *Proc. Br. Crop. Prot. Conf-Weeds* **3,** 813–822.
33. Werck-Reichhart, D., Hehn, A., and Didierjean, L. (2000) Cytochromes P450 for engineering herbicide tolerance. *Trends Plant Sci.* **5,** 116–123.
34. Loutre, C., Dixon, D. P., Brazier, M., Slater, M., Cole, D. J., and Edwards, R. (2003) Isolation of a glucosyltransferase from Arabidopsis thaliana active in the metabolism of the persistent pollutant 3,4-dichloroaniline. *Plant J.* **34,** 485–493.
35. Marrs, K. A. (1996) The functions and regulation of glutathione S-transferases in plants. *Annu. Rev. Plant Physiol.* **47,** 127–158.
36. Schröder, P., Nathaus, F., Lamoureux, G. L., and Rusness, D. G. (1993) The induction of glutathione S-transferase and C-S lyase in the needles of spruce trees. *Phyton* **32,** 127–131.
37. Wolf, A. E., Dietz, K. J., and Schröder, P. (1996) A carboxypeptidase degrades glutathione conjugates in the vacuoles of higher plants. *FEBS Lett.* **384,** 31–34.
38. Theodoulou, F. L., Clark, I. M., He, X. L., Pallett, K. E., Cole, D. J., and Hallahan, D. L. (2003) Co-induction of glutathione-S-transferases and multidrug resistance associated protein by xenobiotics in wheat. *Pest. Manag. Sci.* **59,** 202–214.
39. Kellner, D. G., Maves, S. A., and Slingar, S. G. (1997) Engineering cytochrome P450s for bioremediation. *Curr. Opin. Biotechnol.* **8,** 274–278.

40. Rogers S. M. D. (2003) Tissue culture and wetland establishment of the freshwater monocots *Carex, Juncus, Scirpus,* and *Typha*. In vitro cellular and developmental biology. *Plant* **39,** 1–5.
41. Scheunert, I. and Schröder, P. (1998) Formation, characterization and release of non-extractable residues of [14 C]-labeled organic xenobiotics in soils. *ESPR* **5,** 238–244.
42. May, R. G., Schröder, P., and Sandermann, H. (1997) An *ex-situ* process for treating PAH contaminated soil with *Phanerochaete chrysosporium. Environ. Sci. Technol.* **31,** 2626–2633.

21

Searching for Genes Involved in Metal Tolerance, Uptake, and Transport

Viivi H. Hassinen, Arja I. Tervahauta, and Sirpa O. Kärenlampi

Summary

Despite the recent exploitation of high-throughput methodologies such as cDNA microarrays, the overall picture of plant metal tolerance, accumulation, and translocation is far from complete. Understanding of this network would be beneficial for the optimization of the phytoremediation technique. This chapter compiles the key approaches in the search for novel genes from model plant species as well as other organisms, and briefly describes the genes known thus far to be involved in metal homeostasis in plants. In addition to unravelling the genes, the functional connections between genes, proteins, metabolites, and mineral ions should be understood. Thus, to get a full understanding of the processes, different analytical methods are needed. The main focus of this chapter is in the "omic" technologies, such as transcriptomics, proteomics, and metabolomics, and their potential in the discovery or analysis of the molecules that may play a significant role in metal tolerance, accumulation, and translocation in plants.

Key Words: Omics; metal tolerance; hyperaccumulation; phytoremediation; phytotechniques.

1. Introduction

There is an increasing demand to develop environmentally sound technologies for remediating contaminated sites. Phytoremediation is emerging as one such "green" technology. Although there are cases where this technology has been successfully exploited, it has not yet reached its full potential. Clean up of metal-contaminated land by phytoremediation could be done more cost effectively by using plants with improved metal uptake, translocation, and tolerance, and which also grow faster and produce high biomass *(1,2)*. Even though to date only a few field studies on GM plants—some of them possibly marginally related to phytoremediation—have been conducted there has been interesting development at the

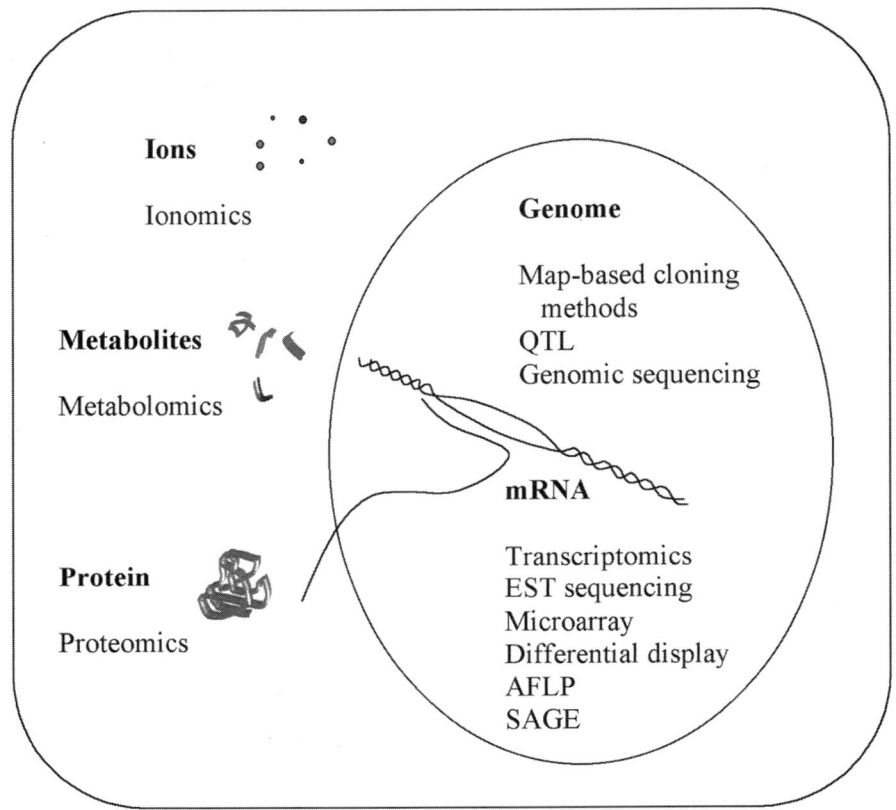

Fig. 1. Modern tools for searching mechanisms behind differential responses to environmental stimuli, such as heavy metal stress.

laboratory level *(3)*. Because our understanding of the tightly regulated metal homeostasis network in plants is not complete *(4)* it is essential to find the genes involved in this network, and particularly in metal tolerance or uptake. This chapter compiles some of the key approaches in the search for these genes (**Fig. 1**) but it should be useful also for readers tackling other biological problems.

2. Where to Search for the Genes?

When searching for genes with desirable characteristics, all the kingdoms—archaea, bacteria, animals, protozoans, plants, and fungi—are at our disposal. The most logical places to start the search of species with extraordinary capabilities of metal tolerance or uptake are extreme environments highly contaminated with metals. As an alternative, well-known model organisms can be improved by mutagenization or other means of genetic modification. However, the ultimate method is to partly, or completely, design synthetic genes by exploiting the fast-developing bioinformatic tools. In the following, a summary and examples are given of the groups of organisms in which interesting findings have been made concerning metal tolerance or uptake.

2.1. Bacteria

Bacteria resistant to metals have been isolated from metal-enriched environments such as polluted soils or even feces of piglets given high amounts of $CuSO_4$ for growth-promoting purposes *(5)*. Both Gram-negative and Gram-positive bacteria (all eubacterial groups from *Escherichia coli* to *Streptomyces*) have resistance systems for metals and metalloids. Bacteria use both chromosomal and plasmid-encoded mechanisms for sequestration and detoxification of metals. Metal efflux ATPases, like ArsA on an *E. coli* plasmid that confers arsenic resistance and CadA of *Staphylococcus aureus* that confers cadmium resistance, are used to remove toxic ions from the cells. Intracellular ions can be sequestered, for example, with metallothioneins as in *Synechococcus* or enzymatically converted to less toxic forms, for example by mercuric reductase, MerA, and arsenate reductase, ArsC, from *E. coli (6)*. The first clear indication that genetic engineering may improve a plant's capacity to phytoremediate metal-polluted soils was achieved using microbial genes. Mercuric ion reductase (*merA*) and organomercury lyase (*merB*) genes from *E. coli* were used to enhance detoxification and phytovolatilization of methylmercury in *Arabidopsis*. Codon usage of the original bacterial sequences was optimized for plants, which increased gene expression and improved functionality *(7)*.

2.2. Mammals

Mammalian genes may be useful for developing novel metal-tolerant or metal-translocating plants. Examples are metallothioneins, which mainly influence tolerance to, and reduce the uptake of, metal ions. Native (human, rodents) or modified genes *(8)* have been transferred into plants *(1)*. In humans, copper deficiency and toxicity disorders have provided information about metal homeostasis in general. Hereditary disorders like Menkes and Wilson diseases with severe symptoms caused by changed transmembrane transport and intracellular distribution of Cu, result from changed Cu-translocating (efflux) P-type ATPases *(9,10)*. A homolog to these mammalian P-type ATPases, RAN1, was found from *Arabidopsis* and transports Cu^{2+} ions to make functional hormone receptors for ethylene signaling *(11)*.

2.3. Plants

2.3.1. Exploiting the Natural Diversity of Plants

Perhaps the best sources of genes are plants with extraordinary capabilities of metal tolerance or uptake. Some environments, such as the serpentine soils, have naturally high concentrations of metals. Mining and industrialization have also led to soils with increased metal contents.

Currently there are a few model plant species, which are used for molecular biological studies on metal tolerance or accumulation. One is *Silene vulgaris*, for which naturally selected metal-tolerant populations are known *(12)*. There are several plant species that not only tolerate large quantities of metals but hyperaccumulate them. Hyperaccumulators are defined as plants that can accumulate 10,000 µg/g Zn or Mn, 1000 µg/g Ni, Co, As, Se, or Cu, or 100 µg/g Cd *(13)*. They have gained great interest as potential sources of genes for developing plants for phytoremediation *(1)*. Nowadays, about 400 plant species are known to hyperaccumulate metals, a majority of them accumulating Ni. Most of the hyperaccumulators belong to the family of Brassicaceae. The hyperaccumulator species most intensively studied are the Ni accumulators *Alyssum lesbiacum* *(14)* and *Alyssum bertolonii* *(15)*, the Ni/Cd accumulator *Thlaspi goesingense* *(16,17)*, the Zn/Cd accumulator *Arabidopsis halleri* *(18–20)*, and the Zn/Cd/Ni accumulator *Thlaspi caerulescens* *(21,22)*. More recently, the accumulation of arsenic in many fern species like *Pteris vittata* *(23)* has been characterized.

Of the hyperaccumulators, *A. halleri* and *T. caerulescens* are interesting because they have populations that grow in noncontaminated habitats. *T. caerulescens* has many advantages over the other hyperaccumulator species. It is widely distributed geographically and there is high variability between the populations. *Thlaspi* is small plant with a fairly short generation time, and it is related to *A. thaliana*, the genome of which is the best characterized among plants. It has been demonstrated recently that *Thlaspi* is also amenable to *Agrobacterium*-mediated transformation *(24)*. *Thlaspi* is a good candidate for a model hyperaccumulator species for large-scale genomic strategies *(24,25)*. It must be kept in mind, however, that different hyperaccumulating species may have evolved different mechanisms.

In addition to metal hyperaccumulating or tolerant species, *Arabidopsis* can also be a useful model. Several *Arabidopsis* ecotypes with altered metal homeostasis are known *(26)* and the natural variation in *Arabidopsis* may be underestimated as a source of genetic information *(27)*. The complex traits associated with each other can be studied using genome-wide analysis such as quantitative trait locus (QTL) mapping (*see* **Subheading 3.1.2.**). When studying metal homeostasis processes, also the green alga *Chlamydomonas reinhardtii* has been used *(28)*. The advantage of using this alga as a model is that it is a unicellular species.

2.3.2. Increasing Plant Diversity in the Laboratory

Besides exploiting the diversity found in nature, mutant plants can be produced to give rise to new characteristics. Seeds can be treated with a mutagen and allowed to germinate in metal-containing dishes. Using chemical mutagenesis with ethyl methyl sulfonate, an *ars1* mutant with increased tolerance to arsenate was found from *Arabidopsis (29)*. In classical forward genetics, plants are mutagenized and selected for desired phenotypes or lack of phenotype. After crossing, positional cloning is used to isolate the particular gene resulting in that trait. The mutagen can be physical (irradiation), chemical (ethyl methyl sulfonate), or biological (transposons, T-DNA). Mutagenization with DNA usually inactivates genes by insertion. The advantage of DNA mutagenization is that DNA can be targeted more precisely and traced easily, whereas chemical or physical mutagenization produces random point mutations. In forward genetics, high-density genetic maps and physical-mapping resources are required.

2.3.3. Choosing the Right Plants

Several aspects should be taken into consideration when selecting a plant for more detailed studies. The less that is known about the plant, the greater the challenge. Genetic studies on crosses of interesting populations have been used to predict the number of genes involved in metal tolerance *(12)*. For some hyperaccumulators, populations differing in their metal-uptake characteristics are available. Crossing these populations allows one to investigate the segregation of interesting genes and do QTL mapping. What still slows down investigations significantly is that full-scale genomic information is available only for *Arabidopsis*, rice, and maize.

Availability of suitable control plants is critical for success in finding the right genes. This is fairly simple in the case of mutant plants where the wild-type/parent can be used as a control. However, problems may arise in all other cases. One way to circumvent the problem is to compare the expression levels (RNA, protein) of control and metal-exposed plants. This approach would reveal only a subset, i.e., up- and downregulated proteins. For example, Zn transporter in the hyperaccumulator *T. caerulescens* is not transcriptionally affected by Zn treatment *(21,22)*. However, the transcript levels are higher in *T. caerulescens* than in the nonaccumulator *T. arvense*, resulting in increased uptake *(21,22)*. Other nonaccumulator relatives used as controls are *T. perfoliatum* for *T. caerulescens*, and *A. lyrata* for *A. halleri*. The suitability of nonaccumulator relatives as controls depends on the particular approach. In microarray studies it may be possible to make use of sequence similarity to *A. thaliana*, whereas for proteomics a very close relative is needed.

2.3.4. Studying Metal Uptake and Tolerance Characteristics of Chosen Plants

The most obvious way to study metal tolerance and uptake in plants is to place them in metal-containing soil, the ultimate matrix in phytoremediation, which is either naturally contaminated with metals or supplemented with metal salts. A problem in interpreting the results is that the soil is not a uniform matrix, and soil pH and other physical and biological parameters influence metal bioavailability. In hydroponic culture, the problem of metal bioavailability is overcome because the metals are uniformly dispersed throughout the medium. Hydroponics is, however, very time-consuming and labor-intensive. Plate assays have been developed, in which survival of seedlings, root growth, and other parameters can be monitored in increased metal concentrations.

2.4. Fungi

Genes controlling plant heavy-metal uptake and tolerance are not well known partly because the pathways are complicated, involving different tissues and organelles. Compared with plants, metal homeostasis mechanisms in baker's yeast (*Saccharomyces cerevisiae*) are much better understood. Many genes in plant metal homeostasis networks have been found from cDNA expression libraries by complementation of metal-sensitive yeast mutants in metal excess or depletion. For example, the ZIP transporter gene family (ZRT, IRT-related proteins) has been isolated using yeast complementation. The *IRT1* (iron-regulated transporter) gene was isolated from *Arabidopsis (30)*, and it was found to complement the iron transport-deficient *S. cerevisiae fet3 fet4* mutant under iron-limiting conditions. Based on sequence similarity to IRT1, ZRT1, and ZRT2 were cloned from yeast, and mediate high- and low-affinity Zn transport, respectively *(31)*. The *zrt1 zrt2* yeast mutant was used, in turn, to clone *Arabidopsis* ZIP1-3 transporters *(32)*. ZIP1 and ZIP3 are expressed mainly in roots and are induced under Zn-deplete conditions, suggesting a role in Zn uptake from the soil. *Thlaspi* ZNT1 was cloned using *zrt1 zrt2* yeast mutant, and shown to be a homolog of *Arabidopsis* ZIP4. *T. caerulescens* ZNT1 mediates high-affinity Zn and low-affinity Cd uptake. The expression of ZNT1 was higher in *T. caerulescens* compared with *T. arvense*, and was found not to be downregulated by elevated Zn levels, probably resulting in increased uptake of Zn *(21,22)*.

The yeast *Schizosaccharomyces pombe* is an interesting model because, unlike in *S. cerevisiae*, the major Cd detoxification route is the phytochelatin-dependent pathway. The results from a nonplant model organism need to be interpreted with caution because, for example, metal transporters have different metal specificities in different organisms. As an example, the *ZRT1* gene encodes a Zn transport protein in *S. cerevisiae (31)*, whereas the plant homolog *IRT1* encodes an iron transporter in *A. thaliana (30)*.

3. Searching for Genes Involved in Metal Tolerance, Uptake, and Transport

3.1. The DNA Level

3.1.1. Genomics

Large-scale genome sequencing projects such as the Arabidopsis Genome Initiative have been of great importance for plant molecular biology research

(33). High-quality sequences of *A. thaliana* can be used for the identification and annotation of genes and proteins from related species like *A. halleri* and *T. caerulescens*.

3.1.2. Map-Based Cloning or Positional Cloning

Positional cloning is a method to identify the genetic basis of a differing phenotype by following markers whose physical locations in the genome are known. An advantage of map-based cloning is that it is done without any prior assumptions, and mutations can be found anywhere in the genome, even in intergenic regions. Using good-quality, high-density maps the localization and identification of the gene conferring a specific trait is quite straightforward. To create high-density maps, markers have been produced by many different techniques, e.g., restriction fragment-length polymorphism, amplified fragment-length polymorphism, single-nucleotide polymorphism, microsatellite sequences, and expressed sequence tag (EST) sequences. For *Arabidopsis*, tens of thousands of randomly distributed genetic markers (single-nucleotide polymorphism; insertion–deletion differences [indels] when one ecotype has an insertion of a number of nucleotides compared with another ecotype) are available in the Cereon Arabidopsis Polymorphism Collection *(34)*. How is it possible to determine if genome polymorphisms (markers) are associated with specific qualitative and quantitative traits? Linkage disequilibrium occurs when haplotype combinations of alleles at different loci occur more frequently than would be expected from random association *(35)*. Linkage disequilibrium has mainly been used for *Arabidopsis* and maize *(36,37)*.

A more recently applied technique for association mapping is to analyze QTL, the genetic variation underlying quantitative phenotypes *(38)*. The chromosomal regions that contribute to variation in complex traits can be statistically identified. Association analysis in plants has been made with crosses of pedigrees of known phenotype, such as F2 populations. These populations unfortunately show only a limited number of recombination events, which means poor resolution for quantitative traits. Therefore, large recombinant inbred line-mapping populations or introgression lines have been produced *(35)*. When the association of a specific locus to the trait of interest is established, the gene itself should be localized and sequenced by genome walking, preferably from a large-insert genomic library instead of plant chromosomal DNA.

Map-based cloning was used to find the *ILR2* gene that affects indole-3-acetic acid (IAA)–leucine resistance in the *Arabidopsis ilr2-1* mutant. It encodes a protein that is polymorphic among *Arabidopsis* accessions and it was found to modulate a metal transporter, thus providing a link between auxin-conjugate metabolism and metal homeostasis *(39)*. The *fer* gene was also isolated by map-based cloning and was subsequently found to take part in the control of root physiology and development at the transcriptional level in response to iron supply and may thus be the first identified regulator for iron nutrition in plants *(40)*.

QTL mapping has been used to study the Al-resistance trait in several plant species like rice, rye, barley, and *Arabidopsis*. In rice, several QTLs have been detected in various chromosomes. The diverse loci in different studies might be explained by the differences in the experiments: populations, phenotypic parameters, Al concentrations, and the time of exposure have all differed. One QTL for Al resistance was common for all studies *(41–44)*. In *Arabidopsis* the Ler and Col ecotypes have also been studied for QTLs of Al resistance. Using crosses of these ecotypes two QTLs were found, one from chromosome 1 and the other from chromosome 4 *(45)*. QTLs were also searched for Al resistance from recombinant inbred lines and two QTLs, in chromosomes 1 and 5, were found to cosegregate with Al-activated release of malate exudation from the roots. There are about 700 predicted genes within those loci, and to dissect the genes of interest a cDNA microarray was made. Altogether, 15 Al-inducible genes, many of them nonannotated, were located in these QTL regions. The function of all genes needs to be verified to find the genes for the QTLs *(46)*.

3.2. The RNA Level

3.2.1. Differential Display

There are several methods for searching mRNAs, the levels of which are altered in metal exposure. One of the methods is differential display (DD), originally developed by Liang and Pardee in the early 1990s *(47)*. In DD, total cDNA is amplified with arbitrary and poly-T primers to produce cDNA pools that are separated on polyacrylamide gel. Differentially expressed fragments can be isolated and sequenced. The method is sensitive, does not require expensive machinery, and has potential for screening all transcripts. A drawback is that the technique is labor intensive and time consuming, because tens of primer combinations are needed to screen the transcripts. Being a PCR-based method, it is also prone to errors. With this method, differentially expressed genes have been isolated from Cd-exposed *Arabidopsis* *(48)* and Zn-exposed *T. caerulescens* populations (Hassinen et al., unpublished).

3.2.2. Amplified Fragment-Length Polymorphism

Differences in gene expression can also be analyzed with amplified fragment-length polymorphism cDNA, in which the cDNA is digested with restriction enzymes, and synthetic adapters are ligated to the ends. The cDNAs are amplified with primers complementary to the adapter sequences, and the PCR products can be displayed on sequencing gels, compared, and differentially expressed fragments can be isolated *(49)*.

3.2.3. Subtractive Hybridization

A classical method to enrich mRNAs present in excess in a sample is to carry out subtractive hybridization *(50)*. Suppressive subtractive hybridization was used to isolate Cd-induced cDNAs from Cd-tolerant *Datura innoxia (51)*, and mercuric ion-induced genes from *A. thaliana (52)*.

3.2.4. Serial Analysis of Gene Expression

Serial analysis of gene expression is another method for profiling differences at the transcriptional level *(53)*. In serial analysis of gene expression, a short sequence tag (10–14 bp) is used to uniquely identify a transcript. Sequence tags are linked together to form a long molecule that can be cloned and sequenced, and the number of times a particular tag is observed provides the expression level of the corresponding transcript. This method has been used to profile transcript levels in *Arabidopsis* roots exposed to 2,4,6-trinitrotoluene *(54)*.

3.2.5. ESTs

ESTs are unedited, automatically processed, single-read sequences produced from cDNAs. Metal-related ESTs have been sequenced from Al-exposed rye and from Fe-deficient barley *(55,56)*. Over 3 million sequences from approx 200 species, including 178,544 *Arabidopsis* EST sequences, are available in the public EST sequence databases *(57)* and should be useful for the identification of genes and proteins in related species.

3.2.6. Transcriptomics

Understanding the functional connections between genes, proteins, metabolites, and mineral ions is one of biology's greatest challenges in the postgenomic era. The "omic" technologies are used as nontargeted approaches in system biology to monitor simultaneously all biological processes operating as an integrated system. Through the study of whole systems, one can visualize how individual pathways or metabolic networks are interconnected. Transcriptomics, proteomics, and metabolomics together may be used to reveal new patterns in specific biological phenomena.

cDNA microarray is one of the potential technologies that can be used to explore the interactions between gene expression and the environment *(58)*. A pool of nucleic acid molecules are isolated from the tissue sample of interest and hybridized to a large number of immobilized DNA molecules arrayed on a solid surface. The DNA fragments on the solid surface can include expressed gene sequences from public databases, oligonucleotides, and ESTs. The gene chip technology produces a wealth of information, and full advantage should be taken of the data using bioinformatic predictors.

Several research groups use microarrays as a tool to understand the metal hyperaccumulation phenomenon. The Affymetrix *Arabidopsis* high-density oligonucleotide array has been widely used for studying differential expression of genes in various stresses in *A. thaliana*, but has also been applied to related species like *A. halleri*. The expression of homologs to metal-responsive genes like ZIP transporters, a putative P-type ATPase (*AtHMA3*), cation-diffusion facilitator *ZAT/AtCDF1*, and the nicotianamine synthase were found to be increased in Zn exposure using the GeneChip microarray that contained 8300 *Arabidopsis* genes *(59,60)*. The genes are known to be related to metal detoxification and transport, and may have a role in metal tolerance and hyperaccumulation.

A significant amount of *Arabidopsis* microarray data (114 microarray datasets in the beginning of 2004) has been made publicly available via TAIR (http://www.arabidopsis.org/). There are data from nutrient effects and of chemical exposures. Five of the datasets are related to metals, i.e., Zn, Al, Ni, and Cd exposures and Fe deficiency. Oxidative stress and ethylene-regulated gene expression has also been studied in *Arabidopsis* *(61,62)*. The genes responding to overall oxidative stress and not specifically to metals might thus be differentiated from each other. Microarrays with selected gene families or groups are useful to study specific routes and processes. An example is transcriptomic analysis done on root transporters in response to cation stress *(63)*. Studies on *S. cerevisiae* under Cu excess and deficiency revealed two transcriptional activators, Ace1 and Mac1. Ace1 mediates copper-induced gene expression in cells exposed to high levels of copper salts (metallothioneins *CUP1* and *CRS5*, *FET3* and *FTR1* in the iron-uptake system), whereas Mac1 activates a subset of genes under copper-deficient conditions (*CTR1*, *CTR3*, *FRE1*, *FRE7*, YFR055w, YJL217w) *(64)*.

The main advantage of the microarray technique is the ability to examine thousands of expressed genes simultaneously. As a high-throughput screening method, it is not aimed at replacing other molecular techniques but is to be used in conjunction with them to confirm and extend the findings. Different methods also produce different results. As demonstrated by Mandaokar et al. *(65)*, additional differentially expressed genes were found with DD compared with results obtained with microarray. Thus, to get full understanding of the processes, different methods should be used.

3.3. Protein Level: Proteomics

Proteins are the products of mRNA translation but the levels of proteins and mRNAs in a sample do not often correlate, partly because of large differences in mRNA and protein turnover *(66)*. Many proteins are also posttranslationally modified, the signal peptides being removed and the peptides being phosphorylated, glycosylated, or glutathiolated. These modifications play an important role in the activity and subcellular localization of the proteins. Protein profiling (proteomics) provides information on the amount of a great number of proteins at a given time in specific plant organ or in response to a given treatment.

Proteomics is traditionally based on two-dimensional electrophoresis (2DE), which separates polypeptides according to their isoelectric point in the first dimension and to their molecular weight in the second dimension (sodium dodecyl sulfate-polyacrylamide gel electrophoresis) *(67)*. Hundreds of proteins can be visualized on a single gel, thus permitting large-scale studies of gene expression and genetic variation at the protein level. Both qualitative and quantitative changes in protein expression can be detected, and statistical analysis can be used to find the most significant differences between samples. Interesting proteins can then be identified using mass spectrometry (MS). Because the peptide masses are compared with masses predicted from gene and protein sequences in databases, identification of proteins from organisms with known genomic sequences such as *Arabidopsis* is fairly straightforward. Clearly more difficult is the identification of previously unknown proteins translated from unknown genes from less-characterized species such as *T. caerulescens*. Cross-species identification is only possible for proteins with high sequence identity. About half of *Thlaspi* proteins differentially expressed in Zn exposure could be identified, mainly based on *Arabidopsis* and *Brassica* sequences *(68)*. One advantage of proteomics is that posttranslational modifications can be detected, but extensive modifications may also cause failure to yield good matches. To identify an unknown protein, sequence data need to be generated; this can be achieved with modern mass spectrometers. The gene can then be searched from a cDNA library, and the sequence compared with databases to predict the possible function of the protein.

The proteomics approach can be used for total proteins but also for subcellular fractions (plastids, mitochondria, nuclei, membranes, and so on). Metal-binding proteins of *Arabidopsis* mitochondrial subproteome have been studied by mobility shifts in the presence of divalent cations with 2D diagonal sodium dodecyl sulfate-polyacrylamide gel electrophoresis *(69)*. Among the proteins shifted, known metal-binding proteins but also proteins without known metal binding properties and several unknown proteins, were found. Because one-third of plant proteins are metalloproteins or metal-binding proteins, subcellular fractionation of proteins is crucial for mobility-shift analysis. One important factor in the success of the 2DE approach is the solubility of the proteins. The IEF (isoelectric focusing) strips in the first dimension are optimal for separation of soluble proteins, but not of hydrophobic proteins like membrane proteins. Also, the mass spectrometric analysis of hydrophobic peptides is more difficult because they are not easily extracted from gel matrices and their ionization is not effective *(70,71)*. Millar et al. *(72)* studied the proteome of mitochondrial-carrier proteins of *Arabidopsis*. Genomic databases indicated 45 putative genes with mitochondrial-carrier features. However, 2DE separation in gel revealed no carrier proteins, presumably because of the fact that they are hydrophobic and basic proteins. Using 1DE of integral protein fraction combined with tandem MS-based sequencing, six mitochondrial-carrier proteins were identified *(73)*. New methods for proteome profiling, such as protein microarray, have been developed to monitor thousands of binding events *(74)*. For plants, no high-throughput protein microarrays have yet been applied.

Large-scale proteomic studies have been done on *Arabidopsis*, rice, and maize *(75–81)*. Many other species have also been subjected to proteomics, but the shortage of genomic data has impeded the identification of the proteins. Several subcellular proteomic databases have been published, especially for *Arabidopsis (72,82–84)*. The effects of metals at the proteome level are still almost unknown, and extensive proteomic studies to solve the mechanisms of metal uptake and tolerance in plants are yet to be published. A Cu-inducible protein, PR-10c, was found from Cu- and Zn-tolerant birch by proteomics *(85)* but no connection to metal tolerance has been established *(86)*. The protein has ribonuclease activity and is posttranslationally modified by glutathione, also shown by using a proteomic approach *(87)*. A comparative proteome profiling of Zn-exposed *T. caerulescens* populations has been recently accomplished *(68)*. Soluble proteins, apparently not including, for example, membrane metal transporters or small-molecular-weight proteins like metallothioneins, were analyzed. Different *Thlaspi* populations were found to have statistically significant differences in protein expression. Among those were proteins for basic metabolism, gene regulation, and signal transduction. Proteomics has also been used to identify changes in mycorrhiza-inoculated pea roots exposed to Cd *(88)*. Symbiosis modulated the expression of several proteins.

3.4. The Metabolite Level: Metabolomics

Metabolomics is the nontargeted profiling of metabolites in biological samples. The methodology includes gas and liquid chromatography coupled to MS (GC–MS, LC–MS) for high sample throughput to identify and quantify small-molecular-weight metabolites. Nuclear magnetic resonance spectroscopy (NMR) can also be used for discrimination of metabolic fingerprints *(89)*.

The genetic and gross metabolic basis for metal tolerance in plants is poorly understood. *Silene cucubalus* is known to respond to Cd through chelation of metal ions by a family of peptide ligands called phytochelatins *(90–92)*. Extracts of Cd-exposed, tolerant *S. cucubalus* were analyzed by ^{1}H NMR spectroscopy to find responses at the metabolite level *(93)*. Statistically significant differences were found between exposed and control plants. Organic acids, such as citric and malic acids, were increased in Cd exposure, whereas glutamine and branched amino acids were decreased.

Root exudates released into soil have important functions in mobilizing micronutrients and causing selective enrichment of beneficial soil micro-organisms that colonize the plant rhizosphere. For analyzing plant-root exudates, chromatography of selected compounds is a common approach. Multinuclear and 2D NMR with GC–MS and high-resolution MS has been recently used to provide *de novo* identification of a number of components directly from crude-root exudates of different plant types *(94)*. The technique was applied in barley and wheat to examine the role of ligands for metal ions in the exudate in the acquisition of Cd and transition metals. The exudation of mugineic acids and malate was enhanced by Fe deficiency, which in turn led to an increase in the tissue content of Cu, Mn, and Zn.

3.5. The Elemental Level: Ionomics

Lahner et al. *(95)* coined the term "ionome" to include all mineral nutrients and trace elements found in an organism. They used inductively coupled plasma spectroscopy to quantify 18 elements, including essential macro- and micronutrients and various nonessential elements in the shoots of 6000 mutagenized M2 *A. thaliana* plants. Altered elemental profiles were found in 51 mutants, e.g., in 8, 14, and 16 mutants to Cu, Zn, and Cd, respectively. This was the first approach to large-scale ionome profiling in plants. To get full benefit of the data, extensive genomic profiling and targeted searching is still needed to find the key genes affecting the ion content of the plants.

3.6. In Silico *Approaches*

Assessing the function of a protein is difficult *in silico* without any annotated homology in databases. Databases and tools available are presented in a special issue of *Nucleic Acids Research (96)*. Computing and algorithms are needed in every step of molecular biological studies. Homologies are searched from databases to identify and characterize the genes and proteins of interest. If only poor homology is found, searching using common sequence motifs may provide further information. The sequence may reveal signal peptides and give information about protein targeting in the cell. Cation-binding motifs can also be searched for. New profiling techniques, such as microarray and proteomics, provide a vast amount of data of the up- and downregulated genes and proteins. By clustering them by the expression, new groups of similarly regulated genes sharing similar transcription factor-binding sites, or genes related to specific pathways, may be found. Efforts to predict interactions of proteins with genes and other proteins can be made.

4. Confirmation of the Desired Function of Candidate Genes

The greatest challenge in the postgenomic era is the functional characterization of the numerous genes found using various techniques. The definite function of a particular gene will be established by expression studies. Reverse genetics is an essential strategy to determine a gene's function by studying the phenotypes of individuals with alterations in the gene of interest. Large insertion mutant collections suited for this purpose have been established for *Arabidopsis* *(97)*. However, at the moment mutants are not available for all possible genes. Developing T-DNA mutants for small genes is a particularly challenging task.

The gene of interest can be introduced into the plant as sense or antisense constructs, and the change in function or phenotype is monitored. The use of the double-stranded RNAi technique *(98)* and specific vectors can aid in producing an efficient gene silencing in plants *(99)*. Yeast complementation is used to characterize the function of genes in an in vivo model. Being a eukaryote, it can process plant proteins correctly, and it has an additional advantage of being a unicellular organism. The yeast mutated in metal homeostasis or detoxification pathways are particularly useful.

In addition to the characterization of structural genes involved in metal homeostasis, it is equally important to know how these genes are regulated. Some of the promoters, e.g., ones that confer metal-inducible or organelle-specific expression, might also be useful when plants are designed for phytoremediation. As an example, the expression of soybean ferritin, driven by endosperm-specific promoter, led to higher Fe and Zn levels in transgenic rice grains *(100)*. The promoter regions of the genes of interest can be analyzed by developing a deletion series of the promoter and fusing it with a reporter gene like *GUS*. In this way, the expression of several metallothionein genes was analyzed under Cu exposure *(101)*. Strategies for the identification of new promoter elements involved in plants stress response was recently discussed *(102)*.

5. Tools for Modification of Plants for Phytoremediation: Current Status

If the starting point is to modify fast-growing, high-biomass-producing plants for phytoremediation, metal tolerance and uptake are most probably the key traits to be introduced. The starting point could equally well be a metal-tolerant hyperaccumulator plant, and the goal is to modify it to grow fast and produce high biomass. However, the latter case is out of the scope of the present review. There are basically two ways of improving metal tolerance and/or uptake in the plant: (1) by introducing new traits from other organisms or (2) by augmenting the network already present in the plant. The *mer (7,103)* and the *ars (104,105)* systems are examples of the former approach. The latter approach is restricted by the inadequate understanding of these processes in the plants. Some of the genes known to be involved in these systems are summarized in **Table 1**. One apparently important group is metal transporters. Both essential and nonessential metals enter the plant via transport systems. It was assumed previously that nonessential heavy metals such as Cd are capable of entering the cells via the

Table 1
Genes and Molecules Possibly Involved in Heavy Metal Detoxification/Homeostasis in Plants

Classification	Gene/molecule	Proposed function
Transporters		
P-type ATPase	AtHMA1-6	Resemble bacterial heavy metal pumps (4, 110, 111)
	RAN1	Cu transport (11)
	PAA1	Cu transport system to chloroplasts (112)
Nramp family	AtNramp1,3,4	Iron homeostasis, Cd uptake? (113, 114)
CDF family	ZAT1 ZTP1, *T. caerulescens* TgMTP1, *T. goesingense*	Intracellular sequestration of Zn (115, 22, 17)
ATP-binding cassette (ABC) transporters	AtMRPs	Cd, Pb transport across the tonoplast? (116, 117)
Divalent cation/proton Antiporters	CAX1-2	CAX1 vacuolar Ca accumulation, CAX2 Cd? (118, 119)
	AtMHX1	Vacuolar membrane exchanger of protons with Mg and Zn (120)
ZIP transporters	IRT1	Iron, possibly Mn, Zn, and Cd uptake (30, 121)
	ZIP1-4	Zn uptake (32)
	ZNT1-2, *T. caerulescens*	High-affinity Zn and low-affinity Cd uptake (21, 22)
	MtZIP2, *Medicago truncatula*	(122)
Other transporters	COPT1	Cu-influx protein (123)
	LCT1	Rb, Na, Ca, and Cd uptake (124, 125)
	AtDX1	Detoxifying Cd efflux carrier? (126)
	NtCBP4, *N. tabacum*	Plasma membrane calmodulin-binding cyclic nucleotide-gated channel (127)
Other	Metallothioneins (MTs)	Cytoplasmic metal buffering (128–130)
	Phytochelatin (PC) pathway; AtPCS1	Cytoplasmic metal buffering (131, 132), long-distance metal trafficking? (133)
	Organic acids; citrate, malate	Increased Al resistance correlates with citrate or malate release from the roots (44)

same transporters as the essential heavy metals like Zn. However, there is also evidence for specific transporters, e.g., the Cd transporter from *T. caerulescens* ecotype Ganges *(106,107)*. Besides metal transporters, tools to enhance metal-buffering capacity are well-established candidates in the improvement of the plants.

6. Designing Plants for Phytoremediation

Several different contaminants typically occur in contaminated soils. Plants should thus take up, or at least tolerate, high levels of more than one contaminant. For this, possibly several genes need to be introduced, and it will be of crucial importance to understand how these genes act in concert. Finally, the possible risks and benefits of the genetically modified plants designed for phytoremediation should be considered in every step of the developmental process to gain real, viable options to the presently employed remediation techniques *(3,108,109)*.

Acknowledgments

This work was funded by the E. C. 5th framework project "PHYTAC" (QLRT-2001-00429) and by the Academy of Finland (project 53885). V. H. was funded by the Finnish Graduate School for Environmental Science and Technology (EnSTe).

References

1. Kärenlampi, S., Schat, H., Vangronsveld, J., et al. (2000) Genetic engineering in the improvement of plants for phytoremediation of metal polluted soils. *Env. Poll.* **107,** 225–231.
2. Krämer, U. and Chardonnens, A. (2001) The use of transgenic plants in the bioremediation of soils contaminated with trace elements. *Appl. Microbiol. Biotechnol.* **55,** 661–672.
3. Kärenlampi, S. (2002) Risk of GMO's: general introduction, political issues, social and legal aspects. In: *Risk Assessment and Sustainable Land Management Using Plants in Trace Element-Contaminated Soil,* (Mench, M. and Mocquot, B., eds.), COST Action 837th WG2 Workshop, Bordeaux 2002. INRA, Centre Bordeaux-Aquitaine, Villenave d'Ornon cedex, France, pp. 157–162.
4. Clemens, S., Palmgren, M. G., and Krämer, U. (2002) A long way ahead: understanding and engineering plant metal accumulation. *Trends Plant. Sci.* **7,** 309–315.
5. Williams, J. R., Morgan, A. G., Rouch, D. A., Brown, N. L., and Lee, B. T. (1993) Copper-resistant enteric bacteria from United Kingdom and Australian piggeries. *Appl. Environ. Microbiol.* **59,** 2531–2537.
6. Silver, S. (1998) Genes for all metals: a bacterial view of the periodic table. The 1996 Thom Award Lecture. *J. Ind. Microbiol. Biotechnol.* **20,** 1–12.
7. Bizily, S. P., Rugh, C. L., and Meagher, R. B. (2000) Phytodetoxification of hazardous organomercurials by genetically engineered plants. *Nat. Biotechnol.* **18,** 213–217.

8. Pan, A., Tie, F., Yang, M., et al. (1993) Construction of a multiple copy of alpha-domain gene fragment of human liver metallothionein IA in tandem arrays and its expression in transgenic tobacco plants. *Protein Engineering* **6**, 755–762.
9. Petris, M. J., Mercer, J. F., Culvenor, J. G., Lockhart, P., Gleeson, P. A., and Camakaris, J. (1996) Ligand-regulated transport of the Menkes copper P-type ATPase efflux pump from the Golgi apparatus to the plasma membrane: a novel mechanism of regulated trafficking. *EMBO J.* **15**, 6084–6095.
10. DiDonato, M., Narindsrasorasak, S., Forbes, J. R., Cox, D. W., and Sarkar, B. (1997) Expression, purification, and metal binding properties of the N-terminal domain from the Wilson disease putative copper-transporting ATPase (ATP7B). *J. Biol. Chem.* **272**, 33,279–33,282.
11. Hirayama, T., Kieber, J. J., Hirayama, N., et al. (1999) RESPONSIVE-TO-ANTAGONIST1, a Menkes/Wilson disease-related copper transporter, is required for ethylene signalling in Arabidopsis. *Cell* **97**, 383–393.
12. Schat, H., Vooijs, R., and Kuiper, E. (1996) Identical major gene loci for heavy metal tolerances that have independently evolved in different local populations and subspecies of *Silene vulgaris*. *Evolution* **50**, 1888–1895.
13. Baker, A. J. M. and Brooks, R. R. (1989) Terrestrial higher plants which hyperaccumulate metallic elements. A review of their distribution, ecology and phytochemistry. *Biorecovery* **1**, 81–126.
14. Kerkeb, L. and Krämer, U. (2003) The role of free histidine in xylem loading of nickel in *Alyssum lesbiacum* and *Brassica juncea*. *Plant Physiol.* **131**, 716–724.
15. Boominathan, R. and Doran, P. M. (2003) Organic acid complexation, heavy metal distribution and the effect of ATPase inhibition in hairy roots of hyperaccumulator plant species. *J. Biotechnol.* **101**, 131–146.
16. Lombi, E., Zhao, F. J., Dunham, S. J., and McGrath, S. P. (2000) Cadmium accumulation in populations of *Thlaspi caerulescens* and *Thlaspi goesingense*. *New Phytol.* **145**, 11–20.
17. Persans, M. W., Nieman, K., and Salt, D. E. (2001) Functional activity and role of cation-efflux family members in Ni hyperaccumulation in *Thlaspi goesingense*. *Proc. Natl. Acad. Sci. USA* **98**, 9995–10,000.
18. Bert, V., Macnair, M. R., de Laguerie, P., Saumitou-Laprade, P., and Petit, D. (2000) Zinc tolerance and accumulation in metallicolous and nonmetallicolous populations of *Arabidopsis halleri* (Brassicaceae). *New Phytol.* **146**, 225–233.
19. Macnair, M. R. (2002) Within and between population genetic variation for zinc accumulation in *Arabidopsis halleri*. *New Phytol.* **155**, 59–66.
20. Bert, V., Meerts, P., Saumitou-Laprade, P., Salis, P., Gruber, W., and Verbruggen, N. (2003) Genetic basis of Cd tolerance and hyperaccumulation in *Arabidopsis halleri*. *Plant Soil* **249**, 9–18.
21. Pence, N. S., Larsen, P. B., Ebbs, S. D., et al. (2000) The molecular physiology of heavy metal transport in the Zn/Cd hyperaccumulator *Thlaspi caerulescens*. *Proc. Natl. Acad. Sci. USA* **97**, 4956–4960.
22. Assunção, A. G. L., Da Costa Martins, P., De Folter, S., Vooijs, R., Schat, H., and Aarts, M. G. M. (2001) Elevated expression of metal transporter genes in three

accessions of the metal hyperaccumulator *Thlaspi caerulescens*. *Plant Cell Environ.* **24**, 217–226.
23. Ma, L. Q., Komar, K. M., Tu, C., Zhang, W., Cai, Y., and Kennelley, E. D. (2001) A fern that hyperaccumulates arsenic. *Nature* **409**, 579.
24. Peer, W. A., Mamoudian, M., Lahner, B., Reeves, R. D., Murphy, A. S., and Salt, D. E. (2003) Identifying model metal hyperaccumulating plants: germplasm analysis of 20 Brassicaceae accessions from a wide geographical area. *New Phytol.* **159**, 421–430.
25. Assunção, A. G. L., Schat, H., and Aarts, M. G. M. (2003) *Thlaspi caerulescens*, an attractive model species to study heavy metal hyperaccumulation in plants. *New Phytol.* **159**, 351–360.
26. Murphy, A. and Taiz, L. (1995) A new vertical mesh transfer technique for metal-tolerance studies in Arabidopsis: ecotypic variation and copper-sensitive mutants. *Plant Physiol.* **108**, 29–38.
27. Alonso-Blanco, C. and Koornneef, M. (2000) Naturally occurring variation in Arabidopsis: an underexploited resource for plant genetics. *Trends Plant Sci.* **5**, 22–29.
28. Hanikenne, M. (2003) *Chlamydomonas reinhardtii* as a eukaryotic photosynthetic model for studies of heavy metal homeostasis and tolerance. *New Phytol.* **159**, 331–340.
29. Lee, D. A., Chen, A., and Schroeder, J. I. (2003) ars1, an *Arabidopsis* mutant exhibiting increased tolerance to arsenate and increased phosphate uptake. *Plant J.* **35**, 637–646.
30. Eide, D., Broderius, M., Fett, J., and Guerinot, M. L. (1996) A novel iron-regulated metal transporter from plants identified by functional expression in yeast. *Proc. Natl. Acad. Sci. USA* **93**, 5624–5628.
31. Zhao, H. and Eide, D. (1996) The yeast ZRT1 gene encodes the zinc transporter protein of a high- affinity uptake system induced by zinc limitation. *Proc. Natl. Acad. Sci. USA* **93**, 2454–2458.
32. Grotz, N., Fox, T., Connolly, E., Park, W., Guerinot, M. L., and Eide, D. (1998) Identification of a family of zinc transporter genes from Arabidopsis that respond to zinc deficiency. *Proc. Natl. Acad. Sci. USA* **95**, 7220–7224.
33. The Arabidopsis Genome Initiative. (2000) Analysis of the genome sequence of the flowering plant *Arabidopsis thaliana*. *Nature* **408**, 796–815.
34. Jander, G., Norris, S. R., Rounsley, S. D., Bush, D. F., Levin, I. M., and Last, R.L. (2002) Arabidopsis map-based cloning in the post-genome era. *Plant Physiol.* **129**, 440–450.
35. Flint-Garcia, S. A., Thornsberry, J. M., and Buckler, E. S. (2003) Structure of linkage disequilibrium in plants. *Annu. Rev. Plant Biol.* **54**, 357–374.
36. Nordborg, M., Borevitz, J. O., Bergelson, J., et al. (2002) The extent of linkage disequilibrium in *Arabidopsis thaliana*. *Nat. Genet.* **30**, 190–193.
37. Remington, D. L., Thornsberry, J. M., Matsuoka, Y., et al. (2001) Structure of linkage disequilibrium and phenotypic associations in the maize genome. *Proc. Natl. Acad. Sci. USA* **98**, 11,479–11,484.

38. Paran, I. and Zamir, D. (2003) Quantitative traits in plants: beyond the QTL. *Trends Genet.* **19,** 303–306.
39. Magidin, M., Pittman, J. K., Hirschi, K. D., and Bartel, B. (2003) ILR2, a novel gene regulating IAA conjugate sensitivity and metal transport in *Arabidopsis thaliana*. *Plant J.* **35,** 523–534.
40. Ling, H. Q., Bauer, P., Bereczky, Z., Keller, B., and Ganal, M. (2002) The tomato *fer* gene encoding a bHLH protein controls iron-uptake responses in roots. *Proc. Natl. Acad. Sci. USA* **99,** 13,938–13,943.
41. Wu, P., Liao, C. Y., Hu, B., et al. (2000) QTLs and epistasis for aluminum tolerance in rice (*Oryza sativa L.*) at different seedling stages. *Theor. Appl. Genet.* **100,** 1295–1303.
42. Nguyen, V. T., Nguyen, B. D., Sarkarung, S., Martinez, C., Paterson, A. H., and Nguyen, H. T. (2002) Mapping of genes controlling aluminum tolerance in rice: comparison of different genetic backgrounds. *Mol. Genet. Genomics* **267,** 772–780.
43. Ma, J. F., Shen, R., Zhao, Z., et al. (2002) Response of rice to Al stress and identification of quantitative trait loci for Al tolerance. *Plant Cell. Physiol.* **43,** 652–659.
44. Ma, J. F. and Furukawa, J. (2003) Recent progress in the research of external Al detoxification in higher plants: a minireview. *J. Inorg. Biochem.* **97,** 46–51.
45. Kobayashi, Y. and Koyama, H. (2003) QTL analysis of Al tolerance in recombinant inbred lines of *Arabidopsis thaliana*. *Plant Cell. Physiol.* **43,** 1526–1533.
46. Hoekenga, O. A., Vision, T. J., Shaff, J. E., et al. (2003) Identification and characterization of aluminium tolerance loci in Arabidopsis (Landsberg erecta x Columbia) by quantitative trait locus mapping. A physiologically simple but genetically complex trait. *Plant Physiol.* **132,** 936–948.
47. Liang, P. and Pardee, A. B. (1992) Differential display of eukaryotic messenger RNA by means of the polymerase chain reaction. *Science* **257,** 967–971.
48. Suzuki, N., Koizumi, N., and Sano, H. (2001) Screening of cadmium-responsive genes in *Arabidopsis thaliana*. *Plant Cell Environ.* **24,** 1177–1188.
49. Bachem, C. W., van der Hoeven, R. S., de Bruijn, S. M., Vreugdenhil, D., Zabeau, M., and Visser, R. G. (1996) Visualization of differential gene expression using a novel method of RNA fingerprinting based on AFLP: analysis of gene expression during potato tuber development. *Plant J.* **9,** 745–753.
50. Sageström, C. G., Sun, B. I., and Sive, H. L. (1997) Subtractive cloning: past, present, and future. *Annu. Rev. Biochem.* **66,** 751–783.
51. Louie, M., Kondor, N., and DeWitt, J. G. (2003) Gene expression in cadmium-tolerant *Datura innoxia*: Detection and characterization of cDNAs induced in response to Cd^{2+}. *Plant Mol. Biol.* **52,** 81–89.
52. Heidenreich, B., Mayer, K., Sandermann, H. J., and Ernst, D. (2001) Mercury-induced genes in *Arabidopsis thaliana*: Identification of induced genes upon long-term mercuric ion exposure. *Plant Cell Environ.* **24,** 1227–1234.
53. Velculescu, V. E., Zhang, L., Vogelstein, B., and Kinzler, K. W. (1995) Serial analysis of gene expression. *Science* **270,** 484–487.

54. Ekman, D. R., Lorenz, W. W., Przybyla, A. E., Wolfe, N. L., and Dean, J. F. D. (2003) SAGE analysis of transcriptome responses in Arabidopsis roots exposed to 2,4,6-trinitrotoluene. *Plant Physiol.* **133,** 1397–1406.
55. Milla, M. A., Butler, E., Huete, A. R., Wilson, C. F., Anderson, O., and Gustafson, J. P. (2002) Expressed sequence tag-based gene expression analysis under aluminum stress in rye. *Plant Physiol.* **130,** 1706–1716.
56. Negishi, T., Nakanishi, H., Yazaki, J., et al. (2002) cDNA microarray analysis of gene expression during Fe-deficiency stress in barley suggests that polar transport of vesicles is implicated in phytosiderophore secretion in Fe-deficient barley roots. *Plant J.* **30,** 83–94.
57. Rudd, S. (2003) Expressed sequence tags: alternative or complement to whole genome sequences? *Trends Plant Sci.* **8,** 321–329.
58. Schena, M., Shalon, D., Davis, R. W., and Brown, P. O. (1995) Quantitative monitoring of gene expression patterns with a complementary DNA microarray. *Science* **270,** 467–470.
59. Becher, M., Talke, I. N., Krall, L., and Krämer, U. (2004) Cross-species microarray transcript profiling reveals high constitutive expression of metal homeostasis genes in shoots of the zinc hyperaccumulator *Arabidopsis halleri*. *Plant J.* **37,** 251–268.
60. Weber, M., Harada, E., Vess, C. V., Roepenack-Lahaye, E., and Clemens, S. (2004) Comparative microarray analysis of *Arabidopsis thaliana* and *Arabidopsis halleri* roots identifies nicotianamine synthase, a ZIP transporter and other genes as potential metal hyperaccumulation factors. *Plant J.* **37,** 269–281.
61. Desikan, R., Mackerness, S. A.-H., Hancock, J. T., and Neill, S. J. (2001) Regulation of the Arabidopsis transcriptome by oxidative stress. *Plant Physiol.* **127,** 159–172.
62. Van Zhong, G. and Burns, J. K. (2003). Profiling ethylene-regulated gene expression in *Arabidopsis thaliana* by microarray analysis. *Plant Mol. Biol.* **53,** 117–131.
63. Maathuis, F. J. M., Filatov, V., Herzyk, P., et al. (2003) Transcriptome analysis of root transporters reveals participation of multiple gene families in the response to cation stress. *Plant J.* **35,** 675–692.
64. Gross, C., Kelleher, M., Iyer, V. R., Brown, P. O., and Winge, D. R. (2000) Identification of the copper regulon in *Saccharomyces cerevisiae* by DNA microarrays. *J. Biol. Chem.* **275,** 32,310–32,316.
65. Mandaokar, A., Kumar, V. D., Amway, M., and Browse, J. (2003) Microarray and differential display identify genes involved in jasmonate-dependent anther development. *Plant Mol. Biol.* **52,** 775–786.
66. Gygi, S. P., Rochon, Y., Franza, B. R., and Aebersold, R. (1999) Correlation between protein and mRNA abundance in yeast. *Mol. Cell Biol.* **19,** 1720–1730.
67. O'Farrell, P. H. (1975) High resolution two-dimensional electrophoresis of proteins. *J. Biol. Chem.* **250,** 4007–4021.
68. Tuomainen, M. H., Nunan, N., Lehesranta, S. J., et al. (2006) Multivariate analysis of protein profiles of metal hyperaccumulator *Thlaspi caerulescens* accessions. *Proteomics* **6,** 3696–3706.

69. Herald, V. L., Heazlewood, J. L., Day, D. A., and Millar, A. H. (2003) Proteomic identification of divalent metal cation binding proteins in plant mitochondria. *FEBS Lett.* **537,** 96–100.
70. Ferro, M., Seigneurin-Berny, D., Rolland, N., et al. (2000) Organic solvent extraction as a versatile procedure to identify hydrophobic chloroplast membrane proteins. *Electrophoresis* **21,** 3517–3526.
71. Molloy, M. (2000) Two-dimensional electrophoresis of membrane proteins using immobilized pH gradients. *Anal. Biochem.* **280,** 1–10.
72. Millar, A. H., Sweetlove, L. J., Giege, P., and Leaver, C. J. (2001) Analysis of the *Arabidopsis* mitochondrial proteome. *Plant Physiol.* **127,** 1711–1727.
73. Millar, A. H. and Heazlewood, J. L. (2003) Genomic and proteomic analysis of mitochondrial carrier proteins in *Arabidopsis*. *Plant Physiol.* **131,** 443–453.
74. Walter, G., Büssow, K., Cahill, D., Lueking, A., and Lehrach, H. (2000) Protein arrays for gene expression and molecular interaction screening. *Curr. Opin. Microbiol.* **3,** 298–302.
75. Tsugita, A., Kawakami, T., Uchiyama, Y., Kamo, M., Miyatake, N., and Nozu, Y. (1994). Separation and characterization of rice proteins. *Electrophoresis* **15,** 708–720.
76. Kamo, M., Kawakami, T., Miyatake, N., and Tsugita, A. (1995) Separation and characterization of *Arabidopsis thaliana* proteins by two-dimensional gel electrophoresis. *Electrophoresis* **16,** 423–430.
77. Tsugita, A., Kamo, M., Kawakami, T., and Ohki, Y. (1996) Two-dimensional electrophoresis of plant proteins and standardization of gel patterns. *Electrophoresis* **17,** 855–865.
78. Komatsu, S., Muhammad, A., and Rakwal, R. (1999) Separation and characterization of proteins from green and etiolated shoots of rice (*Oryza sativa* L.): towards a rice proteome. *Electrophoresis* **20,** 630–636.
79. Chang, W. W., Huang, L., Shen, M., Webster, C., Burlingame, A. L., and Roberts, J. K. (2000) Patterns of protein synthesis and tolerance of anoxia in root tips of maize seedlings acclimated to a low-oxygen environment, and identification of proteins by mass spectrometry. *Plant Physiol.* **122,** 295–318.
80. Rakwal, R. and Komatsu, S. (2000) Role of jasmonate in the rice (*Oryza sativa* L.) self-defense mechanism using proteome analysis. *Electrophoresis* **21,** 2492–2500.
81. Gallardo, K., Job, C., Groot, S. P. C., et al. (2001) Proteomic analysis of *Arabidopsis* seed germination and priming. *Plant Physiol.* **126,** 835–848.
82. Kruft, V., Eubel, H., Jansch, L., Werhahn, W., and Braun, H. P. (2001) Proteomic approach to identify novel mitochondrial proteins in Arabidopsis. *Plant Physiol.* **127,** 1694–1710.
83. Werhahn, W. and Braun, H. P. (2002) Biochemical dissection of the mitochondrial proteome from *Arabidopsis thaliana* by three-dimensional gel electrophoresis. *Electrophoresis* **23,** 640–646.
84. Zabrouskov, V., Giacomelli, L., Van Wijk, K. J., and McLafferty, F. W. (2003) A new approach for plant proteomics: characterization of chloroplast proteins of *Arabidopsis thaliana* by top-down mass spectrometry. *Mol. Cell Proteomics* **2,** 1253–1260.

85. Utriainen, M., Kokko, H., Auriola, S., Sarrazin, O., and Kärenlampi, S. (1998) PR-10 protein is induced by copper stress in roots and leaves of Cu/Zn tolerant clone of birch, *Betula pendula*. *Plant Cell Environ.* **21,** 821–828.
86. Koistinen, K. M., Hassinen, V. H., Gynther, P. A. M., et al. (2002) Birch PR-10c is induced by factors causing oxidative stress but appears not to confer tolerance to these agents. *New Phytol.* **155,** 381–391.
87. Koistinen, K. M., Kokko, H. I., Hassinen, V. H., Tervahauta, A. I., Auriola, S., and Kärenlampi, S. O. (2002) Stress-related RNase PR-10c is post-translationally modified by glutathione in birch. *Plant Cell Environ.* **25,** 707–715.
88. Repetto, O., Bestel-Corre, G., Dumas-Gaudot, E., Berta, G., Gianinazzi-Pearson, V., and Gianninazzi, S. (2003) Targeted proteomics to identify cadmium-induced protein modifications in *Glomus mosseae*-inoculated pea roots. *New Phytol.* **157,** 555–567.
89. Weckwerth, W. (2003) Metabolomics in systems biology. *Annu. Rev. Plant Biol.* **54,** 669–689.
90. Grill, E., Winnacker, E.-L., and Zenk, M. (1985) Phytochelatins: the principal heavy metal complexing peptides of higher plants. *Science* **230,** 674–676.
91. Zenk, M. (1996) Heavy metal detoxification in higher plants: a review. *Gene* **179,** 21–30.
92. Cobbett, C. S. (2000) Phytochelatin biosynthesis and function in heavy metal detoxification. *Curr. Opin. Plant Biol.* **3,** 211–216.
93. Bailey, N. J. C., Oven, M., Holmes, E., Nicholson, J. K., and Meinhart, H. Z. (2003) Metabolomic analysis of the consequences of cadmium exposure in *Silene cucubalus* cell cultures via ^{1}H NMR spectroscopy and chemometrics. *Phytochemistry* **62,** 851–858.
94. Fan, T. W.-M., Lane, A. N., Shenker, M., Bartley, J. P., Crowley, D., and Higashi, R. M. (2001) Comprehensive chemical profiling of gramineous plant root exudates using high-resolution NMR and MS. *Phytochemistry* **57,** 209–221.
95. Lahner, B., Gong, J., Mahmoudian, M., et al. (2003) Genomic scale profiling of nutrient and trace elements in *Arabidopsis thaliana*. *Nat. Biotechnol.* **21,** 1215–1221.
96. Nucleic Acids Research (2004) Database issue. *Nucl. Acids Res.* **32,** 1–599.
97. Sessions, A., Burke, E., Presting, G., et al. (2002) A high-throughput Arabidopsis reverse genetics system. *Plant Cell* **14,** 2985–2994.
98. Chuang, C. F. and Meyerowitz, E. M. (2000) Specific and heritable genetic interference by double-stranded RNA in *Arabidopsis thaliana*. *Proc. Natl. Acad. Sci. USA* **97,** 4985–4990.
99. Wesley, S. V, Helliwell, C. A., Smith, N. A., et al. (2001) Construct design for efficient, effective and high-throughput gene silencing in plants. *Plant J.* **27,** 581–590.
100. Vasconcelos, M., Datta, K., Oliva, N., et al. (2003) Enhanced iron and zinc accumulation in transgenic rice with the ferritin gene. *Plant Sci.* **164,** 371–378.
101. Guo, W.-J., Bundithya, W., and Goldsbrough, P. B. (2003) Characterization of the Arabidopsis metallothionein gene family: tissue-specific expression and induction during senescence and in response to copper. *New Phytol.* **159,** 369–381.

102. Aarts, M. G. M. and Fiers, M. W. E. J. (2003) What drives plant stress genes? *Trends Plant Sci.* **8,** 99–102.
103. Rugh, C. L., Wilde, H. D., Stack, N. M., Thompson, D. M., Summers, A. O., and Meagher, R. B. (1996) Mercuric ion reduction and resistance in transgenic *Arabidopsis thaliana* plants expressing a modified bacterial merA gene. *Proc. Natl. Acad. Sci. USA* **93,** 3182–3187.
104. Dhankher, O. P., Li, Y., Rosen, B. P., et al. (2002) Engineering tolerance and hyperaccumulation of arsenic in plants by combining arsenate reductase and gamma-glutamylcysteine synthetase expression. *Nat. Biotechnol.* **20,** 1140–1145.
105. Dhankher, O. P., Shasti, N. A., Rosen, B. P., Fuhrmann, M., and Meagher, R. B. (2003) Increased cadmium tolerance and accumulation by plants expressing bacterial arsenate reductase. *New Phytol.* **159,** 431–441.
106. Lombi, E., Zhao, F. J., McGrath, S. P., Young, S. D., and Sacchi, G. A. (2001) Physiological evidence for a high-affinity cadmium transporter highly expressed in a *Thlaspi caerulescens* ecotype. *New Phytol.* **149,** 53–60.
107. Zhao, F. J., Hamon, R. E., Lombi, E., McLaughlin, M. J., and McGrath, S. P. (2002) Characteristics of cadmium uptake in two contrasting ecotypes of the hyperaccumulator *Thlaspi caerulescens. J. Exp. Bot.* **53,** 535–543.
108. Linacre, N. A., Whiting, S. N., Baker, A. M., Angle, J. S., and Ades, P. K. (2003) Transgenics and phytoremediation: the need for an integrated risk assessment, management, and communication strategy. *Int. J. Phytorem.* **5,** 181–185.
109. Linacre, N. A. (2003) Making decisions on the release of GM crops. *ISB News Report,* October, 1–4.
110. Axelsen, K. B., and Palmgren, M. G. (2001) Inventory of the superfamily of P-type ion pumps in Arabidopsis. *Plant Physiol.* **126,** 696–706.
111. Cobbett, C. S., Hussain, D., and Haydon, M. J. (2003) Structural and functional relationships between type 1B heavy metal-transporting P-type ATPases in Arabidopsis. *New Phytol.* **159,** 315–321.
112. Shikanai, T., Muller-Moule, P., Munekage, Y., Niyogi, K. K., and Pilon, M. (2003) PAA1, a P-type ATPase of Arabidopsis, functions in copper transport in chloroplasts. *Plant Cell* **15,** 1333–1346.
113. Curie, C., Alonso, J. M., Le Jean, M., Ecker, J. R., and Briat, J. F. (2000) Involvement of NRAMP1 from *Arabidopsis thaliana* in iron transport. *Biochem. J.* **347,** 749–755.
114. Thomine, S., Wang, R., Ward, J. M., Crawford, N. M., and Schroeder, J. I. (2000) Cadmium and iron transport by members of a plant metal transporter family in Arabidopsis with homology to Nramp genes. *Proc. Natl. Acad. Sci. USA* **97,** 4991–4996.
115. van der Zaal, B. J., Neuteboom, L. W., Pinas, J. E., et al. (1999) Overexpression of a novel Arabidopsis gene related to putative zinc-transporter genes from animals can lead to enhanced zinc resistance and accumulation. *Plant Physiol.* **119,** 1047–1055.
116. Tommasini, R., Vogt, E., Fromenteau, M., et al. (1998) An ABC-transporter of *Arabidopsis thaliana* has both glutathione-conjugate and chlorophyll catabolite transport activity. *Plant J.* **13,** 773–780.

117. Bovet, L., Eggmann, T., Meylan-Bettex, M., et al. (2003) Transcript levels of AtMRPs after cadmium treatment: Induction of AtMRP3. *Plant Cell Environ.* **26,** 371–381.
118. Hirschi, K. D., Zhen, R. G., Cunningham, K. W., Rea, P. A., and Fink, G. R. (1996) CAX1, an H+/Ca2+ antiporter from Arabidopsis. *Proc. Natl. Acad. Sci. USA* **93,** 8782–8786.
119. Hirschi, K. D., Korenkov, V. D., Wilganowski, N. L., and Wagner, G. J. (2000) Expression of Arabidopsis CAX2 in tobacco. Altered metal accumulation and increased manganese tolerance. *Plant Physiol.* **124,** 125–133.
120. Shaul, O., Hilgemann, D. W., de-Almeida-Engler, J., Van Montagu, M., Inz, D., and Galili, G. (1999) Cloning and characterization of a novel Mg(2+)/H(+) exchanger. *EMBO J.* **18,** 3973–3980.
121. Korshunova, Y. O., Eide, D., Clark, W. G., Guerinot, M. L., and Pakrasi, H. B. (1999) The IRT1 protein from Arabidopsis thaliana is a metal transporter with a broad substrate range. *Plant Mol. Biol.* **40,** 37–44.
122. Burleigh, S. H., Kristensen, B. K., and Bechmann, I. E. (2003) A plasma membrane zinc transporter from *Medicago truncatula* is up-regulated in roots by Zn fertilization, yet down-regulated by arbuscular mycorrhizal colonization. *Plant Mol. Biol.* **52,** 1077–1088.
123. Williams, L. E., Pittman, J. K., and Hall, J. L. (2000) Emerging mechanisms for heavy metal transport in plants. *Biochim. Biophys. Acta* **1465,** 104–126.
124. Schachtman, D. P., Kumar, R., Schroeder, J. I. and Marsh, E. L. (1997) Molecular and functional characterization of a novel low-affinity cation transporter (LCT1) in higher plants. *Proc. Natl. Acad. Sci. USA* **94,** 11,079–11,084.
125. Clemens, S., Antosiewicz, D. M., Ward, J. M., Schachtman, D. P., and Schroeder, J. I. (1998) The plant cDNA LCT1 mediates the uptake of calcium and cadmium in yeast. *Proc. Natl. Acad. Sci. USA* **95,** 12,043–12,048.
126. Li, L., He, Z., Pandey, G. K., Tsuchiya, T., and Luan, S. (2002) Functional cloning and characterization of a plant efflux carrier for multidrug and heavy metal detoxification. *J. Biol. Chem.* **277,** 5360–5368.
127. Arazi, T., Sunkar, R., Kaplan, B., and Fromm, H. (1999) A tobacco plasma membrane calmodulin-binding transporter confers Ni^{2+} tolerance and Pb^{2+} hypersensitivity in transgenic plants. *Plant J.* **20,** 171–182.
128. Zhou, J., and Goldsbrough, P. B. (1995) Structure, organization and expression of the metallothionein gene family in Arabidopsis. *Mol. Gen. Genet.* **248,** 318–328.
129. Murphy, A., Zhou, J., Goldsbrough, P. B., and Taiz, L. (1997) Purification and immunological identification of metallothioneins 1 and 2 from *Arabidopsis thaliana*. *Plant Physiol.* **113,** 1293–1301.
130. van Hoof, N. A., Hassinen, V. H., Hakvoort, H. W., et al. (2001) Enhanced copper tolerance in *Silene vulgaris* (Moench) Garcke populations from copper mines is associated with increased transcript levels of a 2b-type metallothionein gene. *Plant Physiol.* **126,** 1519–1526.
131. Vatamaniuk, O. K., Mari, S., Lu, Y. P., and Rea, P. A. (1999) AtPCS1, a phytochelatin synthase from Arabidopsis: Isolation and in vitro reconstitution. *Proc. Natl. Acad. Sci. USA* **96,** 7110–7115.

132. Gisbert, C., Ros, R., De Haro, A., et al. (2003) A plant genetically modified that accumulates Pb is especially promising for phytoremediation. *Biochem. Biophys. Res. Commun.* **303,** 440–445.
133. Gong, J., Lee, D. A., and Schroeder, J. I. (2003) Long-distance root-to-shoot transport of phytochelatins and cadmium in Arabidopsis. *Proc. Natl. Acad. Sci. USA* **100,** 10,118–10,123.

22

Manipulating Soil Metal Availability Using EDTA and Low-Molecular-Weight Organic Acids

Longhua Wu, Yongming Luo, and Jing Song

Summary

Soils can be contaminated with heavy metals from various human activities and a number of *ex situ* and *in situ* techniques have been developed to remove heavy metals from contaminated soils. Phytoremediation is a developing technology that aims to extract or inactivate metals, metalloids, and radionuclides in contaminated soils, and chemical enhancements have been used to enhance soil heavy-metal availability to plants. This chapter focuses on synthetic chelates and low-molecular-weight organic acids, in particular induced phytoextraction of heavy metals, the successful cases, the mechanisms of enhancement, and the disadvantages of the method.

Key Words: Bioavailability; heavy metal; phytoremediation; organic acid; synthetic chelate.

1. Introduction

Soils can be contaminated with heavy metals from various human activities including mining, smelting and metal-treatment operations, vehicle emissions, and deposition or leakage of industrial wastes. Because of the potential toxicity and persistence of heavy metals, the clean up of contaminated soils is one of the most difficult tasks for environmental engineering. A number of *ex situ* and *in situ* techniques have been developed to remove heavy metals from contaminated soils.

Phytoremediation is a developing technology that aims to extract or inactivate metals, metalloids, and radionuclides in contaminated soils *(1)*. There are two basic strategies under development. The first is the use of hyperaccumulator plants that have the capacity to hyperaccumulate heavy metals, and the second is chemical chelate-enhanced phytoextraction *(2)*. The major problem

From: *Methods in Biotechnology, vol. 23: Phytoremediation: Methods and Reviews*
Edited by: N. Willey © Humana Press Inc., Totowa, NJ

hindering plant remediation efficiency is that the phytoextraction rate is limited by solubility and diffusion to the root surface, but the metals can be immobile and unavailable in soil. So, chemical enhancements have been used to overcome this problem *(3–7)*. Synthetic chelates, including EDTA, and low-molecular-weight organic acids (LMWOAs) such as citric acid, oxalic acid, and malic acid, have also been tested for their effectiveness in chelate-induced phytoextration of metals. In this chapter, we focus on synthetic chelate- and LMWOA-induced phytoextraction of heavy metals, in particular the successful cases, the mechanisms of enhancement, and the disadvantages of the method.

2. Solubilization and Mobilization of Metals in Soil

2.1. Effects of Chelates on Solubilization of Heavy Metals in Soil

Many chelates have been used in phytoremediation processes. The most promising application of this technology is for the remediation of Pb-contaminated soils. It has been reported that the addition of chelates to Pb-contaminated soil (total soil Pb was 2500 mg/kg) increased shoot Pb concentrations of corn and pea from less than 500 mg/kg to more than 10,000 mg/kg *(4)*. It was supposed that the surge of Pb accumulation in these plants was associated with the surge of Pb level in the soil solution resulting from the addition of chelates to the soil. It has also been found that concentrations of 1.5% Pb in the shoots of *Brassica juncea* could be obtained from soils containing 600 mg of Pb/kg amended with EDTA *(3)*. Enhancement of phytoremediation by EDTA addition has been reported for other heavy-metal contamination such as Cd *(8,9)*, Cr *(10)*, Cu *(6,11)*, and Zn *(5)*. Other aminopolycarboxylic acids have also been tested, but they were all less efficient than EDTA *(4,6,12,13)*.

LMWOAs, such as citric acid, malic acid, oxalic acid, acetic acid, histidine, and malonic acid, are another kind of compound used in phytoremediation. Information is also available about heavy-metal accumulation following the application of these natural organic acids to contaminated soils. For example, it has been reported that the addition of citric acid and its salts selectively increased uranium mobility in soil and subsequently plant uptake *(14,15)*. Nigram et al. *(16)* found that the Cd accumulation by corn after applying the carboxylic acids to a Cd-spiked soil (3.5 μM/kg or 0.39 mg/kg) was enhanced, and it was also higher than with the amino acid aspartic acid or glycine. The organic acids were applied in the same molar concentrations as Cd to the soil, and the Cd concentrations in the corn shoots were more than doubled with citric acid (to 19 mg/kg) and also significantly increased with malic acid (to 15 mg/kg).

2.2. Mechanisms of Soil Metal Mobilization by Chelates

The fundamental mechanism of metal mobilization is a change in the balance of pollutants between soil solution and solid phase. When added to soil, chelates form soluble complexes with metals in soil solution and thus mobilize metals from the solid phase. Excess chelate may exist in free form. The mechanisms of chelate-induced metal solubilization include dissolution of soil minerals via ligand-exchange reaction and remobilization of metals adsorbed onto the solid phase *(17)*. Although being operationally defined, different sequential extraction schemes have been widely employed to investigate the effect of chelate on the solubility of metals associated with different soil fractions. A review of different sequential extraction procedures for fractionation of heavy metals in contaminated soil and sediment has recently been published *(18)*. Elliott and Shastri *(19)* proposed that metals that can be mobilized by EDTA were mainly from the nondetrital soil components (exchangeable fraction, organic matter, and carbonate-bound fractions), and EDTA is ineffective in solubilizing metals from the detrital fractions (metals in oxides and residual-bound fractions). However, other studies *(20,21)* seem to show a complex picture of metal release from different fractions after EDTA addition. For instance, after EDTA extraction, acetic acid-extractable soil Pb (exchangeable + carbonatic fractions) significantly increased, coupled with marked decrease in oxides and organically bound fractions *(21)*. In addition, EDTA seems to be able to release a certain amount of silicate-bound Pb but the effects of EDTA were different among different metals and soils *(21)*.

Although it is difficult to generalize about which fraction is more mobile than the others, the metal-mobilizing effect of chelate can be indicated by changes in metal concentration in soil pore water *(22–24)* or in labile metal fractions determined by soil extraction with water *(7,9)*, 1 M NH$_4$OAc *(25)*, 0.1 M NaNO$_3$ *(12,26)*, 0.1 M CaCl$_2$ *(27)*, 0.01 M CaCl$_2$ *(28)*, or 1 M NH$_4$NO$_3$ *(9,11)*. For example, marked increases of metals in soil pore water within 24 h after application of 2.7 mmol/kg EDTA (added as salt), with soluble Zn increased from 2.4 to 104 mg/L, observed in UK soil, and soluble Pb increases from 0.1 to 36 mg/L in French soil *(23)*. A 1500-fold increase of 1 M NH$_4$NO$_3$-extractable Pb in 5.4 mmol/kg EDTA-treated soil (added as solid EDTA), relative to the control in a field lysimeter study, has also been reported *(11)*.

2.3. Factors Influencing Soil Metal Mobility by Chelates

Factors that affect the balance of metal ions between soil solution and soil solid phase will change metal mobility. Influences on the effectiveness of a given chelate to solubilize soil metals include: metal species and distribution among soil fractions, metal-to-ligand ratio, formation constant of metal–ligand complexes, presence of competing cations, soil pH, and so on.

A study by Epstein et al. *(7)* showed that when sufficient EDTA was added, all PbCO$_3$ spiked in the soil can be solubilized. The authors also indicated that to obtain the same level of soluble Pb from Pb in more recalcitrant forms, more

EDTA is needed. Because of the extreme variability of metal forms and soil property, it is difficult to accurately predict all the competing reactions and conditions *(29)*. Much of the work, therefore, relies on empirical data to determine the dosage used in chelate-induced phytoextraction. Despite large differences in metal species, concentrations, soil properties, and type of chelates, the application rates reported in the literature range from 0 to 10 mmol/kg soil, and the desorption rate of soil heavy metals always increases with the increasing dosage of chelates. Chelate effects on desorption of soil Pb have been examined with soils using a consecutive desorption approach *(12)*. It was found that the average amounts and percentages of total Pb desorbed for the 0, 0.2, 2.0, and 20 mmol/kg chelates were 28, 32, 131, and 948 mg/kg. In the pot experiment conducted by Wenzel et al. *(11)*, metal mobilization was affected by application of EDTA at rates between 0.21 and 1.65 g/kg soil was indicated by corresponding changes of the labile (1 M NH_4NO_3-extractable) fractions in the soil. The largest addition of EDTA increased the labile metal fraction to 34% (Cu), 11% (Pb), and 17% (Zn) of the soil total concentrations. Results from incubation experiments have also showed that concentrations of heavy metals in soil pore water were increased with the EDTA addition rates *(24)*. Normally, the amount of chelate applied was about 3 mmol/kg (1 g/kg), but unusually high application rates could lead to low phytoextraction efficiency and high environmental risks. Plant dry matter yield was significantly affected by the application of the EDTA. Plant growth in untreated or the 0.1 mmol/kg treated soil produced nearly twice the biomass of the plants receiving the 10 mmol/kg EDTA *(3)*.

Kim et al. *(30)* suggested that occlusion of Pb in the Fe oxides may reduce EDTA extraction efficiency of soil Pb. Although the formation constant for 1:1 metal–EDTA complexes follows the order: Cu>Pb>Zn>Fe>Ca, major cations such as Fe and Ca present in the soil may compete for active sites of EDTA. A soil washing *(30)* showed that Fe most probably competed strongly with Pb for EDTA-ligand sites at pH less than 6.0. In a multimetal-contaminated soil (pH <6.0), Cu and Zn may potentially compete with Pb for EDTA ligand sites. Huang et al. *(15)* found a close correlation between the U and the Fe and Al concentrations in the soil solution after the addition of citric acid, which they explained by dissolution of Fe and Al sequioxides and hence release of U from soil material to the soil solution. The abundant Ca^{2+} in calcareous soil could have displaced heavy metals from their EDTA complexes, leading to the formation of CaH_2EDTA and insoluble metal carbonates *(27)*. Metal behavior will be changed while soil pH is changed, but little work has investigated this aspect because soil chemical characters will be different, whereas soil pH changes. The efficiency of organic agents on metal accumulation by plants could, however, be enhanced by lowering the soil pH *(3)*.

3. Uptake of Chelated Metals by Plants

3.1. Mechanism of Chelated Metal Uptake by Plants

The success of chelate enhancement requires a better understanding of the biological mechanisms involved. The predominant theory for chelate-induced metal uptake is the split-uptake mechanism, where only free metals dissociated from the metal–ligand complex will be absorbed by plant roots, leaving the ligand in the soil solution. Another theory suggests that metal–ligand complexes are absorbed by plant roots, transported through the plant via the xylem, and accumulate in the shoots. Recent studies using an isotope tracer ($[^{14}C]$ EDTA) *(31)* and analysis with liquid-scintillation counter seemed to support the latter theory. In addition, it has been hypothesized that it is free protonated EDTA that enters the roots, subsequently forming metal complexes that enhance metal transport to shoots *(11)*. For this to happen, the kinetics of metal–EDTA complex formation in real soil systems need to be slow enough to allow for the diffusion of free protonated EDTA to the root surfaces *(11)*.

Regarding the pathways through which metal–ligand complexes enter xylem tissue, recent ultrastructural studies imply that both apoplastic and symplastic transport pathways may exist, depending on the plant species and chelate used. Transmission electron microscopy has revealed that in the shoots of *Chamaecytisus palmensis* (a fast growing evergreen leguminous tree), Pb chelated with H-EDTA appears to follow an apoplastic route, as indicated by Pb deposition around the intercellular space and within the cell walls *(32)*. On the other hand, EDTA-chelated Pb follows a symplastic path, as indicated by occurrence of Pb within structures in the cytoplasm *(32)*. In the needles of *Pinus radiata* (a fast growing conifer), Pb–ligand complexes were transported exactly in the opposite manner. Pb chelated with H-EDTA appears to follow a symplastic route, whereas EDTA-chelated Pb follows an apoplastic path *(33)*.

Conventional theory suggests that once in the xylem vessel, long-distance transportation of metal–ligand complexes from root to shoot will be driven by the transpiration stream *(3,31,34)*. However, study by Epstein et al. *(7)* implied that transpiration appeared not to be a critical factor for the uptake of Pb and EDTA. The majority of Pb and EDTA uptake may have occurred rapidly (within hours) after the EDTA application before the measured transpiration decreased.

3.2. Factors Influencing Chelate-Induced Metal Accumulation by Plants

Early studies reported EDTA-induced hyperaccumulation of Pb by Indian mustard *(3)*, maize *(8)*, pea, and corn *(4)*. Although many authors have recently reported enhanced metal uptake after addition of other chelates, in a few cases metal hyperaccumulation was achieved *(11,13,22,35–38)*. Recently we studied the effect of EDTA and the biodegradable chelate EDDS (an isomer of EDTA) on metal uptake by *B. juncea* grown on a multimetal-contaminated soil collected near smelters. In treatments with 3 or 6 mmol/kg EDDS, regardless of the mode of application (single or split), average Cu concentration in leaves was much greater than 1000 mg/kg. Bioconcentration factors up to 7.7 for leaf and 2.5 for stem were achieved (unpublished data). The effectiveness of EDDS decreased in the order Cu>Zn> Pb>Cd, which follows the stability constant of EDDS–metal complexes reported in the literature *(39)*. Compared with Cu, the bioconcentration factor of Pb was unsatisfactory in our study. A relatively low formation constant of Pb–EDDS may partly be the reason why Pb hyperaccumulation has not been achieved by using EDDS, even at high dosages (10 mmol/kg).

Vassil et al. *(31)* indicated that a threshold concentration of EDTA is required to obtain accumulation of Pb in plant shoots. They suggested that at high concentrations chelates may physiologically damage the root membranes that would normally impede the uptake of intact metal–ligand complexes. Synthetic chelates could also disrupt the normal function of cell membrane by removing Zn and Ca ions that are involved in the stabilization of plasma membranes *(31)*. Using data from the literature, McGrath et al. *(1)* plotted extractable Pb against Pb in shoots of different plant species induced by different chelates. The wide-span (three orders of magnitude) of data seemed to fit a linear relationship. However, it should be noted that there might be huge differences in plant species regarding their response to elevated metal–ligand concentration in the growth media. For instance, a parabolic relationship between water-extractable Pb and Pb concentration in the shoot of *B. juncea* has been reported *(7)*. Robinson et al. *(40)* found that addition of chelates (NTA, DTPA, and EDTA) increased 1 M NH_4OAc-extractable Ni in an artificial serpentine substrate, but Ni uptake by *Berkhaya codii* decreased about 50% relative to the control. The authors attributed decreased Ni concentration in shoots to competition with the plant's own nickel-binding agents, thereby causing the nickel to diffuse downward to the plant's root system *(1)*. In addition, free protonated EDTA may lead to phytotoxicity *(30)* and pose potential risks to the environment (*see* **Subheading 4.**). Therefore, it is necessary to maintain a proper concentration range of metals in soluble form to achieve a high bioconcentration factor while minimize potential risks associated with chelate addition.

4. Drawbacks of Chelate-Induced Phytoextraction

The purpose of EDTA or other chelates application is to promote heavy-metal mobility, thereby increasing plant heavy-metal uptake and enhancing phytoremediation efficiency. EDTA mobilizes metals rapidly and then their concentration decreases slowly over a long time period, which may assist plant metal uptake. However, when plants are harvested and phytoremediation ends, high concentrations of heavy metals chelated by EDTA can remain in the soil.

We found good agreement between the soil solution total organic carbon concentrations and total molar concentrations of Cu, Zn, Pb, Cd in the soil solution *(41)*. This indicates that most of the heavy metals were complexed by EDTA. Similar observations were made by Lombi et al. *(23)*. Thus, possible side effects of EDTA application should be considered. If low concentrations of EDTA are resistant to degradation, the chelate may persist and affect heavy-metal behavior in the soil over long time periods *(42)*. So, as exogenous substances, chelates carry the risk of negative effects on the environment when applied to soils. Here, we discuss the major risks reported in the literature (e.g., adverse effect on plant and soil biota, metal leaching to groundwater) as well as possible countermeasures. Thus, despite the success of chelate enhancement of heavy-metal mobility, the potentially serious consequential effects should always be taken into account in phytoremediation schemes.

Reported negative effects of chelate on plant growth include foliar necrosis, leaf wilt and abscission, shoot desiccation, reduced transpiration, and biomass *(7,13,22,23,31,38)*. Soils contaminated with phytotoxic levels of Cu, Zn, or Cd present a challenge for phytoremediation. A lysimeter study suggested that toxicity of Cu to canola might be alleviated from complexation with EDTA *(11)*. A hydroponic study suggested that EDTA-induced foliar necrosis may be attributable to the presence of free protonated EDTA in leaves, as no phytotoxicity was observed in treatment with equal molarities of Pb and EDTA *(31)*. HEDTA always proved to be more phytotoxic than EDTA at the same concentration *(32)*. To avoid toxic effects of high concentration of EDTA, it is suggested that EDTA should be applied at rates that minimize the availability of free chelate *(11)*.

The persistence of soluble EDTA–metal complexes in soil can cause prolonged negative effects upon soil microfauna and plant growth. To reduce the long-term negative effects of recalcitrant chelates, a biodegradable EDTA structural isomer EDDS has recently been tested *(36–38,43)*. In a pot trial using multimetal-polluted soils, we compared the effect of EDTA and EDDS on plant growth and metal uptake by *B. juncea*. The third day after application, leaves of all four replicates in the 6 and 3 mmol/kg EDDS treatment started to wilt. While in soil treated with 3 mmol/kg EDTA and lower dosages of EDDS, symptoms of leaf wilt occurred a few days later and not on all four replicates (unpublished data). It has been shown that increasing the concentration of EDTA caused significant reduction in shoot water content *(31)*. Rapid senescence of cabbage shoots occurred in treatments receiving single and weekly additions of 10 mmol/kg EDTA *(38)*. Transpiration rate is not a critical factor for translocation of metal–ligand complexes from root to shoot *(31)*, therefore, decreased transpiration rate will not be a matter of concern in the context of chelate-induced phytoextraction. However, wilted leaves may present a problem as leaf abscission may occur a few days after wilt. To avoid loss of metal-rich plant material, harvest should be done before leaf abscission.

Chelate-induced effects on plant biomass may be largely explained by the mode of application, dosage as well as plant tolerance. The same amount of chelates can be added (1) in a single dose after plants accumulate enough biomass or (2) in a single dose before transplanting, and (3) gradually added at several lower dosage during the growth period. In the first case, plant biomass will less likely be affected by chelate addition as plants will normally be harvested several days after application. Plants may die of phytotoxicity as a result of high concentration of soluble metal–ligand complexes and/or free chelates. In the context of phytoextraction, dead plants are not a big problem. However, excessive soluble metals may inhibit plant growth in the follow-up croppings *(23)*. The inhibition effect will be more pronounced for recalcitrant chelates (e.g., EDTA) than biodegradable chelates (e.g., EDDS). A bioassay with red clover was performed to evaluate the posttreatment toxicity of EDTA- and EDDS-treated soil. The results showed that biomass of red clover shoots was significantly reduced in soil that received 5 and 10 mmol/kg weekly EDTA additions as compared with control and corresponding EDDS treatments *(39)*. More importantly, in cases where recalcitrant chelate is used, highly soluble metals will remain for a long time (e.g., several months). Leaching of soluble metals is then likely to occur (to be discussed next).

In the second approach, when chelates are added before transplanting, plant growth may be inhibited to an extent that reduction in plant biomass cannot be compensated by increased metal concentration in the shoots, thereby decreasing net metal removal *(13)*. The third approach aims to provide maximum soluble metal available for removal while reducing the risk of metal leaching. Studies have suggested that longer exposure of canola plants to toxic EDTA levels in the split application treatments may limit shoot biomass production *(11)*. To determine the time interval between two applications, $0.1\ M$ $NaNO_3$ extraction has been used to indicate the change in a phytoavailable pool of soil metals *(26)*.

Microbial biomass, respiration, nitrogen mineralization, microbial diversity, and functional groups of soil fauna (e.g., nematodes) are well-recognized indicators for evaluating soil quality *(44)*. Römkens et al. *(22)* reported that addition of EGTA resulted in an increase of microbial biomass, that bacterial activity measured as ^{14}C-leucine incorporation was slightly lower in EGTA treatment, whereas bacterial activity measured as ^3H-thymidine incorporation was not affected by EGTA. The net effect of EGTA addition on bacterial growth was, therefore, limited. The effect of EGTA on the number of soil nematodes was found to be dependent on the type of nematodes and plant species. EGTA significantly reduced bacterivores (up to 90% reduction) and fungivores under three crops (grass, lupin, and yellow mustard), but had no direct effect on herbivores. The authors also indicated that EGTA might exert its influence on the soil ecosystem through its effect on plant growth, which in turn affected the number, activity, and type of soil microbes and nematodes that are critical to the function of the soil ecosystem *(22)*. In a laboratory study to assess the potential toxicity of EDTA and [S,S]-EDDS on soil microbes, Kos and Leštan *(37)* observed increased glucose-induced microbial respiration with increasing rate of [S,S]-EDDS presumably because of microbial use of [S,S]-EDDS as an additional carbon or energy source. In contrast, high concentrations of EDTA decreased glucose-induced respiration. However, the authors also suggested that increased Pb leaching in 10 mmol/kg EDDS treatment was likely because of toxic effects on soil microbes, which are capable of degrading EDDS *(37)*. A study by Grěman et al. *(38)* showed that mycorrhizal infection of red clover seemed not to be affected by soil pretreatment with 5 and 10 mmol/kg EDDS or EDTA, although the biomass of red clover was significantly reduced in the 5 and 10 mmol/kg EDTA treatment. Using the PLFA (phospholipid fatty acid) technique, the same authors further demonstrated that EDDS addition was less toxic to soil fungi than EDTA and caused less stress to soil microbes *(36)*. It should be noted that the observed effects of chelate addition on soil microbes and soil biota are joint effects of metal–ligand complexes, free chelates, and existing plants.

As the amount of metals mobilized is normally far beyond the amount that plants can take up in the growing season, solubilized metals may leach down the soil profile to groundwater during rain events. Unusually high metal concentration in leachates collected in column *(20,41,43)* and field lysimeter studies *(11,22)* provided further evidence of metal leaching induced by the addition of EDTA and EGTA. The metal concentrations in the leachates after EDTA addition were clearly related to the rate of EDTA applied *(11,41)*. The order of leaching response to EDTA application is largely consistent with the corresponding formation constant of metal–EDTA complexes (Cu, 20.5; Pb, 19.8; Zn, 18.3) *(20)*. The effect of EDTA addition on metal concentration in leachates was reported to be sustained for several months after EDTA application *(11)*. In contrast to the large percentage of Pb, Cd, and Zn leached through the soil profile, Grěman et al. *(38)* found that when applied at the same dosage, biodegradable [S, S]-EDDS caused much less loss of Pb and Cd –22.7 and 39.8% of initial Pb and Cd leached in 10 mmol/kg EDTA treatments, as compared with 0.8 and 1.5% in EDDS treatment. Leaching of Zn (about 6.2% of initial total concentration) was comparable with the EDTA treatment *(38)*. Recently, a couple of approaches were proposed to facilitate metal mobilization and plant uptake while reducing metal leaching. Using soil column experiments, Kos and Leštan *(36)* tested the effectiveness of increasing field soil water-holding capacity by using acrylamide hydrogel. The idea is to retain chelate solution in the top soil.

Their results showed that acrylamide hydrogel was not particularly of use as a soil conditioner. In another column study, the same authors showed that permeable barriers (consisting of nutrient-enriched vermiculite, peat, or hydrogel in combination with apatite placed underneath polluted soil cores were effective in reducing total Pb leached after the addition of 10 mmol/kg biodegradable EDDS) *(37)*. Although it was environmentally safe, Pb concentration achieved (463 ± 112 mg/kg) in the study was far from the concentration required for efficient phytoextraction of Pb (1%) within a reasonable time frame. Furthermore, the high cost of EDDS ($7800 per ton) and the cost required for installation of permeable barriers, may limit the use of this technique in field application.

Coupled with leaching of heavy metals, chelate addition may also cause loss of macronutrients such as Fe, Ca, and Mg. In a laboratory-leaching experiment using soil columns, Wu et al. *(41)* found the amount of Fe lost increased to 163 mg/kg in the 12 mmol/kg EDTA treatment as compared with 3.37 mg/kg in the control. The effects of volume and pH of simulated rainfall were minimal. On the other hand, EDTA treatment, volume, and pH of simulated rainfall seemed to have no significant effects on the loss of Ca and Mg. The authors also suggested that rainfall pH may play a role in the long term.

5. Final Remarks on Potential Application

Successful chelate-enhanced phytoextraction relies on interactions among soil–metal–chelate–plant. This complex interaction will be affected by a variety of factors such as soil properties (e.g., cation-exchange capacity, buffer capacity, and penetratability), pollution characteristics (e.g., metal species, distribution), chelate application (e.g., type, rate, and mode), plant species used, plant growth stage, and even the weather conditions when the experiment is conducted. To adjust these factors for one situation to make phytoremediation efficiency as high as possible, is a very difficult task. There are always compromises to be made between maximal effectiveness, maximal environmental merits, lowest risks, and lowest costs. Therefore, it is rather difficult to generalize on the prospects of chelate-enhanced phytoextraction. More fundamental research is needed to investigate the mechanism of metal solubilization by selected chelates, the biogeochemistry of metal–ligand complexes, the mechanisms of plant uptake of chelated metals, the effects of agronomic practices, and countermeasures to reduce negative effects of chelates. Such knowledge will be of great help in optimizing the processes involved in chelate-induced phytoextraction and to allow practitioners to customize the processes to site-specific conditions.

The potential risk of negative effects on the environment (e.g., plant health, soil quality, metal leaching) will also depend on climatic conditions (e.g., temperature, rainfall, its pH, and so on), and site hydrogeology (e.g., depth of water table). The dosage, use method, time, and other factors must be carefully considered before chelates, and especially EDTA, are applied. As regards the disadvantages of chelates on soil properties and potential environmental risk, the biodegradable, low-toxicity, synthetic, organic, metal-specific high efficiency, and environmentally friendly chemicals will be introduced in this area. Importantly, the cost of the chelate must also be considered. [S,S] EDDS is highly efficient at increasing soil metal mobility, but its price is higher than EDTA, so direct use of EDDS may not to be economical.

References

1. McGrath, S. P., Zhao, F.J., and Lombi, E. (2002) Phytoremediation of metals, metalloids, and radionuclides. *Adv. Agron.* **75**, 1–56.
2. Salt, D. E., Smith, R. D., and Raskin, I. (1998) Phytoremediation. *Ann. Rev. Plant Phys. Plant Mol. Biol.* **49**, 643–668.
3. Blaylock, J. M., Salt, D. E., Dushenkov, S., et al. (1997) Enhanced accumulation of Pb in Indian mustard by soil-applied chelating agents. *Environ. Sci. Technol.* **31**, 860–865.
4. Huang, J. W., Chen, J., Berti, W. R., and Cunningham, S. D. (1997) Phytoremediation of lead-contaminated soils: role of synthetic chelating in lead phytoeatraction. *Environ. Sci. Technol.* **31**, 800–805.
5. Ebbs, S. D., Lasat, M. M., Brady, D. J., Cornish, J., Gordon, R., and Kochian, L. V. (1997) Phytoextraction of cadmium and zinc from a contaminated soil. *J. Environ. Qual.* **26**, 1424–1430.
6. Wu, L. H., Luo, Y. M., Christie, P., and Wong, M. H. (2003) Effects of EDTA and low molecular weight organic acids on soil solution properties of a heavy metal polluted soil. *Chemosphere* **50**, 819–822.
7. Epstein, A. L., Gussman, C. D., Blaylock, M. J., et al. (1999) EDTA and Pb-EDTA accumulation in *Brassica juncea* grown in Pb-amended soil. *Plant Soil* **208**, 87–94.
8. Bricker, T. J., Pichtel, J., Brown, H. J., and Simmons, M. J. (2001) Phytoextraction of Pb and Cd from a superfund soil: effects of amendments and croppings. *J. Environ. Sci. Health A* **36**, 1597–1610.
9. Jiang, X. J., Luo, Y. M., Zhao, Q. G., Baker, A. J. M., Christie, P., and Wong, M. H. (2003) Soil Cd availability to Indian mustard and environmental risk following EDTA addition to Cd-contaminated soil. *Chemosphere* **50**, 813–818.
10. Shahandeh H. and Hossner, L. R. (2000) Plant screening for chromium phytoremediation. *Int. J. Phytorem.* **2**, 31–51.
11. Wenzel, W. W., Unterbrunner, R., Sommer, P., and Sacco, P. (2003). Chelate-assisted phytoextraction using canola (*Brassica napus* L.) in outdoors pot and lysimeter experiments. *Plant Soil* **249**, 83–96.
12. Cooper, E. M., Sims, J. T., Cunningham, S. D., Huang, J. W., and Berti, W. R. (1999) Chelate-assisted phytoextraction of lead from contaminated soils. *J. Environ. Qual.* **28**, 1709–1719.
13. Chen, H. and Cutright, T. (2001) EDTA and HEDTA effects on Cd, Cr, and Ni uptake by *Helianthus annuus*. *Chemosphere* **45**, 21–28.

14. Ebbs, S. D., Norvell, W. A., and Kochian, L. V. (1998) The effect of acidification and chelating agents on the solubilization of uranium from contaminated soil. *J. Environ. Qual.* **27,** 1486–1494.
15. Huang, J. W., Blaylock, M. J., Kapulnik, Y., and Ensley, B. D. (1998) Phytoremediation of uranium contaminated soils: Role of organic acids in triggering uranium hyperaccumulation in plants. *Environ. Sci. Technol.* **32,** 2004–2008.
16. Nigam R., Srivatava S., Prakash S., and Srivastava M. M. (2001) Cadmium mobilisation and plant availability: the impact of organic acids commonly exuded from roots. *Plant Soil* **230,** 107–113.
17. Nowack, B. (2002) Environmental chemistry of aminopolycarboxylate chelating agents. *Environ. Sci. Technol.* **36,** 4009–4016.
18. Gleyzes, C., Tellier, S., and Astruc, M. (2002) Fractionation studies of trace elements in contaminated soils and sediments: a review of sequential extraction procedures. *Trends Anal. Chem.* **21,** 451–467.
19. Elliott, H. A. and Shastri, N. L. (1999) Extractive decontamination of metal-polluted soils using oxalate. *Water, Air Soil Pollut.* **110,** 335–346.
20. Sun, B., Zhao, F. J., Lombi, E., and McGrath, S. P. (2001) Phytoextraction of cadmium with *Thlaspi caerulescens*. *Environ. Pollut.* **113,** 111–120.
21. Barona, A., Aranguiz, I., and Elias, A. (2001) Metal associations in soils before and after EDTA extractive decontamination: implications for the effectiveness of further cleanup procedures. *Environ. Pollut.* **113,** 79–85.
22. Römkens, P., Bouwman, L., Japenga, J., and Draaisma, C. (2002) Potentials and drawbacks of chelate-enhanced phytoremediation of soils. *Environ. Pollut.* **116,** 109–121.
23. Lombi, E., Zhao, F. J., Dunham, S. J., and McGrath, S. P. (2001) Phytoremediation of heavy metal-contaminated soils: natural hyperaccumulation versus chemically enhanced phytoextraction. *J. Environ. Qual.* **30,** 1919–1926.
24. Wu, L. H., Luo, Y. M., Song, J., Christie, P., and Wong, M. H. (2003) Changes in soil solution heavy metal concentrations over time following EDTA addition to a Chinese paddy soil. *Bull. Environ. Contam. Toxicol.* **71,** 706–713.
25. Robinson, B. H., Brooks, R. R., Gregg, P. E. H., and Kirkman, J. H. (1999) The nickel phytoextraction potential of some ultramafic soils as determined by sequential extraction. *Geoderma* **87,** 293–304.
26. Gupta, S. K., Herren, T., Wenger, K., Krebs, R., and Hari, T. (2000) *In situ* gentle remediation measures for heavy metal-polluted soils. In: *Phytoremediation of Contaminated Soil and Water* (Terry, N. and Bañuelos, G., eds.), Lewis Publishers, Boca Raton, FL, pp. 303–321.
27. Walker, D. J., Clemente, R., Roig, A., and Bernal, M. P. (2003) The effects of soil amendments on heavy metal bioavailability in two contaminated Mediterranean soils. *Environ. Pollut.* **122,** 303–312.
28. Degryse, F., Broos, K., Smolders, E., and Merckx, R. (2003). Soil solution concentration of Cd and Zn can be predicted with a $CaCl_2$ soil extract. *Eur. J. Soil Sci.* **54,** 149–157.
29. Blaylock, M. J. and Huang, J. W. (2000) Phytoextraction of Metals. In: *Phytoremediation of Toxic Metals: Using Plants to Clean Up the Environment* (Raskin, I. and Ensley, B. D., eds.), John Wiley and Son, Inc., New York, pp. 53–70.

30. Kim, C., Lee, Y., and Ong, S. K. (2003) Factors affecting EDTA extraction of lead from lead-contaminated soils. *Chemosphere* **51,** 845–853.
31. Vassil, A. D., Kapulnik, Y., Raskin, I., and Salt, D. E. (1998) The role of EDTA in lead transport and accumulation by Indian mustard. *Plant Physiol.* **117,** 447–453.
32. Jarvis, M. D. and Leung, D. W. M. (2001) Chelated lead transport in *Chamaecytisus proliferus* (L.f.) link ssp *proliferus* var. *palmensis* (H. Christ): an ultrastructural study. *Plant Sci.* **161,** 433–441.
33. Jarvis, M. D. and Leung, D. W. M. (2002) Chelated lead transport in *Pinus radiata*: an ultrastructural study. *Environ. Exp. Bot.* **48,** 21–32.
34. Salt, D. E., Prince, R. C., Pickering, I. J., and Raskin, I. (1995) Mechanisms of cadmium mobility and accumulation in Indian mustard. *Plant Physiol.* **109,** 427–433.
35. Kayser, A., Wenger, K., Keller, A., et al. (2000) Enhancement of phytoextraction of Zn, Cd, and Cu from calcareous soil: The use of NTA and sulfur amendments. *Environ. Sci. Technol.* **34,** 1778–1783.
36. Kos, B. and Lestan, D. (2003a) Induced phytoextraction/soil washing of lead using biodegradahle chelate and permeahle barriers. *Environ. Sci. Technol.* **37,** 624–629.
37. Kos, B. and Lestan, D. (2003b) Influence of a biodegradable ([S,S]-EDDS) and nondegradable (EDTA) chelate and hydrogel modified soil water sorption capacity on Pb phytoextraction and leaching. *Plant Soil* **253,** 403–411.
38. Grčman, H., Vodnik, D., Velikonja-Bolta, Š., and Lestan, D. (2003) Ethylenediaminedissuccinate as a new chelate for environmentally safe enhanced lead phytoextraction. *J. Environ. Qual.* **32,** 500–506.
39. Bucheli-Witschel, M. and Egli, T. (2001) Environmental fate and microbial degredation of amnopolycarboxylic acids. *FEMS Microbiol. Revs* **25,** 69–106.
40. Robinson, B. H., Brooks, R. R., and Clothier, B. E. (1999) Soil amendments affecting nickel and cobalt uptake by *Berkheya coddii*: Potential use for phytomining and phytoremediation. *Annal. Bot.* **84,** 689–694.
41. Wu, L. H., Luo, Y. M., Xing, X. R., and Christie, P. (2004) EDTA-enhanced phytoremediation of heavy metal contaminated soil and associated environmental risk. *Agric. Eco. and Environ.* **102,** 307–318.
42. Schmidt, U. (2003) Enhancing phytoremediation: the effect of chemical soil manipulation on mobility, plant accumulation, and leaching of heavy metals. *J. Environ. Qual.* **32,** 1939–1954.
43. Grčman, H., Velikonja-Bolta, Š., Vodnik, D., Kos, B., and Leštan, D. (2001). EDTA enhanced heavy metal phytoextraction: metal accumulation, leaching and toxicity. *Plant Soil* **235,** 105–114.
44. Schloter, M., Dilly, O., and Munch, J. C. (2003) Indicators for evaluating soil quality. *Agric. Eco. and Environ.* **98,** 255–262.

23

Soils Contaminated With Radionuclides
Some Insights for Phytoextraction of Inorganic Contaminants

Neil Willey

Summary
Soils contaminated with radionuclides provide a particular challenge to soil decontamination and hence a useful perspective on the phytoextraction of inorganic contaminants from soils. As they are potentially potent hazards to the biosphere, radionuclides in soils have attracted a high profile, but this has not yet provoked the development of environmentally benign methods to extract them from soils. Here, radionuclide-contaminated soils are used as a context to discuss phytoextraction development for inorganic contaminants, in particular future legislative pressures, economics, environmental impact, and the potential for manipulating soil availability and plant uptake. It is concluded that radioecologists researching contamination of the soil–plant system have a number of insights that might be useful for the development of phytoextraction not only for some radionuclides but also for other inorganic contaminants.

Key Words: Radionuclides; phytoextraction; transfer factors.

1. The Challenges

Decontamination of radionuclide-contaminated soils presents a particular set of challenges. These challenges, and the knowledge that radioecologists possess that might be helpful in surmounting them, are worth articulating not only because of their potential utility for cleaning radionuclide-contaminated soils but also because of the insights they provide to phytoextraction of inorganic contaminants in general.

Romney et al. *(1)* and Nishita et al. *(2)* reported, using radionuclides in the 1950s, some of the first calculations of the potential of plants to extract inorganic contaminants from soils. A single crop of *Trifolium repens* (clover) was able to remove 4.42% of added ^{90}Sr, and nine crops over 520 d removed 24%.

They suggested that this might make phytoextraction of ^{90}Sr possible and contrasted this with the limited potential for ^{137}Cs. Since then there have been a number of phytoextraction trials for radionuclides including those at Brookhaven National Laboratory, Upton, NY, United States *(3–5)*, the Chernobyl exclusion zone *(6,7)*, Argonne National Laboratory, Argonne, IL, United States *(8)*, and Bradwell Nuclear Power Station, United Kingdom *(9)*. Recent retrospectives *(10–12)* have concluded that phytoremediation for radionuclides could become useful in the near future. However, this is primarily because of the potential for rhizofiltration from effluent or phytostabilization of contaminated soils, rather than phytoextraction from soil. Reviews of potential rehabilitation schemes around Chernobyl have concluded that phytoextraction is not currently useful (www.strategy-ec.org,uk) *(13)*. In fact, effluent filtration or soil stabilization is more efficient for radionuclides than is extraction from soil, with or without plants. The challenge is, therefore, to develop a viable method for extracting radionuclides from soil. Here, I outline a context and suggest research that might help increase the utility of phytoextraction for radionuclides, and use this to make comments on phytoextraction of inorganic contaminants in general.

2. Understanding the Context of Phytoextraction
2.1. The Problem of Radionuclide Contaminated Soil

There is a significant volume of soil contaminated with radionuclides worldwide. It has arisen primarily from nuclear weapons-related activities plus accidents at Chelyabinsk and Chernobyl *(14)*. The concentration of radioactivity in contaminated soils varies widely, and is greatly skewed toward low activity concentrations, but is high enough in some locations to prevent agricultural production or be a potent hazard to human health. This potential to be a potent hazard to humans means that there is great pressure, and in many instances legislation actually in place, to regard soils with any activity concentrations detectably above background as contaminated and legally requiring decontamination. It seems unlikely that legislation on radioactively contaminated soils will become less stringent anywhere in the world in the near future, ensuring that there is very significant pressure to decontaminate large volumes of soil of radionuclides.

The total volume of soils contaminated with nonradioactive inorganics is much larger than that contaminated with radioactive but there is, at present, often much less pressure to decontaminate them. Thus, in addition to the very immediate necessity to deal with those radioactively contaminated soils that are hazardous, the pressure to decontaminate soils of radioactivity provides an interesting case study for those interested in phytoremediation. For example, it is probably sensible to imagine what the soil decontamination challenge might be if public and legislative pressure on heavy-metal-contaminated soils became, as is at least possible, more like that on radionuclide-contaminated soils. Much phytoextraction research is prompted by the perceived problems arising from soil contamination. Such perception is a sociopolitical construction and is variable and changeable. I suggest that the perception of contaminated soils is likely, overall, to change toward the current perception of radioactively contaminated soils and that decontamination efforts for such soils are, therefore, particularly instructive for the phytoremediation community.

2.2. Methods of Soil Decontamination for Radionuclides

Despite there being very significant pressure to decontaminate soils of radioactivity there are relatively few instances of it having been carried out. Perhaps the most thorough decontamination operations for radionuclide-contaminated soils were those at Chelyabinsk and the Bikini Atoll *(14,15)*. For terrestrial ecosystems they primarily involved physical removal of large volumes of contaminated soil plus some regolith reconstruction. Regolith conditions now only approximate, at best, what they were before contamination occurred *(15)*. The method of removing contaminated soil, mostly to waste repositories, is sometimes referred to as an "established" decontamination option but it has a limited track record of successful implementation, primarily for small volumes of highly contaminated soil. It is very doubtful that it is a strategy that could be applied to anything like the majority of radioactively contaminated soils, not least because of restricted volumes in currently approved waste repositories *(9)*, and certain that if it could be it would produce a large regolith reconstruction problem. Rehabilitation of soils contaminated with lower levels of radioactivity is currently achieved almost entirely through other means *(16)*.

With the exception of small volumes of highly contaminated soil, it is at least debatable whether there is an established method that can deal with the volumes of radioactively contaminated soils that currently exist. When faced with the challenge of dealing with soil contaminated with inorganics, it is tempting to think that, if necessary, ultimately it can be capped or dug up and washed or just buried. Radioactively contaminated soils have been in a position of great pressure for clean up for years, perhaps revealing that the problem of contaminated soil is actually less tractable than is often believed. Radioactively contaminated soils may show that it might be sensible to consider that phytoextraction should be viewed within a context in which there are not really any "established" decontamination options for inorganics, rather than a context in which it has to be cheaper and more efficient than "established" decontamination options. If there are established, economically viable decontamination options it is difficult to account for the persistence of so much contaminated soil.

2.3. Radioecological Transfer Factors and Phytoextraction

The rate of phytoextraction of inorganic contaminants depends on the net soil-to-plant transfer rate. Radioecologists have long measured concentration ratios and transfer factors (TFs), and long argued about their potential utility. They have infrequently noted, however, how useful the concepts underpinning TFs are for phytoextraction. The discussions and experience of radioecologists in using TFs might be of quite wide utility in phytoextraction of inorganics.

TFs were developed primarily as part of "empirical" environmental models at the dawn of the nuclear age more than 50 yr ago. These models were among the very first for the behavior of contaminants in the environment and there is much experience of using them over long time periods in the radioecological community. TFs were primarily used to model transfers between ecosystem compartments. Some of these models helped to reveal the limitations of "empirical" environmental models, i.e., their undynamic nature and the difficulty of transferring them between environments. Although physicochemical models that are theoretically more dynamic and more widely applicable than "empirical" models have been constructed for many years for radionuclides, TFs still prove to be extremely useful, especially when urgent responses are necessary. Particularly in the early stages of developing phytoextraction systems, the lessons learned about TFs by radioecologists might be very useful for phytoextraction of inorganics.

Soil-to-plant TFs for radionuclides taken either for a single species on many soils, many species on a single soil, or many species on many soils, have been shown empirically by radioecologists, to be very variable *(17)*, lognormally distributed *(18)*, time dependent *(19)*, and concentration dependent *(20)*. In addition, soil availability can clearly be no more of a guarantee of significant soil-to-plant transfer than high plant uptake because some plants have extremely low uptake of radionuclides just as some soils have very low availability. As it is the net soil-to-plant transfer that determines the utility of phytoextraction, it is the overall behavior of this transfer that needs to be the ultimate aim of phytoextraction research. The TF concept and the experience of radioecologists in applying it is an essential reminder that overall soil-to-plant transfer can be complex on wide spatial and temporal scales and, as the ultimate empirical expression of relevant system behavior, needs analysis in its own right. I am not aware of any such analysis for phytoextraction of nonradioactive inorganic contaminants. The extensive experience of radioecologists in application of TFs might provide a foundation for such analysis.

2.4. The Economics and Environmental Impact of Phytoremediation

Phyoremediation is frequently touted as a cheap and environmentally benign decontamination method *(21)*. This might turn out to be true but very few rigorous economic or environmental assessments of the technology have been reported. Because of the unique challenges of radionuclide-contaminated soils, radioecologists have highlighted some aspects of the technology that merit specific attention in economic and environmental assessments.

Plants used in phytoextraction trials for radionuclides become radioactive waste *(22)*. Rigorous assessment of their potential disposal highlights problems seldom mentioned in discussions of phytoextraction of inorganic contaminants *(23)*. For example, for most waste disposal streams, radioactive or not, fresh plant material, and especially that from fast-growing herbaceous plants, can be problematic and dried material is preferable. Drying of large amounts of plant material to requisite water content can be costly. It is tempting, therefore, to drip feed fresh phytoextraction waste into conventional waste streams. However, many operators are not only loathe to deal with contaminated waste but also unable to because of emission or leakage restrictions. Experience with radioactive waste is not directly transferable to other inorganic contaminants but it seems likely that disposing of large volumes of contaminated fresh plant material is likely to be more costly than is often envisaged. There are also some genuine difficulties in demonstrating how much has been gained environmentally if contamination is extracted from the soil in one place and buried in the ground or dispersed into the air or sea at another.

With phytoextraction of radioactively contaminated soil there are also significant impacts of operator protection *(23)*. Although clearly a potentially low-maintenance option, phytoextraction, and particularly harvesting and waste disposal, does require some operator time. With radioactivity the potential for contamination can necessitate quite expensive protective measures. For some other inorganics, such as Cd, at high concentrations similar protective measures are already necessary in some countries and such legislation is tightening rapidly worldwide. Operating costs seem mostly to be ignored in assessments of the potential for phytoremediation but might be of some significance over the course of years.

3. Increasing Soil Availability of Radionuclides

For those involved in phytoextraction research, not only might the general context for radionuclides previously outlined be useful, but also advances in understanding radionuclide behavior in the soil–plant system. Specific aspects of radionuclide behavior that might be of interest to phytoextraction research are outlined next.

3.1. Radionuclides With High Soil-to-Plant Transfer

There are a number of reports of a single cropping almost completely removing ^{99}Tc from contaminated soils *(24)*. ^{99}Tc is not radiologically significant in terrestrial ecosystems, although there is increasing interest as its disposal in permanent terrestrial waste repositories becomes more necessary *(24)*, but it demonstrates phytoextraction's potential for ions that are available in soil. ^{36}Cl is another radionuclide that is highly available and targeted for terrestrial waste repositories. A number of assessments have noted the potential for ^{36}Cl to be moved up into vegetation even from depths in the soil significantly below the rooting zone *(25)*. There seems little doubt that increases of soil availability of ^{99}Tc or ^{36}Cl are not necessary and that contamination could be attacked using phytoextraction. Therefore, although they have not thus far been significant in terrestrial systems, ^{99}Tc or ^{36}Cl provide useful examples of the potential of phytoextraction.

Sr isotopes behave very similarly in the soil–plant system to the nutrient Ca, often having high soil-to-plant transfer *(26)*. ^{90}Sr is among the most radioecological significant isotopes and is a major contributor to doses at the most radioactively contaminated places on Earth, such as the environs of Chelyabinsk and Chernobyl *(14)*. It has long been noted that ^{90}Sr transfer from soil to plant is close to being high enough for significant phytoextraction to be a reality. ^{35}S, a radioisotope of the plant nutrient S, can also have high soil-to-plant transfer. For ^{90}Sr and ^{35}S it seems very likely that the detailed understanding of Ca and S in the soil–plant system could enable the design of useful phytoextraction systems, probably without specific emphasis on increasing soil availability. It might, however, be salutary to analyze the case of ^{90}Sr more closely. ^{90}Sr is of great radiological significance, established methods involve great environmental disruption and there is potential for improving soil-to-plant transfer to levels that would be considered suitable for phytoextraction, yet there is little sign of phytoextraction being useful for ^{90}Sr in the near future.

3.2. Radionuclides With Low Soil-to-Plant Transfer

There are radionuclides of great radioecological concern, e.g., ^{137}Cs, ^{238}U, and ^{239}Pu, that are generally considered to have the lowest soil-to-plant transfer rates. They represent one of the greatest challenges for phytoextraction of inorganic contaminants. In general, ^{137}Cs is so tightly bound by illitic clays, both on frayed-edge and interlayer sites, that it is very useful for tracing patterns of erosion in soils with such minerals *(27)*. Soil-to-plant transfer from soils with just traces of illite can be very low *(28)*.

However, even for ^{137}Cs there are some angles of attack that give hope. Over the course of years ^{137}Cs concentrations in waters show that ^{137}Cs adsorption in illitic soils is not irreversible but that there is slow leakage *(29)*. This is a long way from sufficient to give soil-to-plant transfer necessary for significant phytoextraction but shows that there are equilibria that might be manipulated. NH_4^+ has long been known to desorb ^{137}Cs from clays *(30)*, primarily nonillitic, and some authors have noted that the presence of NH_4^+ and/or nitrification inhibitors increases ^{137}Cs uptake *(31)*. NH_4^+ from illitic clay interlayers is accessed by plants in significant quantities *(32)* and although the Cs adsorption properties in clay interlayers lead to collapse, interlayers can be opened by molecules such as oxalate *(33)*. Competitor-binding agents such as Na tetraethlyborate can draw K out of clay interlayers *(34)* and other binding agents such as Norbidine A *(35)* might also be able to do so. Recent research has also reported that changes in octahedral Fe in the rhizosphere can increase the availability of NH_4^+ from interlayer sites *(36)*. Thus it seems possible that ^{137}Cs might be removable from illitic clay interlayers, although this is some way from realization at present and achieving it in the field is quite another challenge. There has been much radioecological focus on illitc clay binding of ^{137}Cs because many nuclear facilities are located in areas with such soils. This is, however, not true of all nuclear facilities and there are very significant areas of the planet in which ^{137}Cs is quite available in the soil. For example, organic soils including histosols and spodosols, lateritic soils including oxisols, and allophanitic soils including andosols can all produce high soil-to-plant concentration factors *(37–39)*. Thus, even for a recalcitrant contaminant such as ^{137}Cs, phytoextraction should not perhaps be dismissed as a technology without potential. Radioecologists have a very detailed knowledge of its binding to soils and this is potentially very useful in manipulating its transfer from soil to plant.

U and Pu isotopes are generally very unavailable to plants *(40)*. Further, many of the areas of most attention for U are mining spoils in which a significant proportion of the regolith is composed of U. In many instances there seems little hope of phytoextracting U or Pu in useful quantities. However, both U and Pu have complex soil chemistries and are very available to plants under certain circumstances *(41)*. Organic acids have been demonstrated to make U highly available to plants *(42)*. Similar effects might be achievable for Pu because of its solubility under certain conditions. Thus, phytoextraction is not without potential for both U and Pu. Some of the most formidable phytoextraction challenges, ^{137}Cs, ^{238}U, and ^{239}Pu, are therefore not without hope. It is possible that the chinks of light might be the basis of the development of phytoextraction systems that might be useful in some instances at least. In the absence of other truly effective methods of extracting Cs, U, and Pu from soils this is significant.

4. Methods for Increasing Plant Uptake of Radionuclides
4.1. Ion Availability and Root Exudates

The physicochemical availability of an ion in the soil solution is frequently not what a plant experiences—many plants actively manage the availability of ions in the rhizosphere through symbioses or root exudates. There have been recent advances in engineering root exudates for the breakdown of organic contaminants but they also have great potential for manipulating the availability of inorganic ions.

Phosphate is poorly available in many soils and was probably a major limitation to the colonization of the land by plants. Mycorrhizal associations and proteoid roots have long been known to increase phosphate uptake by plants *(43)*. Soils in which $Fe_2(PO_4)_3$ predominates have extremely low concentrations of phosphate but plants such as *Cajanus cajun* (pigeon pea) have evolved exudates based on piscidic acid that can dissolve the highly insoluble $Fe_2(PO_4)_3$ for uptake *(44)*. Lack of soluble Fe limits plant growth on perhaps 35% of the world's soils *(45)* but many plants exude a variety of phytosiderophores based on mugineic acid to mobilize it *(45)*. Rice plants modified to exude phytosiderophores, which mobilize Fe from soils in which pedological processes render it unavailable, have been produced *(46)*. Root exudates also play a key role in controlling As availability to plants in anaerobic soils. Thus, unavailable ions are mobilized by plants *(47)* and manipulating this process is widely considered to be part of the solution to nutrient limitations in agriculture *(48)*. As yet, there has been very little consideration of this phenomenon for mobilizing radionuclides.

4.2. Manipulating Ion Uptake by Plants

The importance of the concentration of nutrient and toxic ions to agricultural production and to food and forage quality ensures that ion uptake by plants is a vibrant research topic. Plant breeders have succeeded in altering uptake of ions by crop plants and the genetic engineering of ion uptake is now a reality. Plant breeding and genetic engineering strategies to manipulate Cs uptake by plants are now within the realm of the possible. There has been less progress in research focused on manipulating plant uptake of other radionuclides but there is every reason to think that it might be possible.

The molecular biology of K uptake by plants is now advanced *(49)* and has had a great impact on our understanding of Cs uptake by plants. Although it has long been known that Cs and K probably have at least some common modes of entry into plants *(50)* it was not until relatively recently that electrochemical models were used to implicate voltage-independent cation channels in Cs transport *(51)*. Recent research has focused on other types of channel (Corinna Hampton, personal communication), at least partly because knockout mutants have eliminated some types of K transporter as possible Cs transporters *(52)*. It seems unlikely that there will be a single, or even a few, transporters that might be engineered to manipulate Cs uptake by plants but, as is already the case for other nutrients such as S *(53)*, manipulating K uptake by plants is likely to be possible soon.

It now seems likely that crop breeding might play a significant role in manipulating ion uptake by plants. Quantitative trait loci, which identify areas of a genome associated with a phenotype, have been determined for Cs concentration in plants *(54)*, the biodiversity available for breeding assessed *(17)*, and field tests with crop lines carried out (M. Broadley personal communication). All these studies suggest that crops with significantly decreased or increased uptake of Cs might be possible, and suggest considerable potential for manipulating plant uptake of other radionuclides.

5. Conclusions

For radionuclide-contaminated soils, and perhaps for soils with other inorganic contaminants, a first phase of trials revealed phytoremediation systems to have limited utility, especially for phytoextraction. Certainly, phytoremediation is successfully being used in certain circumstances, but there seems little prospect of it being widely used to make significant inroads into the problem of contaminated soils in the near future. It is tempting, therefore, to be pessimistic about the prospects for phytoextraction in particular but this is precipitate. Experience gained from attempts at phytoextraction of radionuclide-contaminated soils provide insights that might help spur further phytoextraction research.

^{137}Cs and ^{90}Sr, neither of which are routinely phytoextracted from soils, might serve as examplars for the challenges facing the development of phytoextraction research. There is very significant pressure to cleanse soils of them and much long-term knowledge of their behavior in the soil–plant system, in particular their TFs. In many soils, ^{137}Cs is bound in collapsed illitic interlayer sites and is probably as difficult to phytoextract as any inorganic contaminant. In contrast, ^{90}Sr is certainly available enough to plants, and there is enough knowledge of manipulating Ca transfer from soils to plants, for phytoextraction systems to have been perceived to be possible for nearly half a century *(2)*. I think that identifying the barriers preventing the utilization of phytoextraction for these radionuclides might provide useful insights for phytoextraction of inorganic contaminants.

Phytoextraction of ^{137}Cs and ^{90}Sr is not prevented by the existence of other cheap, environmentally benign decontamination methods. Certainly, there are soils contaminated with these radionuclides that are dealt with but only with great environmental disruption and expense. The largest soil decontamination projects for radionuclides, perhaps for any inorganic contaminants, are for ^{137}Cs at Bikini and ^{90}Sr around Lake Karachay near Chelyabinsk. In both cases urgent clean up was necessary and soil removal the only option but neither case identified a good cheap, environmentally friendly method of soil decontamination. It seems likely that similar removal of large soil volumes for other inorganic contaminants will become less acceptable as the importance of sustainable environmentally benign decontamination increases. Thus, there is good news for phytoextraction—the competition is not as stiff as it is often made out to be and it will probably get even less stiff. Perhaps there are currently no methods for extracting inorganic contaminants from soils—just damaging ways to remove soil or to wash it. In this light, the challenge is to come up with any method for extracting contaminants from soils—it does not necessarily have to be cheap and it just has to make more environmental sense than soil washing. The management of waste might play a key role in determining whether or not these criteria can be met.

For ^{137}Cs in particular, increases in soil-to-plant transfer are necessary and present a major challenge. I suggest that it might be possible to affect such increases in soil-to-plant transfer—certainly the soil decontamination challenge is big enough, and scientific knowledge great enough, to at last seriously attempt to effect it. At present there are probably enough possible avenues for research from both soil and plant science to make research worthwhile. However, it is also very relevant that ion transfer from soil to plant is a discipline that is undergoing a period of very rapid advancement and that many of these advances will be very useful for phytoextraction research. Clearly, sustainable human existence on Earth depends crucially on sustainable food production systems *(55)* but even before this is achieved there is a global micronutrient crisis in food to be solved *(56)*. For these reasons, managing nutrient transfer from soils to plants underpins at least 2 of the top 10 challenges that environmentalists *(57)* now identify for the planet. Much research focus and investment can therefore be expected into management of the soil–plant system in the next 50 yr. Phytoextraction research needs to be sustained not least so that it can benefit from, and contribute to, global research into management of the soil–plant system.

References

1. Romney, E. M., Neel, J. W., Nishita, J., Olafson, J. H., and Larson K .H. (1957) Plant uptake of Sr-90, Y-91, Ru-106, Cs-137 and Ce-144 from soils. *Soil Sci.* **83,** 369–376.
2. Nishita, H., Steen, A. J., and Larson, K. H. (1958) The release of Sr-90 and Cs-137 from Vina loam upon prologued cropping. *Soil Sci.* **86,** 195–201.
3. Lasat, M. M., Norvell, W. A., and Kochian L. V. (1997) Potential for phytoextraction of ^{137}Cs from a contaminated soil. *Plant Soil* **195,** 99–106.
4. Lasat, M. M., Fuhrmann, M., Ebbs, S. D., Cornish, J. E., and Kochian L. V. (1998) Phytoremediation of a radiocaesium contaminated soil: evaluation of cesium-137 bioaccumulation in the shoots of three plant species. *J. Environ. Qual.* **27,** 165–169.
5. Fuhrmann, M., Lasat, M. M., Ebbs, S. D., Kochian, L. V., and Cornish, J. (2002) Uptake of cesium-137 and strontium-90 from contaminated soil by three plant species: Application to phytoremediation. *J. Environ. Qual.* **31,** 904–909.

6. Dushenkov, S., Mikheev, A., Prokhnevsky, A., Ruchko, M., and Sorochinsky, B. (1999) Phytoremediation of radiocaesium contaminated soil in the vicinity of Chernobyl, Ukraine. *Environ. Sci. Technol.* **33**, 469–475.
7. Victorova, N., Voitesekhovitch, O., Sorochinsky, B., Vandenhove, H., Konoplev, A., and Konopleva, I. (2000) Phytoremediation of Chernobyl contaminated land. *Rad. Protec. Dos.* **92**, 59–64.
8. Negri, M. C. and Hinchman, R. R. (2000) The use of plants for the treatment of radionuclides. In: *Phytoremediation of Toxic Metals: Using Plants to Clean Up the Environment*, (Raskin, I. and Ensley B. D. eds.), John Wiley and Sons, Chichester, UK, pp. 107–132.
9. Willey, N. J., Hall S. C., and Mudiganti, A. (2001) Assessing the potential of phytoextraction at a site in the UK contaminated with ^{137}Cs. *Int. J. Phytorem.* **3**, 321–333.
10. Dushenkov, S. (2003) Trends in phytoremediation of radiouclides. *Plant Soil* **249**, 167–175.
11. Zhu, Y. G. and Shaw, G. (2000) Soil contamination with radionuclides and potential remediation. *Chemosphere* **41**, 121–128.
12. Glass, D. (2000) Economic potential of phytoremediation. In: *Phytoremediation of Toxic Metals—Using Plants to Clean Up the Environment*, (Raskin, I. and Ensley, B. D., eds.), John Wiley and Sons, Chichester, UK, pp. 15–32.
13. Hampton, C. R., Bowen, H. C., Broadley, M. R., et al. (2004) Cesium toxicity in Arabidopsis. *Plant Phys.* **136**, 1–14.
14. Karavaeva, Y. N., Kulinov, N. V., Molchanova, I. V., Pozolotina, V. N., and Yushkov, P. I. (1994) Accumulation and distribution of long-lived radionuclides in the forest ecosystems of the Kyshtym accident zone. *Sci Tot. Env.* **157**, 147–151.
15. Robison, W. L., Bogen, K. T., and Conrado, C. L. (1997) An updated dose assessment for resettlement options at Bikini Atoll—a US nuclear test site. *Health Phys.* **73**, 100–114.
16. Firsakova, S. K., Zhuchenko, Y. M., and Voigt G. (2000) An example of rehabilitation strategies for radioactive contaminated areas in Belarus. *J. Environ. Radioac.* **48**, 23–33.
17. Broadley, M. R., Willey, N. J., and Meade, A. (1999) A method to assess taxonomic variation in Cs concentrations among flowering plants. *Environ. Pollut.* **106**, 341–349.
18. Sheppard, S. C. and Evenden, W. G. (1988) The assumption of linearity in soil and plant concentration ratios: an experimental evaluation. *J. Environ. Radioac.* **7**, 221–247.
19. Bunzl, K. and Kracke, W. (1989) Seasonal variation in soil to plant transfer of K and fall-out Cs-137,134 in peatland vegetation. *Health Phys.* **57**, 593–600.
20. Broadley, M. R., Willey, N. J., Phillipidis, C., and Dennis, R. (1999) A comparison of Cs uptake kinetics in eight species of grass. *J. Environ. Radioact.* **46**, 225–236.
21. Lasat, M. M. (2002) Phytoextraction of toxic metals: a review of biological mechanisms. *J. Environ. Qual.* **31**, 109–120.
22. Hall, S. and Watt, N. (2002) The potential of Phytoextraction to remediate caesium-137 contaminated ground on nuclear licensed sites. *Nuclear Eng.* **43**, 27–31.

23. Watt, N. (2004) *Assessing the potential of phytoextraction to remediate land contaminated with ^{137}Cs at nuclear power station sites.* PhD thesis, University of the West of England, Bristol, UK.
24. Bennett, R. and Willey, N. (2002) Soil availability, plant uptake mechanisms and soil to plant transfer of ^{99}Tc — a review. *J. Environ. Radioac.* **65,** 215–231.
25. White, P. J. and Broadley, M. B. (2001) Chloride in soils and it uptake and movement within the plant: a review. *Ann. Bot.* **88,** 967–988.
26. Nisbet, A. F. and Woodman, R. F. M. (2000) Soil-to-plant transfer factors for radiocaesium and radiotrontium in agricultural systems. *Health Phys.* **78,** 279–288.
27. Quine, T. A. and Walling, D. E. (1991) Rates of soil erosion on arable fields in Britain: Quantitative data from Cs-137 measurements. *Soil Use Manage.* **7,** 169–176.
28. Cheshire, M.V. and Shand, C. (1991) Translocation and availability of radiocaesium in an organic soil. *Plant Soil* **134,** 287–296.
29. Smith, J. T., Comans, R. N. J., Beresford, N. A., Wright, S. M., Howard, B. J., and Camplin W. C. (2000) Chernobyl's legacy in food and water. *Nature* **405,** 141.
30. Evans, D. W., Alberts, J. J., and Clark, R. A. (1982) Reversible ion-exchange fixation of Cs-137 leading to mobilisation from reservoir sediments. *Geochem. Cosm. Acta.* **47,** 1041–1049.
31. Evans, E. J. and Dekker, A. J. (1969) Effect of nitrogen on cesium-137 in soils and its uptake by oat plants. *Can. J. Soil Sci.* **49,** 349–355.
32. Mengel, K., Horn, D., and Tributh, H. (1992) Availability of interlayer ammonium as related to root vicinity and mineral type. *Soil Sci.* **149,** 131–137.
33. Wendling, L. A., Harsh, J. B., Palmer, C. D., Hamilton, M. A. and Flury, M. (2004) Cesium sorption to illite as affected by oxalate. *Clays, Clay Min.* **52,** 375–381.
34. Cox, A. E. and Joern, B. C. (1997) Release kinetics of non-exchangeable potassium in soils using sodium tetraphenylboron. *Soil Sci.* **162,** 588–598.
35. Gauradée, S., Elhabiri, M., Kalney, D., et al. (2002) Allosteric effects in norbadione A. A clue for the accumulation process of ^{137}Cs in mushrooms? *Chem. Comm.* **9,** 944–945.
36. Scherer, H. W. and Zhang, Y. S. (2002) Mechanisms of fixation and release of ammonium in paddy soils after flooding III. Effect of the oxidation state of octahedral Fe on ammonium fixation. *J. Plant Nutr. Soil Sci.* **165,** 185–189.
37. Bergeijk, K. E. van, Noordijk, H., Lembrechts, J., and Frissel, M. J. (1992) Influence of soil pH, soil type and soil organic matter content on soil-to-plant transfer of radiocaesium and strontium as analysed by a non-parametric method. *J. Environ. Radioac.* **15,** 265–276.
38. Fredrikkson, L. (1970) Plant uptake of fission products IV. Uptake of ^{90}Sr and ^{137}Cs from some tropical and sub-tropical soils. *Lantbruk. Ann.* **36,** 61–89.
39. Ban-nai, T. and Muramatsu, Y. (2002) Transfer factors of radioactive Cs, Sr, Mn, Co and Zn from Japanese soils. *J. Environ. Radioac.* **63,** 251–264.
40. Frissel, M. J. (1992) An update of the recommended soil-to-plant transfer factors of Sr-90, Cs-137 and transuranics. In: *8th Report of the IUR Working Group on Soil Plant Transfer.* I.U.R. Banlan, Belgium, pp. 16–25.

41. Shahandeh, H. and Hossner, L. R. (2002) Role of soil properties in phytoaccumulation of uranium. *Water, Air, Soil Pollut.* **141,** 165–180.
42. Huang, J. W., Blaylock, M. J., Kapulnik, Y., and Ensley, B. D. (1998) Phytoremediation of uranium contaminated soils: role of organic acids in triggering uranium hyperaccumulation in plants. *Environ. Sci. Technol.* **32,** 2004–2008.
43. Lamont, B. B. (2003) Structure, ecology and physiology of root clusters: a review. *Plant Soil* **248,** 1–19.
44. Ae, N., Arihara, J., Okada, K., Yoshihara, T., and Johansen, C. (1990) Phosphorus uptake by pigeon pea and its role in cropping systems of the Indian subcontinent. *Science* **248,** 477–480.
45. Marschner, H. (1995) *Mineral Nutrition of Higher Plants, Second ed.* Academic Press. London, UK.
46. Takahashi, M. T., Nakanishi, H., Kawasaki, S., Nishizawa, N. K., and Mori S. (2001) Enhanced tolerance of rice to low iron availability in alkaline soils using barley nicotianamine aminotransferase genes. *Nat. Biotechnol.* **19,** 466–469.
47. Dakora, F. D. and Phillips, D. A. (2002) Root exudates as mediators of mineral acquisition in low-nutrient environments. *Plant Soil* **245,** 35–47.
48. Grotz, N. and Guerinot, M. L. (2002) Limiting nutrients: an old problem with a new solution? *Curr. Opin. Plant Biol.* **5,** 158–163.
49. Chérel, I. (2004) Regulation of K^+ channel activities in plants: from physiological to molecular aspects. *J. Exp. Bot.* **55,** 337–351.
50. Broadley, M. R. and Willey, N. J. (1997) A comparison of Cs uptake in 30 taxa of plants. *Environ. Pollut.* **97,** 11–15.
51. White, P. J. and Broadley, M. R. (2000) Mechanisms of caesium uptake by plants. *New Phytol.* **147,** 241–256.
52. Broadley, M. R., Escobar-Gutiérrez, A. J., Bowen, H. C., Willey, N. J., and White, P. J. (2001) Influx and accumulation of Cs^+ by the *akt1* mutant of *Arabidopsis thaliana* (L.) Heynh. lacking a dominant K^+ transport system. *J. Exp. Bot.* **52,** 839–844.
53. Hawkesford, M. J. (2000) Plant responses to sulphur deficiency and the genetic manipulation of sulphate transporters to improve S-utilisation efficiency. *J. Exp. Bot.* **51,** 131–138.
54. Payne, K. A., Bowen, H. C., Hammond, J. P., et al. (2004) Natural genetic variation in caesium (Cs) accumulation by *Arabidopsis thaliana*. *New Phytol.* **162,** 535–548.
55. Nature Insight (2002) Food and the future. *Nature* **418,** 645–691.
56. Welch, R. M. and Graham, R. D. (2004) Breeding micronutrients in staple food crops from a human nutrition perspective. *J. Exp. Bot.* **55,** 353–364.
57. Phillips, P. W. B. (2002) Biotechnology in the global agri-food system. *Trends Biotech.* **20,** 376–381.

24

Assessing Plants for Phytoremediation of Arsenic-Contaminated Soils

Nandita Singh and Lena Q. Ma

Summary

Arsenic (As) is a pollutant of major concern throughout the world, and causes serious environmental problems in many areas including, for example, West Bengal, Bangladesh, and Vietnam. Phytoremediation is potentially a cost-effective and environmentally benign method of extracting pollutants from soils for which there have been significant recent advances for As. In particular, the discovery of As-hyperaccumulating ferns and on-going research into their biochemistry, ecology, and agronomy are rapidly increasing their potential utility for phytoextraction of As. Here, we review the latest research into (1) the biochemistry of As in plants, (2) plant hyperaccumulation of inorganics including As, (3) the phenomenon of As hyperaccumulation in ferns, and (4) the enhancement of As phytoavailability. We conclude by identifying some technical barriers that need to be overcome to fulfill the great potential for phytoextraction of As from soils.

Key Words: Arsenic (As); hyperaccumulation; phytoremediation; *Pteris vittata*.

1. Introduction

Contamination of the biosphere by heavy metals has increased sharply at the beginning of the 20th century, posing major environmental and human health problems worldwide *(1)*. Among all metals, arsenic (As) has received much attention recently partially because of the well-publicized crises in southeast Asia including West Bengal, Bangladesh, and Vietnam *(2)*. As is a group V_A element and a metalloid, possessing properties of both metals and nonmetals *(3)*. It is a ubiquitous trace metalloid and is present in virtually all environmental media. It is a known human carcinogen, with cancers related to As in drinking water being reported in Taiwan, Argentina, Chile, Bangladesh, and India *(4)*. Because of its extensive contamination in the environment and its carcinogenic toxicity to humans and animals, clean up of As-contaminated sites has received increasing attention.

Unlike organic contaminants, As cannot be eliminated from the environment by chemical or biological transformation *(5)*. Phytoremediation, the use of plants to clean up contaminated soils, has been steadily gaining acceptance in both academia and industry over the last few years *(6–10)*. Among phytoremediation technologies, phytoextraction is one of the most recognized and researched. To successfully apply phytoextraction to As-contaminated soils, it is important to understand the basic biochemistry of As in plants.

2. Biochemistry of As in Plants

The success of phytoremediation, as an effective remediation technology for As-contaminated soils, depends on several factors including the extent of soil contamination, As bioavailability in soil, and plant ability to intercept, absorb, and accumulate As in shoots *(11,12)*. Ultimately, the potential for phytoremediation depends on the interactions between soil, As, and plant. This underlines the importance of understanding the mechanisms and processes that govern As uptake and accumulation in plants.

2.1. As: A Phosphate Analog

To date, research has shown that arsenate is taken up by plants via the phosphate transport systems *(2,13,14)*. This raises the question of how phosphate and arsenate interact during uptake by roots and translocation from roots to shoots. Specifically, how does a plant acquire and maintain sufficient P nutrition under high arsenate stress?

As is toxic, whereas phosphorus is essential for plants. They are both group V_A elements and, thus, have similar electron configurations and chemical properties. Therefore, arsenate competes with phosphate for both soil-sorption sites and uptake by biota. In soil, competition between arsenate and phosphate for soil-sorption sites results in a reduction in their sorption by soil and an increased concentration in soil solutions *(15–17)*. Such competition may help to alleviate arsenate toxicity via improved phosphate nutrition *(18)*.

Arsenate competition with phosphate for the phosphate uptake system has been observed in many organisms—angiosperms *(19)*, mosses *(20)*, lichens *(21)*, fungi *(22)*, and bacteria *(23)*. As a result, arsenate has been reported to inhibit phosphate uptake in yeast *(24)*, phytoplankton *(25,26)*, and terrestrial angiosperms *(27)*. Because the plant uptake system has a higher affinity for phosphate than arsenate, only mild inhibition of phosphate uptake by arsenate has been observed. However, some studies have shown that at low levels, arsenate can increase phosphate uptake *(28,29)*. The authors assumed that this uptake of phosphate may be because of the physiological deficiency of phosphorus caused by low arsenate, because arsenate can substitute for phosphate within the plants but is unable to carry out phosphate's role in energy transfer.

Phosphate has been reported to suppress plant arsenate toxicity in hydroponic systems. Meharg and Macnair *(30)* observed an As uptake reduction of 75% at 0.5 m*M* phosphate in both tolerant and nontolerant plant genotypes of grass (*Holcus lanatus*). Arsenate concentrations in alfalfa (*Medicago sativa* L.) shoots are also reduced by phosphate *(31)*. For Indian mustard (*Brassica juncea* L.), grown in 0.5 m*M* arsenate and 1 m*M* phosphate, a 55–72% reduction of arsenate uptake over the control has been reported *(32)*.

Because of the chemical similarity between arsenate and phosphate, the P/As ratio in plants is important in regulating plant arsenate uptake and toxicity. To adequately protect plants against arsenate toxicity, a P/As molar ratio of at least 12 is proposed by Walsh and Keeney *(33)*. A hydroponic study by Sneller et al. *(18)* shows that, with P/As being constant at 12, arsenate is less toxic at high phosphate levels because more arsenate is taken up by the plants at low phosphate levels. In soil, the influence of phosphate on arsenate varies. This is because soil properties affect the availability of both arsenate and phosphate. Addition of up to 9.7 mmol/kg phosphate does not influence arsenate toxicity when a silt loam soil is spiked with 1.1 mmol/kg arsenate *(34,35)*. This can be explained as the silt loam soil having a high phosphate-fixation capacity and, thus, available phosphate probably does not increase much after phosphate addition. However, in a sandy soil, the same concentration of phosphate enhances arsenate toxicity through displacement of the sorbed arsenate from the soil by phosphate. When P/As molar ratio is ≥ 16, phosphate improves plant yields *(35)*. At high levels of arsenate (30 mmol arsenate/kg), however, phosphate does not overcome arsenate toxicity even at a molar P/As ratio of 24. Arsenate/phosphate uptake can be suppressed in plant roots if the plants are supplied with sufficient P *(36–38)*. This suppression is because of a feedback regulation of the arsenate/phosphate transporter, i.e., reduced arsenate uptake through the suppression of the high-affinity uptake system *(27)*.

2.2. As Toxicity

In soils, As occurs mainly as inorganic species, mostly arsenate (As[V], AsO_4^{3-}), but can also bind to organic matter. Under oxidizing conditions, in an aerobic environment, arsenate is the stable species and is often strongly sorbed onto clays, iron and manganese oxides/hydroxides, and organic matter. Under reducing conditions, in an anaerobic environment, arsenite (As[III], AsO_2^-) is the predominant As species. Inorganic As compounds can be methylated in soils by micro-organisms, producing monomethyl As acid (MMA), dimethylAs acid (DMA), and trimethylarsine oxide under oxidizing conditions *(39)*.

Arsenate, being the predominant form of As present in most soils, means that plants take up As mostly as arsenate. As such, studies on the kinetics of plant

As uptake have focused almost entirely on arsenate *(2,40)*. As a chemical analog of phosphate, arsenate can uncouple the oxidative phosphorylation by displacing phosphate in ATP synthesis *(41,42)*. In addition to arsenate, plants can also take up arsenite from soils. Unlike arsenate, arsenite reacts with sulfydryl groups of enzymes and tissue proteins of plants, leading to the inhibition of cellular function and death *(2)*. Arsenite is generally more toxic than arsenate, and both of them are more toxic than organic As compounds *(43,44)*. The As concentration in soils tolerated by plants varies from 1 to 50 mg/kg *(3)*.

2.3. As Uptake

Different As species have different solubilities and mobilities, and thus different bioavailability to plants. Under hydroponic conditions, the availability of As to a cord grass (*Spartina alterniflora* L.) followed the trend DMA<< MMA<As(V)<As(III) *(38)*. As availability to rice (*Oryza sativa* L.) followed a slightly different order of DMA<As(V)<MMA<As(III) *(45)*. Upon absorption by rice, DMA is readily translocated to the plant shoot, whereas As(III), As(V), and MMA accumulate primarily in the roots *(45)*. In tomato (*Lycopersicon esculentum* Mill.) plants, both MMA and DMA had a greater upward translocation than arsenite and arsenate *(46)*. The presence of other ions also affected As availability and phytotoxicity *(47)*.

The transfer of As from soil to plant is low for most plant species. This may be for several reasons: (1) low bioavailability of As in soil, (2) restricted uptake by plant roots, (3) limited translocation of As from roots to shoots, and (4) As phytotoxicity at relatively low concentrations in plant tissues.

2.4. As Distribution

As absorption by plants is influenced by many factors including plant species *(33)*, As concentrations and forms in the soil *(48)*, soil properties such as pH and clay content *(49)*, and the presence of other ions *(50)*. Most plants grown on As-contaminated soils contain elevated As levels *(51–53)*. For most plants, the highest concentrations of As are found in roots, with the aboveground vegetative parts (leaves and stems) being lower, and the lowest levels found in fruit and seeds *(33,52,53)*. Porter and Peterson *(54)* determined As distribution in both arsenate tolerant and nontolerant clones of *Agrostis capillaris* (syn. *Agrostis tenuis*) after exposing them to different levels of arsenate for 7 d. Tolerant plants accumulated less arsenate than nontolerant plants but it is unclear if nontolerant plants ceased active uptake of arsenate in these experiments. The greatest proportion of accumulated As remained in the roots, with 6–25% of the As being transported to the shoots.

A study of As uptake in As tolerant and nontolerant *H. lanatus (30)* showed that there are considerable differences between the two. In the tolerant plants about 75% of assimilated As is transported to the shoots, whereas only 50% in nontolerant plants. In a study to examine As distribution in Indian mustard (*B. juncea*), a plant that is not tolerant to As, Pickering et al. *(32)* found that after As uptake by the roots, only a small fraction of As is transported to the shoot via the xylem. This suggests that a plant's ability to translocate As from roots to shoots may constitute one of its tolerance mechanisms.

2.5. As Speciation

As speciation is important to the study of its environmental behaviors, as oxidation state and chemical form both affect As uptake, transport, and toxicity in plants *(55)*. To date, several studies have been completed on As speciation in terrestrial plants. Koch et al. *(56)* reported that the predominant As species in terrestrial plants are inorganic forms, of which up to 50% are present as arsenite, whereas only small amounts of the methylated As compounds are present in some plants. Another study showed that following plant uptake of arsenate, considerable amounts of arsenite are found in pine seedlings (*Pinus halepensis*, *Pinus pinea*, and *Pinus radiata*), corn (*Zea mays*), melon (*Cucumis melo*), pea (*Pisum sativum*), and tomato (*L. esculentum*) *(57)*. Large amounts of inorganic As have also been found in carrots (*Daucus carota*) and other vegetables grown in As-contaminated soil but no methylated As species are detected *(58,59)*.

In a study to understand the effect of different levels of arsenate on As speciation in *H. lanatus*, Quaghebeur and Rengel *(60)* reported inorganic As as the predominant species present in the plant, although low levels (<1%) of organic species are detected in shoots. With increasing levels of arsenate in the nutrient medium, the %As(III) observed generally increases in roots, whereas it decreases in shoots. The study further revealed that As(V) can be reduced to As(III) in both roots and shoots.

These studies suggest that inorganic As, including both arsenate and arsenite, is the predominant form of As present in terrestrial plants. This is further supported by a study of Mattusch et al. *(61)* who showed As in 10 herb plants grown on As-contaminated soils is present primarily in inorganic forms, predominantly arsenite. They suggested that the herb plants protect themselves from interruption of oxidative phosphorylation by reducing and storing the As as arsenite. Oxidation of As(III) has rarely been reported in the plants, but it has been seen in soil bacteria *(62)* and mineral-leaching bacteria *(63)*. In general, two pathways are involved in the production of different As species in plants *(64)*. One is oxidation/reduction of As species in the plant; another is direct uptake of different As species from the environment. Some marine algae, which are constantly exposed to arsenate in seawater, have the biochemical capability to intercellularly convert arsenate to harmless organo-As compounds *(65)*.

2.6. As Detoxification

Significant adverse effects have been observed in plants growing on As-contaminated soils *(66,67)*. To protect themselves from As poisoning, plant cells must have developed a mechanism by which As, entering the cytosol of the cell, is immediately complexed and inactivated, thus preventing it from inactivating catalytically active or structural proteins. The biochemistry and toxicology of As is complicated by its ability to convert between oxidation states and organo-metalloidal forms, both in the environment and in biological systems. These processes affect the relative tissue-binding affinities of the various As species, which determine both toxicity and detoxification mechanisms. Finally, As may form stable complexes with organic ligands, such as thiols.

Reactive oxygen species (ROS) may be generated through the conversion of arsenate to arsenite in plants, resulting in damage to not only DNA, but protein and lipids as well *(68)*. The scavenging system controlling ROSs comprises nonenzymatic antioxidants (e.g., glutathione [GSH], ascorbate, and carotenoids) and enzymatic antioxidative systems (e.g., superoxide dismutase [SOD], peroxidase, catalase [CAT], and GSH reductase). Recently, the role of polyamines as antioxidants has also been postulated *(69–71)*. In arsenate-treated *H. lanatus*, increases in lipid peroxidation, and SOD activity correlated with increasing arsenate concentrations *(72)*. Mylona et al. *(73)* reported an increase in CAT, GSH-S-transferase, and SOD in maize after exposure to arsenate and arsenite.

As also can trigger the formation of phytochelatins (PCs), which are thiol (SH)–rich peptides induced by heavy metals including Cd, Cu, and Zn *(74–76)*. Rapid induction of metal-binding PCs in response to inorganic As occurs in cell suspension cultures of *Rouwolfia serpentine*, in *Arabidopsis* seedlings, and enzyme preparations of *Silene vulgaris* *(77)*. In a paper examining As toxicity in *S. vulgaris*, Sneller et al. *(76)* reported that the short-term PC accumulation (over a 3-d period) was positively correlated with As exposure. Isolation of peptide complexes from prolonged As exposure, and subsequent analysis by HPLC-AFS showed that PC_2, PC_3, and PC_4 are present, although the latter is present only after 3 d. As coeluted mainly with PC_2 and PC_3. It is concluded that PCs provide a detoxification mechanism in *S. vulgaris* through the binding of As to their SH groups. An approx 3:1 ratio of the sulfhydryl groups from PCs to As is compatible with reported As–GSH complexes *(77)*. Bleeker et al. *(78)* studied the arsenate tolerance in *Cytisus striatus* in both an As-enriched gold mine population and nonmetalliferous population. In their study, the variation in total PC-SH accumulation seems to have been largely governed by the variation in As accumulation. The root PC-SH:As ratios are not considerably different, i.e., 2.7 in the nonmetalliferous plants and 2.9 in the mine plants. However, the PC chain length distributions are consistently different at all arsenate exposure levels and at both P levels studied, which may be attributable to differential temporal patterns of As accumulation, such as those found in *H. lanatus (76)*.

There is considerable evidence that at least part of the As accumulated by plants is coordinated with GSH and/or PCs, and the complexation is independent of the inorganic As species originally present *(79)*. Although PCs are formed in some plants upon exposure to As and bind with As to some extent, the ratio of the complexes to total As present in the plants is small *(77,80)*, indicating that the majority of As is in non-PC bound forms. The localization of arsenic–PC complexes within plant tissue is as yet unknown. As–PC complexes are not stable at neutral to alkaline pHs, but are stable under the acidic conditions present in the vacuole *(76,77)*. If arsenic–PC complexes are also transported into the root vacuole, then under the acidic conditions present, they might remain stable, thereby preventing reoxidation of As(III) and allowing accumulation of high concentrations of As–PC complexes in arsenate-resistant plants *(2)*.

3. Hyperaccumulators
3.1. Definition

Plants that can accumulate metals to exceptionally high concentrations in their shoots are called hyperaccumulators *(81–84)*. The definition of hyperaccumulators has been discussed by many authors. It is generally defined as those plants containing >1000 µg/g (0.1%) metal (100 µg/g for Cd) in aboveground plant biomass. This term generally represents a concentration much greater (e.g., about 100 times) than that the highest values to be expected in nonaccumulating plants. One of the most distinctive features of hyperaccumulator plants is their ability to accumulate high concentrations of metals in their shoot biomass even when these metals are present at low concentrations in the bulk soil *(85,86)*. This has been observed both in field observations and laboratory studies.

In addition to metal concentration in shoots, two other important factors also need be considered when evaluating a metal hyperaccumulator: bioaccumulation factor (BF) and translocation factor (TF) *(87)*. The BF is defined as the ratio of metal concentrations in plant biomass to those in soils. This is an important factor in determining the efficiency of a plant to remove metals from soil. TF is defined as the metal concentrations in plant shoots to that in the roots. This is used to determine the efficiency of a plant to translocate metals from the root to the shoot *(87)*, which has been used to characterize accumulators and excluders by Baker *(88)*.

According to Ma et al. *(84,87)*, a hyperaccumulator is one that not only tolerates high concentrations of metals, but also accumulates metals to high levels in plant aboveground biomass, as compared with those in soils. They defined As hyperaccumulators as those that have BF >1 and TF >1 as well as being able to accumulate >1000 µg/g As in plant biomass. This definition can be applied to other metal hyperaccumulators as well except for Cd (>100 µg/g). There are examples where a plant can survive in soil with high concentration of a metal but they are not hyperaccumulators of it. Benson et al. *(89)* reported that *A. tenuis* growing on As mine wastes in England, contained 3470 µg/g As, whereas the soil had 26,500 µg/g. This species is "hypertolerant" to As but is not an As "hyperaccumulator" as reported by Francesconi et al. *(90)*.

3.2. Characteristics

The most recent report lists 163 plant taxa belonging to 45 families as being metal hyperaccumulators *(91)*, whereas Reeves and Baker *(83)* listed 418 species of vascular plants. Regardless, hyperaccumulation is a comparatively rare phenomenon because it constitutes only a small fraction of the more than 300,000 known vascular plant species. The majority of hyperaccumulating taxa are endemic to soils that have high metal concentrations, either naturally as a result of mineralization of the parent rocks, or as a result of human activities. These species have been termed "edaphic endemics" *(92,93)*, and ecologically they can be regarded as "strict metallophytes" restricted only to metal-rich soils. On the other hand, a few hyperaccumulating species are "facultative metallophytes" with populations on both metalliferous and nonmetalliferous soils, e.g., *Thlaspi caerulescens* *(94)*, *Arabidposis halleri* *(95)*, and *Pteris vittata* *(84)*.

At present, research efforts to understand plant hyperaccumulation covers many aspects, including screening for new hyperaccumulating taxa *(96)*, studies of the taxonomy and phylogeny of hyperaccumulation *(97,98)*, experiments on ecology and selective advantages of hyperaccumulation *(99)*, and work on the physiological mechanisms by which metals are taken up, transported, sequestered, and tolerated *(84,100)*. The majority of hyperaccumulating species have been described as accumulating only a single metal, i.e., they are metal specific. In some respects, hyperaccumulating plants can be regarded as one subset of a larger category of metal-tolerant plants. The various mechanisms through which plants tolerate potentially toxic metals in the soils have been the subjects of intensive study for many years *(101–104)*. The exact relationship between metal tolerance and hyperaccumulation has not yet been fully resolved. Broadly speaking, the physiological mechanisms of metal tolerance have been categorized as either exclusion, i.e., blocking the movement of metals at the soil–root or root–shoot interface, or accumulation, allowing uptake of metals into aerial parts and rendering them nontoxic through chemical binding or intercellular sequestration *(88,103,105)*.

The studies conducted to date have not consistently revealed a positive correlation between a plant's ability to hyperaccumulate and metal concentration in the soil for a given population. This may suggest that hyperaccumulation is not controlled by the same mechanism as tolerance *(99)*. Studies that have measured both tolerance and hyperaccumulation ability *(95,106,107)* generally suggest that although tolerance is positively correlated with soil metal concentration, a plant's ability to hyperaccumulate shows a significant positive correlation with neither. If hyperaccumulation is not a direct function of soil metal content, this may indicate lower fitness costs associated with metal hyperaccumulation compared with metal tolerance, or an adaptive value of the hyperaccumulator phenotype independent of soil metal concentration. Boyd and Martens *(108)* proposed five hypotheses regarding the possible adaptive benefits of metal hyperaccumulation: (1) a mechanism for increased tolerance, (2) metal-based allelopathy, (3) drought resistance, (4) an effective cation uptake mechanism, or (5) defense against herbivore and pathogen attack.

4. As Hyperaccumulators

Successful application of phytoremediation is determined by two stipulations, i.e., identification of plants with great metal hyperaccumulating efficiency and understanding the factors regulating the growth of the hyperaccumulating plants for maximum metal removal from contaminated sites *(109)*. For a hyperaccumulating plant to be used successfully in remediating As-contaminated sites, it should have sufficient biomass along with efficient extraction of As from the soil. Currently, several As hyperaccumulators are available for such purposes. Numerous As-tolerant plants have been reported in the literature, including those growing on mine wastes from various sites in the United Kingdom *(110)* and on smelter wastes in northeast Portugal *(111)*, however, As hyperaccumulators are a recent phenomenon. The first known As hyperaccumulator, *P. vittata* (also known as Chinese brake fern), was discovered independently by different research groups *(84,112–115)*. Though the discovery was made as early as 1998 *(112)*, it was not well known until after Ma et al. *(84)* reported the discovery in *Nature*.

4.1. The First Known As Hyperaccumulator

P. vittata, the first-known As-hyperaccumulating plant, was found on an abandoned wood-preservation site in north-central Florida *(84,112,113)*. This site was in operation from 1952 to 1962, pressure-treating lumber using an aqueous solution of chromate-copper-arsenate. The site is contaminated with As, with soil As concentration being as high as 361 µg/g *(113)*. This is high in comparison to an average concentration of just 0.42 µg/g in Florida soils *(116)*. *P. vittata* accumulated as much as 14,500 µg/g As in the fronds growing on this chromate-copper-arsenate site *(113)*, which is also exceptionally high. This indicates that *P. vittata* is capable of concentrating as much as 60 times more As in the biomass than the soil where it grows.

Along with As-contaminated sites, *P. vittata* has been seen to accumulate significant concentrations of As from uncontaminated sites *(113)*. A soil with just 0.47 mg As/kg produced a fern with 136 mg As/kg in its fronds *(113,117)*. Under greenhouse conditions, *P. vittata*, growing in a soil spiked with 1500 mg/kg As, effectively accumulates As into its aboveground biomass, in a short period of time, to a level as high as 22,630 mg/kg (>2.3% in dry weight) *(84,113)*. *P. vittata* not only has an exceptional ability to accumulate As, but it also grows rapidly with a large biomass. Moreover, it has a wide distribution and easy adapts to different conditions. These results are consistent with the observations of Chen et al. *(114)* and Visoottiviseth et al. *(115)* who made observations of As hyperaccumulation by *P. vittata* independent of our research.

4.2. Other As Hyperaccumulators

The first-known As hyperaccumulator *P. vittata* was discovered by three research groups all from field studies *(84,114,115)*. It is possible that As hyperaccumulation may occur in other fern species besides *P. vittata*. A second As hyperaccumulator *Pteris cretica* was identified by Ma et al. *(118)* in a screening study of 17 fern plants (15 fern species) for their As-accumulation capability. All three varieties of *P. cretica* (var. Mayii, var. parkeri, and var. albo-lineata) are As hyperaccumulators. After growing in a soil spiked with 50 mg As/kg for 2 wk, As concentrations in the fronds of *P. cretica* reaches 1355 to 3338 mg/kg. Similar to *P. vittata (115)*, a third As-hyperaccumulating fern, *Pityrogramma calomelanos* (silver fern) was found in the field, in the Ron Phibun district of southern Thailand *(119)*. Unlike the first two As hyperaccumulators, it is not from the *Pteris* genus. This fern accumulates higher As concentrations (2760–8350 mg/kg) in the fronds and lower (88–310 mg/kg) in the roots growing in a soil containing 135–510 mg As/kg. The fourth and fifth As hyperaccumulators, *P. longifolia* and *P. umbrosa*, are identified by Zhao et al. *(120)* in a screening study. In the study, they also identified two new varieties of *P. cretica* (var. albo-lineata and var. wimsetti) as As hyperaccumulators. The *Pteris* species and *Pityrogramma calomelanos* all belong to the order Pteridales *(121)*.

Research on As hyperaccumulators may suggest that As hyperaccumulation is not just a character selected in these species on As-contaminated soils, it appears to be constitutive as in the case of Zn and Cd hyperaccumulators *Arabidopsis halleri* and *Thlaspi caerulescens (86)*. The uniqueness in As-hyperaccumulating ferns is that they are widely distributed both geographically and ecologically. Based on the phylogeny of primitive ferns and their allies (Psilotales, Equisetales, Selaginales), Meharg *(122)* postulated that the early land flora would have evolved in subarial hot spring environments rich in As.

Tolerance mechanisms may have been lost as plants spread out from the hot springs into non-As-contaminated environments, with members of the Pteridales retaining this primeval trait as it conferred some subsequent advantage. To address the phylogenetic basis of As hyperaccumulation, Meharg *(123)* screened 45 ferns and their allies, *Equisetum* (five genus) and *Selaginella* (two genera) and *Psilotum nudum* for As hyperaccumulation under standard growth conditions. Five cultivars of *P. cretica* (chilsii, crista, mayii, parkerii, and rowerii) were also included in this study. He concluded that As hyperaccumulators arrive relatively late in terms of fern evolution, and that this character is not exhibited by primitive ferns nor their allies. He also reported that not all members of the *Pteris* genus are As hyperaccumulators—*P. straminae* and *P. tremula* are two examples. In addition, *P. ensiformis* and *P. dentata* do not hyperaccumulate As either *(123a)*.

5. As Hyperaccumulation by *P. vittata*

Among the five known As hyperaccumulators, *P. vittata* is the most well known and therefore has received the most attention.

5.1. As Uptake by *P. vittata*

Though *P. vittata* is highly tolerant to As, Tu and Ma *(124)*, in their experiment with soil containing different forms of As, reported that it suffers As toxicity at ≥500 mg As/kg as arsenate and 50 mg As/kg as DMA. The plants died 4 d after transplanting with 50 mg As/kg as DMA. As a herbicide, DMA is apparently more readily translocated to the shoots than inorganic arsenicals or MMA and thus is more phytotoxic *(38)*. Arsenite is considered at least twice as phytotoxic as arsenate *(42)*. At 50 mg As/kg, such a difference is not observed in *P. vittata*. This may be because of the rapid oxidation of arsenite to arsenate in soils.

In all plant species studied thus far, it has been shown that arsenate is taken up by plants via the phosphate transport systems *(2)* and *P. vittata* is no exception. This was demonstrated by Tu and Ma *(125)* in a hydroponic experiment, which showed P inhibited As uptake at different levels of As (0.67, 2.67, and 5.34 mM) and P (0.8, 1.6, and 3.2 mM) in the growth medium. In a long-term hydroponic experiment by Wang et al. *(126)*, As concentrations in the plants are reduced by increasing phosphate concentration in the nutrient solution. If phosphate transporters are responsible for arsenate uptake, then arsenate uptake should be enhanced in P-deficient plants. Such an effect is demonstrated clearly in *P. vittata*. Phosphate starvation for 8 d increases the maximum net influx rate of As 2.5-fold, suggesting an increased density of phosphate/arsenate transporters on the plasma membranes in root cells *(126)*. In contrast, the K_m (plant affinity for an ion) for arsenate is not significantly affected by P starvation. In the same experiment K_m for *P. vittata* (0.5–1.0 µM) appears to be much lower than the values reported for other plant species (6–25 μM). This difference may explain the highly efficient uptake of As by *P. vittata* from low-As soils *(84)*. The uptake of arsenite, unlike that of arsenate, has no competition from phosphate in the medium, nor does P starvation enhance arsenite influx, indicating that arsenite uptake by *P. vittata* does not share the same transport systems as phosphate *(126,127)*.

Because arsenate and phosphate are chemical analogs, plant P/As molar ratios should be important for plant growth. Generally, the P/As molar ratio in this fern are higher in the roots *(3–29)* than in the fronds (0.5–3.8), as most of the As taken up by the plant is transported to the fronds *(109)*. A ratio of >1.2 in the fronds seems necessary for normal growth of the fern *(109)*, which is much lower than the ratio of 12 required by other plants *(33)*. As an As hyperaccumulator, *P. vittata* shows a greater affinity for As than other plants as discussed earlier *(126,109)*. However, compared with phosphate, it does not necessarily have a greater affinity for As *(109)*. Bioaccumulation preference of arsenate to phosphate measures a plant's selectivity in taking up arsenate from soil as compared with phosphate, i.e., a higher number indicates a greater preference for arsenate uptake *(109)*. Values between 0.1 and 0.4 are observed for the roots of *P. vittata* after exposure to different concentrations of arsenate and phosphate. These low values may suggest that *P. vittata* has a greater affinity for phosphate than arsenate.

5.2. As Distribution in P. vittata

Using excised parts of *P. vittata*, Tu et al. *(127)* showed that excised pinnae, fronds, and roots of the fern all effectively accumulate As(III), As(V), and MMA from solution with their capacity in the order of pinnae>fronds>>roots. This is consistent with the pattern of its As distribution where 83% of As is distributed in the fronds *(84)* and 96% of the total As in the aerial parts is found in the pinnae *(128)*. As concentrations increased from 3000 mg/kg in the most apical pinnae to 6000–9000 mg/kg in the basal pinnae. Energy dispersive X-ray microanalyses showed As is significantly ($p < 0.001$) more abundant in the upper and lower epidermal cells (18 and 13 mM, respectively) than in the palisade and spongy mesophyll *(128)*. The rapid accumulation of As by the excised plants may have relevance for phytoremediation of As-contaminated water, i.e., excised plants can be used to clean up As-contaminated water by simply floating them.

Tissue As concentrations alone may not be a good indicator for comparing As uptake by plants from soils because tissue concentrations do not take into account soil As concentration. The BF, as defined earlier, can be used to compare the effectiveness of a plant in concentrating As from soil into its biomass. A BF value as high as 200 in the fronds of *P. vittata* clearly demonstrates its effectiveness in As hyperaccumulation *(84)*. A high BF value in the fronds requires not only efficient plant As uptake, but also efficient plant As translocation, which is measured by TF. In addition to removing significant amounts of As from soils (BF), *P. vittata* efficiently translocated As from roots to fronds *(84)*. The TF shows that As concentrations in aboveground biomass were 4–25 times greater than those in roots, and are much greater than those for most plants because the highest As concentrations for typical plants are generally found in roots.

5.3. As Speciation

Arsenate can be readily reduced to arsenite both enzymatically and nonenzymatically through reactions with GSH *(2)*. Ma et al. *(84)* have shown that As in *P. vittata* is predominantly present as inorganic arsenate and arsenite. Analysis of As speciation in *P. vittata* showed that 60–74% of the As in the fronds is present as As(III) compared with only 8% in the roots *(129)*. However, this does not rule out the presence of other noncovalently bond organo-As compounds (mainly organic complexes) in *P. vittata*, which may decompose or hydrolyze into simple inorganic As species during the course of extraction.

X-ray absorption near edge spectroscopy analyses by Zhao et al. *(120)* showed that about 75% of the As in the fronds is present as As(III) with the remaining as As(V). This confirms the findings of Ma et al. *(84)* who found similar proportions of As(III) and As(V) in *P. vittata*. Wang et al. *(126)* showed that >85% of the As extracted from the fronds of *P. vittata* is in the form of arsenite, and the remaining mostly as arsenate. Singh and Ma *(129a)* have observed a reduction of As(V) to As(III) both in the rhizomes and fronds of *P. vittata* but the study showed a greater proportion of As(III) in the fronds. This provides evidence that As(V) can be reduced to As(III) in the rhizome as well as the fronds.

Tu et al. *(127)* have observed that in the roots, 30–39% of As is present as As(V) when exposed to As(III), whereas 24–34% As is present as As(III) when exposed to As(V), suggesting that both As(III) oxidation and As(V) reduction occur in the roots. The percentage of arsenate is higher in the older fronds of *P. vittata (130)*, which is indicative of reoxidation of arsenite to arsenate, possibly as a result of a decline in the levels of reductants. Reduction of arsenate to arsenite in plants appears to be related to mechanisms of As tolerance because of the interference of arsenate with P-mediated processes and metabolism *(32,131,132)*. This probably constitutes one of the As detoxification mechanisms in *P. vittata*. However, arsenite is also toxic to plants because of its reaction with sulfydryl groups of enzymes and proteins *(133)*. Therefore, *P. vittata* must have developed additional mechanisms of As detoxification.

5.4. As Detoxification by P. vittata

Understanding the As detoxification mechanism of *P. vittata* is of critical importance in optimizing its use in As phytoremediation. As stored as arsenite can be highly disruptive to metabolic processes in cytoplasm, and hence has to be detoxified. In the leaflets of *P. vittata* As appears to be localized mainly in the vacuoles of epidermal cells *(128)*. Vacuolar sequestration is likely to be the primary mechanism of As detoxification in this hyperaccumulator. There is considerable evidence that exposure to inorganic As results in the generation of ROS in *P. vittata* (Singh et al., 2006). The reduction of arsenate to arsenite, a process that readily occurs, may contribute to the synthesis of enzymatic antioxidants such as SOD and CAT *(134)*. The accumulation of H_2O_2 is minimized in the cell by the ascorbate–GSH cycle where APX reduces it to H_2O *(134)*. Induction of these antioxidative enzymes in turn quenches the concentration of ROS, thus enhancing continuous accumulation of As.

In addition to the effects of As on enzymatic antioxidants, As also affects a wide range of nonenzyme antioxidants (e.g., GSH, ascorbate) (Singh et al., 2006), which may also contribute to the fern's oxidative defense. Acid soluble and total SHs increased in leaflets after As exposure *(135)*. Among the SHs synthesized, acid-soluble SHs are the major component, suggesting that small SH-containing compounds, such as sulfur-containing polypeptides, may be involved in As accumulation and detoxification in *P. vittata*. There is some evidence for the role of PCs in As tolerance but only a few studies have been conducted in the case of *P. vittata*. Zhao et al. *(136)* showed that exposure to arsenate induces the synthesis of PC_2 in the roots and shoots in *P. vittata*. The concentration of PC_2 is higher in the shoots than in the roots, which may be related to a higher concentration of As in the shoots than in the roots. However, *P. vittata* differs from nonhyperaccumulators in two respects: (1) only PC_2 is found in *P. vittata*, whereas other plants studied contain PCs of longer chain length and (2) the concentration of PCs in *P. vittata* is considerably lower than reported in other plants. The results suggest that *P. vittata* has a rather limited capacity to accumulate PCs in response to As exposure. Further research is ongoing to explore the role of GSH, ascorbate, and PCs in As detoxification by *P. vittata*.

6. Enhancement of As Phytoavailability

A key to effective phytoremediation, especially phytoextraction, is to enhance metal phyto-avaiability and to sustain adequate metal concentrations in the soil solution for plant uptake *(137)*. Various soil amendments have been used to aid plant uptake and accumulation of metals *(138–140)*.

6.1. Application of Biosolids

Robertson et al. *(141)* suggested that the decrease in soil pH following biosolid application is the major reason for the greater mobility of heavy metals in biosolid-treated soil. The incorporation of carbon-rich biosolids into soils has been shown to increase the amount of dissolved organic matter in soils *(142,143)*. Dissolved organic matter can facilitate metal transport in soil by acting as a carrier through formation of soluble metal–organic complexes *(144,145)*. Darmody et al. *(146)* also noted that many metals are mobile in a silt loam soil receiving heavy biosolids application. Cao et al. *(147)* have studied the effect of municipal solid waste and biosolid composts on As uptake by the hyperaccumulator *P. vittata*. Their study showed that both composts increase As uptake from the As-contaminated soil, mainly because of the increase in water-soluble As and the transformation of As(V) to As(III), which has a higher solubility and therefore higher availability *(38,148)*. As(III) increased from 9.7 to 20–24% in the soil solution for the compost-amended treatments.

6.2. Chelating Agents

Several chelating agents, such as citric acid, EDTA, CDTA, DTPA, EGTA, EDDHA, and NTA, have been studied for their ability to mobilize metals and increase metal accumulation in different plant species *(149,150)*. Different metals have been targeted; however, the most promising application of this technology is for the remediation of Pb-contaminated soils using Indian mustard in combination with EDTA *(151)*. Despite the success of this technology, some concerns have been expressed regarding the enhanced mobility of metals in soil and their potential risk of leaching to groundwater *(150)*. However, no detailed study regarding As–EDTA complexes in contaminated soils has been conducted.

6.3. pH Change

Rhizosphere pH may differ considerably from that of the bulk soil. Factors affecting rhizosphere pH include the source of nitrogen supply, nutritional status of plants (e.g., Fe and P), excretion of organic acids, CO_2 production by root and rhizosphere micro-organisms, and the buffering capacity of the soil *(64)*. Studies using soil and pure Fe hydroxides generally suggest that As(V) solubility increases on pH increase within pH ranges commonly found in soil (pH 3.0–8.0), whereas As(III) tends to follow the opposite pattern *(152–154)*. Hence, an increase of rhizosphere pH would favor mobilization of labile and exchangeable As(V) fractions in the root vicinity and consequently enhance plant uptake. *P. vittata* prefers calcareous soils *(84)*. This implies that changes of rhizosphere pH would be no prerequisite for As hyperaccumulation owing to the high pH-buffer power of calcareous soils. But *P. vittata* has also been found in acidic soils and mine tailings in Thailand. An increase in the rhizosphere pH could potentially increase As(V) solubility and plant As uptake. On the other hand, very low pH values may dissolve As sorbents such as Fe oxides/hydroxide *(155)*.

6.4. Nutrient Addition

6.4.1. Nitrogen

Fertilization of plants grown on As-contaminated soil with nitrate as the N source would potentially increase rhizosphere pH, and thus possibly enhance As accumulation in plant tissues *(155)*. The results of nitrogen nutrients are further confirmed by Singh and Ma (unpublished) while studying the nitrogen metabolism of As-exposed *P. vittata* with and without nitrogen supply. Further work is needed in this direction to optimize *P. vittata* to hyperaccumulate As.

6.4.2. Phosphorus

Phosphate plays a prominent role in anion–As interactions as a result of its physicochemical similarity to As *(156)*. Arsenate is taken up via the phosphate uptake system and consequently interferes with plant P nutrition. Phosphorus additions at high rates enhance As leaching *(157,158)*, increase extractable fractions of As *(159)* and reduce sorption of As(V) and As(III) onto soils *(160)*. Most studies of As–P interaction in *P. vittata* are carried out using spiked-soil or under hydroponic conditions. Tu and Ma *(125)* examined the interactions of pH, As, and P in *P. vittata*. They noted that P inhibited As uptake at all concentrations of As and P in the growth medium, although the results of Cao et al. *(147)* showed that for *P. vittata* growing on As-contaminated soil, amendment with phosphate rock significantly enhanced plant As uptake with frond As concentration increasing up to 265% relative to the control. The P from phosphate rock may have replaced As from soils, therefore increasing As availability to plants. The differences in As availability between soil and hydroponic systems are partially responsible for the different observation *(2)*.

6.4.3. Iron

Iron can indirectly affect As uptake by plants. *P. vittata* prefers to grow on calcareous soil. It has been reported that root exudates (oxalic and citric acid) of acidifuge plants, effectively mobilize P and Fe from limestone *(161)*. Porter and Peterson *(110)* found a highly significant correlation ($p < 0.001$) between As and Fe in several As-tolerant plants from different mine sites in the United Kingdom. More research is needed to investigate the role of Fe in As hyperaccumulation by *P. vittata*.

6.5. Root Exudates

It has been reported that P-deficient plants show an enhanced exudation of carboxylic acids, such as citric and malic acid. Carboxylate exudation could play a role in the mobilization of As in the rhizosphere and enhance As uptake by plants. Tu et al. *(162)* found that the root exudates from *P. vittata* contain significant amounts of oxalic acid and small amounts of citric acid. These organic acids possess the capabilities of strong proton donation and ion complexation. They can effectively extract As from soil minerals and may therefore play a role in enhancing soil As availability. Tu et al. *(162)* also noted that large amounts of phytic acid also existed in root exudates. Little is known about the role of phytic acid in plant nutrient uptake or metal hyperaccumulation. Phytic acid has been seen to release significant amounts of As from the As minerals and the contaminated soil. The amount of As released increased with phytic acid concentration and extraction time *(162)*.

6.6. Mycorrhizal Associations

Mycorrhizae have been reported in plants growing on heavy-metal-contaminated sites *(163–165)* indicating that these fungi have evolved a tolerance to heavy metals and that they may play a role in the phytoremediation of the site. For arbuscular mycorrhizae (AM) the results are conflicting. Some reports indicate higher concentrations of heavy metals in plants because of AM, even resulting in toxic levels in plants *(164–167)*. Although others have found reduced plant concentrations of Zn and Cu in mycorrhizal plants *(168–170)*. Meharg et al. *(171)* investigated an As-tolerant phenotype of *H. lanatus* and showed that 11% had higher P status and a 34% higher AM-infection rate of roots. Wright et al. *(172)* conducted a field experiment using clones of tolerant and nontolerant *H. lanatus* populations. Though no difference in AM mycorrhization were observed, tolerant plants did accumulate more P in shoots.

The role of mycorrhizae in As hyperaccumulation is not yet known. Fitz and Wenzel *(155)* found that *P. vittata* grown in pots are colonized by AM fungi. In a recent experiment Al Agely et al. (2005) studied the role of mycorrhizal symbiosis in *P. vittata* in plant growth and As and P association. They concluded that mycorrhizal fungi increase As transfer as well as plant biomass. This result conflicts with the result of Leyval et al. *(173)*, who reported that AM-limited pollutant transfer to the host plant. The prospect of symbionts existing in *P. vittata* has important implications for phytoremediation. Mycorrhizal associations increase the absorptive surface area of the plant because of extramatrical fungal hypae exploring rhizospheres beyond the root-hair zone. The protection and enhanced capacity of greater uptake of minerals result in greater biomass production, a prerequisite for successful remediation.

7. Conclusion

As is a pollutant of major concern throughout the world. Phytoremediation has emerged as a cost-effective and environment-friendly technology in cleaning up contaminated soils. For decontamination of As-polluted sites *P. vittata* is the most suitable plant at the present time. It is equipped with all the properties required by an ideal hyperaccumulating plant for phytoremediation purposes, i.e., versatility and hardiness, large biomass, fast growth, extensive root system, high As accumulation in fronds, perennial, resistance to disease and pests, diverse ecological niches, and mycorrhizal associations.

However, phytoremediation of As-contaminated soils using *P. vittata* is still in the research phase and some technical barriers need to be addressed. This includes the optimization of the process, greater understanding of how the plant absorbs, translocates, and metabolizes As, the identification of the genes responsible for As uptake, and disposal of the As-laden biomass. Nevertheless, much progress has been made toward the understanding of As tolerance and hyperaccumulation by *P. vittata* during the past a few years. More research efforts are needed to make phytoremediation using *P. vittata* a practical technology to clean up As-contaminated soils.

Acknowledgment

One of the authors (N. S.) is thankful to the US Department of States and USEFI for the Fulbright Scholarship to work at the University of Florida. The authors thank Mr. Tom Luongo for proofreading the manuscript.

References

1. Nriagu, J. O. (1979) Global inventory of natural and anthropogenic emissions of trace metals to the atmosphere. *Nature* **276,** 409–411.
2. Meharg, A. A. and Hartley-Whitaker, J. (2002) Arsenic uptake and metabolism in arsenic resistant and nonresistant plant species. *New Phytol.* **154,** 29–43.
3. Bondada, B. R. and Ma, L. Q. (2002) Tolerance of heavy metals in vascular plants: arsenic hyperaccumulation by Chinese Brake Fern (*Pteris Vittata* L.). In: *Pteridology in the New Millennium,* (Chandra, S. and Srivastava, M., eds.), Kluwer Academic Publishers, Dordrecht, The Netherlands.
4. WHO (2001) Arsenic in drinking water. http://www.who.int/int-fs/en/fact210.html. Fact sheet No. 210. May 30, 2000.
5. Cunningham, S. D. and Ow, D. W. (1996) Promises and prospects of phytoremediation. *Plant Physiol.* **110,** 715–719.
6. Comis, D. (1995) Metal-scavenging plants to cleanse the soil. *Agric. Res.* **43,** 4–9.
7. Salt, D. E., Smith, R. D., and Raskin, I. (1998) Phytoremediation. *Ann. Rev. Plant Physiol. Plant Mol. Biol.* **490,** 643–668.
8. Prasad, M. N. V. and Freitas, H. (1999) Feasible biotechnological and bioremediation strategies for serpentine soils and mine spoils. *Elec. J. Biotechnol.* **2,** 35–50.

9. Prasad, M. N. V. and Freitas, H. (2000) Removal of toxic metal from the aqueous solution by the leaf, stem and root phytomass of *Quercus ilex* L.(Holly Oak) *Environ. Pollut.* **110,** 277–283.
10. Prasad, M. N. V. (2004) Phytoremediation of metals and radionuclides in the environment: the case for natural hyperaccumulators, metal transporters, soil-amending chelators and transgenic plants. In: *Heavy Metal Stress in Plants: From Biomolecules to Ecosystems,* Springer-Verlag, Heidelberg, Germany, pp. 351–375.
11. Ernst, W. H. O. (2000) Revolution of metal hyperaccumulation and phytoremediation. *New Phytol.* **146,** 357–358.
12. Lasat, M. M. (2002) Phytoextraction of toxic metals: a review of biological mechanisms. *J. Environ. Qual.* **31,** 109–120.
13. Dixon, H. B. F. (1997) The biochemical action of arsenic acids especially as phosphate analogues. *Adv. Inorg. Chem.* **44,** 191–227.
14. Wang, J. R., Zhao, F. J., Meharg, A. A., Raab, A., Feldmann, J., and McGrath, S. P. (2002) Mechanisms of arsenic hyperaccumulation in *Pteris vittata*: arsenic species uptake kinetics and interaction with phosphate. *Plant Physiol.* **130,** 1552–1561.
15. Liversey, N. T. and Huang, P. M. (1981) Adsorption of arsenate by soils and its relation to selected chemical properties and anions. *Soil Sci.* **131,** 88–94.
16. Manning, B. A. and Goldberg, S. (1997) Arsenic (III) and arsenic (V) adsorption on three California soils. *Soil Sci.* **162,** 886–895.
17. Smith, E., Naidu, R., and Alston, A. M. (2002) Chemistry of arsenic in soils: II Effect of phosphorus, sodium and calcium on arsenic sorption. *J. Environ. Qual.* **31,** 557–563.
18. Sneller, F. E. C., Van Heerwaarden, L. M., Kraaijeveld-smit, F. J. L., et al. (1999) Toxicity of arsenate in *Silene vulgaris,* accumulation and degradation of arsenate-induced Phytochelatins. *New. Phytol.* **144,** 223–232.
19. Asher, C. J. and Reay, P. F. (1979) Arsenic uptake by barley seedlings. *Aust. J. Plant Physiol.* **6,** 459–466.
20. Wells, J. M. and Richardson, D. H. S. (1985) Anion accumulation by the moss *Hylocomium splendens*: uptake and competition studies involving arsenate, selenate, phosphate, sulphate and sulphite. *New Phytol.* **101,** 571–583.
21. Nieboer, E., Padovan, D., and Vavoie, P. (1984) Anion accumulation by lichens II. Competition and toxicity studies involving arsenate, phosphate, sulphate and sulphite. *New Phytol.* **96,** 83–94.
22. Beever, R. E. and Burns, D. W. J. (1980) Phosphorus, uptake, storage and utilization by fungi. *Adv. Bot. Res.* **8,** 127–219.
23. Silver, S. and Misra, T. K. (1988) Plasmid-mediated heavy metal resistances. *Annu. Rev. Microbiol.* **42,** 717–743.
24. Rothstein, A. and Donovan, K. (1963) Interaction of arsenate with the phosphate-transporting system of yeast. *J. Gen. Physiol.* **46,** 1075–1085.
25. Blum, J. J. (1966) Phosphate uptake by phosphate-starved *Euglena. J. Gen. Physiol.* **49,** 1125–1137.
26. Planas, D. and Healey, F. P. (1978) Effects of arsenate on growth and phosphorus metabolism of phytoplankton. *J. Phycol.* **14,** 337–341.

27. Meharg, A. A. and MacNair, M. R. (1992) Suppression of the high affinity phosphate uptake system: a mechanism of arsenate tolerance in *Holcus lanatus* L. *J. Exp. Bot.* **43**, 519–524.
28. Burlo, F., Guijarro, I., Barrachina, A. A. C., and Vlaero, D. (1999) Arsenic species: effects on and accumulation by tomato plants. *J. Agric Food Chem.* **47**, 1247–1253.
29. Carbonell Barrachina, A. A., Aarabi, M. A., Delaune, R. D., Gambrell, R. P., and Patrick, W. H. Jr. (1998a) Arsenic in wetland vegetation: Availability, Phytotoxicity, uptake and effects on plant growth and nutrition. *Sci. Total Environ.* **217**, 189–199.
30. Meharg, A. A. and MacNair, M. R. (1991) Uptake, accumulation and translocation of arsenate in arsenate-tolerant and non-tolerant *Holcus lanatus* L. *New Phytol.* **117**, 225–231.
31. Khattak, R. A., Page, A. L. Parker, D. R., and Bakhtar, D. (1991) Accumulation and interactions of arsenic, selenium, molybdenum and phosphorus in alfalfa. *J. Environ. Qual.* **20**, 165–168.
32. Pickering, I. J., Prince, R. C., George, M. J., Smith, R. D., George, G. N., and Salt, D. E. (2000) Reduction and coordination of arsenic in Indian mustard. *Plant Physiol.* **122**, 1171–1177.
33. Walsh, L. M. and Keeney, D. R. (1975) Behaviour and phytotoxicity of inorganic arsenicals in soils. In: *Arsenical Pesticides,* (Woolson, E. A., ed.), ACS, Washington, DC, pp. 35–52.
34. Jacobs, L. W. and Keeney, D. R. (1970) Arsenic phosphorus interactions on corn. *Commun. Soil. Sci. Plant Anal.* **1**, 85–93.
35. Woolson, E. A., Axley, J. H., and Kearney, P. C. (1973) The chemistry and phytotoxicity of arsenic in soils: II. Effects of time and phosphorus. *Soil. Sci. Soc. Am. Proc.* **37**, 254–259.
36. Buresh, R. J., De Laune, R. D., and Patrick, W. H. Jr. (1980) Nitrogen and phosphorus distribution and utilization by *Spartina alterniflora* in a Louisiana Gulf coast marsh. *Estuaries* **3**, 111–121.
37. De Laune, R. D. and Pezeshki, S. R. (1988) Relationship of mineral nutrients to growth of *Spartina alterniflora* in Louisiana salt marshes. *North East Gulf Science* **10**, 55–60.
38. Carbonell-Barrachina, A. A., Aarabi, M. A., De Laune, R. D., Gambrell, R. P., and Patrich, W. H., Jr. (1998) The influence of arsenic chemical form and concentration on *Spartina patens* and *Spartina alterniflora* growth and tissue arsenic concentration. *Plant Soil* **198**, 33–43.
39. Mandal, B. K. and Suzuki, K. T. (2002) Arsenic round the world: a review. *Talanta* **58**, 201–235.
40. Abedin, M. J., Feldmann, J., and Meharg, A. A. (2002) Uptake kinetics of arsenic species in rice (*Oryza sativa* L.) plants. *Plant Physiol.* **128**, 1120–1128.
41. Ernst, W. H. O. (1997) Wirkungen erhohter Bodengehalte an Arsen, Blei und Cadmium auf Pflanzen, in *Beurteilung von Schwermetallen in Boden von Ballungsgebieten: Arsen, Blei and Cadmium.* DECHEMA, pp. 319–355.

42. Terwelle, H. F. and Slater, E. C. (1967) Uncoupling of respiratory chain phosphorylation by arsenate. *Biochem. Biophys. Acta* **143**, 1–17.
43. Sachs, R. M. and Michaels, J. L. (1971) Comparative phytotoxicity among four arsenical herbicides *Weed Sci.* **19**, 558–564.
44. Lepp, N. W. (1981) Effect of heavy metal pollution on plants. In: *Effects of Trace Metals on Plant Function, Vol 1.* Applied Science Publishers, London, UK 111–143.
45. Marin, A. R., Masscheleyn, P. H., and Patrick, W. H., Jr. (1992) The influence of chemical form and concentration of arsenic in rice growth and tissue arsenic concentration. *Plant Soil* **139**, 175–183.
46. Burlo, F., Guijarro, I., Barrachina, A. A. C., and Vlaero, D. (1999) Arsenic species: effects on and accumulation by tomato plants. *J. Agric. Food Chem.* **47**, 1247–1253.
47. Fowler, B. A. (1983) *Biological and Environmental Effects of Arsenic.* Elsevier Sci. Publ. Amsterdam, The Netherlands.
48. National Academy of Sciences (1977) *Arsenic.* The National Research Council. National Academy of Sciences, Washington, DC.
49. Von Endt, D. W., Kearney, P. C., and Kaufman, D. D. (1968) Degradation of monosodium methanearsonic acid by soil microorganisms. *J. Agri. Food. Chem.* **16,** 17–20.
50. Khattak, R. A., Page, A. L., Parker, D. R., and Bakhtar, D. (1991) Accumulation and interactions of arsenic, selenium, molybdenum and phosphorus in alfalfa. *J. Environ. Qual.* **20**, 165–168.
51. Wauchope, R. D. and McWhorter, C. G. (1977) Arsenic residues in soybean seed from simulated MSMA spray drift. *Bull. Environ. Contam. Toxicol.* **17**, 165–167.
52. Carbonell-Barrachina, A. A., Burtiõ, F., and Mataix, J. (1995) Arsenic uptake, distribution and accumulation in tomato plants: effect of arsenite on plant growth and yield. *J. Plant Nut.* **18**, 1237–1250.
53. Carbonell-Barrachina, A. A., Burtiõ, F., Burgos-Hernãndez, A., López, E., and Mataix, J. (1997) The influence of arsenite concentration on arsenic accumulation in tomato and bean plants. *Sci. Hortic.* **71**, 167–176.
54. Porter, E. K. and Peterson, P. J. (1977) Arsenic tolerance in grasses growing on mine waste. *Environ. Pollut.* **14**, 255–265.
55. Masscheleyn, P. H., DeLaune, R. D., and Patrick, W. H., Jr. (1991) A hybrid generation atomic absorption technique for arsenic speciation. *J. Environ. Qual.* **20,** 96–100.
56. Koch, I., Wang, L., Ollson, C. A., Cullen, W. R., and Reimer, K. (2000) The predominance of inorganic arsenic species in plants from Yellowknife, Northwest Territories, Canada. *Environ. Sci. Technol.* **34**, 22–26.
57. Nissen, P. and Benson, A. A. (1982) Arsenic metabolism in freshwater and terrestrial plants. *Physiol. Plant.* **54**, 446–450.
58. Pyles, R. A. and Woolson, E. A. (1982) Quantitation and characterization of arsenic compounds in vegetables grown in arsenic acid treated soil. *J. Agric. Food Chem.* **30**, 866–870.

59. Helgesen, H. and Larsen, E. H. (1998) Bioavailability and speciation of arsenic in carrots grown in contaminated soil. *Analyst* **123**, 791–796.
60. Quaghebeur, M. and Rengel, Z. (2003) The distribution of arsenate and arsenite is shoots and roots of *Holcus lanatus* in influenced by arsenic tolerance and arsenate and phosphate supply. *Plant Physiol.* **132**, 1600–1609.
61. Mattusch, J., Wennrich, R., Schmidt, A. C., and Reisser, W. (2000) Determination of arsenic species in water, soil and plants. *Fresen J. Anal. Chem.* **366**, 200–203.
62. Philips, S. E. and Taylor, M. L. (1976) Oxidation of arsenite to arsenate by *Alcaligenes foecalis. Appl. Environ. Microbiol.* **32**, 392–399.
63. Sehlin, H. M. and Lindström, E. B. (1992) Oxidation and reduction of arsenic by *Sulfololus scidocaldarius* strain BC. *FEMS Microbial Lett.* **93**, 87–92.
64. Marschner, H. (1995) *Mineral Nutrition of Higher Plants, 2nd ed.* Academic Press, London, UK.
65. Edmonds, J. S. and Francesconi, K. A. (1987) Transformations of arsenic in the marine environment. *Experientia* **43**, 553–557.
66. Sachs, R. M. and Michaels, J. L. (1971) Comparative phytotoxicity among four arsenical herbicides. *Weed Sci.* **19**, 558–564.
67. Galbraith, H., Lejeune, K., and Lipton, J. (1995) Metal and arsenic impacts to soils, vegetation communities and wildlife habitat in southwest Montana uplands contaminated by smelter emissions. 1. Field evaluation. *Environ. Toxicol. Chem.* **14**, 1895–1903.
68. Elstner, E. F. (1982) Oxygen activation and oxygen toxicity. *Ann. Rev. Plant Physiol.* **33**, 73–96.
69. Benavides, M. P., Gallego, S. M., Comba, M. E., and Tomaro, M. L. (2000) Relationship between polyamines and paraquat toxicity in sunflower leaf discs. *J. Plant Growth Regul.* **31**, 215–224.
70. Chang, C. J. and Kao, C. H. (1997) Paraquat toxicity is reduced by polyamines in rice leaves. *J. Plant Growth Regul.* **22/3**, 163–168.
71. Borrell, A., Carbonell, L., Farras, R., Puig-Parellada, P., and Tiburcio, A. F. (1997) Polyamines inhibit lipid peroxidation in senescing oat leaves. *Physiol. Plant.* **99**, 385–390.
72. Hartley-Whitaker, J., Ainsworth, G., and Meharg, A. A. (2001) Copper and arsenate-induced oxidative stress in *Holcus lanatus* L. clones with different sensitivity. *Plant Cell Environ.* **24**, 713–722.
73. Mylona, P. V., Polidoros, A. N., and Scandalios, J. G. (1998) Modulation of antioxidant responses by arsenic in maize. *Free Radical Biol. Med.* **25**, 576–585.
74. Grill, E., Winnacker, E.-L., and Zenk, M. H. (1987) Phytochelatins, a class of heavy–metal–binding peptides from plants are functionally analogous to metallothioneins. *Proc. Natl. Acad. Sci. USA* **84**, 439–443.
75. Maitani, T., Kubato, H., Sato, K., and Yamada, T.(1996) The composition of metals bound to class III metallothione in (phytochelatins and its desglycyl peptide) induced by various metals in root cultures of *Rubia tinctorum, Plant Physiol.* **110**, 1145–1150.

76. Sneller, F. E. C., Van Heerwaarden, L. M., Kraaijeveld-Smit, F. J. L., et al. (1999) Toxicity of arsenate in *Silene vulgaris*, accumulation and degradation of arsenate-induced phytochelatins. *New Phytol.* **144,** 223–232.
77. Schmöger, M. E. V., Oven, M., and Grill, E. (2000) Detoxification of arsenic by phytochelatins in plants. *Plant Physiol.* **122,** 793–801.
78. Bleeker, P. M., Schat, H., Vooijs, R., Verkleij, J. A. C., and Ernst, W. H. O. (2003) Mechanisms of arsenate tolerance in *Cytisus striatus*. *New Phytol.* **157,** 33–38.
79. Delnomdedieu, M., Basti, M. M., Otvos, J. D., and Thomas, D. J. (1994) Reduction and binding of arsenate and dimethylarsinate by glutathione a magnetic-resonance study. *Chem-Biol. Interact.* **90,** 139–155.
80. Zhao, F. J., Wang, J. R., Barker, J. H. A., Schat, H., Blecker, P. M., and McGrath, S. P. (2003) The role of phytochelatins in arsenic tolerance in the hyperaccumulator *Pteris vittata*. *New Phytol.* **159,** 403–410.
81. Baker, A. J. M. and Brooks, R. R. (1989) Terrestrial higher plants which hyperaccumulate metallic elements: a review of their distribution, ecology and phytochemistry. *Biorecovery* **1,** 81–126.
82. Baker, A. J. M., McGrath, S. P., Reeves, R. D., and Smith, J. A. C. (2000) Metal hyperaccumulator plants: a review of the ecology and physiology of a biological resource for phytoremediator of metal-polluted soils. In: *Phytoremediation of Contaminated Soil and Water,* (Terry, N. and Banuelos, G., eds.), Lewis Publishers, Boca Raton, FL. pp. 85–107.
83. Reeves, R. D. and Baker, A. J. M. (2000) Metal accumulating plants. In: *Phytoremediation of Toxic Metals: Using Plants to Clean-Up the Environment,* (Raskin, I. and Ensley, B. D., eds.), John Wiley and Sons, New York, NY, pp. 193–230.
84. Ma, L. Q., Komar, K. M., Tu, C., Zhang, W., Cai, Y., and Kennelley, E. D. (2001) A fern that hyperaccumulates arsenic. *Nature* **409,** 579.
85. Salt, D. E. and Krämer, U. (2000) Mechanisms of metal hyperaccumulation in plants. In: *Phytoremediation of Toxic Metals,* (Raskin, I. and Ensley, B. D., eds.), John Wiley, New York, NY, pp. 231–246.
86. Baker, A. J. M. and Whiting, S. N. (2002) In search of the Holy Grail: a further step in understanding metal hyperaccumulation? *New Phytol.* **155,** 1–7.
87. Ma, L. Q., Komar, K. M., Tu, C., Zhang, W., and Cai, Y. (2001) A fern that hyperaccumulates arsenic-addendum. *Nature* **411,** 438.
88. Baker, A. J. M. (1981) Accumulators and excluders-strategies in the response of plants to heavy metals. *J. Plant Nutr.* **3,** 643–654.
89. Benson, A. A., Cooney, R. V., and Harrera-Lasso, J. M. (1981) Arsenic metabolism in algae and higher plants. *J. Plant Nutr.* **3,** 285–292.
90. Francesconi, K., Visoottiviseth, P., Sridokchan, W., and Goessler, W. (2002) Arsenic species in an arsenic hyperaccumulating fern, *Pityrogramma calomelanos*: a potential phytoremediator of arsenic-contaminated soils. *Sci. Tot. Environ.* **284,** 27–35.
91. Prasad, M. N. V. and Freitas, H. (2003) Metal hyperaccumulation in plants: biodiversity prospecting for phytoremediation technology. *Elec. J. Biotech.* **6,** 285–321.

92. Kruckeberg, A. R. and Rabinowitz, D. (1985) Biological aspects of endemism in higher plants. *Annu. Rev. Ecol. Syst.* **16,** 447–479.
93. Kruckeberg, A. R. (1986) An essay: the stimulus of unusual geologies for plant speciation. *Syst. Bot.* **11,** 455–463.
94. Reeves, R. D., Schwartz, C., Morel, J. L., and Edmondson, J. (2001) Distribution and metal accumulating behaviour of *Thlaspi caerulescens* and associated metallophytes in France. *Int. J. Phytorem.* **3,** 145–172.
95. Bert, V., Macnair, M.R., DeLaguerie, P., Saumitou-Laprade, P., and Petit, D. (2000) Zinc tolerance and accumulation in metallicolous and non-metallicolous populations of *Arabidopsis halleri* (Brassicaceae). *New Phytol.* **146,** 225–233.
96. Reeves, R. D., Kruckeberg, A. R., Adiguzel, N., and Krämer, U. (2001) Studies on the flora of serpentine and other metalliferous areas of western Turkey. *S. Afr. J. Sci.* **97,** 513–517.
97. Broadley, M. R., Willey, N. J., Wilkins, J. C., Baker, A. J. M., Mead, A., and White, P. J. (2001) Phylogenetic variation in heavy metal accumulation in angiosperms. *New Phytol.* **152,** 9–27.
98. Meharg, A. A. (2003) Variation in arsenic accumulation – hyperaccumulation in ferns and their allies. *New Phytol.* **157,** 25–31.
99. Pollard, A. J., Dandridge, K. L., and Jhee, E. M. (2000) Ecological genetics and the evolution of trace element hyperaccumulation. In: *Plants in Phytoremediation of Contaminated Soil and Water,* (Terry, N. and Banuelos, G., eds.), Lewis Publishers, Boca Raton, FL, pp. 251–264.
100. Clemens, S., Palmgren, M. G., and Krämer, U. (2002) A long way ahead: understanding and engineering plant metal accumulation. *Trends Plant Sci.* **7,** 309–315.
101. Antonovics, J., Bradshaw, A. D., and Turner, R. G. (1971) Heavy metal tolerance in plants. *Adv. Ecol. Res* **7,** 1–85.
102. Baker, A. J. M. (1987) Metal tolerance. *New Phytol.* **106,** 93–111.
103. Ernst, W. H. O., Assunção, A. G. L., Verkleij, J. A. C., and Schat, H. (2002) How important is apoplastic zinc xylem loading in *Thlaspi Caerulescens? New Phytol.* **155,** 4–5.
104. Meharg, A. A. (1994) Integrated tolerance mechanisms: constitutive and adaptive plant responses to elevated metal concentrations in the environment. *Plant Cell Environ.* **17,** 989–993.
105. Baker, A. J. M. and Walker, P. L. (1990) Ecophysiology of metal uptake by tolerant plants. In: *Heavy Metal Tolerance in Plants: Evolutionary Aspects,* (Shaw, A. J., ed.), CRC Press, Boca Raton, FL, pp. 155–177.
106. Escarré, J., Lefèbvre, C. Gruber, W., et al. (2000) Zinc and cadmium hyperaccumulation by *Thlaspis caerulescens* from metalliferous and non metalliferous sites in the Mediterranean area: implications for phytoremediation. *New Phytol.* **145,** 429–437.
107. Assunção, A. G. L., Martins, P. Da. C., De Folter, S., Vooijs, R., Schat, H., and Aarts, M. G. M. (2001) Elevated expression of metal transporter genes in three accessions of the metal hyperaccumulator *Thlaspi caerulescens.* *Plant Cell Environ.* **24,** 217–226.

108. Boyd, R. S. and Martens, S. N. (1992) The *raison d'etre* for metal hyperaccumulation by plants. In: *The Vegetation of Ultramafic (Serpentine) Soils,* (Baker, A. J. M., Proctor, J., and Reeves, R. D. eds.), Intercept Ltd., Andover, UK, pp. 279–289.
109. Tu, C. and Ma, L. Q. (2003) Effects of arsenate and phosphate on their accumulation by an arsenic hyperaccumulator. *Pteris vittata* L. *Plant Soil* **249,** 373–382.
110. Porter, E. K. and Peterson, P. J. (1975) Arsenic accumulation by plants on mine waste (United Kingdom). *Sci. Total Environ.* **4,** 365–371.
111. De Koe, T. (1994) *Agrostis castellana* and *Agrostis delicatula* on heavy metal and arsenic enriched sites in NE Portugal. *Sci. Total Environ.* **145,** 103–109.
112. Komar, K. M., Ma, L. Q., Rockwood, D., and Syed, A. (1998) Identification of arsenic tolerant and hyperaccumulating plants from arsenic contaminated soils in Florida. *Agronomy Abstract* p 343.
113. Komar, K. M. (1999) *Phytoremediation of arsenic contaminated soils: plant identification and uptake enhancement.* MS University of Florida, Gainesville, FL.
114. Chen, T., Wei, C., Huang, Z., Huang, Q., Lu, Q., and Fan, Z. (2002) Arsenic hyperaccumulator Pteris Vittata L. and its arsenic accumulation. *Chin. Sci. Bull.* **47,** 902–905.
115. Visottiviseth, P., Francesconi, K., and Sridokchan, W. (2002) The potential of Thai indiginous plant species for the phytoremediation of arsenic contaminated land. *Environ. Pollut.* **118,** 453–461.
116. Chen, M., Ma, L. Q., Harris, W. G., and Hornsby, A. (1999) Background concentrations of 15 trace metals in Florida soils. *J. Environ. Qual.* **28,** 1173–1181.
117. Cai, Y. and Lena, Q. M. (2003) Metal tolerance, accumulation, and detoxification in plants with emphasis on arsenic in terrestrial plants. In: *ACS symposium Series 835. Biogeochemistry of Environmentally Important Trace Elements,* (Cai, Y. and Braids, O. C., eds.), American Chemical Society, pp. 95–114 Washington, DC.
118. Ma, L. Q., Komar, K. M., and Kennelley, E. D. (2001) Methods for removing pollutants from contaminated soils materials with a fern plant. USA Patent US Patent No. 6,280,500. Issue date 8/28/01.
119. Francesconi, K., Visooltiviseth, P., Sridokchan, W., and Goessler, W. (2002) Arsenic species in an arsenic hyperaccumulation fern, *Pityragramma calomelanos*: a potential phytoremediator of arsenic-contaminated soils. *Sci. Tot. Environ.* **284,** 27–35.
120. Zhao, F. J., Dunham, S. J., and McGrath, S. P. (2002) Arsenic hyperaccumulation by different fern species. *New Phytol.* **156,** 27–31.
121. Jones, D. L. (1987) *Encyclopaedia of Ferns: An Introduction to Ferns, Their Structure, Biology, Economic Importance, Cultivation and Propogation.* Lothian Publ. Company, Melbourne, Australia.
122. Meharg, A. A. (2002) Arsenic and old plants. *New Phytol.* **156,** 1–4.
123. Meharg, A. A. (2003) Variation in arsenic accumulation-hyperaccumulation in ferns and their allies. *New Phytol.* **157,** 25–31.
123a. Luonge, T. and Ma, L. Q. (2005) Characteristics of arsenic accumulation by Pteris and non-Pteris ferns. *Plant Soil* **277,** 117–126.

124. Tu, C. and Ma, L. Q. (2002) Effects of arsenic concentrations and forms on arsenic uptake by the hyperaccumulator ladder brake. *J. Environ. Qual.* **31,** 641–647.
125. Tu, S. and Ma, L. Q. (2003) Interactive Effects of pH, As and P on growth and As/P uptake in hyperaccumulator *Pteris vittata*. *Environ. Exp. Bot.* **50,** 1–9.
126. Wang, J., Zhao, F.-J., Meharg, A. A. P. Raab, A., Feldmann, J., and McGrath, S. P. (2002) Mechanisms of arsenic hyperaccumulation in *Pteris vittata*. Uptake kinetics, interactions with phosphate, and arsenic speciation. *Plant Physiol.* **130,** 1552–1561.
127. Tu, S., Ma, L. Q., McDonald, G. E., and Bondada, B. (2003) Arsenic absorption, speciation and thiol formation in excised parts of *Pteris Vittata* in the presence of phosphorus. *Environ. Exp. Bot.* **51,** 121–131.
128. Lombi, E., Zhao, F., Fuhrmann, M., Ma, L. Q., and McGrath, S. P.(2002) Arsenic distribution and speciation in the fronds of the hyperaccumulator *Pteris Vittata*. *New Phytol.* **156,** 195–203.
129. Zhang, W., Cai, Y., Tu, C., and Ma, L. Q. (2002) Arsenic speciation and distribution in an arsenic hyperaccumulating plant. *Sci. Total Environ.* **300,** 167–177.
129a. Singh, N., and Ma, L. Q. (2006) Arsenic specification, and arsenic and phosphate distribution in arsenic hyperaccumulator *Pteris vittata* and non-hyperaccumulator *Pteris ensiformis*. *Enivorn. Pollut.* **141,** 238–246.
130. Tu, C., Ma, L. Q., Zhang, W., Cai, Y., and Harris, W. G. (2003) Arsenic species and leachability in the fronds of the hyperaccumulator Chinese brake (*Pteris vittata* L.). *Environ. Pollut.* **124,** 223–230.
131. Delnomdedieu, M., Basti, M. M., Orvos, J. D., and Thomas, D. J. (1994) Reduction and binding of arsenate and dimethylarsenate by glutathione: a magnetic resonance study. *Chem-Biol. Interact.* **90,** 139–155.
132. Maltusch, J., Wennrich, R., Schmidt, A. C., and Reisser, W. (2000) Determination of arsenic species in water, soils and plants. *Fresen. J. Anal. Chem.* **366,** 200–203.
133. Ullrich-Eberius, C. I., Sanz, A., and Novacky, J. (1989). Evaluation of arsenic and vandate associated changes of electrical membrane potential and phosphate transport in *Lemma gibba* G1. *J. Exp. Bot.* **40,** 119–128.
134. Cao, Y., Ma, L. Q., and Tu, C. (2003) Antioxidative responses to arsenic in the arsenic-hyperaccumulator Chinese brake fern (*Pteris Vittata* L.). *Environ. Pollut.* **128,** 317–325.
135. Cai, Y., Su, J., and Ma, L. Q. (2004) Low molecular weight thiols in arsenic hyperaccumulation *Pteris Vittata* upon exposure to arsenic and other trace elements. *Environ. Pollut.* **129,** 69–78.
136. Zhao, F. J., Wang, J. R., Barker, J. H. A., Schat, H., Bleeker, P. M., and McGrath, S. P. (2003) The role of phytochelatins in arsenic tolerance in the hyperaccumulator *Pteris Vittata*. *New Phytol.* **159,** 403–410.
137. Lombs, E., Zhao, F. J., Dunham, S. J., and McGrath, S. P. (2001) Phytoremediation of heavy metal contaminated soils: natural hyperaccumulation versus chemically enhanced phytoextraction. *J. Environ. Qual.* **30,** 1919–1926.
138. Heeraman, D. A., Claassen, V. P., and Zasos Ki, R. J. (2001) Interaction of lime, organism matter and fertilizer on growth and uptake of arsenic and mercury by Zorro fescue (*Vulpra myuros* L.). *Plant Soil.* **234,** 215–231.

139. Peryea, F. J. (1998) Phosphate starter fertilizer temporarily enhances soil arsenic uptake by apple trees grown under field conditions. *Hort. Sci.* **33**, 826–829.
140. Zhou, L. X. and Wong, J. W. C. (2001) Effect of dissolved organic matter from sludge compost on soil copper sorption. *J. Environ. Qual.* **30**, 878–883.
141. Robertson, W. K., Lutrick, M. C., and Yuan, T. L. (1982) Heavy applications of liquid-digested sludge on three ultisols: I. effect on soil chemistry. *J. Environ. Qual.* **11**, 278–282.
142. Baham, J. and Sposito G. (1983) Chemistry of water-soluble, metal complexing ligands extracted from an anaerobically digested sewage sludge. *J. Environ. Qual.* **12**, 731–737.
143. Lamy, I., Bourgeois, S., and Bermond, A. (1993) Soil cadmium mobility as a consequence of sewage sludge disposal. *J. Environ. Qual.* **22**, 731–737.
144. McCarthy, J. F. and Zachara, J. M. (1989) Subsurface transport of contaminants. *Environ. Sci. Technol.* **23**, 496–502.
145. Temminghoff, E. J. M., VanderZee, S. E. A. T. M., and De Haan, F.A.M. (1997) Copper mobility in a copper-contaminated sandy soil as affected by pH and solid and dissolved organic matter. *Environ. Sci. Technol.* **31**, 1109–1115.
146. Darmody, R. G., Foss, J. E., Mcintosh, M., and Wolf, D. C. (1983) Muncipal sewage sludge compost—amended soils: some spatio temporal treatment effects. *J. Environ. Qual.* **12**, 231–236.
147. Cao, X. Ma, L. Q., and Shiralipour, A. (2003) Effects of compost and phosphorus amendments on arsenic mobility in soils and arsenic uptake by the hyperaccumulator, *Pteris vittata* L. *Environ. Pollut.* **126**, 157–167.
148. Marin, A. R., Masschelyn, P. H., and Patrick, H. W., Jr. (1992) The influence of chemical form and concentration of arsenic on rice growth and tissue arsenic concentration. *Plant Soil* **139**, 175–183.
149. Huang, J. W. W., Chen, J. J., Berti, W. R., and Cunningham, S. D. (1997) Phytoremediation of lead contaminated soils-role of synthetic chelates in lead phytoextraction. *Environ. Sci. Technol.* **31**, 800–805.
150. Cooper, E. M., Sims, J. T., Cunningham, S. D., Huang, J. W., and Berti, W. R. (1999) Chelate-assisted phytoextraction of lead from contaminated soils. *J. Environ. Qual.* **28**, 1709–1719.
151. Blaylock, M. J. (2000) Field demonstration of phytoremediation of lead contaminated soils. In: *Phytoremediation of Contaminated Soil and Water,* (Terry, N. and Banuelos, G. eds.), Lewis, Publ., Boca Raton, FL, pp. 1–12.
152. Manning, B. A. and Goldberg, S. (1997) Arsenic (III) and arsenic(V) adsorption on three California soils. *Soil Sci.* **162**, 886–895.
153. Smith, E., Naidu, R., and Alston, A. M. (1999) Chemistry of arsenic in soils: I. Sorption of arsenate and arsenite by four Australian soils. *J. Environ. Qual.* **28**, 1719–1726.
154. Tyler, G. and Olsson, T. (2001) Concentration of 60 elements in the soil solution as related to the soil acidity. *Eur. J. Soil Sci.* **52**, 151–165.
155. Fitz, W. J. and Wenzel, W. W. (2002) Arsenic transformations in the soil-rhizosphere plant system: fundamentals and potential application to phytoremediation. *J. Biotechnol.* **99**, 259–278.

156. Adriano, D. C. (2001) *Trace Elements in the Terrestrial Environment.* Springer, New York, NY.
157. Woolson, E. A., Axley, J. H., and Kearny, P. C. (1973) The chemistry and phytotoxicity of arsenic in soils: II. Effects of time and phosphorus. *Soil Sci. Soc. Am. Proc.* **37,** 254–259.
158. Peryae, F. J. and Kammereck, R. (1995) Phosphate enhanced movement of arsenic out of lead arsenate-contaminated top-soil and through uncontaminated sub-soil, *Water Air Pollut.* **93,** 243–254.
159. Peryae, F. J. (1991) Phosphate-induced release of arsenic from soils contaminated with lead arsenate *Soil Sci. Soc. Am. J.* **55,** 1301–1306.
160. Smith, E., Naidu, R., and Alston, A. M. (2002) Chemistry of arsenic in soils: II. Effect of phosphorus, sodium and calcium on arsenic sorption. *J. Environ. Qual.* **31,** 557–563.
161. Strom, L., Olsson, T., and Tyler, G. (1994) Differences between calcifuge and acidifuge plants in root exudation of low molecular organic acids. *Plant Soil* **167,** 239–245.
162. Tu, S., Ma, L. Q., and Luongo T. (2004) Root exudation and its role in arsenic hyperaccumulation of *Pteris vittata. Plant Soil* **258,** 9–19.
163. Sneltty, P. K., Hetrick, B. A. D., Figge, D. A. H., and Schwab, A. P. (1994) Effects of mycorrhizae and other soil contaminated mine spoil. *Environ. Pollut.* **86,** 181–188.
164. Weissenhorn, I. and Leyval, C. (1995) Root colonization of maize by a Cd-sensitive and a Cd- tolerant *Glomus mosseae* and cadmium uptake in sand culture. *Plant Soil* **175,** 233–238.
165. Chaudhary, T. M., Hayes, W. J., Khan, A. G., and Khoo, C. S. (1998) Phytoremediation-focusing on accumulator plants that remediate metal contaminated soils. *Austr. J. Ecotoxicol.* **4,** 37–51.
166. Killham, K. and Firestone, M. K. (1986) Vesicular arbuscular mycorrhizal mediation of grass response to acid and heavy metal deposition. *Plant Soil* **72,** 39–48.
167. Joner, E. J. and Leyval, C. (1997) uptake of ^{109}Cd by roots and hypae of a *Glomus mosseae/ Trifolium subterraneum* mycorhiza from soil amended with high and low concentration of cadmium. *New Phytol.* **72,** 39–48.
168. Schuepp, H., Dehn, B., and Sticher, H. (1987) Interaktionen zwischen VA-Mycorrhizen and Schwermetall bela stungen. *Agnew Botanik* **61,** 85–96.
169. El-Kherbawy, M., Angle, J. S., Heggo, A., and Chaney, R. L. (1989) Soil pH, rhizobia and vesicular mycorrhizae inoculation effects on growth and heavy metal uptake of alfalfa (*Medicago sativa* L.). *Biol. Fertil. Soils.* **8,** 61–65.
170. Heggo, A., Angle, J. S., and Chaney, R. L. (1990) Effects of vesicular arbuscular my corrhizal fungi on heavy metal uptake by soybeans. *Soil Bio. Biochem.* **22,** 865–869.
171. Meharg, A. A., Bailey, J., Breadmore, K., and Macnair, M. R. (1994) Biomass allocation, phosphorus nutrition and vesicular arbuscular mycorrhiza infection in clones of Yorkshire Fog, *Holcus lanatus* L. (Poaceae) that differ in their phosphate uptake kinetics and tolerance to arsenate. *Plant Soil* **160,** 11–20.

172. Wright, W., Fitter, A., and Meharg, A. A. (2000) Reproductive biomass in *Holcus lanatus* clones that differ in their phosphate uptake kinetics and mycorrhizal colonization. *Plant Soil* **146,** 493–501.
173. Leyval, C., Turnau, K., and Haselwandter (1997) Effects of heavy metal pollution on mycorrhizal colonization and function: physiological, ecological and applied aspects. *Mycorrhiza* **7,** 139–153.

IV

CONTEXTS AND UTILIZATION OF PHYTOREMEDIATION

25

Phytoremediation in China
Inorganics

Shirong Tang

Summary

There is an old saying in China "food is heaven for people while soil is the mother for food." Soil is considered to be one of the most important natural resources that people are dependent upon. However, more and more anthropogenic and natural factors are speeding up soil contamination, which poses a potential threat to human health and the environment. The situation is getting worse in some parts of the People's Republic of China although the trend has been slowing down in recent years. The main contaminants causing degradation of the soil in China include heavy metals, radionuclides, and organic pollutants. In this chapter, the focus is on soil contamination by heavy metals and other inorganic pollutants in China, and the state-of-the-art green remediation technologies for them.

Key Words: Phytoremediation; heavy metals; radionuclides; soil degradation.

1. General Review of Soil and Water Contamination With Inorganics in China

Soil contamination with heavy metals has become an environmental concern in China in the past decade. Statistics show that the acreage of arable lands contaminated with heavy metals such as Cd, As, Cr, and Pb to various degrees has been increasing in recent years, now totaling 2×10^7 ha, i.e., about one-fifth of the country's arable land. The most serious problem with heavy-metal contamination occurs in the soils around city suburbs, and mining and wastewater irrigation areas. It has been reported that the soil on the outskirts of Shanghai City is contaminated with Cd and Hg. About 9.5% of the arable land in the suburbs of Guangzhou City is reportedly polluted by Cd, Pb, and As. The soil in Tianjin (Tientsin) City had Cd and Hg concentrations 5 and 60 times, respectively, higher than the background values, with about 23,000 ha arable land contaminated by waste water irrigation there *(1)*. A survey made of several key areas in Liaonin Province in the northeastern part of China showed that hazardous heavy metals were ubiquitous in the soil in the province, with some areas being considerably above the background values, especially in waste water-irrigated agricultural lands, and soils around cities and mining sites. The reported contaminated land in the province was estimated to be more than 2×10^5 km^2 *(2)*, with pollutants being mainly Cd, Hg, Cu, As, Pb, Cr, and Zn. Soils at Zhang Shiguan situated in the suburb of the provincial capital city Shengyan in the northeastern part of China were irrigated with waste water for about 20 yr, and about 2500 ha land there was contaminated as a result of that irrigation practice. The Cd concentration in some rice fields ranged from 5 to 7 mg/kg in the soil.

Soil contamination in some areas of the Zhejiang Province in the Southeastern part of China is serious, too. It has been reported that about 333,300 ha arable land has been contaminated by the "three wastes" (i.e., industrial wastewater, hazardous materials, and waste air), taking up more than 20% of the total arable land in the Province *(3)*. Analysis of soil samples taken from three vegetable-producing areas in the suburb of the provincial capital city Hangzhou showed that the concentration of Hg in the tilled layers of the vegetable-producing land was 2.6 times that of the soil far away from contamination sources and free from application of sewage sludge, with Pb, Cu, As, and Zn concentrations being 69, 58, 19, and 19%, higher, respectively, than noncontaminated soils nearby. It was found that about half of the areas (180 km^2) in the northern part of Xiaoshan District, Hangzhou City, which are labeled as a National Modern Agriculture Development Base, were contaminated to various degrees because of industrial waste water discharge. In the suburb of Fuyang City (about 60-km southwest away from Hangzhou City) and along both banks of the Fuchun River, Zhejiang Province, there are reportedly several hundred hectares of contaminated arable land because of local mining and refinery activities, with reports of human and domestic animals being poisoned. Leqing City, a city located near Wenzhou where the Chinese private township enterprises originated, also has a serious environmental pollution problem, with soil contamination being caused mainly by polluted water irrigation (**Table 1**) *(4)*. Water sampling investigations made at Yandang town of the city showed that the water in the irrigation furrow of the arable lands has a pH value of 3.27, and that the average heavy-metal concentrations of Cu, Zn, Pb, Cd, and Ni in the water were 182, 9.5, 0.24, 0.06, and 108.0 mg/L, respectively. Soil surveys made at Lou village and Fuao village of the city showed that the average concentrations of Cu, Zn, Pb, Cd, and Ni were 711, 84.0, 51.80, 1.25, 196.0 mg/kg in the soil of Lou village, and 148, 60.1, 33.50, 0.99, 35.8 mg/kg in the soil of Fuao village, respectively, suggesting accumulation of heavy metals in the soils. It was also

Table 1
Concentrations of Heavy Metals in the Vegetable-Producing Lands in Leqing City, Zhejiang Province (mg/kg) (4)[a]

Sampling sites	Beipeixiang	Liushi	Lecheng	Hongqiao
Total Cu	29.30	36.21	26.90	29.48
Bioavailable Cu	5.23	7.52	4.35	6.27
Total Cd	0.932	0.943	1.125	0.970
Bioavailable Cd	0.193	0.231	0.248	0.170
Total Zn	104.44	108.60	103.95	103.28
Bioavailable Zn	7.652	14.630	9.732	6.010
Total Pb	42.70	39.13	42.65	42.30
Bioavailable Pb	5.122	13.149	6.205	9.890
Total Ni	28.40	29.91	35.55	31.92
Bioavailable Ni	2.862	2.616	1.515	1.960
Total Cr	28.9	29.9	18.4	38.0

[a]The background values of Cu, Cd, Zn, Pb, Ni, and Cr in the vegetable-producing lands of Leqing City, Zhejiang Province was 18.83, 0.107, 83.18, 33.05, 29.53, and 59.9 mg/kg with standard deviation being 4.77, 0.054, 10.96, 9.32, 7.29 and 12.2 mg/kg, respectively.

Table 2
The Average Contents of Heavy Metals in Vegetable-Producing Soils in Zhongqing City and Their Corresponding Background Contents in Soils (mg/kg) (5)

Metals	Cd	Hg	As	Cu	Pb	Cr	Zn	Ni
Soil in Zhongqing City	0.285	0.056	6.757	23.044	37.036	52.117	82.274	33.436
Background contents in soils	0.141	0.037	6.76	21.96	22.20	48.55	79.47	35.69

documented that the main vegetable-producing lands there were contaminated with Cd, and secondarily with Cu and Zn. Li et al. (5) reported that the soils used for growing vegetables in Zhongqing City of the southwest part of China was contaminated to a variable degree, mainly by Cd, Hg, and Pb (**Table 2**) with Cd contamination being the most serious.

Sources of heavy metals that cause soil contamination in China are various, including agricultural irrigation with metal-containing wastewater, agricultural utilization of municipal sewage sludge, application of agricultural chemicals, mining and refinery, application of organic and phosphorus fertilizers, and deposition of atmospheric polluted particles. Irrigation of agricultural land with waste water is one of the major contributions to heavy metal increase in Chinese soils. It has been estimated that the wastewater discharged in China adds up to

400 million tons per year, among which there were about 2700 tons of heavy metals such as Cd and Hg. Most of the contaminants discharged through wastewater have entered the agricultural- and livestock-farm environment. A survey has shown that about 40% of underground water could not meet the standards for agricultural irrigation, influencing more than 3.2 hundred million mu (a Chinese unit of area, 1 hectare = 15 mu) of arable land. There are reportedly about 8 million mu of agricultural land irrigated with wastewater. On 70% of the land, waste water is used as a main or sole source of irrigation water. Liaonin Province of the northeastern part of China is one of the typical examples, where the average fresh water resource is only 900 m^3 per person. The main water source for irrigation in the province is the Liao River of which some sections were contaminated to various degrees. Waste water irrigation resulted in deposition of heavy metals such as Cd, Hg, Cu, and Zn in the soils that could potentially be taken up subsequently by agricultural crops. There are about 10 million ha of arable land polluted by the "three wastes." The amount of the "three wastes" discharged by Chinese industries, especially by township enterprises, was tremendous, with treatment rate of the "three wastes" being less than 30% *(6)*. It has been estimated that China has 3.3 million ha of land that was irrigated with waste water, among which 64.8% of the land was contaminated with heavy metals, 46.7% was slightly polluted, 9.7% intermediately polluted, and 8.4% heavily polluted *(7)*. Statistics from the National Environmental Protection Bureau showed that the arable land contaminated by the industry "three wastes" added up to 7 million ha, resulting in crop reduction of more than 1×10^{11} kg. A second survey of the national waste water irrigation showed that among 57 typical waste water irrigation areas 14.6% of them was contaminated by heavy metals, with the acreage polluted by Cd taking up 56.9% of the total investigated areas. Relative to the results revealed by the first survey conducted from 1977 to 1982, the contaminated acreage more than doubled. It was reported that waste water irrigation resulted in contamination of about 2700 ha arable land in the suburbs of Guangzhou City. A survey conducted in one of the waste water-irrigated areas of the capital city Beijing in the middle 1980s showed that about 60% of the soil and 36% of the rice produced were contaminated by heavy metals to some degree.

Besides heavy-metal contamination in soils, nitrate contamination is also a serious problem in areas where intensive agricultural systems are prosperous. Investigation showed that the vegetables produced in some big cities such as Beijing, Tientsin, Shenyang, Urumchi, Shanghai, Zhongqing, Nanjing, Hangzhou, Ninbo, Guangzhou, and Fuzhou were contaminated with nitrate to various degrees, posing a potential threat to human health *(8)*.

Radionuclide contamination of the soil in China has mainly arisen from a discharge of radioactive waste materials from nuclear facilities, including ^{85}Kr,

^{133}Xe, ^{131}I, ^{60}Co, ^{137}Cs, ^{134}Cs, and ^{3}H. They can be discharged into environments during the normal operating practice of nuclear power stations or by nuclear accidents. Application of radioactive fertilizers is another major contribution of radionuclide increase in soils *(9)*. A survey made on fertilizers produced in China showed that all phosphorus and potassium fertilizers contained natural radionuclides such as ^{238}U, ^{230}Th, and ^{40}K in varying amounts. Another contribution to the increase of radionuclides in soils is coal and coal-fired power stations because they discharge some radionuclides such as ^{230}Th, $^{232, 238}$U, ^{226}Ra, and ^{40}K. Mining also contributes a lot to the increase of radionuclides such as ^{226}Ra and $^{232, 238}$U in soil *(10)*.

The previously reviewed data show that soil contamination with inorganics is a serious problem facing China, particularly as it is reducing the yield and quality of crops and thus posing a great threat to the country's sustainable development. For example, it has been estimated that the contamination resulting from heavy metals causes annual reduction of agricultural crops of 100 million tons with 12 million tons of food being contaminated with heavy metals and economical loss up to 20 billion RMB Yuan. Soil contamination with heavy metals also reduces the quality of agricultural crops. The situation in some areas of the country is so bad that so-called "cadmium rice," "lead rice," or "copper rice" is produced. For example, the rice grown in soils around Shengyang City of the northeastern part of China usually contains 0.4–1.0 mg/kg Cd. In one of the counties of Jiangxi Province where 44% of the arable land is contaminated with cadmium, there are 670 ha producing "cadmium rice" *(11)*.

2. Field and Experimental Investigation of Heavy-Metal Accumulators and Hyperaccumulators in China

The study of uptake of heavy metals by plants started a long time ago. The early phase of investigation focused on uptake of heavy metals by agricultural crops *(12)*. In the early 1990s, with the introduction of the concept of green clean-up technologies to the country, studies of plant uptake of heavy metals in terms of phytoremediation began. Tang *(13)* first introduced the concept "hyperaccumulator" to his national colleagues, followed by the concept of "phytoremediation" *(14)*. Since then, dozens of review papers have been published on different levels of peer-reviewed journals focusing on the state-of-the-art phytoremediation research in China and abroad *(15–43)*, mainly focusing on summarizing the development of different aspects of phytoremediation.

Over the years, progress in the field of phytoremediation has been made in the following aspects (1) screening plants with a strong ability to take up large amount of heavy metals for phytoremediation purposes on the basis of field sampling and experimental data; (2) screening soil additives for enhancing metal bioavailability; (3) understanding how rhizospheric environments function, and how mycorrhiza affect plant uptake of heavy metals; and (4) CO_2-induced hyperaccumulation of zinc and copper in Indian mustard and sunflowers.

2.1. Screening Metal-Accumulating Species on the Basis of Field Sampling and Experimental Data

In conjunction with the field sampling and greenhouse experiments, Chinese scientists have conducted extensive investigations of potential heavy-metal-accumulating plants native to China. The goal of these studies was to identify plant species with a strong ability to take up heavy metals from artificial- and natural-contaminated soils. These investigations were based on the assumption that plants growing naturally in historically contaminated areas, such as mining spoils and soils adjacent to smelting operations found in China, may have adapted to high metal concentrations by absorbing the metals. Such field-sampling investigations conducted by Chinese scientists have characterized the distribution of accumulators and hyperaccumulators in wild contaminated environments and their occurrences in various mining habitats. For example, Tang et al. *(44,45)* reported three species *Elsholtzia haichowensis* Sun. (**Fig. 1**), *Commelina communis* Linn. (**Fig. 2**), and *Rumex acetosa* Linn. (**Fig. 3**) dominantly and vastly growing over copper-mining-spoil heaps and copper-contaminated soils in the areas along the middle and lower streams of the Yangtze River. These three species, representative of the low altitude copper flower association, could take up copper in their organs to various degrees. The highest concentration of copper was found in *R. acetosa* with the leaf copper concentration ranging from 340 to 1102 mg/kg and averaging 601 mg/kg (on a dry weight basis). *C. communis* also contained a high concentration of copper in its leaves ranging from 19 to 587 mg/kg and averaging 157 mg/kg. Field investigations made by Dr. Shu and his colleagues from the School of Life Science, Zhongshan University, showed that *C. communis* could accumulate copper in its aerial shoots up to 1000 mg/kg *(46)*. *E. haichowensis* has the lowest copper concentration in its leaves ranging from 18 to 391 mg/kg and averaging 102 mg/kg, but it contains quite a high concentration of copper in its roots. The ability to accumulate mixed heavy metals by *C. communis* and *E. haichowensis* using hydroponic and pot experiments was tested by Tang et al. *(47)*. *C. communis* was also found to hyperaccumulate chromium in hydroponic culture *(48)*. Tang et al. *(49)* reported another group of copper flowers growing on copper-mining spoil in Yunnan Province, southwest China. *Polygonum microcephalum* D. Don (**Fig. 4**) and *Rumex hastatus* D. Don (**Fig. 5**) were found to grow extensively on copper-mining spoils in Yunnan Province as representatives of typical high-altitude copper flowers. Analytical results showed that *P. microcephalum* has a high concentration of copper in the roots, stems, and leaves ranging from 36 to 2854 mg/kg, 14 to 244 mg/kg, and 9 to 332 mg/kg, respectively

Phytoremediation in China: Inorganics 357

Fig. 1. *Elsholtzia haichowensis* Sun., dominantly and vastly growing over the copper mining spoil heaps and copper-contaminated soil of the areas along the middle and lower streams of the Yangtze River. (Photo by Dr. Shirong Tang.)

(on a dry weight basis), and averaging 491 ± 782 mg/kg, 110 ± 72 mg/kg, and 133 ± 94 mg/kg, respectively, and that *R. hastatus* contained lower concentration of copper in its organs with the copper concentration ranging from 4 to 74 mg/kg,

Fig. 2. *Commelina communis* Linn., dominantly and vastly growing over the copper-mining spoil heaps and copper-contaminated soil of the areas along the middle and lower streams of the Yangtze River. (Photo by Dr. Shirong Tang.)

Fig. 3. *Rumex acetosa* Linn., dominantly and vastly growing over the copper-mining spoil heaps and copper-contaminated soil of the areas along the middle and lower streams of the Yangtze River. (Photo by Dr. Shirong Tang.)

Fig. 4. *Polygonum microcephalum* D. Don., dominantly and vastly growing over the copper-mining spoil heaps and copper-contaminated soils in high elevation areas in Yunnan Province, P. D. China. (Photo by Dr. Shirong Tang.)

Fig. 5. *Rumex hastatus* D. Don, dominantly and vastly growing over the copper-mining spoil heaps and copper-contaminated soils in high elevation areas in Yunnan Province, the People's Republic of China. (Photo by Dr. Shirong Tang.)

3 to 145 mg/kg, and 2 to 130 mg/kg, and averaging 33 ± 23 mg/kg, 42 ± 42 mg/kg, and 45 ± 32 mg/kg in the roots, stems, and leaves, respectively. The ability to accumulate Cu by three species *R. acetosa*, *P. microcephalum*, and *R. hastatus* was experimentally tested under different Cu treatments with seed germination and quartz sand culture *(50)*.

Wang and Shen *(51)* used hydroponic and pot cultures to investigate Cd uptake by mung bean (*Phaseolus aures* Poxb), cabbage (*Brassica campestris* ssp. Chinensis L. Mark), and wheat (*Triticum aestivum* L.). They found that mung bean had the lowest Cd concentration in the shoots and roots among the three species, and that applications of pig manure (30 g/kg soil) and lime (3 g/kg soil) could significantly decrease the concentration of Cd in mung bean shoots. Huang et al. *(52)* investigated the uptake of Cd, As, Pb, and Cr by 11 weeds growing near a smelter in Zhejiang Province, the People's Republic of China. They showed that *Poa annua*, *Salvia anthemifolia*, *Lepidium virginicum*, and *Bidens frondosa* had strong ability to take up copper, whereas *Plantago virfinica*, *S. anthemifolia*, and *Veronica perefrina* could take up large amounts of Cd and As. *S. anthemifolia* and *V. perefrina* also contained high concentrations of Pb. *V. perefrina* and *P. virfinica* had high uptake of Cr. Wang et al. *(53)* studied the differences in uptake of Cd by 13 varieties of oilseed rapes and concluded that some varieties, including Zhongyouza no. 1, Chuxuanxiaoyoucai, Maoanhuazhe, and Zhongyou no. 119, could hyperaccumulate Cd. It was also found that the potential for utilization of Zhongyouza no. 1 to remediate Cd-contaminated soils was higher than the model plant species India mustard. Yang et al. *(54)* reported a zinc accumulator *Sedum alfredii*. It was found that *S. alfredii* cannot only tolerate high zinc but also accumulate this element, with Zn concentration ranging from 4134 to 5000 mg/kg and averaging 4515 mg/kg. He et al. *(55)* showed that *S. alfredii* could take up large amounts of lead when grown in hydroponic culture supplied with different concentrations of $Pb(NO_3)_2$ and could be used as a potential plant species for phytoremediation of lead-contaminated soils. Xue et al. *(56)* reported a new Mn-hyperaccumulator plant called *Phytolacca acinosa* Roxb with leaf concentration of Mn up to 19,299 mg/kg dry weight. Wang et al. *(57)* reported that *Polygonum hydropiper* growing on contaminated soils in a sewage pond accumulated 1061 mg/kg of Zn in its shoots, and that *R. acetosa* L. growing near a smelter accumulated more than 900 mg/kg of Zn in both its shoots and roots. It was concluded that both species have some potential for phytoremediation of metal-contaminated sites. Li et al. *(58)* reported that two species from the Asteraceae growing vigorously on copper-mining spoils with large biomass, *Artemisia argi* Levl.et Vant and *Artemisia scoparia* Waldst.et Kit, contained high concentrations of copper in their organs, and both could have potential for phytoremediation. Su and Wong *(59)*, using pot culture methods to investigate uptake of Cd by mustard-type oilseed rapes

Table 3
Arsenic Bioconcentration and Translocation of *Pteris cretica* L. Under Field Conditions in Southern China's Hunan Province (61)[a]

Sample no.	As in soils (mg/kg)	As in plants (mg/kg) Fronds	Roots	BFs	TFs
0011SM01	299	694	552	2.32	1.28
0011SM03	261	560	215	2.15	2.60
0011SM15	123	338		2.75	
001SM21	39	258	184	6.62	1.40
001SM23	252	401	403	1.59	1.00
001SM26	131	635	277	4.85	2.29
001SM29	111	149	126	1.34	1.18
001SM30	124	307		2.48	

[a]BFs, ratio of As concentration in fronds to that in soils; TFs, ratio of As concentration in fronds to that in roots.

found that the variety Xikou Huazi has markedly high shoot biomass and Cd uptake, and concluded that this species could be used to remediate Cd-contaminated soils.

There are some reports showing that ferns can take up a suite of elements up to high concentrations. Chen et al. *(60)* showed that *Pteris vittata* L. endemic to China can hyperaccumulate As in above-ground tissues. This species contained As as high as 5070 mg/kg in the leaves under experimental pot culture. Another fern called Cretan Brake (*Pteris cretica* L.) was subsequently reported by Wei et al. *(61)* to have the ability to accumulate a large amount of As in its organs (**Table 3**). Besides uptake of As, some Chinese ferns have the ability to take up large amount of other trace elements. Wei et al. *(62)* investigating the uptake of rare Earth elements (REEs) by ferns concluded that *Dicranopteris linearis* was an important REE-accumulating plant. This species is widely distributed on acid soils to the south of the Yangtze River, China and naturally colonizes forest land repeatedly deforested or burned, abandoned farmlands, and mining areas, and often forms the dominant vegetation. The concentrations of REEs in *D. linearis* reported by Wei et al. *(62)* ranged from 134 to 1754 mg/kg in root, 107 to 632.9 mg/kg in stem, 51 to 102 mg/kg in stipes, and 977 to 2272 mg/kg in fronds. These are concentrations that in the case of other uncommon elements like Cd would define the plants as hyperaccumulators. Hong et al. *(63)* investigated the distribution pattern of La, Ce, Nd, Tb, Dy in *Dicranopteris dichotoma* and found that light and moderately heavy REEs were easily selectively absorbed and then accumulated by the fern.

There are some other plants in China that were found to have the ability to take up salts. Sheng et al. *(64)* reported some species from the Chenopodiaceae and

Table 4
Some Species Shown to Accumulate Salts in China *(64)*

	Cation (% wt)					Anion (% wt)		
Species	K^+	Na^+	Ca^{2+}	Mg^{2+}	Total	Cl^-	SO_4^{2-}	Total
Suaeda glauca	1.663	3.983	0.291	0.461	6.368	3.144	0.624	3.768
Suaeda corniculata	1.037	6.596	0.537	0.414	8.584	2.601	0.011	2.612
Kochia sieversiana	2.319	4.369	0.554	0.331	7.574	2.738	0.415	3.153
Polygonum sibiricum	1.893	2.120	0.941	0.425	5.379	1.728	0.062	1.790

Polygonaceae with strong ability to accumulate salts, such as *Polygonum sibiricum*, *Kochia sieversiana*, *Suaeda glauca*, and *Suaeda corniculata* (**Table 4**). It was concluded that these species could be used for phytoremediation of alkali salt-contaminated soils.

2.2. Achievement in Screening Soil Additives for Enhancement of Bioavailability

Some research in China during recent years was on the application of soil amendments to enhance phytoremediation efficiency *(65–70)*. Jiang et al. *(71)* investigating the role of EDTA in releasing Cd from the soil and transferring it into the shoots of Indian mustard, concluded that Cd concentration increased in soil solution after the addition of EDTA. Chen et al. *(72)* using pot culture methods investigated the potential of Indian mustard for phytoremediation of Pb-contaminated soil under the addition of EDTA. It was concluded that the addition of EDTA could enhance the uptake of Pb by Indian mustard, with the highest concentration of Pb in the shoot up to 1.4%. Wu et al. *(67)* reported that the addition of EDTA (3.15 mmol/kg) remarkably raised the water-soluble copper concentration from 0.18 mg/kg at control to 22.5 mg/kg at treatment 10 d before harvest. Sheng et al. *(70)* showed that the addition of EDTA and DTPA into the culture solution could remarkably decrease the uptake of Zn, Cu, and Mn by *Thlaspi caerulescens*. Similar results were reported by Zhang et al. *(68,69)*. The differences in responses of plants to chelating agents may reflect different mechanisms by which plants accumulate heavy metals, possibly related to the differences in plant genetic makeup. Wu et al. *(65)* showed that the mobilization of heavy metals in soils with chelating agents mainly started at the beginning of application. With time passing, heavy-metal concentrations in soil solution decreased sharply, being in agreement with the trend in EDTA-induced variation. They suggested that total organic carbon variation characteristics be used as an indicator of EDTA variation in soil solution. It was shown that EDTA degradation followed the equation: total organic carbon (mg/L) = $601.4e^{-0.0603t}$, when *t* equals days after EDTA application *(66)*.

2.3. Study of the Effect of Rhizospheric Environments and Mycorrhizae on Plant Uptake of Heavy Metals

Wei et al. *(73)* reviewed the state of the art in the field of roles of rhizosphere in remediation of contaminated soils and its mechanisms. There were a few papers published by Chinese scientists during recent years that dealt with how rhizospheric environments function and how mycorrhiza affect plant uptake of heavy metals *(74,75)*. Huang et al. *(74)* showed that inoculation of maize (*Zea mays* L.) with VA-mycorrhiza could increase exchangeable Cu in rhizospheric soil significantly, but decreased the exchangeable Cd compared with the control soil. They also found that in the rhizosphere of vesicular arbuscular (VA)-mycorrhizal-inoculated maize, the amounts of Cu, Zn, and Pb bound to organic matter were significantly higher than those in the rhizosphere of control maize, whereas the four tested metals bound to carbonates and to iron and manganese oxides were constant in the rhizosphere of mycorrhizal and nonmycorrhizal maize. Their results suggested that the plant roots could have greater influence on the distribution and dynamics of metal forms in the rhizosphere for mycorrhizal plants than for nonmycorrhizal plants. The study from Wu et al. *(75)* showed that the treatment with high concentration of Cd in the mixed contaminated soil has a negative inhibiting effect on the growth of bacteria and actinomycetes, whereas the addition of copper (250 mg/kg) could stimulate the growth of bacteria.

2.4. CO_2-Triggered Hyperaccumulation of Zn and Cu in Indian Mustard and Sunflowers and its Possible Application in Phytoremediation

One of the bottlenecks in phytoremediation that needs solving is how to enhance the uptake of metals by plants to increase absolute phytoremediation efficiency and, meanwhile, to increase the biomass production to increase relative phytoremediation efficiency *(21)*. A review of the literature showed that more than 400 plants have been identified to have the ability to uptake and absorb unusually large amounts of metals, but the majority of them have very low biomass production in their native habitats *(76,77)*. Thus, the possibility of increasing the plant biomass production, enhancing the uptake of metals by plants, while creating no secondary contamination has intrigued many scientists for many years. A survey of literature showed that three major ways were proposed to make a breakthrough in this regard: inoculation of micro-organisms *(78,79)*, application of soil amendments *(80–83)*, and transgenic plants *(76, 84–89)*. Although any of these techniques might help scientists achieve improvement of phytoremediation efficiency, each still has its disadvantages. For example, inoculation of plants with micro-organisms was reported to facilitate phytoremediation of heavy-metal contamination in some cases in a greenhouse study.

One disadvantage with this technique is that the inoculated micro-organisms cannot survive well in the field sites because of their weak competitiveness with native microbial communities for local niches. Traditional phytoremediation relies heavily on soil amendments/chelating agents to mobilize otherwise unavailable metals from contaminated soils. A problem with amendment/chelating agent application technology for phytoextraction is the concern over the potential effects of repeated application of amendments such as EDTA on the environment, including the toxicity of soil amendments to soil microbiota, leaching of the soil additives down to the underground water, and suddenly increased availability of essential nutrients that could create a toxic environment for plant growth. As for the use of transgenic plants to remediate heavy metal contaminated soil, there are potential problems with transgenic pollen and seed escape. Populating extensive areas of heavy-metal pollution with metal-accumulating transgenic trees could lead to widespread problems. Although the previously mentioned methods have proven to work to some extent in certain cases, there is still doubt about the contribution of any of the techniques to improve the efficiency of the designed phytoremediation systems.

Are there any other alternative ways to improve phytoremediation efficiency? Tang et al. *(77)* showed that enriching air with CO_2 could increase biomass production of Indian mustard (*Brassica juncea* [L.] Czern.) and sunflower (*Helianthus annuus* L.), improve copper tolerance of these metal-accumulating plant species to high levels of Cu, and trigger Cu hyperaccumulation in plants. This finding is of great significance. Tang et al. showed that all Indian mustard and sunflower seedlings growing at elevated CO_2 showed better growth than the CO_2 control ones, whereas those growing at ambient CO_2 level showed poorer growth at high levels of Cu, suggesting an improvement of growth following the application of CO_2 (**Fig. 6**). The biomass production of shoots increased at elevated CO_2 levels with the average dry shoot weight increases up to 200% for Indian mustard and sunflowers (**Table 5**). It was also found that with CO_2 enrichment in the air, Indian mustard and sunflower grew higher and larger, and had more and thicker leaves, and larger leaf areas compared to the plants growing under ambient CO_2 level. This is a significant finding, because the increase of plant biomass resulting from CO_2 application could suggest that more metals be taken up from the contaminated growth media and that the tolerance to metal toxicity be improved. Obviously, this could help metal accumulators survive on the metal stress conditions and shorten the time needed for clean up of the contaminated sites, and, therefore, increase relative phytoremediation efficiency.

More importantly, Tang et al. also found much more accumulation of Cu by Indian mustard and sunflower growing under elevated CO_2 levels than at ambient atmospheric CO_2 levels. All plants growing in pots treated with Cu and at enriched CO_2 levels exhibited hyperaccumulation of Cu, with Cu concentration

CO2   1200 µL L-1   800 µL L-1   350 µL L-1

Fig. 6. Increase of the biomass of sunflowers growing in pots treated with 200 mg copper soil/kg (dry weight) and under different levels of CO_2.

being more than 1000 mg/kg in the plant tissues on a dry weight basis (**Table 6**). The bioaccumulation factor, calculated as a ratio of Cu concentration in leaf to Cu concentration in soil, increased with increasing CO_2 (**Table 6**). CO_2 application also altered the leaf/root ratios of Cu in plants (**Table 6**). With increasing CO_2 levels in the growth chambers, plants exhibited a significant increase in the ratios. The changes of leaf/root ratios in plants with enriching CO_2 may suggest that the types of plant–soil relationship *(90)* alter, possibly from excluder to accumulator or even hyperaccumulator. This finding is of great significance, because that the leaf/root ratios of Indian mustard and sunflower increased with CO_2 enrichment may suggest translocation of more Cu from root to shoot when exposed to higher CO_2 levels.

3. The Study of Plant Uptake of Radionuclides in China

China has almost completed four nuclear power stations, including Qinshan Nuclear Power Station situated in Zhejiang Province, Dayawan Nuclear Power Station, Lingao Nuclear Power Station located in Guangdong Province, and Tianwan Nuclear Power Station located in Jiangsu Province. It is reported that there will be more than 30 nuclear power stations built by the year 2020. With rapid development of such nuclear technologies in China, environmental concern is increasing, especially the potential for contamination of environments by long-lived radionuclides such as ^{137}Cs and ^{90}Sr. In recent years, the Chinese government has paid more attention to the development of novel remediation technologies to avert risk to humans or the environment from

Table 5
Dry Weight (g/pot) of Indian Mustard and Sunflower Grown in Pots and Exposed to Different Levels of CO_2 (77)[a]

Plant species	Cu content (mg/kg)	CO_2 concentration (µL/L)	Root (average ± SD)	Shoot (average ± SD)
Sunflower	0	350	2.96 ± 0.44 A, a, a'	8.84 ± 0.53 A, a, a'
		800	3.63 ± 1.9 A, a, a'	9.01 ± 1.19 A, a, b'
		1200	3.04 ± 0.73 A, a, a'	9.28 ± 0.48 A, a, c'
	100	350	2.84 ± 1.46 A, b, a'	7.66 ± 3.4 A, b, a'
		800	2.23 ± 0.25 A, b, a'	8.11 ± 0.50 A, b, b'
		1200	2.51 ± 0.52 A, b, a'	9.29 ± 0.87 A, b, c'
	200	350	0.38 ± 0.23 A, b, a'	2.13 ± 0.86 A, c, a'
		800	0.36 ± 0.08 A, b, a'	3.14 ± 1.02 A, c, b'
		1200	0.66 ± 0.35 A, b, a'	4.30 ± 0.46 A, c, c'
Indian mustard	0	350	0.83 ± 0.82 B, a, a'	2.35 ± 0.07 B, a, a'
		800	0.81 ± 0.22 B, a, a'	2.49 ± 0.49 B, a, b'
		1200	0.73 ± 0.23 B, a, a'	3.61 ± 0.71 B, a, c'
	100	350	0.89 ± 0.96 B, b, a'	2.36 ± 1.57 B, b, a'
		800	0.83 ± 0.35 B, b, a'	3.24 ± 0.44 B, b, b'
		1200	0.85 ± 0.43 B, b, a'	4.42 ± 0.60 B, b, c'
	200	350	0.52 ± 0.12 B, c, a'	1.83 ± 1.03 B, c, a'
		800	0.43 ± 0.19 B, c, a'	1.94 ± 0.72 B, c, b'
		1200	0.47 ± 0.07 B, c, a'	2.81 ± 0.83 B, c, c'

[a]Within each column, values followed by the same letter are not significantly different as determined by Dunnett's test ($p < 0.005$) for all values. A, a, and a' represent plant species, copper treatments, and CO_2 treatments, respectively).

radionuclide contamination, and scientists are already aware of the potential value of phytoremediation for cleanup of sites contaminated with low levels of long half-life radionuclides. Despite this, little research on this technology has been done in the country so far compared with other countries in the field. Many plants with the ability to abnormally accumulate radiocesium have been well documented outside China (*91–104*) but only a few species have been reported within China (*105*). Zhu and Qiu (*106*) studied the uptake of ^{90}Sr by 10 species and ^{137}Cs by 7 species, respectively (**Table 7**). They showed that two species from the Cucurbitaceae family have the strongest ability to take up radiostrontium, followed by two species *Boehmeria nivea* (L.) Gaud and *Salsola collina* Pall. from the Urticaceae and Chenopodiaceae families, respectively. *B. campestris* L. and *Cucurbita moschata* Duch. Ex Poiret had the highest transfer factors among the species investigated (**Table 7**). Wang et al. (*107*) determined the concentrations of uranium and radium in four species

Table 6
Influence of CO_2 on Copper Concentration in Roots, Stems, and Leaves of Indian Mustard and Sunflower (77)[a]

Species	Cu added (mg/kg)	CO_2 (μL/L)	Leaf	Stem	Root	Leaf/root ratio	BF
Indian mustard	0	350	154 ± 44	245 ± 31	199 ± 23	0.49	7.0
		800	527 ± 125	621 ± 129	287 ± 52	2.56	101.3
		1200	186 ± 3	121 ± 46	197 ± 60	0.83	37.8
	100	350	423 ± 123	361 ± 171	765 ± 152	0.11	0.6
		800	4586 ± 263	13,696 ± 1853	2301 ± 1751	1.79	23.7
		1200	1587 ± 173	831 ± 175	672 ± 245	2.03	10.7
	200	350	538 ± 87	443 ± 187	977 ± 114	1.27	4.1
		800	2277 ± 325	1091 ± 282	2270 ± 166	2.24	10.3
		1200	1382 ± 236	957 ± 342	1362 ± 503	1.49	5.0
Sunflower	0	350	51 ± 10	42 ± 11	106 ± 25	0.78	21.1
		800	664 ± 214	506 ± 159	290 ± 73	1.84	71.9
		1200	277 ± 27	192 ± 96	333 ± 73	0.94	25.4
	100	350	60 ± 7	69 ± 16	557 ± 28	0.55	3.9
		800	2539 ± 1110	1401 ± 402	1418 ± 507	1.99	42.7
		1200	1567 ± 106	1031 ± 371	564 ± 117	2.36	14.8
	200	350	857 ± 297	558 ± 108	674 ± 51	0.55	2.6
		800	2143 ± 507	1433 ± 442	958 ± 12	1.00	11.0
		1200	1037 ± 149	957 ± 342	696 ± 183	1.01	6.7

[a] BF, copper concentration in leaf/copper concentration in soil. For the copper control pots, copper concentration in soil, copper concentration determined in soil. For the copper-treated pots, copper concentration in soil, the copper concentration and the copper added.

Table 7
^{90}Sr and ^{137}Cs Uptake by Selected Plant Species (106)[a]

Radionuclide	Speices	Family	TFs	URR (%)*
^{90}Sr				
	Triticum aestivum L.	Gramineae	1.2	0.8
	Crotalaria juncea L.	Fabaceae	5.6	4.8
	Cannabis sativa L.	Cannabinaceae	6.8	4.6
	Solanum melongena L. var. esculentum	Solanaceae	6.9	4.1
	Helianthus annuus L.	Asteraceae	7.2	4.8
	Amaranthus mangostanus L.	Amaranthaceae	9.3	8.0
	Salsola collina Pall.	Chenopodiaceae	9.4	11.4
	Boehmeria nivea (L.) Gaud	Urticaceae	9.6	12.0
	Cucumis sativus L.	Cucurbitaceae	13.7	8.5
	Cucurbita pepo L.	Cucurbitaceae	14.0	11.2
^{137}Cs				
	T. aestivum L.	Gramineae	0.05	0.03
	Astragalus adsurgens Pall.	Fabaceae	0.37	0.20
	Solanum tuberosum L.	Solanaceae	0.40	0.20
	Xanthium sibiricum Patrin.	Asteraceae	1.00	0.60
	Beta vulgaris L. var. lutea DC	Chenopodiaceae	0.24	0.70
	Cucurbita moschata Duch. Ex Poiret	Cucurbitaceae	1.88	1.50
	Brassica campestris L.	Brassicaceae	2.31	1.40

[a]URR, unit recovery rate, defined as percentage of radionuclides removed by plants grown in a unit area to the total amount of radionuclides applied to the soil.

Pteridium aquilinum, *Nerium indicum*, *Cynodon dactylon*, and *Eremochloa ophiuroides* growing in a uranium-tailing area at Hengyang, Hunan Province, central China. They found that all four species were tolerant plants with higher concentrations of uranium and radium in aerial parts than underground parts, and could be used as pioneering species for the remediation of uranium-mining spoils and abandoned mine (**Figs. 7** and **8**).

Tang and Willey (105) investigated the uptake of ^{134}Cs by *Lactuca sativa* L., *Silybum marianum* Gaertn., *Centaurea cyanus* L., *Carthamus tinctorius* L. from the Asteraceae, and *Beta vulgaris* L. var. Lutiancai, and *Beta vulgaris* L. var. Hongtiancai from the Chenopodiaceae grown in two widely distributed soils (a paddy soil and a red soil) in south China. The results showed that the plants growing on the paddy soil had a relatively high yield and low ^{134}Cs acitivty concentration, whereas those growing on the red soil showed the opposite trend. The accumulation of ^{134}Cs was dependent on plant species and soil types. For

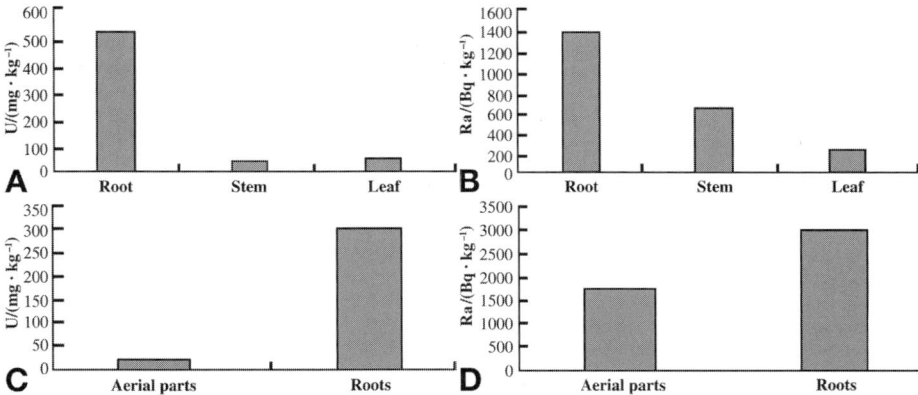

Fig. 7. Concentrations of uranium and radium in the organs of *Nerium indicum* (**A,B**) and *Pteridium aquilinum* (**C,D**) *(107)*.

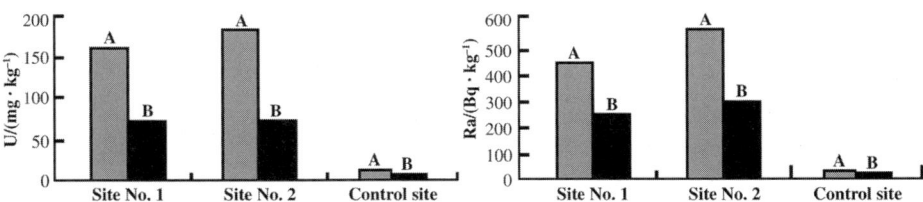

Fig. 8. Concentrations of uranium and radium in *Cynodon dactylon* (**A**) and *Eremochloa ophiuroides* (**B**) *(106)*.

the paddy soil, mean values for ^{134}Cs activity concentration were higher for the species of the Asteraceae (ranging from 165 to 185 Bq/g) than for those of the Chenopodiaceae (less than 140 Bq/g). For the red soil, *S. marianum* and *C. cyanus* of the Asteraceae had high average activity concentrations of ^{134}Cs ranging from 340 to 400 Bq/g, but *L. sativa* and *C. tinctorius* from the same family had low concentrations of ^{134}Cs ranging from 115 to 200 Bq/g on a dry weight basis. *B. vulgaris* L. var. Lutiancai and *Beta vulgaris* L. var. Hongtiancai accumulated from 120 to 231 Bq ^{134}Cs/g of plant shoot. The transfer factor values of ^{134}Cs for the studied species were in general higher in red soil than in paddy soil except for *C. tinctorius*. All plant species from the Asteraceae family growing on the paddy soil had higher transfer factors than the *B. vulgaris* species. *S. marianum*, and *C. cyanus* growing on the red soil had transfer factors >1, being much higher than the *B. vulgaris* species. They concluded that the plant species from the Asteraceae could accumulate a higher concentration of radiocesium than the *B. vulgaris* that has previously been suggested as a candidate for phytoremediation of radiocesium-contaminated soils.

Tang et al. *(108)* investigated the difference of two species with an extreme ability to take up potassium in response to soil additives. It was found that among the 26 soil amendments, $(NH_4)_2SO_4$ was found to be the most effective in desorbing ^{134}Cs from the investigated soil, and that the plant species showed different responses to the $(NH_4)_2SO_4$ addition compared with the control (ammonium free). $(NH_4)_2SO_4$ application decreased the uptake of ^{134}Cs by *A. tricolor* but increased the accumulation of ^{134}Cs by *A. cruentus* growing in pots treated with low-to-medium ^{134}Cs activity. $(NH_4)_2SO_4$ addition also increased the bioaccumulation ratios in *A. cruentus* and *A. tricolor* compared with the control, with the exception of the case where high ^{134}Cs was applied to the soil. Total ^{134}Cs removed and biomass for both species became less in the $(NH_4)_2SO_4$ treatments than in the no-$(NH_4)_2SO_4$ treatments. The results suggest that chemicals with the greatest ability to enhance the desorption of ^{134}Cs might play an unexpected role in transferring the ^{134}Cs to shoots.

Because China is abundant in plant taxa, it is unfortunate if there is little research on screening taxa for phytoremediation from Chinese-endemic species. Being aware of this situation, Dr. Shirong Tang and Dr. Neil Willey investigated uptake of radiocesium by 67 Chinese endemic species using substrate culture methods. It was found that there were dramatic differences in radiocesium-accumulating ability, with some species from the following six families Brassicaceae, Caryophyllaceae, Amaranthaceae, Polygonaceae, Asteraceae, and Phytolaccaceae able to accumulate a large amount of radiocesium from the growing substrate Fission's F2 (**Table 8**). These data are very valuable in screening of plant species for phytoremediation of sites contaminated with radiocesium, and physiological mechanisms for radiocesium accumulation.

4. Field Investigation of Phytoremediation of Environments Contaminated With Inorganic Contaminants in China

Little research on phytoremediation has been conducted on a field scale in the country. Dr. Chen and his colleagues, Institute of Geographical Sciences and Natural Resources Research, Chinese Academy of Sciences, successfully established a field trial where they grow *P. vittata* L. endemic to China to clean up As-contaminated sites near an As mine in south China. There has been no field study of phytoremediation of radionuclide-contaminated sites aimed to test its feasibility in China. However, in some uranium-mining districts, phytostablization technology has already been used to reduce the hazardous effects resulting from mining activities.

5. Brief History and Future Trends in Phytoremediation of Inorganic Contaminants

Research on the uptake of heavy metals by various plants started a long time ago. However, studies of plant uptake of heavy metals in terms of phyto-

Table 8
Radiocesium Concentration Activity in 67 Plant Species Native to China (Bq/g dry weight, according to Tang and Willey, unpublished data)

Species	^{137}Cs in shoots Average	SD	TFs Average	SD	Species	^{137}Cs in shoots Average	SD	TFs Average	SD
Brassica juncea Coss. Var. napiformis Pall. Et Bols	44	35	0.11	0.09	Chenopodium amaramtricolor	86	104	0.73	0.89
Brassica juncea Coss. var. tumida Tsen et Lee	74	110	0.19	0.28	Chenopodium quinoa	275	225	3.09	2.52
Brassica juncea Coss var. folia Bailey	208	160	1.76	1.36	Chenopodium quinoa	275	225	2.34	1.91
Brassic campestris L.	14	5	0.04	0.01	Chenopodium spp.	66	58	0.73	0.63
Melandrium apricum (Turcz.) Rohrb.	198	46	0.44	0.11	Beta vulgaris L. var. "Lutiancai"	538	157	5.6	1.63
Lychnis coronata Thunb	170	126	0.38	0.31	Beta	231	211	1.49	2.11
Lychnis senno Sieb. Et Zucc	68	24	0.17	0.06	Spinacia oleracea L.	25	6	0.21	0.05
Dianthus spp.	61	7	0.15	0.02	Chenopodium album Linn. Var. centroruburm Makino	259	118	2.83	1.29
Dianthus barbatus L.	124	108	0.32	0.28	Cassia occidentalis Linn.	28	5	0.29	0.05
Gypsophila oldhamiana Miq.	341	251	0.89	0.66	Astragalus sinicus L. var. "Zhezi No. 84"	31	17	0.32	0.17
Rumex hastatus	247	87	0.6	0.21	Astragalus sinicus L.var. "Zhezi No. 5"	21	14	0.22	0.15
Rumex patientia X.R. tianschanicus Rumex K-1	17	8	0.04	0.02	Astragalus sinicus L. var. "Changde species"	62	86	0.65	0.9
Polygonum microcephalum D. Don	711	116	1.57	0.28	Glycine max (L.) merr. Var. "Xiangxi No.3"	26	13	0.29	0.15
Gomphrea globosa L.var. alba	433	179	1.13	0.47	Glycine max (L.) merr. Var. "Xiangxi No.3"	26	13	0.22	0.11
Celosia argentea L.	214	311	0.54	0.78	Glycine max (L.) merr. Var. "Xiangxi No.119"	36	35	0.41	0.4
Gomphrea globosa L. cv. Alba	298	209	0.75	0.52	Glycine max (L.) merr. Var. "Xiangxi No.119"	36	35	0.31	0.3
Amaranthus paniculatas L.	544	281	1.37	0.71	Glycine max (L.) merr. Var. "Aijiaohan"	19	9	0.21	0.1
Amarantus tricolour L.	120	44	0.3	0.11	Glycine max (L.) merr. Var. "Aijiaohan"	19	9	0.16	0.07
Amaranthus cruentus L.	445	224	1.04	0.53	Aeschynomene	18	11	0.19	0.12
Cirsium japonicum (DC.) Maxim.	66	14	0.14	0.03	Nicotiana tabacum	218	113	2.09	1.09
Chrysanthemum coroarium L.	506	151	1.32	0.39	Nicotiana tabacum Samsum	374	121	3.6	1.17
Zinnia elegans Jacq.	231	54	0.54	0.13	Datura stramonium L.	402	374	3.41	3.17
Bidens pilosa L.	72	37	0.18	0.09	Nigella damascena L.	15	3	0.16	0.04
Calendula officinalis L.	61	17	0.15	0.04	Aquilegia viridiflora Pall	41	11	0.4	0.11
Centaurea cyanus L	30	5	0.07	0.01	Malva sinensis Cavan	47	12	0.53	0.14
Carthamus tinctorius L.	133	26	0.31	0.06	Malva sinensis Cavan	47	12	0.4	0.1
Silybum marianum Gaertn.	44	30	0.11	0.08	Celosia cristata L.	144	52	1.56	0.56
Lactuca sativa L.	70	34	0.18	0.09	Campanula ymanensis Hong	61	25	0.47	0.34
Pyrethrum pulchrum Ledeb	96	44	0.23	0.1	Pentapetes phoeniceae L.	63	9	0.68	0.09
Phytolacca acinosa Roxb	151	59	0.35	0.14	Lolium perenne L.	14	3	0.16	0.03
Artemisia annua L	75	30	0.2	0.08	Elsholtzia haichowensis Sun ex C.H. Hu	44	16	0.48	0.17
Bellis perennis L.	106	15	0.26	0.04	Salvia farinacea Benth.	52	20	0.57	0.21
Kniphofia varia Hook	56	21	0.12	0.05	Papaver rhoeas L.	35	20	0.38	0.22

remediation began in the early 1990s. Over the years China has developed some international reputation in phytoremediation and a lot of scientific publications in phytoremediation have resulted from research studies conducted by Chinese scientists, an accomplishment attracting the attention of scientists from many other countries in the world. Today, China is one of the few countries in the world investing much money to conduct research in phytoremediation. Because phytoremediation has many benefits over other traditional soil remediation techniques, the Chinese government has been paying more and more attention to environmental remediation in recent years, and more funding resources are available for phytoremediation research across the country. Chief among them are National Natural Science Foundation, the Ministry of Science and Technology, and the Ministry of Land and Resources. With support from different nonprofit organizations, more results will be expected in the coming years. Future studies will focus on the mechanisms, physiological or physiochemical, to better understand how plant species accumulate inorganics. Other interesting research aspects could be (1) further exploration for wild heavy-metal hyperaccumulators from field sites; (2) studies on the mechanisms physiological or physiochemical by which plant species accumulate or hyperaccumulate inorganics; (3) adjusting and controlling mechanisms that influence the bioavailibity of inorganics through soil, soil chemistry, and agrotechnical engineerings; (4) rhizospheric microbial characteristics and their roles in increasing the bioavailability of heavy metals; and (5) application of advanced analytical technologies such as Proton Microprobe, X-ray fluorescence, and X-ray absorption spectroscopy to phytoremediation research.

Because China is one of the biggest countries in the world with a variety of geographical and climatic conditions that need to be remediated, a variety of plant species should be explored for phytoremediation. Plant species native to the local area being restored are most desirable and, therefore, more work is needed to be done on screening species with a strong ability to survive under different geographical and climatic conditions. More information is also needed regarding utilization of endemic species for the phytoremediation purpose. To achieve this, more research should be conducted in this country on screening plant species native to the local contaminated sites in terms of field sampling and phylogenetic characteristics. Results from field sampling studies need to be tested in the greenhouse and field environments to determine their effectiveness.

Acknowledgments

This work was funded by the Ministry of Science and Technology, P. R. China (Grant Number: NKBRSFG 1999011808) and by a Leverhulme Trust, UK, Research Fellowship.

References

1. Dong, Y. H. and Zhang, T. L. (2003) Sustainable management of soil resources for food safety. *Soils* **35**, 182–186.
2. Wang, Y. F., Dong, X., Wang, L., and Han, D. (2003) Prerequisition, current situation and countermeasures for the development of nuisanceless agriculture in Liaonin Province. *Chin. J. Soil Sci.* **34**, 370–373.
3. Jin, L. P. (2001) Twenty percent of arable lands contaminated with three industrial wastes in Zhejiang Province. Zhejiang Daily News of Science and Technology, Nov. 28: A1.
4. Zhao, L. F., Huang, P. W., Zhang, Z. X., and Xie, D. K. (2000) The current state of the arable lands and water resources in Leqing City and improvement strategies. *Zhejiang Agricultural Science* **1**, 25–26.
5. Li, Q. L. and Wang, Y. (2000) Variation of heavy metal concentrations in the vegetable grown in the vegetable producing bases of the Zhongqing City suburbs. *Rural Environ. Dev.* **17**, 42–44.
6. Yang, X. E., Yu, J. D., Nie, W. Z., and Zhu, C. (2002) Quality of agricultural environmental and agricultural food safety. *Rev. Agric. Sci. Technol.* **4**, 3–9.
7. Xie, J. Y. (2002) Investigation and evaluation of the current state of soil contamination by heavy metals in Baoding, Hebei Province. *Bulletin of Hebei Agricultural University* **25**, 38–41.
8. Zhou, Z. Y. (1998) Study of factors resulting in contamination with nitrite and nitrate in Chinese vegetables, and countermeasure strategies. *Environ. Sci. Prog.* **7**, 1–3.
9. Tang, S. R. (2002) Bioremediation of low-level radionuclides in soil-water substrates. *Chin. J. Appl. Ecol.* **13**, 243–246.
10. National Environmental Protection Bureau (NEPB) (1996) *Remediation of Contamination Resulting from Uranium Mining and Refinery*. Chinese Environmental Science Press, Beijing, China.
11. Chen, T. B. (1998) More attention should be paid to soil contamination in China. Daily News of Science and Technology, Dec. 22: A3.
12. Liao, Z. J. (1992) *Environmental Chemistry and Biological Effects of Trace Elements*. China Environmental Science Press, Beijing, P. R. China.
13. Tang, S. R. (1996) Hyperaccumulators. *Agri. Environ. Dev.* **3**, 14–18.
14. Tang, S. R., Huang, C. Y., and Zhu, Z. X. (1996) Using plants to remediate heavy metal contaminated soils. *Adv. Environ. Sci.* **4**, 10–15.
15. Dai, S. G., Liu, X. Q., and Xu, H. (1998) Progress on phytoremediation of contaminated soils. *Shanghai Environ. Sci.* **17**, 25–31.
16. Gong, Y. H., Wang, J. R., and Gao, J. F. (1998) Phytoremediation and its application to environmental protection. *Agro-Environ. Protec.* **17**, 268–270.
17. Shen, D. Z. (1998) Phytoremediation of contaminated soil. *Chin. J. Ecol.* **17**, 59–64.
18. Cheng, Y. C. (1999) Bioremediation of contaminated soils. *Adv. Environ. Sci.* **6**, 7–11.
19. Luo, Y. M. (1999) Phytoremediation of heavy metal contaminated soils. *Soil* **5**, 261–265.

20. Sang, W. L. and Kong, F. X. (1999) Advances in research on phytoremediation. *Adv. Environ. Sci.* **7,** 40–44.
21. Tang, S. R. and Wilke, B. M. (1999) Phytoremediation and agrobiological environmental engineering. *Trans. Chin. Soc. Agric. Eng.* **15,** 21–26.
22. Shen, Z. G. and Chen, H. M. (2000) Bioremediation of heavy metal polluted soils. *Rural Eco-Environ.* **16,** 39–44.
23. Wang, X. C., Shi, W. M., and Cao, Z. H. (2000) Phytoremediation of heavy metals in soil—a green and clean technique. *Acta Agric. Nucl. Sinica* **14,** 315–320.
24. Zhao, A. F., Zhao, X., and Chang, X. L. (2000) Advances in research on phytoremediation of contaminated soil. *Chin. J. Soil Sci.* **31,** 43–46.
25. Zhao, Z. Q., Niu, J. F., and Quan, X. (2000b) Progress in phytoremediation of toxic metals from the environment. *Research of Environ. Sci.* **13,** 54–57.
26. Chu, G. X. and Ren, G. (2001) Advances in phytoremediation and soil pollution by heavy metals. *J. Shihezi University (Natural Sci.)* **5,** 342–346.
27. Liu, X. M., Nie, J. H., and Wang, Q. R. (2001) Advances in research on phytoremediation of heavy metal contaminated soil. *J. Gansu Agric. Uni.* **36,** 8–13.
28. Peng, Z. R., Wang, Y. F., and Xu, B. X. (2001) Chelate-induced phytoremediation of contaminated soils by heavy metals. *Shanghai Chemistry* **17,** 4–7.
29. Tang, S. R. (2001) Distribution of hyperaccumulators in genera and family as well as at time and space. *Rural Eco-Environ.* **17,** 12–16.
30. Wei, C. Y. and Chen, T. B. (2001) Hyperaccumulators and phytoremediation of heavy metal contaminated soil: a review of studies in China and abroad. *Acta Ecol. Sinica* **21,** 1196–1203.
31. Zhou, Q. X. and Song, Y. F. (2001) Technological implication of phytoremediation and its application in environmental protection. *J. Safety Environ.* **13,** 48–53.
32. Zhong, Z. K. and Gao, Z. H. (2001) The mechanism of phytoremediation and its application prospect. *World Fores. Res.* **14,** 23–28.
33. Jiao, F. C., Mao, X., and Li, R. Z. (2002) Strategies and application of clean-up of environmental pollutants by plants. *Agro-Environ. Protec.* **21,** 281–284.
34. Leng, J., Jie, Y. C., and Xu, Y. (2002) The state of the art in utilization of plants to clean up heavy metal contaminated soils and a future development. *Chin. J. Soil Sci.* **33,** 467–470.
35. Liu, G. H. and Shu, H. L. (2002) Phytoremediation of soils contaminated with heavy metals. *Jiangxi Forestry Sci.* **2,** 30–31.
36. Wei, C. Y. and Chen, T. B. (2002) The state of the art in research and application of phytoremediation of heavy metal contaminated soils. *Adv. Earth Sci.* **17,** 833–839.
37. Wu, Z. H. (2002) Advances in phytoremediation of soils contaminated with heavy metals. *J. Yancheng Institute Technol.* **15,** 53–57.
38. Wu, Z. H., Zhang, Y. F., Wang, X. R., and Hu, X. (2002) Application of gene technology in phytoremediation fro contaminated soil by heavy metals. *Agro-Environ. Protec.* **21,** 84–86.
39. Zhou, N. Y. and Wang, R. W. (2002) Phytoremediation—new approach of heavy metal cleanup from heavy metal-polluted soils. *J. Chin. Biotech.* **22,** 53–57.

40. Zhang, K. S. and Liang, J. D. (2002) Roles of rhizosphere in remediation of contaminated soils and its mechanisms. *Chin. J. Appl. Ecol* **14**, 143–147.
41. Zhou, G. H., Huang, H. Z., and He, H. L. (2002) Phytoremediation: a new approach for the remediation of heavy metal-contaminated soils. *Techniq. Equip. Environ. Pollut. Control* **3**, 33–39.
42. Fang, X. H. and Qiu, R. L. (2003) Advaces on study of the role of organic chelators on phytoremediation of nickel-contaminated soil. *Techniq. Equip. Environ. Pollut. Control* **3**, 1–5.
43. Li, F., Zang S., and Luo, Y. (2003) Bioremediation of contaminated soils: a review. *Chin. J. Ecology* **22**, 35–39.
44. Tang, S. R., Huang, C. Y., and Zhu, Z. X. (1997) *Commelina communis* L: copper hyperaccumulator found at Tongling city, Anhui province of China. *Pedosphere* **24**, 10–11.
45. Tang, S. R., Wilke, B. M., and Huang, C. Y. (1999) The uptake of copper by plants dominantly growing on copper mining spoils along the Yangtze River, the People's Republic of China. *Plant Soil* **209**, 225–232.
46. Shu, W. S., Yang, K. Y., Zhang, Z. Q., Yang, B., and Lan, C. Y. (2001) Flora and heavy metals in dominant plants growing on an ancient copper spoil heap on Tonglushan in Hubei Province, China. *Chin. J. Appli. Environ. Biol.* **7**, 7–12.
47. Tang, S. R., Wilke, B. M., and Brooks, R. R. (2001) Heavy-metal uptake by metal-tolerant *Elsholtzia haichowensis* and *Commelina communis* from China. *Comm. Soil Sci. and Plant Anal.* **32**, 895–906.
48. Tang, S. R. and Xi, L. (2002) Accumulation of chromium by *Commelina communis* L. grown in solution supplied with different concentrations of Cr and L-histidine. *Bullet. Zhejiang University (Sciences)* **32**, 232–236.
49. Tang, S. R. and Fang Y. H. (2001) Copper accumulation by *Polygonum microcephalum D. Don* and *Rumex hastatus D.Don* from copper mining spoils in Yunnan Province, P. R. China. *Environ. Geol.* **40**, 902–907.
50. Li, H. Y., Tang, S. R., and Zheng, J. M. (2005) Copper tolerance and accumulation Rumex acetosa Linn., Polygonum microcephalum D. Don, and Rumex hastatus D. Don. Bulletin of Science and Technology, **21(4)**: 480–484 (In Chinese with English summary).
51. Wang, C. C. and Shen, Z. G. (2001) Uptake of Cd by three species of plants and responses of mung bean to Cd toxicity. *J. Nanjing Agric. Uni.* **24**, 9–13.
52. Huang, C. B., Guo, S. L., Chen, X. M., and Huang, P. Y. (2001) Absorption and accumulation of four heavy metals by eleven weeds in Jinhua, Zhejiang. *Agro-Environ. Protec.* **20**, 225–228.
53. Wang, J. Q., Liu, B., and Su, D. C. (2003) Selection of oilseed rapes as a hyperaccumulator cadmium. *J. Agric. Uni. Hebei* **26**, 13–16.
54. Yang, X. E., Long, X. X., Nie, W. Z., and Fu, C. X. (2002) *Sedum alfredii*: a new zinc hyperaccumulator. *Sci. Bulletin* **47**, 1003–1006.
55. He, B., Yang, X. E., Ni, W. Z., Wei, Y. Z., Long, X. X., and Ye, Z. Q. (2002) *Sedum alfredii*: a new lead-accumulating ecotype. *Acta Botanica Sinica* **44**, 1365–1370.

56. Xue, S. G., Chen, Y. X., Lin, Q., Xu, S., and Wang, Y. P. (2003) *Phytolacca acinosa* Roxb. (Phytolaccaceae): a new manganese hyperaccumulator plant from Southern China. *Acta Ecologica Sinica* **23,** 935–937.
57. Wang, Q. R., Cui, Y. S., Liu, X. M., Dong, Y. T., and Christie, P. (2003) Soil contamination and plant uptake of heavy metals at polluted sites in China. *J. Environ. Sci. Health Part A Tox. Hazard Subst. Environ. Eng.* **38,** 823–838.
58. Li, H. Y., Tang, S. R., and Zheng, J. M. (2003) Copper contents in two species plants of Compositae growing on copper mining spoils. *Rural Eco-Environ.* **19,** 53–55.
59. Su, D. C. and Wong, J. W. C. (2002) The phytoremediation potential of oilseed rape (*B. juncea*) as a hyperaccumulator for cadmium contaminated soil. *China Environ. Sci.* **22,** 48–51.
60. Chen, T. B., Wei, C. Y., Huang, Z. C., Huang, Q. F., Lu, Q. G., and Fan, J. L. (2002) Arsenic hyperaccumulator *Pteris vittata* L. and its accumumulation characteristics of arsenic. *Sci. Bullet.* **47,** 207–210.
61. Wei, C. Y., Chen, T. B., Huang, Z. C., and Zhang, X. Q. (2002) Cretan Brake (*Pteris cretica* L.): an arsenic-accumulating plant. *Acta Ecologic Sinica* **22,** 777–778.
62. Wei, Z. G., Yin, M., Zhang, X., et al. (2001) Rare earth elements in naturally grown fern *Dicranopteris linearis* in relation to their variation in soils in South-Jiangxi region (Southern China). *Environ. Pollut.* **114,** 345–355.
63. Hong, F. S., Wei, Z. G., Tao, Y., et al. (1999) Distribution of rare earth elements and structure characterization of chlorophyll-lanthanum in a natural plant fern *Dicranopteris dichotoma*. *Acta Botanica Sinica* **41,** 851–854.
64. Sheng, L. X., Ma, X. F., and Wang, Z. P. (2002) Study on the recovery and control of the alkili-saline lands in Songnen Plain. *J. Northeast Normal University* **34,** 30–35.
65. Wu, L. H., Luo, Y. M., and Wang, H. Z. (2001) Cheleting agents induced phytoremediation of copper contaminated dry red soil. *Chin. J. Appl. Ecol.* **12,** 435–438.
66. Wu, L. H., Luo, Y. M., and Zhang, H. B. (2001) Study of environmental risk with utilization of organic chelating agents to enhance phytoremediation I Efftect of EDTA on TOC in mixed contaminated soil and dynamic variations of heavy metals. *Soil* **33,** 189–192.
67. Wu, L. H., Luo, Y. M., and Lu, R. H. (2000) Study of organic adjustment for phytoremediation of copper contaminated soils II Mobilization of rhizospheric soil copper by organic chemicals. *Soil* **32,** 67–70.
68. Zhang, J. S., Li, H. F., and Yi, C. J. (1999) Effect of organic acids on uptake of cadmium by rice. *Agro-Environ. Protection* **18,** 278–280.
69. Zhang, J. S., Li, H. F., and Yi, C. J. (1999) Effect of organic acids on mobilization of cadmium in soil and uptake of cadmium by wheat. *Acta Pedologica Sinica* **36,** 61–66.
70. Sheng, Z. G., Liu, Y. L., and Cheng, H. M. (1998) Effect of chelating agents on uptake of zinc, copper, manganese, and iron by heavy metal hyperaccumulator *Thlaspi caerulescens*. *Acta Phytophysiol. Sinica* **24,** 340–346.

71. Jiang, X. J., Luo, Y. M., Zhao, Q. G., and Ge, Y. Y. (2003) The role of EDTA in Cd absorption and translocation by Indian mustard. *Acta Pedologica Sinica* **40,** 205–209.
72. Chen, Y. H., Li, X. D., Liu, H. Y., and Shen Z. G. (2002) The potential of Indian mustard (*Brassica juncea* L.) for phytoremediation of Pb-contaminated soils with the aid of EDTA addition. *J. Nanjing Agric. Uni.* **25,** 15–18.
73. Wei, S. H., Zhou, Q. X., Zhang, K. S., and Liang, J. D. (2002) Roles of rhizosphere in remediation of contaminated soils and its mechanisms. *Chin. J. Appl. Ecol.* **14,** 143–147.
74. Huang, Y., Chen, Y., and Tao, S. (2000) Effect of rhizospheric environment of VA-mycorrhizal plants on forms of Cu, Zn, Pb and Cd in polluted soil. *Chin. J. Appl. Ecol.* **11,** 431–434.
75. Wu, S. C., Luo, Y. M., Jiang, X. J., et al. (2000) Study of phytoremediation of heavy metal contaminated soils. *Soil* **2,** 75–78.
76. Karenlampi, S., Schat, H., Vangronsveld, J., et al. (2000) Genetic engineering in the improvement of plants for phytoremediation of metal polluted soils. *Environ. Pollut.* **107,** 225–231.
77. Tang, S. R., Xi, L., Zheng, J. M., and Li, H. Y. (2003) The responses of Indian mustard and sunflower growing on copper contaminated soil to elevated CO_2. *Bull. Environ. Contam. Toxicol.* **71,** 988–997.
78. Entry, J. A., Vance, N. C., and Hamilton M. A. (1996) Phytoremediation of soil contaminated with low concentrations of radionuclides. *Water, Air, Soil Pollut.* **88,** 167–176.
79. Khan, A. G., Kuek, C., and Chaudhry, T. M. (2000) Role of plants, mycorrhizae and phytochelators in heavy metal contaminated land remediation. *Chemosphere* **41,** 197–207.
80. Huang, J. W., Chen, J., Berti, W. R., and Cunningham, S. D. (1997) Phytoremediation of lead-contaminated soils: role of synthetic chelates in lead phytoextraction. *Environ. Sci. Technol.* **31,** 800–805.
81. Huang, J. W., Chen, J., and Cunningham, S. D. (1997) Phytoextraction of lead from contaminated soils. In: *Phytoremediation of Soil and Water Contaminants,* (Kruger, E. I., Anderson, T. A., and Coats, J. R., eds.), ACS Symposium Series No. 664. American Chemical Society, Washington, DC, pp. 283–297.
82. Anderson, C. W. N., Brooks, R. R., Stewart, R. B., and Simcock, R. (1998) Harvesting a crop of gold in plants. *Nature* **395,** 553–554.
83. Blaylock, M. J., Salt, D. E., Dushenkov, S., et al. (1997) Enhanced accumulation of Pb in Indian mustard by soil-applied chelating agents. *Environ. Sci. Technol.* **31,** 860–865.
84. Clemens, S., Palmgren, M. G., and Kramer, U. (2002) A long way ahead: understanding and engineering plant metal accumulation. *Trends Plant Sci.* **7,** 309–315.
85. Kramer, U. and Chardonnens, A. (2001) The use of transgenic plants in the bioremediation of soils contaminated with trace elements. *Appl Microbiol Biotechnol.* **55,** 661–672.

86. Bizily, S. P., Rugh, C. L., and Meagher, R. B. (2000) Phytodetoxification of hazardous organomercurials by genetically engineered plants. *Nature Biotech.* **18,** 213–217.
87. Pilon-Smits, E. and Pilon, M. (2000) Breeding mercury-breathing plants for environmental cleanup. *Trends Plant Sci.* **5,** 235–236.
88. Rugh, C. L., Senecoff, J., Meagher, R. B., and Merkle, S. A. (1998) Development of transgenic yellow poplar for mercury phytoremediation. *Nature Biotech.* **16,** 925–928.
89. Black, H. (1995) Absorbing possibilities: phytoremediation. *Environ. Health. Perspec.* **103,** 1106–1108.
90. Baker, A. J. M. (1981) Accumulators and excluders-strategies in the response of plants to heavy metals. *J. Plant Nutr.* **3,** 643–654.
91. Negri, M. C. and Hinchman, R. R. (2000) The use of plants for the treatment of radionuclides. In: *Phytoremediation of Toxic Metals: Using Plants to Clean Up the Environment,* (Raskin, I. and Ensley, B., eds.), John Willey and Sons, New York, NY, pp. 107–132.
92. Dushenkov, S., Mikheev, A., Prokhnevsky, A., Ruchko, M., and Sorochinsky, B. (1999) Phytoremediation of radiocesium-contaminated soil in the vicinity of Chemobyl, Ukraine. *Environ. Sci. Technol.* **33,** 469–475.
93. Entry, J. A., Vance, N. C., Hamilton, M. A., Zabowski, D., Watrud, L. S., and Adriano, D. C. (1996) Phytoremediation of soil contaminated with low concentrations of radionuclides. *Water, Air, Soil Pollut.* **88,** 167–176.
94. Broadley, M. R. and Willey, N. J. (1997) Differences in root uptake of radiocesium by 30 plant taxa. *Environ. Pollut.* **97,** 11–15.
95. Buysse, J., van Den-Brande, K., and Merckx, R. (1996) Genotypic differences in the distribution of radiocaesium in plants. *Plant Soil* **178,** 265–271.
96. Entry, J. A., Rygiewicz, P. T., and Emmingham, W. H. (1993) Accumulation of cesium 137 and strontium-90 in Ponderosa and Monterey pine seedlings. *J. Environ. Qual.* **22,** 742–745.
97. Lasat, M. M., Fuhrmann, M., Ebbs, S. D., Cornish, J. E., and Kochian, L. V. (1998) Phytoremediation of a radiocesium-contaminated soils: evaluation of cesium-137 bioaccumulation in the shoots of three plant species. *J. Environ. Qual.* **27,** 165–169.
98. Lasat, M. M., Norvell, W. A., and Kochian, L. V. (1997) Potential for phytoextraction of ^{137}Cs from a contaminated soil. *Plant Soil* **195,** 99–106.
99. Salt, C. A. and Mayes, R. W. (1991) Seasonal variations in radiocaesium uptake by reseeded hill pasture grazed at different intensities by sheep. *J. of App. Ecol.* **28,** 947–962.
100. Salt, C. and Mayes, R. B. (1990) Seasonal patterns of ^{134}Cs uptake into hill pasture vegetation. In: *Transfer of Radionuclides in Natural and Semi-natural Environments,* (Desmet, G., Nassimbeni, P., and Belli, M., eds.), Elsevier Applied Science, London, UK, pp. 334–340.
101. Mascanzoni, D. (1990) Uptake of ^{90}Sr and ^{137}Cs by mushroom following the Chernobyl accident. In: *Transfer of Radionuclides in Natural and Semi-natural*

Environments, (Desmet, G., Nassimbeni, P., and Belli, M., eds.), Elsevier Applied Science, London, UK, pp. 459–467.
102. Sawidis, T. (1988) Uptake of radionuclides by plants after the Chernobyl accident. *Environ. Pollut.* **50,** 317–324.
103. Coughtrey, P. J., Jackson, D., and Thorne, M. C. (eds.) (1983) *Radionuclide Distribution and Transport in Terrestrial and Aquatic Ecosystems, Vol. 1.* A. A. Balkema, Rotterdam (for CEC), pp. 1–496.
104. Wallace, A. and Romney, E. M. (1972) *Radioecology and Ecophysiology of Desert Plants at the Nevada Test Site.* Environmental Radiation Division, Laboratory of Nuclear Medicine University of California, Riverside, CA, p. 432.
105. Tang, S. R. and Willey, N. J. (2003) Uptake of ^{134}Cs by four species from the Asteraceae and two species from the Chenopodiaceae grown in two types of Chinese soil. *Plant Soil* **250,** 75–81.
106. Zhu, Y. Y. and Qiu, T. C. (1991) The behaviors of the fission products ^{90}Sr, ^{137}Cs, and ^{144}Ce in the soil-plant system. *China Environ. Sci.* **11,** 266–269.
107. Wang, R. L., Yi, S., Cheng, K. G., and Zhang, X. T. (2002) Uranium and radium accumulation in *Pteridium aquilinum, Nerium indicum, Cynodon dactylon,* and *Eremochloa ophiuroides. J. Xiangtan Normal University (Natural Science Edition)* **24,** 73–77.
108. Tang, S. R., Chen, Z. Y., Li, H. Y., and Zheng, J. M. (2003) Uptake of ^{134}Cs in the shoots of *Amaranthus tricolor* and *Amaranthus cruentus. Environ. Pollut.* **125,** 305–312.

26

Phytoremediation in China

Organics

Shirong Tang and Cehui Mo

Summary

During recent decades contamination of ecosystems by synthetic organic compounds has increased tremendously in China, and now poses a major environmental and human health problem. As in other countries, China so far has not found any effective ways to solve this problem of soil and water contamination by organics. More attention was paid to inorganic pollutants than organic contaminants in the country during the past years but with progress in international cooperation, there is an increasing awareness of the seriousness of environmental contamination by different kinds of organic compounds among Chinese scientists. They are trying, as their international counterparts are, to explore some cost-effective technologies to cope with the problem because they have already been aware of the inadequateness of traditionally used technologies to treat soils contaminated with organic pollutants. Here, we focus both on describing soil and water contaminated with organics in China, and various phytoremediation research activities conducted in the country, including research on plant uptake of organic pollutants for the remediation of contaminated sites. A brief discussion of the application of this approach at a bench- and field- scale is included.

Key Words: Phytoremediation; organic contaminants; pesticides; contaminated soil.

1. Introduction

It has been recognized in China that, as a complement to traditional methods, the currently emerging phytoremediation technology is an alternative way to solve the problem of contamination with synthetic organics because this green technology has many advantages over other remediation techniques. It cannot only inhibit leaching of the organic pollutants but can also metabolize the pollutants into a form that is either not available for the food web or nontoxic.

There have been several important reviews in the Chinese literature in recent years, mostly focusing on understanding and evaluating phytoremediation potentials and techniques, and covering the research on organics conducted outside China *(1–7)*. These reviews of phytoremediation of organic pollutant-contaminated environments are mainly summaries with little critical analysis.

Much research work has been done during the past decade on the use of plants for the removal of organic pollutants and heavy metals from spillage sites, sewage waters, sludges, and polluted areas on bench and field scales outside China *(8–10)*. In contrast, little research into plant uptake of organic pollutants has been conducted within China. It seems that there have been no successful case reports so far in China of phytoremediation of organic-contaminated soils on a field scale. On the other hand, the number of contaminated sites resulting from industrial, village, and town enterprise activities in the country is increasing drastically, which presents a challenge to Chinese scientists involved in research on phytoremediation of organics-contaminated sites.

2. General Review of Water and Soils in China Contaminated With Organics

2.1. The Contaminants

Contamination of the environment by organics in China is serious and has become a major concern in the country, with pollutants being varied and widespread. The situation is worse in coastal areas where industries and intensive agricultural systems are prosperous with a lot of township workshops having produced large amounts of organic pollutants. Wenzhou, a city of Zhejiang Province, in the southeastern part of China from where the Chinese private township enterprises originated, is one of the typical examples. It is considered to be the capital of artificial leathers in the world. The township workshops making artificial leathers are prosperous, and produce large amounts of organic pollutants that have caused serious contamination of the local air, water, and soils.

It has been estimated that more than 200 varieties of organics can be identified in the environment of the country. Many of them are on the US Environmental Protection Agency (EPA) top contaminant list. One of the characteristics with organic contamination in China is that low level but highly toxic organic pollutants in the environment in this country have been dramatically increasing and has become an environmental concern. Typical representatives of organic pollutants include benzene, toluene, dimethyl benzene and ethylbenzene, and pesticides mainly derived from automobile off-gas, coal burning, industrial and life rubbish, street barbecues, agricultural activities, and so on. Long-term exposure to these organic pollutants may influence humans genes and cause serious diseases. Another characteristic of contamination with organic pollutants such as pesticides, oils and their products, solid waste materials and their leachates in China is that the pollutants are massive and regional. There are many organic-contamination sources in this country.

A major source of organic contaminants is pesticides. China is one of the biggest pesticide producers and consumers in the world, with total production second only to the United States. China has more than 2000 pesticide manufacturers with the registered number of products being more than 1500, with annual production of raw pesticides adding up to 40,000 tons and pesticide preparation being more than 1 million tons. Estimation showed that about 80% of the pesticides applied have entered the environment and become organic pollutants of environmental concern *(11)*. The main problems associated with utilization of pesticides in China includes utilization of irrational varieties of pesticides and a large proportion of highly toxic pesticides and preparations, the misuse of methods, all causing serious environmental consequences such as considerable pesticide residues in agricultural products, and human beings and animals being poisoned with pesticides.

Another important contamination source in China is oils *(12)*. This group of organic pollutants has caused serious soil and water contamination (including surface, underground water, and oceanic water contamination). Industrial "three wastes" were also considered to be another important organic-contamination source *(11)*.

2.2. Contaminated Water

Waters contaminated with organics in China include water springs, tap water, municipal streams, natural rivers, and some lakes. The organics found in some water sources and tap water are varied and abundant, posing a serious threat to the local residents' health. Taking source water in Nanjing, Jiangsu Province as an example, research showed that many organic pollutants on the US EPA list of the top 129 contaminants were found in the water samples, suggesting that the source water was polluted to some extent. Another example is the water of the Danjiangkou Reservoir where a lot of trace organic pollutants were identified, including hydrocarbon ring compounds, carboxylic acid and its derivatives, heterocyclic compounds, Polycyclic aromatic hydrocarbons (PAHs), ethanol, ether, and phenol *(13)*. Tian et al. *(14)* reported the identification of more than 100 kinds of organic compounds in source water, with 60 kinds found in the Jialing River and 50 in the Yangtze River, including phthalic acid esters, keton, phenol, benzene, and derivatives.

According to statistics published by the Chinese government in 2000, sections of municipal streams passing through some cities have been contaminated to a various extent. Major sources that cause contamination include residents' daily life wastewater and industrial discharge. Water samples taken at the sites near cities in China were found to contain large amount of dissolved organics. In some industrial discharge wastewater, toxic and hazardous synthetic organic chemicals such as man-made pesticides and dyes were identified.

The contamination situation in rivers and lakes in the country is also astonishing (**Table 1**). Field investigation of sections in 35 rivers showed that a large number of trace organics in the water could be identified, in some cases with chemicals three times above the sanitary standards. In the 1970s and 1980s, 26 species of organics in the Songhua River, one of the biggest rivers in northeast China were identified, of which 14 of the chemicals are on the EPA list of the top contaminants. A report in 1998 says that there are at least 60 nondegradable organics in the sediments of the second Songhua River, among which many pollutants are PAHs. In the Yangtze River Delta areas, south China, most of the rivers are polluted with organics. Investigation in 14 typical river sections in seven stream districts showed that 197 species of organics can be identified, of which 25 chemicals are carcinogenic and 53 are on the EPA list of the top contaminants. The organic varieties in the Shanghai Wangpu River, Shanghai City, ranged from 500 to 700, of which 218 varieties are quantitatively detectable with gas chromatography–mass spectrometry, and 39 are on the US EPA list of the top contaminants. Wang and Liu *(15)* reported serious contamination with different organic contaminants in the Baiyangdian Lake that was nicknamed "Northern China's Shining Pearl." Pan and Xie *(16)* presented data showing 46 kinds of organic pollutants with concentration in the range of ppb to ppm in the Fenhe River of Taiyuan section, Shanxi Province. There were 64 kind of organic pollutants identified in the Huan River, Henan Province, People's Republic of China, including several important organic pollutants on the US EPA priority list *(17)*. The water of the MengJin-Huayuankou section of the Yellow River was contaminated by organic pollutants *(18)*, and organic pollutants included volatile phenols, oils, organic pesticides, PAHs, and phenols. Ma *(19)* reported contamination with organic pollutants mainly consisting of phenols and oils on the Lanzhou section of the Yellow River. An et al. *(20)* reported recognition of 63 kinds of organic pollutants in Kunming Lake's water in Beijing, among which 17 kinds are on the priority list of control pollutants of US EPA. Zhang et al. *(21)* made a survey of polychlorinated biphenyl (PCB) concentration in the water, interstial water, and sediments sampled from Min River, Fujiang Province, south China, and found that PCB concentration in the water, interstial water, and sediments (dry weight) ranged from 0.20 to 2.47 µg/L, 3.19 to 10.86 µg/L, 15.13 to 57.93 µg/L, respectively. Contamination with organics in the water of Baihua and Hongfeng Lakes, Guizhou Province, in the southwest part of China was reported by Liang et al. *(22)* (**Table 2**). All these lines of evidence showed that organic contamination in China has become of environmental concern.

Table 1
Organics Contamination in Some Chinese Rivers and Lakes *(22)*

Locations	Description of organics contamination
Shengzhen River (section between Shengzhen and Hongkong)	Dominant pollutants are organics
Xiangjiang River	Combined contaminants, being dominated by organics
2nd Songhua River (Jiling Province, Northeast China)	374 organic pollutants identified, including aromatics, halogenated hydrocarbons, aldehyde, ketone, alcohols
Huangshi Section of the Yangtze River	100 organic pollutants identified
Taihu	74 organic pollutants identified
Tuojiang River	175 organic pollutants identified
Donghu source water, the Pearl River	241 organic pollutants identified
Sediments in the Suzhou River	102 organic pollutants identified

Table 2
Contamination With Organics in the Water of Baihua and Hongfeng Lakes, Guizhou Province, PR China (μg/L) *(23)*

Organics	Hongfeng lake no. 10				Baihua lake no. 2		
Sampling date	92.1	92.7	96.6	91.6	92.3	92.7	96.6
Naphthalene	0.6	13	5.0	1.2	5.4	2.0	1.6
Biphenyl	2.2	5.7	1.1	0.2	1.8	1.9	1.9
Ethylene naphthalene	1.0	4.4	5.0	1.4	4.7	5.8	1.4
Phenanthrene	8.4	3.5	1.4	5.3	6.5	3.1	5.5
Fluoranthene		2.7	1.7	1.3	2.7	21	11
Benzo (a)pyrene	3.7		1.1	4.5	0.8	0.6	18
n-C12	3.6	3.2	4.7	1.7	4.5	4.6	0.9
n-C20	7.0	7.8	9.3	1.0	13	25	14
Di-iso-butyl phthaiate	230.0	49	310	87	73	78	29
Dibutyl phthalate	16	390	880	700	600	630	170
Dioctyl phthalate	3.1	36	85	70	46	68	47
Dinitrobenzene	5	6.4	44	4.1	32	16	
Ethyl benzene	30	4.6	1.3	11		8.9	

There are some reports about underground water contamination by organics. Mo et al. *(24)* identified 60 kinds of organic pollutants in the underwater of Liantang Town, Jiangxi Province, P. R. China. One hundred and eight kinds of organic pollutants in groundwater in the Beijing sewage irrigation area were identified, of which 20 kinds are on the list of priority pollutants suggested by US EPA *(25)*.

2.3. Contaminated Soil

In addition to water contamination by organic pollutants, soil pollution by organics, especially by pesticides, has become a serious problem. It was estimated that the acreage of the soils contaminated with pesticides in China adds up to more than 6% of the total arable land. There are 1.7 million tons of pesticides annually applied for protection against plant diseases and insect pests, among which 30% is organo-phosphorus, resulting in toxic remains, and posing a potential threat to the environment and human health *(26)*. Taking Liaonin Province of the northeastern part of China as an example, about 25,000 tons of commercial pesticide are applied each year, implying that on average 7.5 kg of pesticide is applied to 1 ha of arable land. Such a high dose of pesticide applied not only poses a hazard to soil environmental quality and agricultural crop quality but also contaminates surface and underground water as well as oceanic environments, being a threat to the environment and human health. Another example is the Pearl River Delta district of south China where the soils were contaminated with organics to various degrees. An investigation of the soils in some vegetable-producing bases of the district conducted by Mo et al. showed that 6 varieties of phthalates and 11 kinds of PAHs were identified, and that the total amount of the two mentioned groups of organic compounds ranged from 3.00 to 45.67 mg/kg and 0.06 to 8.00 mg/kg, respectively, in the soils *(27)*.

The increasing incidence of cancers, and increasing death rate with some bizarre diseases happening here and there were shown to be linked with environmental and food contamination with pesticides *(28)*. In a word, contamination of the environment in China by organics, especially by pesticides, has been becoming a major concern in terms of food safety and human health. More effective countermeasures and remediation strategies are needed to cope with the problem.

3. Studies of Plant Uptake of Organics in China

Some sporadic data are available in the Chinese literature in this regard. During the past decade in China, some research has been done on testing plant species for their potential to take up different organics in terms of phytoremediation. For example, An et al. *(29)* compared the phytoremediation ability of 10 varieties of grasses grown in pot experiments in 5 soil types contaminated with different concentrations of DDT and its main degradation products. The 10 grasses included Kentucky bluegrass (var Nassan, United States), perennial ryegrass (var. taya, Denmark), tall fescue, perennial ryegrass (var. evening shade, United States), Kentucky bluegrass (var. Conni, Denmark), Kentucky bluegrass (var. Merit, United States), perennial ryegrass (var. Manhattan, United States), tall fescue (var. Millennium, Denmark), Kentucky bluegrass (var. Rugby, United States). It was found that the cleanup ability for each species varies depending on variety and that the same variety of grass showed different remediation potential when grown in different contaminated soil (**Table 3**). They also showed that the uptake of DDT and its main degradation products by the grass varieties contributed very little to removal of the contaminants from the soil. The experimental data indicated that the amount of DDT and its main degradation product taken up by grasses ranged from 0.13 to 0.30% of the initial amount of the total DDT applied to the soil, and that there were about 7.10–71.94% of the initial amount of DDT and its main degradation product removed by the grasses from the soil after growth for 3 mo.

Song et al. *(30)* compared the individual and combined responses of phenanthrene, pyrene, 1,2,4-trichlorobenzene, and the extent of combined toxicity effects by determining the inhibition rates of phenanthrene, pyrene, and 1,2,4-trichlorobenzene on higher plants (wheat, Chinese cabbages, and tomatoes) and the toxic effects of combined pollution with these chemicals in meadow brown soils. Their results showed that there was a significant linear or logarithmic relationship between the concentration of phenanthrene, pyrene, and 1,2,4-trichlorobenzene and the inhibition rates of root elongation of plants, and that the inhibition strength on plant elongation was greatest for 1,2,4-trichlorobenzene followed by phenanthrene and pyrene. They also showed that wheat was the most sensitive species to the applied organic pollutants and that there was a synergism of phenanthrene, pyrene, and 1,2,4-trichlorobenzene in the tested soil–plant system.

Because experimental data about plant uptake of organic pollutants are sporadic in Chinese literature, no investigation has been made so far concerning the phylogenetic characteristics of plant uptake of organics in this country.

4. Phytotechnologies Tested for Remediation of Organically Contaminated Environments on a Bench or Field Scale in China

Broadly speaking, phytoremediation of organic-contaminated sites on a bench scale was tested in China during the past years mainly with a focus on remediation of oil-mining sites and areas with diffuse pollution. In the early 1990s, Chinese scientists designed some treatment systems that were used to treat organic-contaminated sites in the field but only a few successful cases were reported. Trees and other plant species have been planted at several locations for the purpose of treatment of oil-contaminated environments *(12,31)* because of their oil tolerance and fast growth. Ji et al. *(12)* used the reed wetland system to treat crude oil pollutants. This system was operated at the Liaohe Oilfield of the northeastern part of China. The experimental results showed that large amounts of oil pollutants could be removed by this system, and that the removal efficiencies of the system increased with increasing levels of the crude oil. It was also shown that crude oil pollutants had little effect on the number of the reed leaves but a positive impact on the reed height.

Table 3
The Concentration of Total DDT in Grasses Grown in Contaminated Soil *(29)*

No.	Grass Common name	Variety name, producer	Total DDT applied (mg/kg, DW)					
			0.215			1.064		
			30 d*	60 d	90 d	30 d	60 d	90 d
1	Kentucky bluegrass	Nassan, USA	0.070	0.955	0.675	14.960	17.755	12.055
2	Perennial ryegrass	Taya, Denmark	0.210	1.210	1.120	5.985	11.790	6.740
3	Tall fescue	Titan, USA	0.735	0.430	0.680	6.210	15.295	3.385
4	Perennial ryegrass	Evening shade, USA	0.320	0.420	0.980	3.605	5.440	5.490
5	Kentucky bluegrass	Conni, Denmark	0.115	0.820	0.775	16.590	7.715	5.410
6	Kentucky bluegrass	Merit, USA	0.610	0.520	1.470	16.320	10.670	5.790
7	Perennial ryegrass	Manhattan, USA	0.325	0.840	3.405	6.690	11.185	5.940
8	Kentucky bluegrass	Midnight, USA	3.875	3.915	1.635	17.865	19.235	9.205
9	Tall fescue	Millennium, Denmark	0.250	0.430	1.670	6.140	17.715	3.305
10	Kentucky bluegrass	Rugby, USA	1.465	0.925	0.635	16.250	25.300	5.405

*Harvest days after period of growth.

5. Future Prospects for the Phytoremediation of Organic Contaminants in China

Although governments at all levels in China have paid more attention to research on phytoremediation of organically contaminated environments at present than in the past, there is still a long way to go before it becomes a commercial technology. More money is urgently needed to invest in this potential technology, and national and international cooperation should benefit its development. The authors believe that progress in the following aspects will be made in the field of phytoremediation of environments contaminated with organic pollutants in China:

1. Development in the field of phytoremediation of organics in this country will progress toward screening plant species with the ability to take up organic pollutants.

With Chinese taxonomists being involved more in phytoremediation research, more research will be conducted on screening species from the viewpoint of phylogenetic characteristics.
2. Screening highly efficient soil additives with a strong ability to make organic pollutants dissolve into bioavailable species will be carried out. Some Chinese scientists have already been aware of this potential technique because there are some applications sent to the Chinese National Natural Science Foundation in the last few years.
3. More work will be done in this country to provide a basis for genetic modification of plants for improved performance. Progress has been made during the last decade in China toward molecular breeding technologies for modification of agronomically important plant traits, such as genetically modified cottons and corns that are already grown in the field on a large scale. The authors believe that molecular biology will allow the production of plants specifically targeted for the needs related to phytoremediation of organically contaminated soil.
4. A national information network will be set up to provide comprehensive information on the respective pollutants, the available plant systems for remediation, their associated micro-organisms and constitutive enzyme sets, possible application of agricultural engineering measures, or other means of inducing metabolic activity in the plants for the degradation of organic pollutants, and so on. Scientists from Tsinghua University have made some progress in compiling data on POPs (www.china-pops.org).

Because phytoremediation still has some drawbacks and limits, integration of this technology into other physical and chemical technologies should be the priority of the future research regarding this potential technology. We believe that more progress in the field of phytoremediation of organically contaminated soil and water will be made in the vast country of China.

References

1. Xia, H. L., Wu, L. H., and Tao, Q. N. (2003) Phytoremediation of organic contaminated environments: a review. *Chin. J. Appl. Ecol.* **14,** 457–460.
2. Yi, X. Y., Dang, Z., and Shi, L. (2002) Phytoremediation of soil polluted by organic contaminants. *Agro-Environ. Protection* **21,** 477–479.
3. Yang, L. C., Zheng, M. H., Liu, W. B., An, F. C., and Mo, H. H. (2002) The study progress of phytoremediation of organic polluted environments. *Tech. Equip. Environ. Pollution Control* **3,** 1–7.
4. Wang, Q. R., Liu, X. M., Cui, Y. S., and Dong, Y. T. (2001) Concept and advances of applied bioremediation for organic pollutants in soil and water. *Acta Ecol. Sin.* **21,** 159–163.
5. Zhou, Q. X. and Song, Y. F. (2001) Technological implications of phytoremediation and its application in environment protection. *J. Safety Environ.* **1,** 48–53.
6. Tang, S. R. and Wilke, B. M. (1999) Phytoremediation and agrobiological environmental engineering. *Trans. Chin. Soc. Agric. Eng.* **15,** 21–26.

7. Tang, S. R., Huang, C. Y., and Zhu, Z. X. (1996) Using plants to clean up heavy metal contaminated soil. *Adv. Environ. Sci.* **4,** 10–15.
8. Schröder, P., Harvey, P. J., and Schwitzguébel, J. P. (2002) Prospects for the phytoremediation of organic pollutants in Europe. *Environ. Sci. Pollut. Res.* **9,** 1–3.
9. Korte, F., Kvesitadze, G., Ugrekhelidze, D., et al. (2000) Review: organic toxicants and plants. *Ecotox. Environ. Safety* **47,** 1–26.
10. Macek, T., Mackova, M., and Kas, J. (2000) Exploitation of plants for the removal of organics in environmental remediation. *Biotechnol. Adv.* **18,** 23–34.
11. Xia, H. H. and Lin, Y. S. (2000) Advances in bioremediation of organics contaminated soil. *Contam. Prev. Rem. Technol.* **13,** 46–47.
12. Ji, G. D., Sun, T. H., Sui, X., and Chang, S. J. (2002) Impact of ground crude oil on the ecological engineering purification system of reed wetland. *Acta Ecol. Sinica* **22,** 649–654.
13. Peng, B., Huang, Z., and Wang, C. H. (1997) Preliminary study of trace organic pollutants in the water of the Danjiangkou Reservoir. *Renmin Changjiang* **28,** 27–29.
14. Tian, H. J., Shu, W. Q., Zhang, X. K., Wang, Y. M., and Cao, J. (2003) Organic pollutants in source water in Jialing River and Yangtze River (Chongqing Section). *Resources Environ. In the Yangtze Basin* **12,** 118–123.
15. Wang, Y. Z. and Liu, J. A. (1995) Qualitative analysis of organic contaminants in the water of Baiyangdian District. *Environ. Chemistry* **14,** 442–448.
16. Pan, S. X. and Xie, J. F. (1992) Monitoring and assessment of the organic pollutant in Fenhe River of Taiyuan Section. *Res. Environ. Sci.* **5,** 34–41.
17. Yang, W. F., Li, J., Yan, B., Li, Y., Ding, Z. A., and Dong, L. P. (2001) Study of trace organic pollutants in the water of Huan River, Henan Province. *Environ. Monitoring in China* **17,** 26–30.
18. Hu, G. H., and Zhao, P. L. (1995) Analysis of organic contamination in the MengJin-Huayuankou section of the Yellow River, and prevention as well as remediation strategies. *People's Yellow River* **11,** 6–9.
19. Ma, H. M. (2002) Regularity and changing trend of contamination with organic pollutants in the Lanzhou section of the Yellow River. *Gangsu Sci. Technol.* **6,** 67–68.
20. An, S. J., Zheng, S. Z., Mao, S. Z., Jin, Z., and Li, S. R. (2000) Detection and removal of organic pollutants in Kunming Lake's water in Beijing. *Environ. Chemistry* **19,** 284–288.
21. Zhang, Z. L., Hong, H. S., and Yu, G. (2002) Preliminary study on persistent organic pollutants (POPs)—PCBs in multi-phase matrices in Minjiang River Estuary. *Acta Scientiae Circumstantiae* **22,** 788–791.
22. Liu, X. R., Feng, H. H., and Zhang, Y. (2002). The state of organic contamination in water environments in China and its control strategies. Technical Supervision in Water Resources, **10(5)**: 58–60(in Chinese).
23. Liang, X. J., Fu, W. J., Zhang, M. S., and Wang, A. M. (1998) Investigation on nutrient elements and organic pollutants of Baihua and Hongfeng Lakes. *Guizhou Sci.* **16,** 311–315.

24. Mo, H. H., An, F. C., Yang, K. W., Wang, T. H., and Chen, H. J. (2001) Preliminary study of organic pollutants in the underwater of Liantang Town, Jiangxi Province, P. R. China. *Series Environ. Sci.* **13,** 38–54.
25. Li, S. Q. and Yuan, M. (2000) Organic pollutants in ground water in the Beijing sewage irrigation area. *Tianjin Construction Sci.* **3,** 32–34.
26. Chen, X. W. (2001) Imperative to develop green industry in China. High-class forum of the first Chinese green industry sustainable development. Beijing.
27. Cai, Q. Y. (2003) Study of the soil-vegetable systems contaminated with organics in South China. *Ph.D. dissertation, South China Agricultural University*, Guangzhou, Guangdong, Province, P. R. China, pp. 1–78.
28. Zheng, F. T. and Zhao, Y. (2003) China food security: problems and policy measures. *China Soft Science* **2,** 16–20.
29. An, F. C., Mo, H. H., Zheng, M. H., and Zhang, B. (2003) Phytoremediation of DDT and its main degradation product-contaminated soil using grass. *Environ. Chemistry* **22,** 19–25.
30. Song, Y. F., Zhou, Q. X., Xu, H. X., Ren, L. P., Song, X. Y., and Gong, P. (2002) Eco-toxicological effects of phenanthrene, pyrene and 1, 2, 4-trichlorobenzene in soils on the inhibition of root elongation of higher plants. *Acta Ecol. Sinica* **22,** 1945–1950.
31. Ji, Z. G., Zhang, D., and Zhu, H. L. (eds.) (1997) *Biodegradation of Pollutants*. East China Huadong Univers. Sci. and Engin. Press, Shanghai, P. R. China, pp. 1–312.

27

Phytoremediation of Arsenic-Contaminated Soil in China

Chen Tong-Bin, Liao Xiao-Yong, Huang Ze-Chun, Lei Mei, Li Wen-Xue, Mo Liang-Yu, An Zhi-Zhuang, Wei Chao-Yang, Xiao Xi-Yuan, and Xie Hua

Summary

Arsenic (As) is a common pollutant of concern in environmental clean up because its contamination is recognized to lead to a variety of cancers, cardiovascular diseases, diabetes, and other health problems. Because *Pteris vittata* L. was discovered to hyperaccumulate As from soils, As hyperaccumulators have been attracting more and more attention and are proposed to be promising for phytoremediation. Although laboratory studies on the tolerance and accumulation of As by the hyperaccumulators are available, little information about field performance of phytoremediation using the plants is available. Here, the research priorities for As-phytoremediation technologies, As accumulation, and the relationships between As and other elements in the plants, are discussed. Primarily, however, results from a pilot field study on phytoremediation of As-contaminated soil in Chenzhou City of Hunan Province, China are summarized. It is concluded that *P. vittata* can effectively phytoextract As from an As-contaminated site under a subtropical climate.

Key Words: Arsenic; *Pteris vittata* L.; contamination; field demonstration; hyperaccumulator; phytoremediation; soil.

1. Introduction

Arsenic (As), which is a metalloid of the Group 5A elements and exhibits both metallic and nonmetallic properties, naturally occurs in the form of sulfides. It has many positive industrial and agricultural applications but is also a pollutant of concern in the environment. Soil As concentrations can become elevated as a result of human activities such as mining, waste discharges, coal burning, and applying arsenical pesticides (*1*). In recent years, soil As has been reported in high quantities as a result of As-containing water and atmospheric deposition around some industrial areas as a result of mining and smelting activities *(2–4)*. Irrigation water may be a major carrier of As in dissolved or adsorbed forms that may be a cause of regional contamination *(5,6)*.

Both geochemical As enrichment and anthropogenic soil contamination have a risk to human health. Long-term exposure to As can cause skin abnormalities, including the appearance of dark and light spots on the skin, which may ultimately progress to skin cancer. Arsenic has also been associated with an increased risk of liver, bladder, kidney, and lung cancer, cardiovascular diseases, diabetes, and other health problems. It has been proven that human disease can result from As in soils *(7)*. The lowest concentration of As in soil at which phytotoxic effects have been observed is 10 mg/kg in green beans, spinach, radishes, cabbage, and lima beans *(8)*. Soil As has caused phytotoxicity to sensitive plants at numerous locations, especially where mine wastes and smelters caused As-contamination of soils *(9–11)*. For example, there are many As-toxic soils in the Hunan and Guangxi Provinces of China, which lead to serious toxicity to humans and plants *(6,12–14)*.

Arsenic is, therefore, a common pollutant of concern in environmental cleanup. However, because it readily changes valence state and reacts to form species with varying toxicity and mobility, its effective treatment is difficult. Eight kinds of technologies applicable to As-contaminated soils and wastes, i.e., solidification/stabilization, vitrification, soil washing/acid extraction, pyrometallurgical recovery, *in situ* soil flushing, electrokinetics, biological treatment, and phytoremediation, are identified by the US Environmental Protection Agency *(15)*. The traditional technologies may not be applicable to soils and wastes containing low concentrations of As, and the new technologies such as phytoremediation have been applied in only a limited number of applications and there is no adequate performance data for application.

Phytoremediation of As-contaminated sites using hyperaccumulators has been attracting more attention. Phytoremediation is usually defined as the use of green plants to remove pollutants, which offers an economical, feasible, and "green" remediating technology for contaminated environments compared with the soil replacement, solidification, and washing means presently used *(16–19)*. Phytoextraction, one of the promising strategies within the field of phytoremediation, refers to the translocation and concentration of metals from soils into harvestable parts by metal-accumulating plants. There have been some successful demonstrations of phytoextraction. After three crops of *Brassica juncea*, approx 50% of the lead was removed from the surface soil with a Pb concentration of 2055 mg/kg in Bayonne *(19)*. *Thlaspi caerulescens* (Brassicaceae) took a relatively long time of continuous cultivation (13–14 yr) to clean a site of Ni and Zn contamination *(20)*. Plants of *B. oleracea*, *Raphanus sativus, T. caerulescens, Alyssum lesbiacum, A. murale*, and *Arabidopsis thaliana* have been shown to extract Zn, Cd, Ni, Cu, Pb, and Cr, respectively, from soils *(21,22)*. The application of phytoextraction is limited to small-scale field trials, but it has been thought of as a very feasible technology for the removal of metal pollutants from the environment. Although there are an increasing number of papers dealing with As hyperaccumulators, especially *P. vittata*, there is no information available about phytoremediation of As under field conditions.

2. Discovery of As Hyperaccumulators

In 1997, Chen Tong-Bin speculated about the possibility of As hyperaccumulators existing in the old mining areas in southern China and began to screen them *(23)*. More than 100 plant species were screened to identify potential hyperaccumulating plants, which are capable of tolerating and accumulating high concentrations of As from the As-contaminated soils. Fortunately, Chinese brake (*Pteris vittata* L.), a fern of the brake family (Pteridaceae), was discovered and verified by Chen's group to be an As-hyperaccumulating plant both in a field investigation and a greenhouse study *(24)*. A primary report about the discovery of the As-hyperaccumulating plant was presented by Chen and Wei at the International Conference of Soil Remediation held in Hangzhou, China in October of 2000 *(25)*. Both Chen's and Ma's groups independently discovered the As hyperaccumulation of *P. vittata* in China and in the United States, respectively, and detailed information on the findings of the As hyperaccumulator were reported correspondingly by Ma et al. *(26)* and by Chen et al. *(24)*.

P. vittata produces large biomass and adapts itself to different conditions. It grows on limestone, calcareous soils, and stone fissures or wall surfaces with a distribution below 2000 m altitude. It is found to grow normally in soils of 50–4030 mg/kg As, and even in tailings with As up to 23,400 mg/kg (**Table 1**), indicating its great As tolerance. Field investigation showed that As concentrations varied from 120 to 1540 mg/kg, from 70 to 900 mg/kg and from 80 to 900 mg/kg in the dried pinnae, petioles, and rhizomes, respectively *(24)*. It is characteristic that As in the roots of common plants grown on normal soil is greater than that in shoots, and that As concentrations vary from less than 0.01 to about 5 mg/kg *(27–29)*. As is accumulated by *P. vittata* with bioaccumulation coefficients greater than one from ordinary As-contaminated soils. From an investigation of field sites around the Shimen arsenic sulphide mine, Hunan Province, China, the rhizomes account for 21–71% of the total As accumulated by Chinese brake the fern and they are a major storage organelle for As, especially in soil with higher As concentrations *(30)*. Therefore, it is proposed that in

Table 1
Arsenic Concentrations in Chinese Brake Grown at Different Field Sites of Hunan Province, China

Growing media	As concentration (mg/kg)			
	Media	Pinna	Petiole	Rhizome
Soils (N = 10)	50–4030	120–1110	70–670	80–220
Tailings (N = 3)	3400–23,400	1540–1530	680–900	~900

addition to the previously known As-storing organ, the pinna, the rhizome is regarded as a major storage organelle for As, which might alleviate toxicity to the frond.

The depth of remediation in phytoextraction, a root-based biotechnology, is limited by the root distribution of the cultivated plant. The elevated As levels found in most soils contaminated as a result of human activities, such as waste discharges of metal-processing plants, burning of fossil fuels, mining of As-containing ores and use of arsenical pesticides, are usually focused in the surface layer *(31–34)*. *P. vittata* roots are mainly distributed in the upper 0–30 cm of the soil profile under field conditions, which suggest that it may be able to phytoremediate the rooted layer of As-contaminated soils *(35)*.

The biomass, growth rate, and metal tolerance and metal concentration of hyperaccumulators are factors determining their phytoremediation effectiveness. The known hyperaccumulators, although demonstrating great accumulation of heavy metals, are low in biomass, slow in growth, and difficult to cultivate, and thus cannot be applied widely in the field *(36)*. Compared with most of the traditional hyperaccumulators previously reported, *P. vittata* can not only accumulate large amounts of As in the fronds but also grow rapidly and produce great biomass. Therefore, it is thought to have favorable prospects for phytoremediation. *P. vittata* is distributed widely in tropical and subtropical zones in China and other countries, suggesting adaptability to a range of climates and soils. All these characteristics suggest favorable prospects for its application in the phytoextraction of As-contaminated soils.

The highest As concentration in the shoot of *P. vittata* is more than 10,000 mg/kg, which is more than 100,000 times greater than that in normal plants, and higher than the P concentrations in normal plants. As is not essential to plant growth, and a excess of As can be very toxic to plants, but *P. vittata* can accumulate a high concentration of As in its tissue and is tolerant to a high concentration of As. This highlights its great value to studies of physiological and biochemical mechanisms such as As absorption, translocation, tolerance, and detoxification.

3. Phytoremediation of As-Contaminated Soils: Greenhouse Studies

3.1. Arsenic Accumulation in P. vittata as Enhanced by N Fertilizers

Plant growth and elemental concentration are closely related to N supply levels and N source. There are substantial differences between plant species when provided with different N sources *(37)*. Although *P. vittata* biomass is greatly enhanced by N supply, fertilization applications depress As transport to shoot *(38)*. The inhibitory effect of As transport resulting from N application does not lessen the As accumulation resulting from the greater increase in its biomass. The biomass of *P. vittata* is enhanced by applying NH_4^+-N fertilizers as an N source compared with the NO_3^--N fertilizers *(38)*. The As accumulation by the plant provided with different N resources follows the trend: NH_4^+-N > Urea-N > NO_3^--N (**Table 2**, *38*). Although application of N fertilizer can enhance As accumulation in the shoot of *P. vittata*, appropriate N source should be considered for a higher efficiency of As removal.

3.2. Arsenic Accumulation in P. vittata Enhanced by P Fertilizers

Because As and P have similarities in chemical properties, they act similarly in many ways in the soil–plant system. It was reported in previous studies that P and As were taken up by the same system in common plants *(39–41)*. However, the results from *P. vittata* grown in a calcareous soil showed that the As concentrations in the fronds were not influenced significantly under lower P concentrations (≤400 mg/kg) and increased sharply under higher P concentrations (>400 mg/kg) (**Fig. 1**) *(42)*. Therefore, no competition between P and As is found in *P. vittata*. The addition of a higher rate of P correlated with a great increase in the total quantity of shoot As. The discovery implies that the efficiency of As removal in phytoremediation using the hyperaccumulating plant can be greatly elevated by application of P fertilizer.

3.3. Calcium and Arsenic in P. vittata

Calcium and As often occur simultaneously in soils, and available As is closely associated with Ca concentration in calcium soils *(43–45)*. *P. vittata* is an indicator plant for calcareous soil and limestone *(26)*. However, Liao et al. *(46)* found that in sand culture, plants treated with 0.03 m*M* of Ca had the highest As concentration (4218 mg/kg) in pinnae, compared with those of 2.5 and 5.0 m*M* Ca. With addition of Ca at rate of 2.5 and 5.0 m*M*, plants accumulated 6019 and 2014 mg/pot of As, respectively, which were 78.6 and 26.3% of the total As accumulation in the plant treated with 0.03 mM of Ca *(47)*. The ratio of As concentration in pinna to root decreased with the increase in Ca concentration, implying that Ca inhibited As translocation. These results indicate that excessive Ca in the growing media may reduce the efficiency of As removal from contaminated soils.

Table 2
As Accumulation in Chinese Brake Applied Different Forms of N Fertilizers in Pot Experiment (38)

Fertilizer	As accumulation (mg/pot)		
	Shoot	Rhizome	Total
Control	3.19 c*	0.17 c	3.36 c
NH_4HCO_3	7.50 a	0.47 ab	7.97 a
$(NH_4)_2SO_4$	6.31 ab	0.52 ab	6.83 ab
KNO_3	4.97 bc	0.33 bc	5.30 bc
$Ca(NO_3)_2$	5.46 ab	0.55 a	6.02 ab
Urea	5.60 ab	0.48 ab	6.60 ab

3.4. Arsenic and Heavy Metals in P. vittata

The cocontamination of As and other heavy metals is observed at many contaminated sites. A series of field investigations and greenhouse experiments were carried out by our group to determine tolerance of *P. vittata* to heavy metals, and whether it can be applied to phytoremediation and revegetation of soils cocontaminated with As and other heavy metals. It is found that *P. vittata* has a great ability to tolerate Cd, and can normally grow in field soils with very high concentrations of Cd (up to 301 mg/kg) and As *(48)*. Cadmium concentration in the shoot could reach up to 186 mg/kg, which is above the Cd-hyperaccumulator level, under field conditions. Although Cd inhibits the plant's growth, the addition of lower concentrations of Cd elevated the As concentrations in fronds and enhanced the translocation of As from root to frond *(48)*. The plant can tolerate a high concentration of Pb, Cu, and Zn, and phytoextract As effectively from soils also contaminated with heavy metals *(49)*. Therefore, *P. vittata* may have the potential of phytoremediating and rehabilitating the sites cocontaminated with As and other heavy metals.

4. Phytoremediation of As-Contaminated Soils Using *P. vittata*: Field Studies

4.1. The Field Site and Plant Propagation

The site is located in Chenzhou City, Hunan Province, China. The soil had been used to grow rice, and frequently alternated between dry and wet before a serious pollution accident, which led to two deaths and nearly 400 persons being hospitalized in the winter of 1999, since then it has gone out of cultivation. Its As concentration, ranging from 24 to 192 mg/kg, was elevated from irrigation with As-containing water from an As smelter. The As accumulated mainly in the surface layer, 0–20 cm, of soil with the As concentrations in the layer of

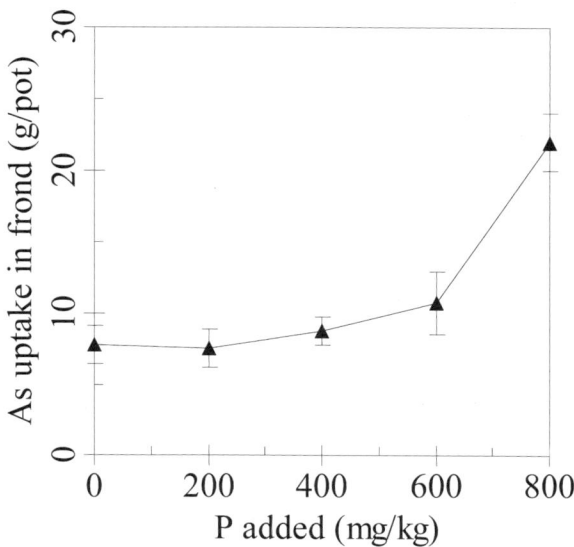

Fig. 1. Arsenic accumulation in Chinese brake as influenced by P added in a pot experiment.

40–80 cm not being greatly affected *(14)*. About 1 ha of the contaminated soil was planted with *P. vittata* to verify the feasibility of phytoremediating the As-contaminated soil under a subtropical climate. The field experiment of phytoremediation has been carried out since 2001.

It is not easy to propagate sporophytes of *P. vittata* directly in the field. Therefore, the source of sporophytes is a problem, which limits the application in *P. vittata* in phytoremediation. Sporophytes and sectioned rhizomes of *P. vittata* can be used as explant material for in vitro culture for mass propagation purposes *(50)*. The plant sporophytes obtained from tissue culture can grow well and hyperaccumulate As from soil. Frequencies of harvesting (e.g., three harvestings per year) of the plant is an important factor related to phytoextraction efficiency using *P. vittata*. It may be an effective measure to enhance the efficiency of As removal if the plant is harvested at suitable frequency *(51)*.

4.2. Application of P Fertilizer

Plants treated with 200 kg/ha of P accumulated maximum As (3.74 kg/ha), in their aboveground part, which was 2.4 times more than the control and 1.2 times more than in plants treated with 600 kg/ha of P *(35)*. After 7 mo of experiment, soil As concentrations were significantly reduced at all treatments compared with those before transplanting. When 200 kg/ha of P was added, efficiency of As removal was the highest, 7.84%, whereas the efficiencies of As

Fig. 2. Field (1 ha of area) of an arsenic-contaminated site phytoremediated with Chinese brake in Chenzhou City, Hunan Province, China.

removal at control and 600 kg/ha treatments were 2.31 and 6.63%, respectively. Moreover, P application can maintain a balance of available As during phytoremediation. Application of P fertilizer is necessary for phytoremediation using As hyperaccumulator and optimization of P application can significantly enhance the efficiency of As removal from contaminated soils.

4.3. Efficiency of As Removal by Harvesting of the Shoot

The As-contaminated site described previously was phytoremediated with *P. vittata* in Chenzhou City, Hunan Province, China (**Fig. 2**). The plant was fertilized with N, P, K, and irrigated as needed. Seven months after transplanting, the plant shoot was harvested for the purpose of remediation. The shoot dry matter weight ranged from 872 to 4767 kg/ha (**Table 3**). The As concentrations in shoot varied from 127 to 3269 mg/kg, which were significantly related to As concentration in initial soils. The efficiency of As removal ranged from 6 to 13%, which proves the ability of *P. vittata* for effectively extracting soil As in the field. As an innovative technology to clean up As contamination, the phytoremediation based on cropping *P. vittata* still requires time for further refinement, and should undergo more detailed and long-term

Table 3
Chinese Brake Growth and As-Hyperaccumulation Under the Field Condition

	Shoot biomass (kg/ha)	As concentration (mg/kg)			Efficiency of As removal (%)
		Shoot	Initial soil	Remediated soil	
R1	872	127	23.9	21.5	10.0
R3	1364	206	28.3	24.6	13.2
R4	1616	211	35.4	31.1	12.0
R8	917	708	48.0	45.1	6.1
R16	1849	2292	123.0	114.6	6.9
R20	4767	3269	192.1	169.5	11.8

studies. From the pilot study carried out in China, it is concluded that *P. vittata* can phytoextract As effectively from an As-contaminated site under a subtropical climate.

Acknowledgments

This work was supported by the National Grant for Excellent Young Scientists (grant no. 40325003), the National High-Tech R & D Program (no. 2001AA6450), the China State Program for Basic Research (no. 2002CCA 03800), and the National Natural Science Foundation of China (grant no. 4023 2022). The authors wish to thank Dr. S. R. Tang of Zhejiang University for his help in preparation of the manuscript.

References

1. Fergusson, J. E. (1990) *The Heavy Elements: Chemistry, Environmental Impacts and Health Effects*. Pergamon Press, Oxford, UK, p. 614.
2. Gidhagen, L., Kahelin, H., Schmidt-Thomé, P., and Johansson, C. (2002) Anthropogenic and natural levels of arsenic in PM10 in Central and Northern Chile. *Atmos. Environ.* **30**, 3803–3817.
3. Lynch, J. A., McQuaker, N. R., and Brown, D. F. (1980) ICP-AES analysis and the composition of airborne and soil materials in the vicinity of a lead/zinc smelter complex. *J. Air Pollut. Control Assoc.* **30**, 257–260.
4. Mitchell, P. and Barr, D. (1995) The nature and significance of public exposure to arsenic: a review of its relevance to south west England. *Environ. Geochem. Health* **17**, 57–82.
5. Pandey, P. K., Yadav, S., Nair, S., and Bhui, A. (2002) Arsenic contamination of the environment: a new perspective from central-east India. *Environ. Int.* **28**, 235–245.
6. Liao, X.-Y., Chen, T.-B., Xie, H., and Liu, Y.-R. (2005) Soil As contamination and its risk assessment in areas near the industrial districts of Chenzhou City, Southern China. *Environ. Int.* **31**, 791–798.

7. Rahman, M. M., Paul, K., Chowdhury, U. K., Biswas, B. K., Lodh, D., Basu, G. K., Roy, S., Das, R., Ahmed, B., Kaies, I., Barua, A. K., Palit, S. K., Quamruzzaman, Q., and Chakraborti, D., (2001) Current status of arsenic pollution and health impacts in West Bengal and Bangladesh. *An International Workshop on Arsenic Pollution of Drinking Water in South Asia and China*, Jinji Roumu Kaikan, Ohsaki, Shinawawa, Tokyo, Japan, March 10, 2001.
8. Woolson, E. A. (1973) Arsenic phytotoxicity and uptake in six vegetable crops. *Weed Sci.* **21,** 524–527.
9. United States Environmental Protection Agency (1997) *Recent Development for In-situ Treatment of Metal Contaminated Soils*. Office of Solid Waste and Emergency Response. EPA-542-R-97-004, p.8.
10. Roychowdhury, T., Uchino, T., Tokunaga, H., and Ando, M. (2002) Arsenic and other heavy metals in soils from an arsenic-affected area of West Bengal, India. *Chemosphere* **49,** 605–618.
11. Smith, E., Naidu, R., and Alston, A. M. (2002) Chemistry of arsenic in soils: II. Effect of phosphorous, sodium and calcium on arsenic sorption. *J. Environ. Qual.* **31,** 557–563.
12. Chen, T.-B. (1990) Arsenic in soil-plant system and its effect on rice growth and development. PhD dissertation, Chinese Academy of Agricultural Sciences, Beijing, China, p. 92.
13. Chen, T.-B., Liu, G.-L., Xie, K.-Y., and Gan, S.-W. (1992) Arsenic contents in soils and crops in high As district of Hunan Province. *Soil Fertil.* **2,** 1–4.
14. Liao, X.-Y., Chen, T.-B., Xiao, X.-Y., Huang, Z.-C., An, Z.-C., Mo, L.-Y., Li, W. X., Chen, H., and Zheng, Y. M. (2003) Spatial distribution charactersistics of arsenic in contaminated paddy soils. *Geog. Res.* **22,** 635–643.
15. United States Environmental Protection Agency (2002) *Arsenic Treatment Technologies for Soil, Waste, and Water*. Office of Solid Waste and Emergency Response. EPA-542-R-02-004.
16. Baker, A. J. M., McGrath, S. P., Sidoli, C. M. D., and Reeves, R. D. (1994) The possibility of in-situ heavy metal decontamination of polluted soils using crops of metal-accumulating plants. *Res. Conserv. Recyc.* **11,** 41–49.
17. Chaney, R. L., Brown, S. L., Li, Y. M., Angle, J. S., Homer, F. A., and Green, C.E. (1995) Potential use of metal hyperaccumulators. *Min. Environ. Mag.* **3,** 9–11.
18. Chaney, R. L., Malik, M., Li, Y. M., Brown, S. L., Angle, J. S., and Baker, A. J. M. (1997) Phytoremediation of soil metals. *Curr. Opin. Biotech.* **8,** 279–284.
19. Blaylock, M. J., Muhr, E., Page, K., Montes, G., Vasudev, D., and Kapulnik, Y. (1996) Phytoremediation of lead contaminated soil at a Brownfield site in New Jersey. *Proceeding of Am. Chem. Soc.*, Birmingham, AL, Sept. 9–11.
20. Salt, D. E., Blaylock M, Kumar, N. P., et al. (1995) Phytoremediation: a novel strategy for the removal of toxic metals from the environment using plants. *Biotechnol.* **13,** 468–474.
21. Baker, A. J. M., Reeves, R. D., and McGrath, S. P. (1991) In situ decontamination of heavy metal polluted soils using crops of metal-accumulating plants: a feasibility study. In: *In Situ Bioreclamation*, (Hinchee, R. E. and Olfenbuttel, R. F., eds.), Butterworth, Heinemann, Boston, MA, pp. 600–605.

22. Brown, S. L., Chaney, R. L., Angle, J. S., and Baker, A. J. M. (1995) Zinc and cadmium uptake by hyperaccumulator *Thlaspi caerulescens* and metal tolerant *Silene vulgaris* grown on sludge-amended soils. *Environ. Sci. Technol.* **29,** 1581–1585.
23. Chen, T.-B. (1997) Ecological study on genetic difference in tolerance of plants to arsenic. A proposal submitted to National Natural Science Foundation of China for grant application, Institute of Geography, Chinese Academy of Sciences, Beijing, China. pp. 15.
24. Chen, T.-B., Wei, C.-Y., Huang, Z.-C., Huang, Q.-F., Lu, Q.-G., and Fan, Z.-L. (2002) Arsenic hyperaccumulator *Pteris vittata* L. and its arsenic accumulation. *Chinese Sci. Bull.* **47,** 902–903.
25. Chen, T.-B. and Wei, C.-Y. (2000) Arsenic hyperaccumulation in some plant species in South China. In: *Proceedings of International Conference of Soil Remediation*, (Luo, Y.-M. et al. eds.), Hangzhou, Zhejiang, China from October 15–19, 2000, pp. 194–195.
26. Ma, L. Q., Komar, K. M., Tu, C., Zhang, W., Cai, Y. and Kennelley, E. D. (2001) A fern that hyperaccumulates arsenic. *Nature* **409,** 579.
27. Liebig, G. F.m Jr. (1973) Arsenic. *Diagnostic Criteria for Plants and Soils* (Chapman, H. D. ed.), Quality Printing Company Inc., TX, Riverside, California, USA, pp. 13–23.
28. Feed Additive Compendium (FAC) (1975) vol. 13. The Miller Publishing Company, Minneapolis, MN, p. 330.
29. Kabata-Pendias, A. and Pendias, H. (eds.) (1991) *Trace Elements in Soils and Plants*. CRC Press, Boca Raton, FL, pp. 309.
30. Liao, X.-Y., Chen, T.-B., Lei, M., Huang, Z.-C., Xiao, X.-Y., and An, Z.-Z. (2004) Root distributions and elemental accumulations of Chinese brake (*Pteris vittata* L.) from As-contaminated soils. *Plant Soil,* 109–111.
31. Allinson, G., Turoczy, N. J., Kelsall, Y., et al. (2000) Mobility of the constituents of chromated copper arsenate in a shallow sandy soil. *New Zeal. J. Agr. Res.* **43,** 149–156.
32. Galasso, J. L., Siegel, F. R., and Kravitz, J. H. (2000) Heavy metals in eight 1965 cores from the Novaya Zemlya Trough, Kara Sea, Russian Arctic. *Mar. Pollut. Bull.* **140,** 839–852.
33. Kalbitz, K. and Wennrich, R. (1998) Mobilization of heavy metals and arsenic in polluted wetland soils and its dependence on dissolved organic matter. *Sci. Total Environ.* **209,** 27–39.
34. Tack, F. M. G., Verloo, M. G., Vanmechelen, M., and Van, R. E. (1997) Baseline concentration levels of trace elements as a function of clay and organic carbon contents in soils in Flanders (Belgium). *Sci. Total Environ.* **201,** 113–123.
35. Baker, A. J. M. and Brooks, R. R. (1989) Terrestrial higher plants which hyperaccumulate metallic elements: a review of their distribution, ecology and phytochemistry. *Biorecovery* **1,** 81.
36. Liao, X.-Y., Chen, T.-B., Xie, H., and Xiao, X.-Y. (2004) Effect of application of P fertilizer on efficiency of As removal in contaminated soil using phytoremediation: Field demonstration. *Acta Scient. Circumst.* **24,** in press.
37. Siddipi, M. Y., Malhotram, B., Xiangjia, M., and Glass, A. D. M. (2002) Effects of ammonium and inorganic carbon enrichment on growth and yield of a hydroponic tomato crop. *J. Plant Nutr. Soil Sci.* **165,** 191–197.

38. Liao, X.-Y., Chen, T.-B. Xiao, X.-Y., Yun, X-L, Zhai, L.-M., and Wu, B., and Xie, H. (2006) Influences of the form of nitrogen fertilization on the removal efficiency of arsenic from soils using Chinese brake. Selecting appropriate forms of nitrogen fertilizer to enhance arsenic removal from soil using Pteris vittata: A new approach in phytoremediation. *Acta Ecol. Sin., chemosphere* submitted.
39. Meharg, A. A., Naylor, J., and Macnair, M. R. (1994) Phosphorus nutrition of arsenate-tolerant and nontolerant phenotypes of velvetgrass. *J. Environ. Qual.* **23**, 234.
40. Brolo, F., Guijarro, I., and Carbonell-Barrachina, A. A. (1999) Arsenic species: effects on and accumulation by tomato plants. *J. Agric. Food. Chem.* **47**, 1247.
41. Sharples, J. M., Meharg, A. A., Chambers, S. M., and Cairney, J. W. G. (2000) Evolution: symbiotic solution to arsenic contamination. *Nature* **404**, 951.
42. Chen, T.-B., Fan, Z.-L., Lei, M., Huang Z.-C., and Wei, C.-Y. (2002) Effect of phosphorus on arsenic uptake by As-hyperaccumulator *Pteris vittata* L. and its implications. *Chinese Sci. Bull.* **47**, 1156–1159.
43. Woolson, E. A., Axley, J. H., and Kearney, P. C. (1971) The chemistry and phytotoxicity of arsenic in soils. I. Contaminated field soils. *Soil Sci. Soc. Am. J.* **35**, 938–943.
44. Onken, B. M. and Hossner, L. R. (1995) Plant uptake and determination of arsenic species in soil solution under flooded conditions. *J. Environ. Qual.* **24**, 373–381.
45. Liao, Z.-J. (ed.) (1992) *Environmental Chemistry and Biological Effect of Trace Elements*. Environmental Science Press, Beijing, P.R. China, pp. 162.
46. Liao, X.-Y., Xiao, X.-Y., and Chen, T.-B. (2003) Effects of Ca and As addition on As, P and Ca uptake by hyperaccumulator *Pteris vittata* L. under sand culture. *Acta Ecol. Sin.* **23**, 2057–2065.
47. Xiao, X.-Y., Liao, X.-Y., Chen, T.-B., and Zhang, Y.-Z. (2003) Effects of arsenic and calcium on metal accumulation and translocation in *Pteris vittata* L. *Acta Ecol. Sin.* **23**, 1477–1487.
48. An, Z.-Z. (2004) Tolerance of *Pteris vittata* L. to cadmium, lead, copper and zinc and effect of phosphate on arsenate, arsenite uptake. Working Report of Postdoctoral Research. Institute of Geographical Sciences and Natural Resources Research, Chinese Academy of Sciences, Beijing, China, p. 90.
49. An, Z.-Z., Chen, T.-B., Lei, M., Xiao, X.-Y., and Liao, X.-Y. (2003) Tolerance of *Pteris vittata* L. to Pb, Cu and Zn. *Acta Ecol. Sin.* **23**, 2594–2598.
50. Ma, L.-Y. (2004) Exploration for improving the ability of arsenic accumulation in Chinese brake fern and in vitro propagation of the fern. Working Report of Postdoctoral Research. Institute of Geographical Sciences and Natural Resources Research, Chinese Academy of Sciences, Beijing, China, p. 90.
51. Li, W.-X. (2004) Studies on the arsenic distribution in Chinese brake and the two ways to improve the phytoextraction efficiency of arsenic. Working Report of Postdoctoral Research. Institute of Geographical Sciences and Natural Resources Research, Chinese Academy of Sciences, Beijing, China, p. 80.

28

Phytoremediation in Portugal

Present and Future

Cristina Nabais, Susana C. Gonçalves, and Helena Freitas

Summary

A specific database concerning the number of sites suitable for phytoremediation, i.e., those sites that contain contaminants in moderate concentrations in near-surface groundwater or in shallow soils, is not available in Portugal and field application of phytoremediation is practically nonexistent. However, there are some projects that have used this remediation technology and suggest its possible benefits concerning environmental pollution. For example, in a former gold mine a small-scale field trial has been carried out since 1998 to test a variety of inexpensive mineral amendments for the *in situ* inactivation of trace metals on the fine-grained spoils, allowing a better restoration of the vegetation cover.

Phytoremediation will have more success if appropriate stress-adapted plants are associated with efficient microbial isolates that can tolerate pollution. Additional research is needed if technologies based on the combined action of plants and the microbial communities they support within the rhizosphere are to be adopted in large-scale remediation actions.

Key Words: Phytoremediation; heavy metals; mycorrhizae; mine spoil; constructed wetlands.

1. Contaminated Areas in Portugal

The first step when considering remediation actions for contaminated sites is to have an inventory of the main spots, with a good description of the type of contaminated media (water, soil, sediment), the type and concentration of contaminants, and an evaluation of the environmental risk. This is the basic information to establish priority areas (based on environmental risk assessment) and the type of remediation that should take place in each contaminated area.

Phytoremediation is the use of vegetation for *in situ* treatment of contaminated soils, sediments, and water. It is applicable at sites containing organic, nutrient, or metal pollutants that can be accessed by the roots of plants and sequestered, degraded, immobilized, or metabolized in place *(1)*. Applications include hazardous waste sites where other methods of treatment are too expensive or impractical, low-level contaminated sites where treatment is required over long periods of time, and sites where phytoremediation is used in conjunction with other technologies such as a final cap and closure *(1)*.

The national potential for phytoremediation could be estimated by first totaling the number of sites that contain organic compounds and metals suitable for phytoremediation, i.e., those sites that contain contaminants in moderate concentrations in near-surface groundwater or in shallow soils *(1)*. Currently, such specific information about hazardous waste sites in Portugal is not available.

The main official institutions responsible for environmental issues in Portugal are: Instituto do Ambiente (IA; Environmental Institute); Instituto Nacional dos Resíduos (INR; National Residues Institute), Instituto Geológico e Mineiro (IGM; Geologic and Mining Institute), and Instituto Nacional da Água (INAG; National Water Institute). These institutions collect information concerning waste production from industry, abandoned mining areas, and water quality.

1.1. Waste Production From Industry

According to data from the INR, in 2001 the estimated total amount of industrial waste was 29×10^6 tons, of which 254×10^3 tons (0.9%) were considered hazardous waste *(2)*. Most of the hazardous waste was produced in the region of Lisboa e Vale do Tejo, followed by the north and center (**Fig. 1**). Of the total hazardous waste production, 48% are oil wastes and 13% are wastes from organic chemical processes.

The contamination of soils in Portugal is mainly caused by inappropriate disposal of waste and an excessive use of fertilizers and pesticides *(3)*. Detailed information about the location of these contaminated sites is not available.

1.2. Abandoned Mining Areas

One important source of environmental contamination is mining activity. The IGM is doing important work to characterize the mining situation in Portugal, particularly the old and abandoned mines of metallic ores, and the associated environmental risks *(4)*. **Figure 2** shows the location and classification of the environmental state of the abandoned mines in Portugal. In the north and center of Portugal the main extractives were coal, Au, Fe, Pb, Ag, Sn, W, Zn, and radioactive minerals *(5)*. In the South the main extractives were the sulfide deposits for Cu, Fe, and S *(5)*.

Phytoremediation in Portugal

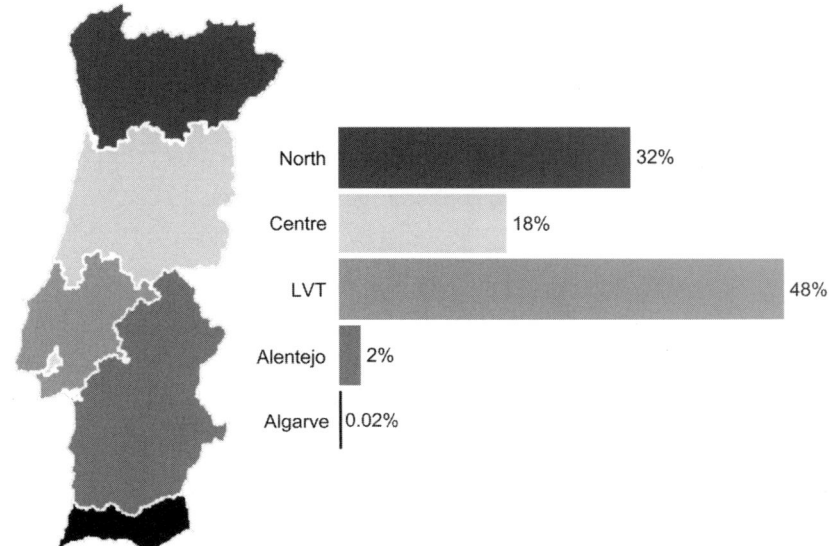

Fig. 1. Percentage of total hazardous waste production in the different regions of Portugal. (Modified after **ref. 2**).

One of the major environmental impacts of mining activity is the occurrence of high concentrations of trace metals (As, Cd, Co, Cr, Cu, Hg, Mn, Ni, Pb, Sb, Se, Zn) in the mine waste materials such as tailings, slags, and heap dumps *(6)*. Certain anions are also important environmental pollutants in mining areas, namely bromite, cianite, iodite, nitrate, nitrite, phosphate, and sulfate *(6)*. In northeast Portugal the presence of radioactive minerals is a significant environmental concern (**Fig. 2**). From the 85 abandoned mines studied, 14% were considered to present high environmental risk, mainly a result of trace metal contamination of water, soil, and sediment.

Remediation actions are already taking place in some of the mining areas. Those actions are mainly engineering interventions to stabilize the tailings and heap dumps. In addition, more detailed studies are being undertaken to characterize the environmental impact of mining using new methodologies, namely Earth observation techniques *(7)*.

1.3. Water Quality

According to a report from the IA *(8)* superficial water resources have low quality (**Fig. 3**). The water-quality parameters that were most often outside the legal limits were organic matter, bacteriological parameters, nutrients, and total suspended solids (**Fig. 4**). To improve the quality of water, a more detailed knowledge is necessary concerning land use.

Location of abandoned mining areas with risk assessment

n°	name	elements	n°	name	elements
1	Castelhão	Sn, W	44	Segura	Ba, Pb, W, Sn
2	Montesinho	Sn	45	Argemela	Sn
3	Guadramil	Fe	46	Escádia Grande	Au, Ag
4	França	Au, Ag	47	Várzea	Pb, Zn
5	Covas	W	48	Mata da Rainha	W, Sn
6	Carris	W, Mo	49	Talhadas	Cu, Pb, Ag
7	Bessa	Sn, W	50	Sarzedas	W, Sb, Au
8	Tuela	Sn	51	Tinoca	Cu
9	Murços	W, Sn	52	Miguel Vacas	Cu
10	Borralha	W	53A	Preguiça	Zn, Pb
11	Ribeira	W, Sn	53B	Vila Ruiva	Zn, Pb
12	S. Martinho	Sn, W	54	Grou	Sb, Au
13	Argozelo	Sn, W	55	Bugalho	Cu
14	Três Minas	Au	56	Mociços	Cu
15	Almendra	W, Sn	57	Chaminé	Au, Cu
16	Adoria	W, Sn	58	Caerinha	Cu
17	Freixeda	Au, Ag	59	Lousal	Pirite
18	Jales	Au	60	Caveira	Pirite
19	Vieiros	Sn	61	Aparis	Cu
20	Vale das Gatas	W, Sn	62	Lagoas do Paço	Mn
21	Fonte Santa	W	65	Orada	Fe
22	Moncorvo	Fe	66	S. Domingos	Pirite
23	S. Pedro da Cova	coal	67	Cortes Pereira	Sb
24	Terramonte	Pb, Zn, Ag	68	Ferrarias	Cu
25	Sta. Leocádia	W, Pb	69	Balôco	Pb, Cu
26	Banjas	Sb, Au	70	Azeiteiros	Cu
27	Várzea de Trevões	Pb, Zn, Ag	71	Sta. Eulália	Sn, W
28	Freixo de Numão	W, Sn	72	Mostardeira	Cu
29	Ribª de Alva	W, Sn	73	Monges	Fe, Py
30	Pintor	W, As	74	Nogueirinha	Fe, Py
31	Penedono	Au	75	Algares	Zn, Pb
32	Tarouca	W, Sn	76	Poço dos Freitas	Au
33	Regoufe	W, Sn	77	Ferragudo	Mn
34	Massueime	W, Sn, Li	78	Balança	Mn
36	Ladeira das Vinhas	W, Sn, Qz., Feld.	79	Montinho	Pirite
37	Raseira	Qz, Feld.	80	Chança	Pirite
38	Peixeiro	W	81	Barrigão	Cu
39	Tapada do Lobo	Qz, Feld.	82	Saramaga	Mn
40	Serra de Bois	Sn	83	Alvito	Fe
41A	Braçal	Pb	84	Juliana	Cu, Mn
41B	Coval da Mó	Pb	85	Cercal - Rosalgar	Mn
41C	Malhada	Pb	86	Defesa das Mercês	Cu
42	Góis - Vale Pião	Sn	87	Botefa	Cu
42A	Góis - Senhora da Guia	W	88	Azenhas	Fe
43	Monfortinho	Au			

Fig. 2. (*Continued*)

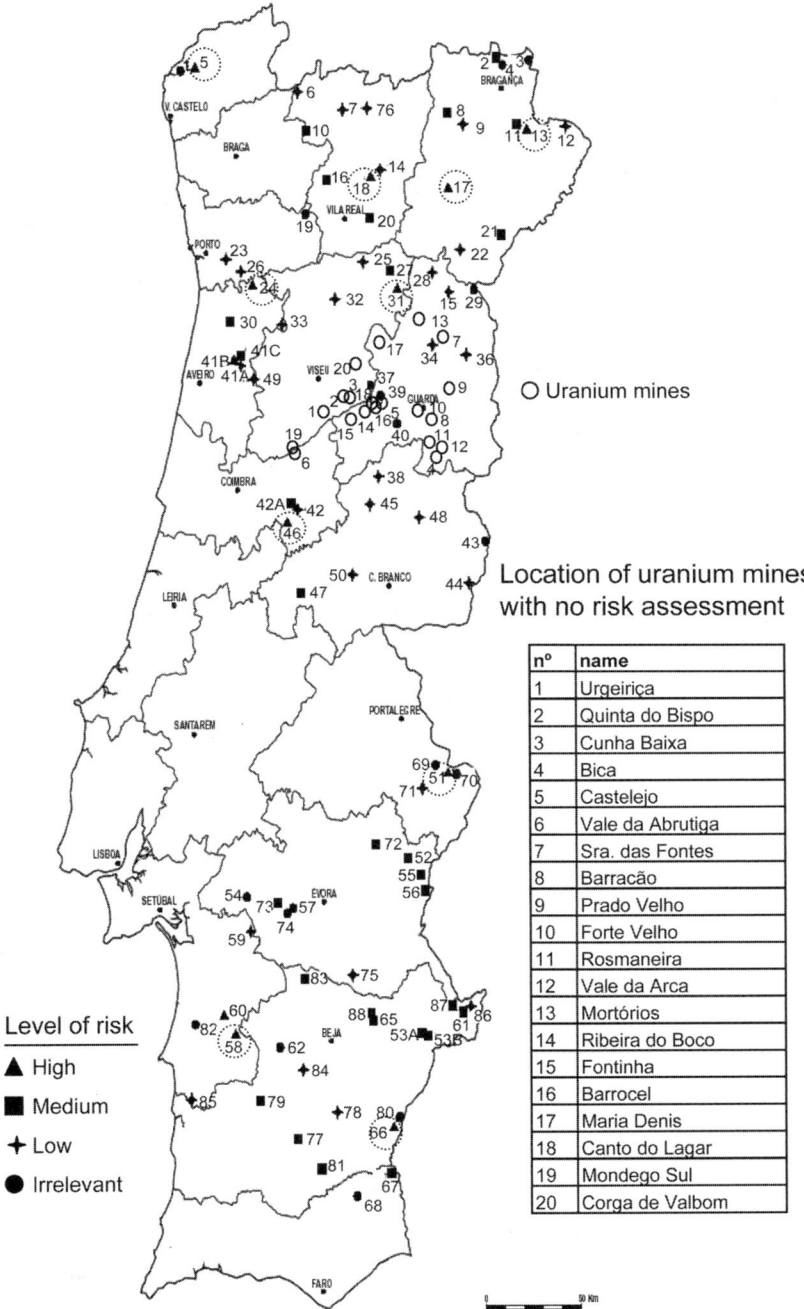

Fig. 2. Location of abandoned mining areas of Portugal (Modified after **ref. 5**).

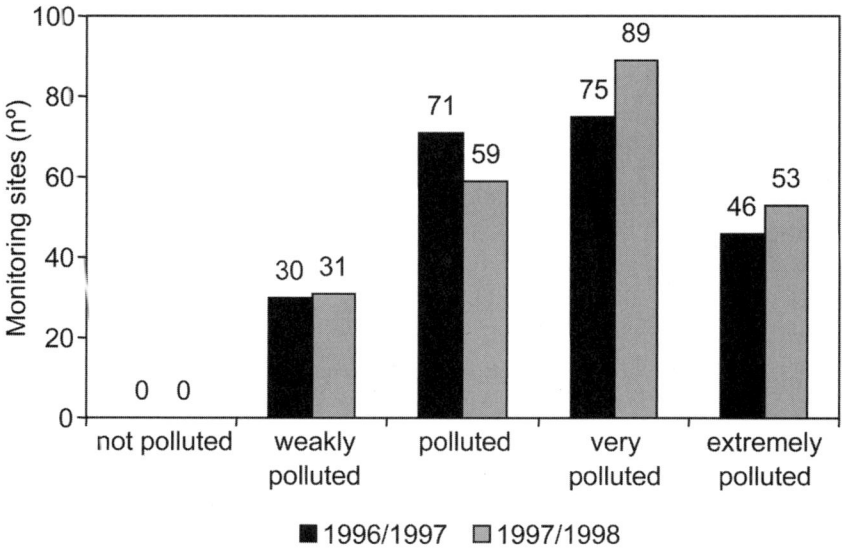

Fig. 3. Classification of water quality in Portugal (Modified after **ref. 8**).

2. Phytoremediation in Portugal: From the Laboratory to the Field

Few evaluations of full-scale phytoremediation have been reported in the literature. It is not enough to simply show that vegetation is growing at a contaminated site to prove its efficacy *(9)*. It is important to show that the plants are able to remediate site contaminants. Long-term, objective field evaluation is critical to understanding how well phytoremediation may work, what the real cost of application will be, and how to build models to predict the interaction between plants and contaminants *(9)*. Monitoring needs to address both the decrease in the concentration of the contaminants in the media of concern, and examine the fate of the contaminants *(1)*. The existing knowledge base is limited, and specific data are needed on more plants, contaminants, and climate conditions.

Limitations of the technology include the potential for introducing the contaminant or its metabolites into the food chain, the long time required for clean up to below-action levels, and toxicity encountered in establishing and maintaining vegetation at waste sites *(9)*. In Portugal, field application of phytoremediation is practically nonexistent. However, there are some projects that used this remediation technology and suggest its possible benefits concerning environmental pollution.

2.1. Constructed Wetlands for Water Treatment

The research group of the Centre for Biological and Chemical Engineering (Technical University of Lisbon) has been studying the effect of constructed

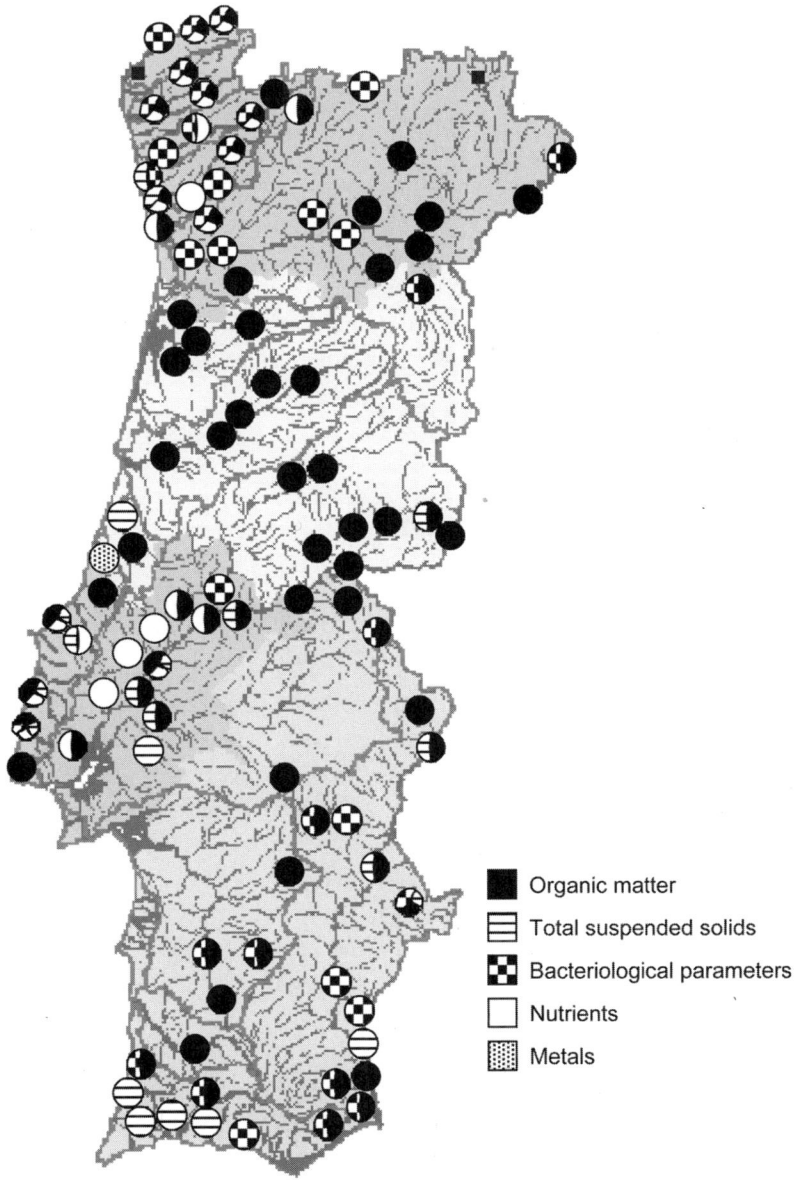

Fig. 4. Problematic water-quality parameters in Portugal (Modified after **ref. 8**).

wetlands on the removal of organic pollutants from industrial wastewaters. The objective was to treat polluted industrial effluents containing nitrogenous aromatic compounds from aniline and nitrobenzene production using reed beds (*Phragmites australis*). With an inlet effluent composition of 10–300 ppm aniline,

10–100 ppm nitrobenzene, and 10–30 ppm nitrophenols, a reduction of aromatic compounds up to 100% was obtained, using a total planted area of 10,000 m^2. The retention time of compounds varied between 1 and 20 d as a function of the size, nature, and charge of the molecule involved. The system has been monitored for about 6 yr *(10)*.

In a strict sense, constructed wetlands are not considered a type of phytoremediation because the primary mechanism of treatment is not by plants. Rather, plants provide the niche for bacteria to flourish and utilize nutrients, degrade organic compounds, and bind or precipitate metals *(9)*. Soil microbial activity is enhanced by the presence of root exudates that are compounds (e.g., sugars, amino acids, organic acids) produced by plants and released from plant roots.

2.2. Phytoremediation in Mining Areas

Metals cannot be degraded so at sites with low contamination levels or vast contaminated areas where a large-scale removal action or other *in situ* remediation is not feasible, stabilizing them *in situ* is sometimes the best alternative *(1)*. The elements As, Cd, Co, Cu, Ni, Pb, Zn can be treated by phytostabilization and phytoextraction; Hg and Se with phytovolatilization, and metals and radionuclides with rhizofiltration *(1)*. In the former gold mine of Jales a small-scale semifield trial has been carried out since 1998 to test a variety of inexpensive mineral amendments for the *in situ* inactivation of trace metals on the fine-grained spoils *(11)*. Compost combined with steelshots (iron grit) and/or beringite additives allowed a better restoration of the vegetation cover *(11)*. The amendments decreased As, Cd, and Zn taken up by plants.

In addition, after human disturbance such as mining, revegetation may occur slowly. Recolonization of contaminated or disturbed ground by plants typically starts at the edges of an impacted area *(1)*. There is already a database concerning the plants growing in and around some of the Portuguese mining areas *(12,13)*. These plants can have an important role in the revegetation of the mining areas.

2.3. Riparian Corridors

In the Alentejo region (south Portugal) research was carried out to understand the origin of phenolic compounds detected in the water of dams. The main phenolic compound detected was 2,4-dinitrophenol, a synthetic compound, indicating that its origin was the pesticides used in agriculture *(14)*. The detection of the phenol was especially prominent after the application of pesticides and was followed by a raining event in the second half of February (**Fig. 5**). In one of the dams studied (mainly surrounded with agriculture), the Santa Vitória river showed lower levels of phenolic compounds compared with other rivers (**Fig. 5**). The major difference of this sampling area was that vegetation

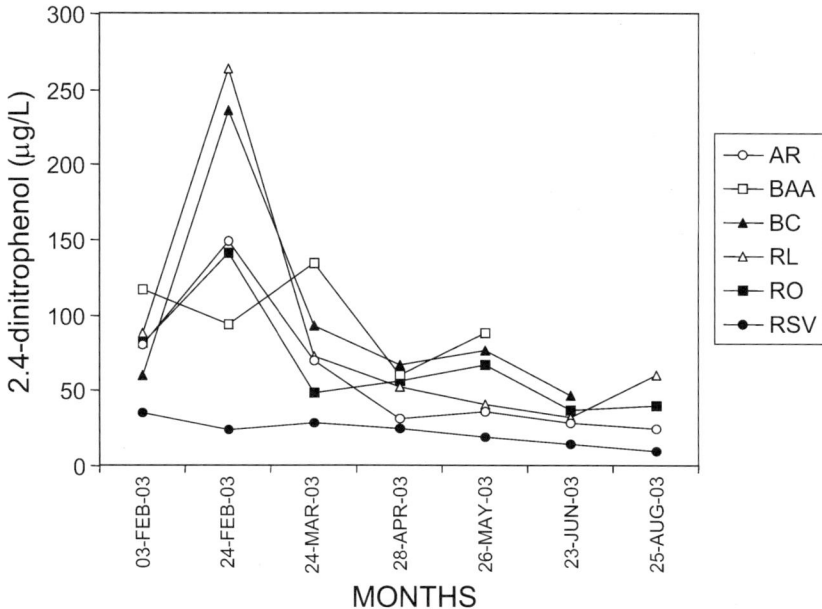

Fig. 5. Seasonal concentration of 2,4-dinitrophenol in some rivers of Roxo dam (Alentejo, S. Portugal). Ar, Albufeira do Roxo; BAA, Barranco da Água Azeda; BC, Barranco dos Castelhanos; RL, Ribeira dos louriçais; RO, Ribeira do Outeiro; RSV, Ribeira de Santa Vitória.

was growing in the margins of the river (e.g., *Juncus*, *Scirpus*, and *Typha* spp.) and probably had an important role in "cleaning" the water.

Although the objective of this research was not phytoremediation, it showed a likely beneficial role of plants in the decontamination of water. It is known that riparian corridors/buffer strips are generally applied along streams and river banks to control and remediate surface runoff and groundwater contamination moving into the river *(1)*.

3. Phytoremediation Research in Portugal
3.1. Trace metals: Metallophytes and Mycorrhizal Fungi in Focus

Worldwide ecological principles are now being considered in trace-metal-contaminated site restoration. Ecological restoration involves returning a site to a state as close as possible to the one existing before, thus requiring high native plant diversity rather than monocultures or introduced species *(15)*. To accomplish this task, metallophytes—the plant species that thrive on metal-rich substrates—should be identified, conserved, and studied in relation to their metal tolerance and ecological function *(16)*.

Freitas et al. *(13)* have sampled plants from populations established in an abandoned copper mine in Mina de São Domingos, southeast Portugal. Sampling resulted in the collection of 24 plant species, representing 16 genera and 13 families. Plant samples were analyzed for total Ag, As, Co, Cu, Ni, Pb, and Zn. The higher concentrations of Pb and As were recorded in the semi-aquatic species *Juncus conglomeratus* with 84.8 and 23.5 mg/kg dry weight (DW), *Juncus effusus* with 22.4 and 8.5 mg/kg DW, and *Scirpus holoschoenus* with 51.7 and 8.0 mg/kg DW, respectively. *Thymus mastichina* also had a high content of As in the above-ground parts, 13.6 mg/kg DW. A concentration of Pb more than 20 mg/kg DW was found in the leaves of three species of *Cistus*, typical mediterranean shrubs known for their tolerance to drought and low nutrient availability *(17)*. Mine restoration could benefit from including plant species belonging to different functional groups as they could perform distinct roles in the remediation process *(18,19)*. Freitas et al. *(12)* have analyzed a serpentine plant community from northeast Portugal to examine the trace metal budget in different tissues of the plants growing in this habitat naturally enriched in metals, such as Cr and Ni. One hundred and thirty five plant species belonging to 39 families were analyzed for total Co, Cr, Fe, Mn, Ni, Pb, and Zn. Although *Alyssum serpyllifolium* was confirmed to be the only hyperaccumulator of Ni, reaching 38,105 mg/kg DW in the above-ground tissue, the uptake of Cr by *Linaria spartea* was remarkably high, 707 mg/kg DW.

Comprehensive surveying of contaminated sites aimed at ecological restoration should also include soil micro-organisms. Among these, mycorrhizal fungi should receive special attention as they represent direct links between plants and soil and may prove essential for revegetation efforts in contaminated soils, mainly through phytoextraction or phytostabilization *(20)*. Evidence of mycorrhizal colonization in serpentinophytes from Portugal was shown by Gonçalves et al. *(21)*. In this study, ectomycorrhizas (ECM) were reported in *Quercus ilex* and the species *Cenococcum geophilum* was detected. This fungus, the dominant ECM morphotype in Portuguese serpentine, has been, ever since, the subject of research *(22–24)*. Interest in *C. geophilum* was prompted because of its wide geographical distribution and broad host range *(25)* and the knowledge that it is also a common partner of the mediterranean shrubs *Cistus* sp. (Gonçalves, personal observation). Moreover, several studies reported that *C. geophilum* is more resilient to stress than other ECM fungi *(26)* and may protect roots from desiccation under severe drought stress when in symbiosis *(27,28)*. Portugal et al. *(22,23,29–31)* analyzed the genetic variability of *C. geophilum* isolates obtained from serpentine and non-serpentine soils in northeast Portugal using PCR/RFLP of the nuclear rDNA, IGS, ITS1, and ITS2 regions, microsatellite-primed PCR, and AFLP profiles. Results confirmed that *C. geophilum* is a very heterogeneous species, showing a high level of polymorphism. UPGMA analysis of the microsatellite data indicated that the isolates from the same morphotype are more similar to each other than they are to the isolates from another morphotype, regardless of the serpentine or nonserpentine origin. From these results, no genetic divergence seems to have occurred based on a serpentine effect *(32)*. Interestingly, preliminary results on the isolates' responses to Ni suggest that different morphotypes also exhibit distinct physiological responses (Gonçalves, unpublished data). More studies are needed to unravel the role of *C. geophilum* to *Q. ilex* in serpentine soils. Highly mycotrophic trees, such as oak, are unlikely to colonize a site that is too toxic to support ECM fungi. Thus, in these soils fungal tolerance to stress may be of great value to the plant, even if the fungus does not reduce trace metal uptake *(33)*.

Arbuscular mycorrhizas (AM) were also observed in the Portuguese serpentine plant species *Anthyllis sampaiana* and *Festuca brigantina (21)*, both belonging to genera with potential for revegetation *(34,35)*. AM were further studied in *F. brigantina (36)*. The colonization was quantified monthly throughout one growing cycle of the species and related to plant phenology, phosphorus nutrition, and Ni concentrations of soil and plant tissues. *F. brigantina* was found systematically colonized during the sampling period, the highest colonization levels preceding the beginning of the plant's reproductive period and maximum shoot investment. An extremely low shoot/root Ni concentration ratio was observed, in agreement with Menezes de Sequeira and Pinto da Silva *(37)*, the first authors devoted to serpentine ecology in Portugal. AM fungi could be directly involved in reducing Ni translocation to the shoot in *F. brigantina*, possibly through Ni binding by hyphal wall components or sequestration within the fungal hyphae *(38)*. However, because no correlation was found between AM colonization and Ni translocation to the shoot, the results from this study did not support this hypothesis. Nevertheless, a significant relationship was found between AM colonization and root phosphorus concentration suggesting an improvement of *F. brigantina* P nutrition by the fungi. As in this study one could not distinguish between the P in the root plant cells from the P in the fungal hyphae, it is possible that the fungi sequester Ni in the form of polyphosphates inside the vacuoles *(39,40)*. Further research is needed to test this hypothesis.

3.2. Organic Pollutants and Trace Metals: Integrating Microbial Aspects

Several studies are being pursued on the microflora of the rhizosphere and roots of plants inhabiting contaminated soils and sediments in Estarreja, northern Portugal *(41)*. The study sites have received the discharge of chemical industry effluents for over 50 yr, including trace metals and organic pollutants.

Phragmites has been used in constructed wetlands designed to treat contaminated effluents in Portugal and elsewhere *(10,42,43)*. A bacterial consortium has been obtained from the rhizosphere of *P. australis*, which colonizes these sites *(44)*. The consortium was capable of utilizing 4-nitrophenol (4-NP) as the sole carbon and energy source. Furthermore, the biodegradation of 4-NP was enhanced in the presence of plant extract obtained from *P. australis*. The mycorrhizal status of *P. australis* was also investigated *(45)*. Root colonization by AM fungi was low but arbuscules peaked during spring and autumn prior to flowering. In this study, the mycorrhizal colonization of *J. effusus* and *Salix atrocinerea*, two other plant species growing in these soils, was also registered. The results confirmed that semiaquatics can become colonized by AM fungi and that colonization status changes during the year depending on soil moisture content and plant phenology. The colonization was apparently not influenced by soil or sediment pH (soils pH 4.1 and 7.1, sediments 12.6). The role of mycorrhizal fungi in these polluted soils is still uncertain but the potential exists to establish a more diverse plant ecosystem during remediation of these areas by management of adapted plants and mycorrhizal fungal partners *(46)*.

More recently, Oliveira et al. *(47)* have focused their attention on the remediation of sediments from the waste of an acetylene and polyvinylchloride factory in the area. These authors presented a model using *Alnus glutinosa* together with inoculated stress-tolerant bacteria and fungal symbionts for the phytoremediation of these sediments. *Alnus* establishes symbiosis with *Frankia* but also with AM and ECM fungi *(48)*. The presence of the symbionts may reduce stress caused by extreme pH (11.0–12.0) and the lack of nutrients in the sediment. The authors conducted a 6-mo greenhouse experiment using *A. glutinosa* seedlings inoculated either with the AM fungus *Glomus intraradices* (BEG 163), *Frankia* sp., or both symbionts. Plants inoculated with both symbionts had significantly greater total leaf area, shoot height and biomass, root collar diameters, and total biomass when compared with the uninoculated controls, the *Frankia* sp. alone, and the *G. intraradices* treatment alone.

4. Future Perspectives for Phytoremediation in Portugal
4.1. Improvement of a National Database of Contaminated Sites

The national official institutions have collected data concerning environmental quality, especially in the last 10 yr. However, more detailed information about the location, origin and extent of soil, sediment, and water contamination is needed as well as an evaluation of the environmental risk. This information is important to decide on the remediation actions to take place and their urgency. Additionally, a good database on contaminated sites is necessary to establish research priority areas on environmental pollution. It would also be important to establish multidisciplinary research groups including biologists, geologists, chemists, and hydrologists to study contaminated areas from complementary disciplines.

The use of phytoremediation in Portugal should be considered more seriously as an alternative/complementary remediation action by the official institutions responsible for environmental quality.

4.2. Research Directions: Some Suggestions

The role of mycorrhizosphere bacteria and fungi in plant ecology can no longer be ignored because of their impacts on plant community structure *(49)*. Phytoremediation will have more success if appropriate stress-adapted plants are associated with efficient microbial isolates that can tolerate pollution. These organisms may prove crucial in plant establishment in degraded ecosystems with high levels of trace metals and/or organic pollutants *(50)*. For instance, a survey of the literature data shows that only one of the 24 plant species studied by Freitas et al. *(13)* belongs to a genus considered nonmycotrophic (*Rumex*).

Additional research is needed if technologies based on the combined action of plants and the microbial communities they support within the rhizosphere are to be adopted in large-scale remediation actions *(51)*. Matching appropriate ecotypes of plants and microbes is an important task. The selection of effective microbial isolates should include investigations of their colonization abilities, mechanisms of microbial tolerance, host specificity, and competitiveness in soil *(20,50)*.

The establishment of a plant cover is important for stabilization, pollution control, and visual improvement. However, the rate of natural revegetation is usually slow because of low levels of nutrients and the lack of organic matter in the soil. The presence of beneficial micro-organisms could speed up the process. However, at least for mycorrhizal fungi, spontaneous establishment is low *(52,53)*. Therefore, stimulation of native microbial populations or inoculation with isolates adapted to contaminants will enhance the rate of restoration. Fungi to be used in inoculation can be isolated from areas that are either naturally enriched in trace metals, like serpentines, or originate from old mine/industry wastes. Serpentine plant communities are a good example of plant adaptation to trace metal toxic levels, the hyperaccumulation trait being the most remarkable *(54)*. Moreover, serpentine biodiversity offers genetic material that might be used in remediation and ecological restoration of trace-metal-contaminated sites. In these habitats, coevolution of plants and micro-organisms is likely to have occurred. This demands investigation and urgent conservation efforts focused on serpentines *(55,56)*.

Anthropogenic contamination can also drive the evolution of stress-tolerant ecotypes. In the contaminated areas of São Domingos and Estarreja, it is reasonable to admit that the plant and micro-organisms found are tolerant ecotypes. Thus, the remediation process could benefit from practices that would enhance microbial populations. It is noteworthy that ECM fungi, and possibly AM fungi, are able to degrade organic pollutants, implying a role for these organisms in phytodegradation *(57,58)*. In Portugal, soils contaminated with uranium also require remediation. Here again, micro-organisms may play an important role *(59)* and demand further investigation.

References

1. EPA (2000) *Introduction to phytoremediation.* Environmental Protection Agency (EPA) Report EPA/R-99/107.
2. INR (2003) *Estudo de inventariação de resíduos industriais.* Relatório de Síntese.
3. Jorge, C. (1999) Contaminação do solo: potenciais zonas em Portugal. *Revista Ambiente Magazine,* 22.
4. Oliveira, J. M. S. (1997) *Algumas reflexões com enfoque na problemática dos riscos ambientais associados à actividade mineira.* Estudos, Notas e Trabalhos 39, Instituto Geológico e Mineiro.
5. Oliveira, J. M. S., Farinha, J., Matos, J. X., et al. (2002) *Diagnóstico ambiental das principais áreas mineiras degradadas do país.* Boletim de Minas, 39, Instituto Geológico e Mineiro.
6. Oliveira, J. M. S., Machado, M. J. C., Pedrosa, M. Y. Ávila, P., and Leite, M. R. M. (1999) *Programa de investigação e controlo ambientais em áreas do país com minas abandonadas: compilação de resultados.* Estudos, Notas e Trabalhos. 41, Instituto Geológico e Mineiro.
7. Quental, L., Bourguignon, A., Sousa, A. J., et al. (2002) *MINEO Southern Europe environment test site. Contamination/impact mapping and modelling: final report.* MINEO IST-1999-10337.
8. DGA (2000) *Relatório do Estado do Ambiente 1999.* Direcção Geral do Ambiente, Ministério do Ambiente e Recursos Naturais, Lisboa, Portugal.
9. Schnoor, J. L. (2002) *Phytoremediation of soil and groundwater.* Ground-Water Remediation Technologies Analysis Center (GWRTAC) Technology Evaluation Report TE-02-01.
10. Dias, S. M. (2000) Nitro-aromatic compounds removal in a vertical flow reed bed case study: industrial wastewater treatment. In: *Intercost Workshop on Bioremediation,* Sorrento, pp. 104–105.
11. Mench, M., Bussière, S., Boisson, J., et al. (2003) Progress in remediation and revegetation of the barren Jales gold mine spoil after *in situ* treatments. *Plant Soil* **249,** 187–202.
12. Freitas, H., Prasad, M. N. V., and Pratas, J. (2004) Analysis of serpentinophytes from north-east of Portugal for trace metal accumulation—relevance to the management of mine environment. *Chemosphere* 54, 1625–1642.
13. Freitas, H., Prasad, M. N. V., and Pratas, J. (2004) Plant community tolerant to trace elements growing on the degraded soils of São Domingos mine in the south east of Portugal: environmental implications. *Environ Int.* **30,** 65–72.

14. Nabais, C., Barrico, M. L., Martins, M. J., Castro, H., and Freitas, H. (2003) *Avaliação do contributo de espécies vegetais para a contaminação das águas das bacias hidrográficas das albufeiras de Santa-Clara e do Roxo por compostos fenólicos—relatório final*. Direcção Regional do Ambiente e Ordenamento do Território do Alentejo (DRAOT-Alentejo).
15. Dobson, A. P., Bradshaw, A. D., and Baker, A. J. M. (1997) Hopes for the future: restoration ecology and conservation biology. *Science* **277**, 515–522.
16. Whiting, S. N., Reeves, R. D., and Baker, A. J. M. (2002) Mining, metallophytes and land reclamation. *Mining Environmental Management* March, 11–16.
17. Correia, O. (2002) Os *Cistus*: as espécies do futuro? In: *Fragmentos em Ecologia*, (Martins-Loução, M. A., ed.), Faculdade de Ciências da Universidade de Lisboa, Escolar Editora, Lisboa, Portugal, pp. 97–119.
18. Hooper, D. U. and Vitousek, P. M. (1997) The effects of plant composition and diversity on ecosystem processes. *Science* **277**, 1302–1305.
19. Hooper, D. U. and Vitousek, P. M. (1998) Effects of plant composition and diversity on nutrient cycling. *Ecol Monogr.* **68**, 121–149.
20. Leyval, C., Joner, E. J., del Val, C., and Haselwandter, K. (2002) Potential of arbuscular mycorrhizal fungi for bioremediation. In: *Mycorrhizal Technology in Agriculture,* (Gianinazzi, S., Shuëpp, H., Barea, J. M., and Haselwandter, K., eds.), Birkhäuser Verlag, Basel, Switzerland, pp. 175–186.
21. Gonçalves, S. C., Gonçalves, M. T., Freitas, H., and Martins-Loução, M. A. (1995) Mycorrhizae in a Portuguese serpentine community. In: *The Proceedings of the Second International Conference on Serpentine Ecology,* (Jaffré, T., Reeves, R., and Becquer, T., eds.), Nouméa, pp. 87–90.
22. Portugal, A., Martinho, P., Vieira, R., and Freitas, H. (2001) Molecular characterization of *Cenococcum geophilum* isolates from an ultramafic soil in Portugal. *S. Afr. J. Sci.* **97**, 617–619.
23. Portugal, A., Gonçalves, S. C., Vieira, R., and Freitas, H. (2003) Characterization of *Cenococcum geophilum* isolates from a serpentine area by microsatellite-primed PCR. A tool for future revegetation programmes. In: *Proceedings of the Fourth International Conference on Serpentine Ecology,* (Baker, A. J. M., Boyd, R. S., and Iturralde, R. B., eds.), in press.
24. Gonçalves, S. C., Portugal, A., Gonçalves, M. T., Vieira, R., Martins-Loução, M. A., and Freitas, H. (2003) *Cenococcum geophilum* isolated from serpentine and non-serpentine soils: genetic variation and in vitro response to Ni and Mg/Ca ratio. In: *Abstracts of the Fourth International Conference on Serpentine Ecology*, Havana, pp. 18–19.
25. Horton, T. R. and Bruns, T. D. (2001) The molecular revolution in ectomycorrhizal ecology: peeking into the black-box. *Mol. Ecol.* **10**, 1855–1871.
26. Mexal, J. and Reid, C. P. P. (1973) Growth of selected mycorrhizal fungi in response to induced water stress. *Can. J. Bot.* **51**, 1579–1588.
27. Pigott, C. D. (1982) Survival of mycorrhiza formed by *Cenococcum geophilum* Fr. in dry soils. *New Phytol.* **92**, 513–517.
28. Jany, J. L., Martin, F., and Garbaye, J. (2003) Respiration activity of ectomycorrhizas from *Cenococcum geophilum* and *Lactarius* sp. in relation to soil water potential in five beech forests. *Plant Soil* **255**, 487–494.

29. Portugal, A., Martinho, P., Freitas, H., and Vieira, R. (2001) Molecular characterization of *Cenococcum geophilum* isolates—a case study. *J. Medit. Ecol.* **2**, 21–30.
30. Portugal, A., Vieira, R., and Freitas, H. (2001) The use of genetic markers in the characterisation of the mycobiont *Cenococcum geophilum* isolates from ultramafic soils of NE Portugal. *Revista de Ciências Agrárias* **24**, 227–238.
31. Portugal, A., Vieira, R., and Freitas, H. (2003) Molecular characterisation by AFLP of ectomycorrhizal fungus *Cenococcum geophilum* isolates from ultramafic rocks derived soils of NE Portugal. *Revista de Ciências Agrárias*, in press.
32. Proctor, J. (1999) Toxins, nutrient shortage and droughts: the serpentine challenge. *Trends Ecol. Evol.* **14**, 334–335.
33. Jentschke, G. and Goldbold, D. L. (2000) Metal toxicity and ectomycorrhizas. *Physiol Plant* **109**, 107–116.
34. Hetrick, B. A. D., Wilson, G. W. T., and Figge, D. A. H. (1994) The influence of mycorrhizal symbiosis and fertilizer amendments on establishment of vegetation in heavy metal mine spoil. *Environ. Pollut.* **86**, 171–179.
35. Requena, N., Perez-Solis, E., Azcon-Aguilar, C., Jeffries, P., and Barea, J. M. (2001) Management of indigenous plant-microbe symbioses aids restoration of desertified ecosystems. *App. Environ. Microbiol.* **67**, 495–498.
36. Gonçalves, S. C., Martins-Loução, M. A., and Freitas, H. (2001) Arbuscular mycorrhizas of *Festuca brigantina*, an endemic serpentinophyte from Portugal. *S. Afr. J. Sci.* **97**, 571–572.
37. Menezes de Sequeira, E. and Pinto da Silva, A. R. (1992) The ecology of serpentinized areas of north-east Portugal. In: *The Ecology of Areas With Serpentinized rocks. A World View* (Roberts, B. A. and Proctor, J., eds.), Kluwer Academic Publishers, Dordrecht, Germany, pp. 169–197.
38. Galli, U., Schuepp, H., and Brunold, C. (1994) Heavy-metal binding by mycorrhizal fungi. *Physiol. Plant* **92**, 364–368.
39. Kunst, L. and Roomans, G. M. (1985) Intracellular-localization of heavy-metals in yeast by X-ray-microanalysis. *Scan. Electron Microsc.* **1**, 191–199.
40. Jones, M. D. and Hutchinson, T. C. (1988) Nickel toxicity in mycorrhizal birch seedlings infected with *Lactarius rufus* or *Scleroderma flavidum*. 1. Effects on growth, photosynthesis, respiration and transpiration. *New Phytol.* **108**, 451–459.
41. Oliveira, R., Carvalho, F., Manaia, C., Dodd, J. C., and Castro, P. (2000) Plants colonising polluted sites: integrating microbial aspects. In: *Abstracts From the Intercost Workshop on Bioremediation*, Sorrento, p. 41.
42. Dias, S. M. (1998) Tratamento de efluentes em zonas húmidas construídas ou leitos de macrófitas. *Boletim de Biotecnologia* **60**, 14–20.
43. Haberl, R., Grego, S., Langergraber, G., et al. (2003) Constructed wetlands for the treatment of organic pollutants. *J. Soil Sed.* **3**, 109–124.
44. Oliveira. R. S., Zarzycki, R., Manaia, C. M., and Castro, P. M. L. (2001) Influence of plant components on the degradation of 4-nitrophenol by a bacterial consortium isolated from the rhizosphere of *Phragmites australis*. *Minerva Biotechnologica* **13**, 27–31.

45. Oliveira, R. S., Dodd, J. C., and Castro, P. M. L. (2001) The mycorrhizal status of *Phragmites australis* in several polluted soils and sediments of an industrialised region of Northern Portugal. *Mycorrhiza* **10**, 241–247.
46. Vangronsveld, J., Colpaert, J. V., and van Tichelen, K. K. (1996) Reclamation of a bare industrial area contaminated by non-ferrous metals: physicochemical and biological evaluation of the durability of soil treatment and revegetation. *Environ. Pollut.* **94**, 131–140.
47. Oliveira, R. S., Castro, P. M. L., Dodd, J. C., and Vosátka, M. (2003) The effect of two microbial symbionts, *Glomus intraradices* and *Frankia* sp. on the growth of *Alnus glutinosa* in extremely alkaline anthropogenic sediments. In: *Abstracts of The Fourth International Conference On Mycorrhiza*, Montréal, Canada, p. 463.
48. Rose, S. L. (1980) Mycorrhizal associations of some actinomycete nodulated nitrogen-fixing plants. *Can. J. Bot.* **58**, 1449–1454.
49. van der Heijden, M. G. A., Klironomos, J. N., Ursic, M., et al. (1998) Mycorrhizal fungal diversity determines plant biodiversity, ecosystem variability and productivity. *Nature* **396**, 69–72.
50. Turnau, K. and Haselwandter, K. (2002) Arbuscular mycorrhizal fungi, an essential component of soil microflora in ecosystem restoration. In: *Mycorrhizal Technology in Agriculture,* (Gianinazzi, S., Shuëpp, H., Barea, J. M., and Haselwandter, K., eds.), Birkhäuser Verlag, Basel, Switzerland, pp. 137–149.
51. Harvey, P., Campanella, B., Castro, P. M. L., et al. (2002) Phytoremediation of polyaromatic hydrocarbons, anilines and phenols. *Environ. Sci. Pollut. Res.* **9**, 29–47.
52. Turnau, K. (1998) Heavy metal content and localization in mycorrhizal *Euphorbia cyparissias* from zinc wastes in southern Poland. *Acta Societatis Botanicorum Poloniae* **67**, 105–113.
53. Turnau, K., Ryszka, P., Gianinazzi-Pearson, V., and van Tuinen, D. (2001) Identification of arbuscular mycorrhizal fungi in soils and roots of plants colonizing zinc wastes in southern Poland. *Mycorrhiza* **10**, 169–174.
54. Baker, A. J. M. and Whiting, S. N. (2002) In search of the Holy Grail: a further step in understanding metal hyperaccumulation? *New Phytol.* **155**, 1–7.
55. Prasad, M. N. V. and Freitas, H. (1999) Environmental biotechnology feasible biotechnological and bioremediation strategies for serpentine soils and mine spoils. *Electr. J. Biotechnol.* **2**, http://www.ejbiotechnology.Info/content/vol2/issue1/full/5/index.html.
56. Prasad, M. N. V. and Freitas, H. (2003) Metal hyperaccumulation in plants—biodiversity prospecting for phytoremediation technology. *Electr. J. Biotechnol.* **6**, http://www.ejbiotechnology.info/content/vol6/issue3/full/6.
57. Meharg, A. A. and Cairney, J. W. G. (2000) Ectomycorrhizas: extending the capabilities of rhizosphere remediation? *Soil Biol. Biochem.* **32**, 1475–1484.
58. Joner, E. J., Johansen, A., Loibner, A. P., et al. (2001) Rhizosphere effects on microbial community structure and dissipation and toxicity of polycyclic aromatic hydrocarbons (PAHs) in spiked soil. *Environ. Sci. Technol.* **35**, 2773–2777.
59. Rufyikiri, G., Thiry, Y., and Declerck, S. (2003) Contribution of hyphae and roots to uranium uptake and translocation by arbuscular mycorrhizal carrot roots under root-organ culture conditions. *New Phytol.* **158**, 391–399.

29

Phytoremediation in Russia

Yelena V. Lyubun and Dmitry N. Tychinin

Summary

Phytoremediation is taking a prominent place in the processes of environmental cleanup from hazardous pollutants, and there is increasing interest among Russian scientists (and, quite importantly, among various organizations) in this technology. This chapter reviews the current state of phytoremediation research in the Russian Federation. Topics addressed are the use of crops and grasses in soil remediation from oil hydrocarbons, chemical warfare agent degradation products, and heavy metals; and the use of algae, duckweed, water hyacinth, and miscanthus in water remediation from oil hydrocarbons and heavy metals.

Key Words: Phytoremediation; oil hydrocarbons; chemical warfare agent degradation products; heavy metals.

1. Introduction

Environmental pollution is a major socio-economic problem facing all the world's nations today. Many natural and man-made factors—in particular the growth of industrial production made possible by scientific and technological progress, population growth, and urban expansion—increasingly affect our environment. Land contamination resulting from human activities remains a priority problem for 48.2% of Russia's territory, including well-developed industrial areas (chemicals and petrochemicals, nonferrous metallurgy), densely populated areas, and areas affected by the Chernobyl disaster. Soil pollution by oil and its products, heavy metals, agrochemicals, and various types of waste (industrial, domestic, agricultural) is of the greatest ecological concern. In 9.6% of Russia's territory (Kemerovo, Smolensk, Penza, and Cheliabinsk Regions, Krasnoyarsk Territory, the Khanty-Mansi and Yamal-Nenets Autonomous Regions), land contamination has reached an ecological crisis point.

Despite economic hardship, environmental protection is on the list of Russia's national priorities. The federal program "Priority Directions for Research, Technology, and Engineering in the Russian Federation," approved by President Vladimir Putin on March 30, 2002, aims specifically at promoting new environmental cleanup technologies. One such technology is phytoremediation—the use of pollutant-hyperaccumulating plants in soil and water remediation. This chapter covers mostly Russian-language publications, which are not readily available to the majority of Western readers.

2. Soil Phytoremediation
2.1. Oil

Russia's fuel and energy complex remains the principal source of environmental pollution, accounting for about 48% of atmospheric emissions and 27% of polluted waste water discharged into surface water bodies. Oil hydrocarbons find their way into the environment during oil extraction, transportation, processing, and storage *(1)*. The dangerous consequences of the pollution become manifest in all ecosystem components, including soil. As a rule, oil pollution leads to considerable decreases in soil fertility, making recultivation necessary. The planting of phytoremediating crops and grasses is essential to soil fertility restoration and is recommended as the final stage of polluted-soil biorecultivation *(2)*. The remedial action of grasses is determined by their ability to produce above- and below-ground biomass and to selectively accumulate specific elements, depending on physiological peculiarities and on ecological conditions *(3)*. Successful soil remediation depends largely on the choice of effective phytoremediating plants.

Oil and oil products are major technogenic pollutants in the Republic of Bashkortostan, eastern Russia. For over 7 yr, scientists at Bashkir State University have been studying the effect of oil-contaminated soils on agricultural plants *(4,5)*. They examined the effect of oil addition on the growth and development of common chickweed (*Stellaria media* L.), couch grass (*Elytrigia repens* L.), and barnyard grass (*Echinochloa crusgalli* L.) at both laboratory and field scales. Addition of oil prolonged the vegetative period, but the plant height in polluted soil did not reach the control value. In polluted (0.5–4%) soil, a decrease in the yield of the above-ground biomass was observed. Under the effect of oil, the protein content of couch grass hay decreased twofold. The herbage of all three grasses contained 3,4-benzo(a)pyrene 10- to 15-fold in excess of the background maximum allowable values, ruling out the possibility that these grasses could be used as fodder crops. However, because couch grass is able to form turf, it can be recommended for use on oil-contaminated lands with a view to recovering soil fertility *(6)*.

Laboratory experiments conducted at Kazan State University *(7)* were aimed at selecting crops resistant to hydrocarbon pollution. The experiments were performed with kerosene-polluted, leached chernozem. The germination of seeds of some unconventional fodder plants (fescue grass [*Festuca*], brome [*Bromopsis*], timothy [*Phleum*], goat's-rue [*Galega officinalis* L.], clover [*Trifolium*], sainfoin [*Onobrychis viciifolia* Scop.], holy thistle [*Silybum*], and corn [*Zea mays*]) at various hydrocarbon concentrations (0, 1, 2, 3, 5, 10, and 15%), the accumulation of above-ground and root biomass, and the residual hydrocarbon content in the soil were determined. At a 1% concentration, a phytoremediation effect was found for sainfoin, corn, goat's-rue, and clover (the soil hydrocarbon content decreased 1.5- to 6-fold, as compared with unplanted soil); at a 2% concentration, for sainfoin and goat's-rue (the decrease was 1.8- to 4.3-fold). Sainfoin, goat's-rue, and clover excel in tolerance with respect to several parameters *(7)*.

Davydova and Pakhnenko-Durynina *(8)* conducted a greenhouse experiment aimed at detecting changes in the functional properties of oil-contaminated soil by the response of agricultural plants with various sensitivities to adverse environmental factors. The experiment used soil from the plowing horizon of leached chernozem. The oil concentration was 10 g/kg of soil—a pollution level 10-fold greater than the oil-concentration threshold in soil *(6)*. The soil had been contaminated with oil before agricultural crops were planted. The plants grown in clean and polluted soil were corn (*Z. mays*), clover (*Trifolium*), ryegrass (*Lolium*), rough meadow grass (*Poa trivialis* L.), fescue grass (*Festuca*), oat (*Avena sativa* L.), and barley (*Hordeum*). The plants' response to the pollution was evaluated visually during vegetative growth and also by the crop-yield level. The authors found that when soil is heavily polluted by crude oil, the crop-yield level depends strongly on the plants' peculiarities. Corn, clover, the lawn grasses, and barley showed a decreased crop capacity (70–85%, as compared with the control), whereas oat was more tolerant (40%, as compared with the control). The authors conclude that ryegrass, meadow grass, fescue grass, and clover are not suitable for biological recultivation at a pollution level of 10 g/kg. It was found that the oil-degradation rate in the soil decreased in the order barley > clover > lawn grasses > oat > corn. The soil-pollution level of 10 g/kg did not lead to an increase in soil toxicity to oat, barley, clover, or the lawn grasses. An aftereffect of the pollution was a considerable stimulation of root growth in these crops, particularly in oat. On the basis of their results, the authors recommend growing oat in chernozem with a pollution level of 10 g/kg during biological recultivation because this crop is the least sensitive to this pollution level *(8)*.

As plants are capable of promoting microbial pollutant degradation in rhizosphere soil, much effort has been directed toward elucidating the mechanisms of plant–microbe interactions that lead to successful phytoremediation, as well as toward searching for plants that best enhance the growth of natural degradative microorganisms in the rhizosphere. In a collaborative Russian–German study, Muratova et al. *(9)* examined the capacity of alfalfa (*Medicago sativa* L.) and reed (*Phragmites australis* [Cav.] Trin. Ex Steud) for stimulating microbial growth in soil polluted by polycyclic aromatic hydrocarbons (PAHs). They conducted laboratory pot experiments by using sandy material obtained by biological cleanup of ground excavated from under railroad tracks near Leipzig, Germany. The contaminated material contained 16 PAHs, among them (mg/kg): naphthalene, 8.71; phenanthrene, 1.67; fluoranthene, 18.17; pyrene, 17.67; and benzo(a)pyrene, 3.37. The total PAH content of the material was 79.80 mg/kg. The experiments allowed the soil-PAH content to be reduced by 74.5% with alfalfa and 68.7% with reed; however, alfalfa was better at stimulating the growth of a PAH-degrading microbial population. Alfalfa and reed are equally effective in removing PAHs from soil: the alfalfa root exudates provide the microflora with organic acids and mineral nitrogen, and the root system and aerenchyma of reed release oxygen into the rhizosphere *(9)*. Both plants enhance microbial activity in soil contaminated with paraffinic bitumen, but the rhizosphere microflora of alfalfa survives the pollution much better than does the rhizosphere microflora of reed *(10)*.

Recent work *(11)* aimed at selecting efficient phytoremediating plants for the cleanup of oil-polluted sites in Saratov Region, southern Russia, involved testing wheat (*Triticum*), rye (*Secale cereale*), broomcorn (*Sorghum vulgare* var. *technicum*), and alfalfa for tolerance to oil-slime pollution (1.6, 6.8, and 13.6 g/kg). Broomcorn survived the best and was considered a candidate plant for phytoremediation of oil-contaminated soils in Saratov Region in view of its high tolerance to arid conditions *(11)*.

2.2. Chemical Warfare Agent Degradation Products

Russia possesses the world's largest chemical weapons stockpile (about 40,000 tons). Under the international Chemical Weapons Convention, signed by Russia in November 1997, destruction means a process by which chemical warfare agents are irreversibly converted to a form unsuitable for their subsequent use as raw material for weapons production *(12,13)*. The storage, transportation, and destruction of vesicant agents, in particular yperite (β,β'-dichlorodiethylsulfide; $S[CH_2CH_2Cl]_2$) and Lewisite (dichloro[2-chlorovinyl]arsine; $Cl\text{-}CH=CH\text{-}AsCl_2$), may become new potential sources of environmental pollution. Within the Federal Program for the Destruction of Chemical Weapons, research is being carried out on the possible use of plants for cleaning soils polluted by the degradation products of the chemical agents.

During their laboratory experiments, Zakharova et al. *(14)* selected seeds of oat (*A. sativa* L.) as being tolerant of the yperite "reaction masses" (the toxic byproducts generated in the destruction process). The plant rhizosphere is rich in root exudates, which serve as carbon and nitrogen sources for rhizospheric organisms. Certain bacterial strains produce phytohormones (e.g., indole-3-acetic acid [IAA]), thereby contributing much to the improvement of plant growth and development. Therefore, the plant seeds were treated with various IAA concentrations. It was found that the uptake of the reaction masses is enhanced considerably after treatment of 3-d-old oat seedlings with IAA solutions. Of the three IAA concentrations used (10^{-7}, 10^{-5}, and 10^{-3} g/L), the best results are obtained with 10^{-5} g/L IAA, with which the soil content of the sulfur-containing products decreased more than 20-fold. The presence of IAA, known to be stimulatory to plant growth and development *(15)*, enhances the uptake of the yperite decomposition products through increased permeability of the plant-root membrane *(14)*.

The destruction of Lewisite gives rise to reaction masses containing sodium arsenite (Na_3AsO_3). Arsenic, not an essential micronutrient for either plant or animal kingdoms, is mainly known for its toxicity. The use of phytoremediation to clean up soils polluted by arsenic salts and the possibility of intensifying the process with natural and synthetic phytohormones were discussed by Lyubun et al. *(16)*. A study involving two phytohormones (IAA and 2,4-dichlorophenoxyacetic acid [2,4-D]) found two- to fivefold enhancement of arsenic uptake by 3-d-old seedlings of common sunflower (*Helianthus annuus* L.) and sugar sorghum (*Sorghum saccharatum* Pers.) after plant treatment with 10^{-5} g/L IAA and 2,4-D. Mixed cropping of sorghum and sunflower was suggested as another way of improving phytoremediation of sodium arsenite-polluted sites. Although sunflower exhibits a higher tolerance for arsenic than does sorghum, together the crops form an effective remediation system. Sorghum is very drought resistant and has an extensive root system, and sunflower is a highly transpiring plant (transpiration coefficient 500–600 L/kg dry weight). On the basis of this research, a method was proposed for soil cleansing from the products of natural and technological destruction of vesicant chemical warfare agents *(17)*. The contribution of phytoremediation to chemical weapons destruction was discussed at international workshops in the "Problems in Chemical Weapons Destruction" series held in Saratov, Russia, in 2000 and 2001 *(18,19)*.

2.3. Heavy Metals

Food chains can be protected against heavy-metal pollution by the use of plants' great potential to absorb and neutralize heavy metals. Some plants have developed stable forms that can survive and thrive in soils containing increased concentrations of heavy metals. Most of the articles published in Russia deal with the effect of heavy-metal doses on the metabolism, growth, and reproductive functions of cultivated plants *(20–22)*. The possibility of using "accumulator" plants in heavy-metal phytoremediation is less studied.

An X-ray fluorescence study *(23)* determined the content of heavy metals in the below- and above-ground parts of dandelion (*Taraxacum officinale* Wigg.), Canada thistle (*Cirsium arvense* [L.] Scop.), common wormwood (*Artemisia absinthium* L.), and yarrow (*Achillea millefolium* L.). It was found that the aerial parts of thistle accumulate lead, zinc, and chromium, whereas the aerial parts of wormwood accumulate zinc and nickel. These plants can be regarded as bioindicators of terrestrial ecosystem pollution and as possible cleanup agents for heavy-metal-contaminated soils *(23)*.

Plants and microbes play an important role in metal transformation through hydrolysis, oxidation, and other processes occurring in natural environments. They give rise to various compounds whose transformation potential and mechanisms have not been adequately studied. Scientists at the Skryabin Institute of Biochemistry and Physiology of Microorganisms (Pushchino, Moscow Region) have studied the possibility of using soil microorganisms to intensify metal phytoremediation. They obtained a strain of the soil bacterium *Pseudomonas* that shows resistance to heavy metals but at the same time is capable of protecting plants against pathogenic fungi *(24)*. They succeeded in cultivating bacteria that can grow and produce antibiotics in the presence of zinc, nickel, cadmium, and cobalt. The pseudomonads containing a protective plasmid do not let heavy metals into their cells. These results show that it is possible to produce genetically modified *Pseudomonas* strains that are resistant to heavy metals and to use them as components of a phytoremediation system *(24)*.

Sizova et al. *(25)* proposed a bacteria-assisted phytoremediation technology for the cleanup of arsenic-contaminated soils. In their experiments, they used sugar sorghum (*S. saccharatum* Pers.), rhizospheric bacteria of the genus *Pseudomonas*, and *Pseudomonas* derivatives harboring an arsenite/arsenate-resistance plasmid, pBS3031. The strains were chosen because they are highly antagonistic to a wide range of phytopathogenic fungi and bacteria, and they can stimulate plant growth. Sorghum inoculation with the *Pseudomonas* strains was found to increase the survivability of the plants growing in sodium arsenite-containing soils, and the presence of the conjugative plasmid offered the derivative strains a selective advantage over plasmidless rhizospheric bacteria, ensuring increased numbers of them on plant roots. Treatment of seeds, roots, and seedlings with such strains may substantially increase the tolerance of plants to metal pollutants in soil and, consequently, the effectiveness of phytoremediation.

Some investigators believe that the introduction of genes coding for the synthesis of metal-binding peptides and proteins is the most convenient strategy for increasing plant tolerance to metals and/or for increasing metal accumulation. Approaches to searching for metal-resistant hyperaccumulator plants include creating genetically modified plants having the necessary properties *(26)*. However, studies of this type are still in an embryonic stage.

3. Water Phytoremediation
3.1. Theoretical Foundations

Russia's water resources are about 4,310 km^3. The amount of waste water poured into surface water bodies is about 54.8 km^3; of this, about 37.7% (20.6 km^3) is rated as "polluted." The most common surface-water pollutants are oil products, phenols, readily oxidizable organic substances, metal compounds, ammonium and nitrite nitrogen, and specific pollutants (lignin, xanthogenates, formaldehyde, and so on) coming mostly from industrial and agricultural effluents. The tributaries to large rivers and the water-storage reservoirs at sites of unchecked waste water discharge show signs of human-caused ecological regress (biodiversity damage, the shortening of food chains). In such objects, a further increase in the anthropogenic load may destroy the biocenosis.

Available artificial systems for soil/water remediation are based on the use of pollutant mineralization and concentration, and they imitate natural systems (standing and flowing water bodies, bogs, and alluvial lands). The theoretical foundations of these cleanup systems were laid by the famous Russian biologists Uspensky *(27)*, Skadovsky *(28)*, Timofeyeva-Resovskaya *(29)*, and Vinberg *(30)*. Phytoremediation is the closest to natural processes and allows the energy resources of the ecosystem being cleaned to be increased with moderate use of organic fertilizers to stimulate microbial activity. Some authors consider phytoremediation as controlled eutrophication of a water body for the degradation of admixtures of abnormally high hydrocarbon concentrations. Considering that even the cleanest (e.g., Lake Baikal's) water contains low concentrations of hydrocarbons and indigenous microbes capable of their degradation, Kvitko et al. *(31)* suggest that phytoremediation should be based on the use of natural associations. Natural associations are an order of magnitude richer in their biogeochemical functions than microbial cultures, traditionally used in biotechnologies for the removal of xenobiotic contaminants. This richness is because such associations include photosynthetic organisms: higher plants, eukaryotic algae, and cyanobacteria. The water fern *Azolla caroliniana* and various duckweed (*Lemna*) species grow rapidly in waste water from agricultural, stockbreeding, and industrial enterprises, removing heavy metals, hydrocarbons, and aromatic and organic compounds. Work is underway to test these cultures for applicability to actual situations under various conditions *(32,33)*.

3.2. The Use of Algae and Duckweed

Some blue-green algae (cyanobacteria of the genera *Phormidium* and *Microcoleus*) can assimilate oil hydrocarbons, combining the heterotrophic and the phototrophic types of metabolism. Scientists at St. Petersburg State University have been investigating whether algae and cyanobacteria can degrade toxic aromatic compounds, and they are using algal strains from the university's collection in the cleanup of oil-polluted waters and industrial effluents. Experiments were run to rehabilitate a heavily fuel-oil-polluted landlocked water body at a power-and-heating plant. Oil-oxidizing biopreparations of the Oleovorin® family (Gossintezbelok company, Moscow), algae of the genus *Chlorella*, and duckweed (*Lemna minor*, *L. trisulca*) were used. One-third of the water body (total water surface area, 700 m^2) was covered with a 7-mm-thick fuel-oil slick, and oil was observed on the stems of the marginal vegetation (cane and reed) *(34)*. The cleanup was performed in the 1995–1996 spring/summer period. The concentration of water-dissolved hydrocarbons was determined, and the microbial population in various parts of the water body was studied. The numbers of heterotrophic bacteria, saprophytic oligotrophs, oligocarbophiles, and alkanotrophs were determined. The qualitative and quantitative ratios between these bacteria allow one to judge the ecological and sanitary state of the water body as a whole. The 2-yr efforts reduced the pollution level from 0.65 to 0.13 mg/L.

In summer 1997, as a result of an accidental spill, a large amount of diesel fuel entered a river in Leningrad Region. Most of the fuel was retained by oil booms and was collected, both manually and by using sorbent materials. The biological cleanup of the riverside and vegetation, as well as of the water surface, continued for less than 1.5 mo. Most effort was directed toward minimizing the amount of water-dissolved oil hydrocarbons. Duckweed application to some sites of the river was first accompanied by the plant's mass death and by sorption of the oil products onto the dead-plant layer. On second application, however, duckweed retained its viability; the sorption and the photosynthesis of the survivor plants intensified the oxidation of oil hydrocarbons in the near-surface water layer. After a 1.5-mo remediation, the pollution level was lowered from 14 to 0.2 mg/L. The use of photosynthetics (algae and floating aquatic plants) as a biofilter component for oil-degrading microorganisms substantially accelerates phytoremediation under natural conditions *(34)*.

3.3. The Use of Water Hyacinth

Water hyacinth (*Eichhornia crassipes*, family Pontederiaceae) has gained wide acceptance in Russia as a phytoremediating agent. The first mention about the benefits of using this exotic plant in Russia is of relatively recent date, but the preliminary results are very promising *(35,36)*. Water hyacinth is exceptionally good at taking up and accumulating heavy metals. It can be used for remediating water from cesium and strontium radionuclides. The radionuclide accumulation by water hyacinth occurs both through assimilation by the plant and through sedimentation of radionuclide-containing suspended material on the plant's roots. Current work is aimed at wider application of water hyacinth to the cleanup of small rivers, reservoirs, and industrial effluents.

A classic aquasystem using water hyacinth as its main plant component was developed for the treatment of waste water. The well-developed root system, the high rate of clonal reproduction, and the rapid biomass accumulation allow this plant to be used to clean up runoff from stockbreeding units and municipal services. The effectiveness of pollutant removal per hectare of biopond per day is 150–200 kg for ammonium nitrogen and 1.6–4.5 kg for oil products. Experiments using runoff from the Pavlodar oil-processing plant showed that water hyacinth effectively removes high concentrations of oil products (23.7 mg/L), and also of sulfides and phenols, with a high (up to 80 mg/L) water content of ammonium nitrogen. The technology was tested at the ZAO Kudryashovskoye (a pig-breeding complex) and at Tolmachyovo Airport (Pavlodar, Russia). The heavy-metal concentrations over 20 d of growth were (in mg/kg dry plant wt): Cu^{2+}, 1,955; Zn^{2+}, 1,809; Pb^{2+}, 414; and Cd^+, 370.

3.4. The Use of Miscanthus

The Far-East grass miscanthus was found to be promising in the recultivation and cleanup of soils and as bank vegetation around man-made water bodies. Miscanthus can grow in dry, wet, high-mineral, and oil-polluted soils with a wide range of organic-matter contents. It does not reproduce by seed, which allows control of its dispersal. It is well adapted to Siberian frosts, grows in April, and increases its per-season numbers 25- to 30-fold. In the Institute of Cytology and Genetics, Siberian Branch of the Russian Academy of Sciences, methods have been developed for the reproduction and preservation of these plants.

In Russia, however (as in many European countries), such treatment facilities may not prove cost effective because of the rigorous climate. They can operate only during a warm season. With early frosts (at the beginning of October), the leaves die or sink to the reservoir's bottom. In winter, uncovered treatment facilities are useless, or they require additional effort and investments to generate the conditions for the plants' adaptation, maintain its viability throughout the year, and choose effective treatment conditions. Since 2000, work on covered facilities has been conducted by the limited-liability company Vektor E under the supervision of the Moscow Government's Department of Nature Management and Environmental Protection. A useful model has been developed ("Covered Constructions for Cultivation of the Plant *Eichhornia*") *(37)*.

4. Conclusions

Russia plays a key role in maintaining the global functions of the biosphere because much of the Earth's biodiversity is represented in its vast territories and variety of natural ecosystems. Phytoremediation is taking a prominent place in the processes of environmental cleanup from hazardous pollutants. Although there has been little large-scale field work, there is increasing interest among scientists (and, quite importantly, among various organizations) in this technology. This interest is attested by the many scientific meetings that include various aspects of phytoremediation research in their programs.

To preserve the stability of natural ecosystems, scientists should expand their research on the functioning of the key component—the plants. Russian scientific schools and research teams working on the basic problems of plants, microorganisms, and plant–microbial associations are increasingly studying pollutant effects on such communities. Work of this type forms a scientific basis for plant introduction and acclimatization, and for the applied aspects of phytoindication and phytoremediation.

References

1. Mazhaisky, Y. A., Davydova, I. Y., Yevtyukhin, V. F., and Yevsenkin, K. N. (1999) The agroecological assessment of oil-polluted land in the territories of IPDS's. In: *Advances in Ecological and Safe-Life Research*. Proc. Fourth All-Russia Scientific–Practical Conf. with Int. Participation, St. Petersburg, Russia, June 16–18, 1999, vol. 1, pp. 396–398.
2. Panchenko, L. V., Turkovskaya, O. V., Dubrovskaya, Ye. V., and Muratova, A. Yu. (2003) *Methodological Recommendations on the Biorecultivation of Oil-Polluted Lands*. Saratov University Press, Saratov, Russia.
3. Novikova, A. F. and Gololobova, A. V. (1976) On the reclamation of solonetz soils of a dark chestnut subzone in Kustanai Region. *Pochvovedenie* **4**, 97–106.
4. Kireeva, N. A., Yumaguzina, K. A., and Kuzyakhmetov, G. G. (1996) The growth and development of oat plants in oil-contaminated soils. *Selskaya Biologiya* **5**, 48–54.
5. Kireeva, N. A., Novosyolova, Y. I., and Kuzyakhmetov, G. G. (1997) The performance of agricultural crops in oil-polluted and recultivated soils. In: *Ecological Problems of the Republic of Bashkortostan*, BSPI Press, Ufa, Russia, pp. 293–299.
6. Kireeva, N. A., Miphtakhova, A. M., and Kuzyakhmetov, G. G. (2001) Growth and development of weeds on the technogene pollution environment. *Vestnik Bashkirskogo Universiteta* **1**, 32–34.
7. Larionova, N. L. (2003) The tolerance of unconventional fodder plants to soil hydrocarbon pollution and the effect of phytoremediation. In: *Abstracts, 7th Pushchino School–Conference of Young Scientists "Biology: The Science of the Twenty-First Century,"* Pushchino, Russia, April 14–18, 2003, abstract no. 71.
8. Davydova, I. Y. and Pakhnenko-Durynina, Y. P. (2003) The response of agricultural plants to soil oil pollution. In: *Territorial Scientific–Practical Conf. of Students, Young Scientists, and Experts "Youth and Science of the Third Millenium,"* Stavropol, Russia, Sept. 15, 2003.

9. Muratova, A., Hübner, T., Tischer, S., Turkovskaya, O., Möder, M., and Kuschk, P. (2003) Plant–rhizosphere-microflora association during phytoremediation of PAH-contaminated soil. *Int. J. Phytorem.* **5,** 137–151.
10. Muratova, A., Hübner, T., Narula, N., et al. (2003) Rhizosphere microflora of plants used for the phytoremediation of bitumen-contaminated soil. *Microbiol. Res.* **158,** 151–161.
11. Muratova, A. Y., Dmitrieva, T. V., and Turkovskaya, O. V. (2003) Study of phytotoxicity of oil-contaminated soil and development of rhizosphere contaminant degrading microbial population. In: *Abstracts, International Symposium "Biochemical Interactions of Microorganisms and Plants with Technogenic Environmental Pollutants,"* Saratov, Russia, July 28–30, 2003, pp. 26–28.
12. Kapashin, V. P. (2002) *Chemical Disarmament. Industrial Environmental Monitoring.* Nauchnaya Kniga Publ., Saratov, Russia.
13. Kapashin, V. P., Sevostyanov, V. P., Shebanov, N. P., and Tolstykh, A. V. (2002) *Chemical Disarmament. Technologies for Destroying Chemical-Warfare Agents.* Nauchnaya Kniga Publ., Saratov, Russia.
14. Zakharova, E. A., Kosterin, P. V., Brudnik, V. V., et al. (2000) Soil phytoremediation from the breakdown products of the chemical warfare agent, yperite. *Environ. Sci. Pollut. Res.* **7,** 191–194.
15. Takahashi, N. (ed.) (1986) *Chemistry of Plant Hormones.* CRC Press Inc., Boca Raton, FL.
16. Lyubun, Ye. V., Kosterin, P. V., Zakharova, E. A., Shcherbakov, A. A., and Fedorov, E. E. (2002) Arsenic-contaminated soils: phytotoxicity studies with sunflower and sorghum. *J. Soils Sediment.* **2,** 143–147.
17. Ignatov, V. V., Fedorov, E. E., Kosterin, P. V., et al. (2002) Remediation method for soils polluted by products of natural and technological destruction of vesicant chemical warfare agents. Patent 2185901. Date issued: July 27, 2002.
18. Tychinin, D. N. and Kosterin, P. V. (2000) Problems in chemical-weapon destruction: Report on the Third International Workshop "Biotechnological Approaches to Chemical-Weapon Destruction" (Saratov, Russia, August 21–22, 2000). *Environ. Sci. Pollut. Res.* **7,** 245–246.
19. Tychinin, D. N. and Kosterin, P. V. (2002) Contribution of biotechnology to chemical-weapons destruction. Report on the Fourth International Workshop "Contribution of Biotechnology to Chemical-Weapons Destruction," (Saratov, Russia, September 6–7, 2001). *Environ. Sci. Pollut. Res.* **9,** 217–218.
20. Obroucheva, N. V., Bystrova, E. I., Ivanov, V. B., Antipova, O. V., and Seregin, I. V. (1997) Root growth responses to lead in young maize seedlings. *Plant Soil* **200,** 55–61.
21. Sobotik, M., Ivanov, V. B., Obroucheva, N. V., et al. (1998) Barrier role of root system in lead-exposed plants. *Angew. Bot.* **72,** 144–147.
22. Talanova, V. V., Titov, A. F., and Boyeva, N. P. (2001) Effect of increasing heavy-metal concentrations on the growth of barley and wheat seedlings. *Fiziol. Rast.* **48,** 119–123.

23. Samkayeva, L. T., Revin, V. V., Rybin, Y. I., Kulagin, A. N., Novikova, O. V., and Pugayev, S. V. (2001) Study of heavy-metal accumulation by plants. *Biotekhnologiya* **1**, 54–59.
24. Boronin, A. M. (1998) Plant growth-promoting rhizobacteria *Pseudomonas*. *Soros Educational Journal* **10**, 25–31.
25. Sizova, O. I., Kochetkov, V. V., Validov, S. Z., Boronin, A. M., Kosterin, P. V., and Lyubun, Y. V. (2002) Arsenic-contaminated soils: genetically modified *Pseudomonas* spp. and their arsenic-phytoremediation potential. *J. Soils Sediment.* **2**, 19–23.
26. Galkin, A. P., Bulko, O. V., Leoshina, L. G., Vasiliev, A. N., and Medvedeva, T. V. (1997) Cleanup of contaminated lands from heavy metals using transgenic plants. In: *In Situ and On-Site Bioremediation, Volume 3 (4-3)*, (Leeson, A. and Alleman, B. C., eds.), Battelle Press, Columbus, OH, pp. 325–330.
27. Uspensky, Y. Y. (1932) On the question of problems and solutions in microbiology in connection with the development of municipal water-supply, particularly during construction of water-storage reservoirs. *Mikrobiologiya* **3**, 107.
28. Skadovsky, S. N. (ed.) (1961) *Epibiotic Biocenoses as Sorbents (Novel Method for the Preliminary Water Cleanup for Water-Supply Purposes)*. Moscow State Univ. Press, Moscow, Russia.
29. Timofeyeva-Resovskaya, Y. A., Agafonov, B. M., and Timofeyev-Resovsky, N. V. (1961) On the soil biological deactivation of water. In: *Proceedings of the Institute of Biology of the Urals Branch of the USSR Academy of Sciences* **13**, 35–48.
30. Vinberg, G. G. and Sivko, T. N. (1956) Phytoplankton as an agent of self-cleaning of polluted waters. *Proceedings of the All-Union Hydrobiological Society* **7**, 5–23.
31. Kvitko, K. V., Iankevitch, M. I., and Dmitrieva, I. A. (1998) The cooperation of algal and heterotrophic components in oil polluted wastewaters. In: *Proceedings of the Workshop "Microbiology of Polluted Aquatic Ecosystems,"* Leipzig, Germany, 4–5 December 1997, UFZ-Bericht no. 10, pp. 174–181.
32. Mikryakova, T. F. (1994) Heavy-metal distribution in the higher aquatic plants of the Uglich Water-Storage Reservoir. *Ekologiya* **1**, 16–21.
33. Gogotov, I. N. (2003) Additional purification of water bodies and waste waters with consorciums of aquatic plants containing microorganisms. In: *Biotechnology: State of the Art and Prospects of Development: Proceedings of the II Moscow International Congress*, Moscow, Russia, Nov. 10–14, 2003, p. 6.
34. Iankevitch, M. I. and Kvitko, K. V. (1998) Bioremediation of oil-polluted water bodies. *Ekologiya i Promyshlennost Rossii* **10**, 21–27.
35. Fedin, E. (1998) The flowers win. *Izobretatel i Ratsionalizator* **6**, 5.
36. Voronina, L. P., Malevannaya, N. N., and Karpova, Y. V. (2001) Search for qualitative characteristics of *Eichhornia crassipes* facilitating realization of its ecological functions. In: *Proceedings of the First Russian Scientific and Applied Conf. "Actual Problems of Innovations in Nonconventional Plant Resources and in Development of Functional Products,"* Moscow, Russia, June 18–19, 2001, Abstract 1vr69.
37. Lyalin, S. V. (2003) Method for growing *Eichhornia* during hydrobotanical wastewater treatment. Patent 2193532. Date issued: Jan. 14, 2003.

30

Phytoremediation in India

M. N. V. Prasad

Summary

In India, urbanization, excessive utilization of natural resources, and population growth are the causes for air, water, and soil contamination and pollution. Major environmental problems in India are land degradation (deforestation, overgrazing, overcultivation, faulty irrigation), destruction of wildlife habitat and erosion of genetic resources (including those of crops and trees, terrestrial animals, and fish), and pollution (air, water, and soil pollution with toxic wastes and other substances). Soil conservation and restoration of degraded soils (wasteland/marginal land) is the most serious environmental concern to India. In India, soil erosion is a serious problem ranging from loss of top soil in 130.5 million ha to terrain deformation in 16.4 million ha. Soil loss under different land-use options has been reported and minimum loss found when trees and grass were grown together in a silvipastoral system. For e.g., Shivaliks (foothills of Himalayas, one of the most fragile ecosystems) has included combinations of eucalyptus-bhabar grass; Acacia catechu-forage grass; Leucaena-Napier grass; teak-Leucaena-Bhabar; Eucalyptus-Leucaena-Turmeric; poplar-Leucaena-Bhabar; and Sesamum-rape seed. Sodic soils of the Indo-gangetic alluvial plain are characterized by high pH, high exchangeable sodium and phosphorus, low infiltration, dispersed soil, low organic matter content, and poor fertility. Special planting techniques have been developed for raising multipurpose tree species in sodic and saline soils. A silvipastoral model comprising *Prosopis juliflora* and *Leptochloa fusca* has been developed, and alkali soils have been standardized. Another serious problem is the physical deterioration of soil because of water logging or submergence/flooding that has affected around 11.6 million ha of land in India. Suitable trees and grass species for such situations are trees (*Eucalyptus tereticornis, Populus deltoids, Terminalia arjuna, Acacia auriculiformis, Syzigium cumini, Albizia lebbek, Dalbergia sissoo,* and *Pongamia pinnata*) and grasses (para grass, cord grass, lemon grass, and *Setaria* grass). Contamination of food and other agricultural products with pesticide residues is a widespread problem in India. India's 15 oil refineries generate a huge amount of oily sludge annually. This also takes a toll on the scarce soil, because land requirements increase with an increase in oil sludge generation. Besides the sludge from oil refineries, crude oil spills too are a cause of environmental degradation. The "Mission Mode" experiment of fly-ash management including using fly ash in forestry systems is one of the important strategies to protect environmental degradation.

Key Words: Phytoremediation in India; restoration of -degraded; salt effected and desertified soils; mine spoil revegetation; fly ash management; oil slick treatment.

1. Introduction

Phytoremediation is a rapidly expanding area of environmental science that holds great promise for cleaning up the polluted and contaminated environment both of inorganics and organics. There have been a number of reports using the native flora and microbes from various research laboratories of India—algae, cyanobacteria, vascular plants, and aquatic macrophytes have been used extensively in laboratory and field conditions *(1–22)*. Terrestrial plants and lichens have also been used to monitor air pollution around industrial sites and in cities. Criteria investigated for potential phytoremediators include seed germination and seedling growth, plant growth, and reproduction *(23)*. Because of its low cost and lack of technical problems, phytoremediation has become very popular not only in India but also globally.

In India considerable effort is being devoted to identifying indigenous plant species that can be used to remediate pollutants such as pharmaceutical wastes, arsenic, fly ash, and metals (**Fig. 1**). Phytoremediation technology is fairly low in cost, does not require extensive equipment, and is appropriate not only for India, Pakistan, Bangladesh, and other Asian nations, but also for advanced nations *(24,25)*. In India, because of urbanization, excessive utilization of natural resources and population growth, air, water, and soil are contaminated and polluted with a variety of xenobiotics that are amenable to phytoremediation. From its independence, India's strategies for coherent ecodevelopment have gained global attention.

The fact sheet about the country's environmental profile is shown in **Table 1**: physical and chemical monitoring of environmental pollutants in the region is a problem because of costs and lack of appropriate equipment and expertise. In terms of air pollutants, which are extensive in India, an alternative technology such as phytoremediation may be quite useful. Low-cost, simple passive samplers have been developed to detect cumulative concentrations of air pollutants, such as sulfur dioxide, nitrogen oxides, and ozone. These can be used to assess concentrations in air that may adversely affect humans and plants. Major environmental problems in India are land degradation (deforestation, overgrazing, overcultivation, faulty irrigation), destruction of wildlife habitat and erosion of genetic resources (including those of crops and trees, terrestrial animals, and fish), and pollution (including air, water, and soil pollution with toxic wastes and other substances) *(26,27)*.

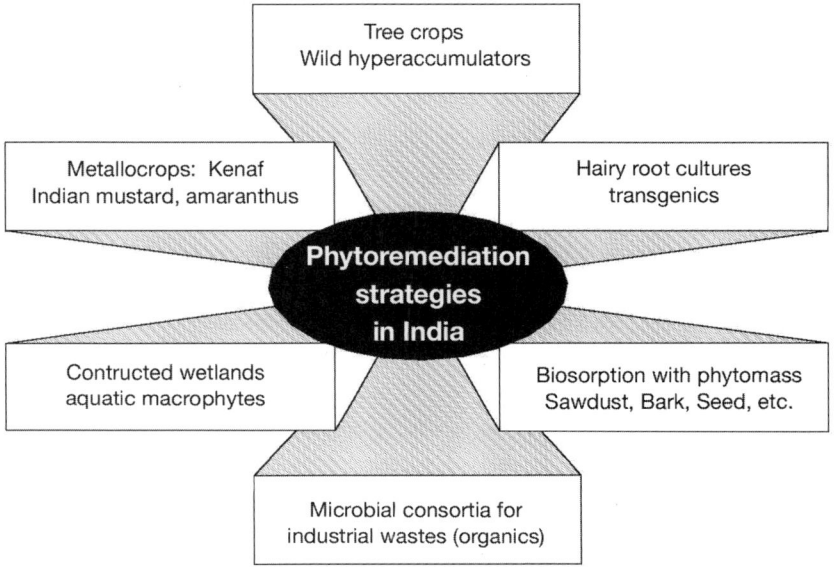

Fig. 1. Feasible phytoremediation strategies in India.

Table 1
Environmental Profile of India

Area	~3.3 million square miles
Arable Land	55%
Population	Nearly 1 billion
Biggest city	Kolkata (Calcutta) ~10–12 million
Climatic zones	From arctic high-mountain areas to desert and tropical humid forests
Agriculture	Mainfood crops: rice, wheat, maize
	Main cash crops: sugar cane, oils seed, cotton, and jute
Livestock (in millions)	Cattle 190, goats 90, sheep 50, buffaloes 70, poultry 200
Health	Infant mortality 10%
Life expectancy	Nearly 60 yr
Forest cover	~20%
Access to safe water	40%
Energy consumption pattern	Firewood 29%, coal 16%, electricity 15%, oil 26%, other 15%
GDP per capita	US$400

Table 2
Estimates of Land Degradation in India (Adapted from [28])

Agency	Estimated extent (million ha)
National Commission on Agriculture (NCA, 1976)	148
Ministry of Agriculture (1978) (Soil and Water Conservation Division)	175
Society for promotion of Wastelands Developments (SPWD, 1984)	130
National Remote Sensing Agency (NRSA, 1985)	53
National Bureau of Soil Survey and Land Use Planning (NBSSLUP, 1994)	108
Ministry of Agriculture (MOA, 1985)	174
Ministry of Agriculture (MOA, 1994)	107
Water erosion	57
Wind erosion	11
Salt-effected and waterlogging	10
Shifting cultivation	2.4
Degraded forests	25
Ravines	2.7
Others (mines quarry, landslide, and acid sulfate soils)	0.34

2. Phytoremediation for Restoration of Degraded Land

Soil conservation and restoration of degraded soils (wasteland/marginal land) is the most serious environmental concern to India (**Table 2**). Wasteland/marginal land is defined as land which is presently degraded and is lying unutilized (except current fallows) because of various constraints. This can be divided into two broad classes *viz.* (1) culturable land and (2) unculturable land. Lands, which in spite of having the potential to support vegetation, are not being properly utilized because of different constraints and could be put under culturable wasteland, whereas land which has no potential to develop vegetation cover, could be termed as nonculturable wastelands.

2.1. Eroded Soils

In India, soil erosion is a serious problem ranging from loss of top soil in 130.5 million ha to terrain deformation in 16.4 million ha. Also there are nearly 3.67 million ha ravine lands of which about 72% are confined to the states of Uttar Pradesh, Madhya Pradesh, Rajasthan, and Gujarat. In addition, every year because of faulty agricultural practices more than 8000 ha of land are converted into ravines.

The best use of ravine land is to put it under suitable permanent vegetation cover. Medium and shallow gullies can be utilized under silvipasture and hortipasture and deep gullies under tree plantations. A list of promising agroforestry species suitable for different purposes are mentioned in **Table 3** *(28)*.

Acacia nilotica is the most promising fuel-wood tree species for ravine rehabilitation. The main grasses suitable for gully stabilization in Rajasthan, Uttar Pradesh, Madhya Pradesh, and Gujarat are *Dichanthium annulatum, Cenchrus ciliaris*, and *Schima nervosum*. By planting and protecting of these grasses, reasonable green fodder yield can be achieved in 2–5 yr. This practice also reduces run-off and soil loss considerably to 6–10 t/ha/yr. Vegetative barriers are cheap and effective as compared with mechanical measures on mild slopes. Living *bunds* (banks) of Guinea (*Panicum maximum*), bhabar (*Eulaliopsis binata*), and khus grass (*Vetivera* spp.) reduced run-off by more than 18% and soil loss by more than 78% as compared with cultivated fallow on a 4% slope in Doon valley *(29,30)*. Guinea grass was found more effective than others grasses. Soil loss under different land use options has been reported *(31)* and minimum loss found when trees and grass were grown together in a silvipastoral system. Similarly, Tejwani et al. *(32)* compared different grasses for their role in soil conservation and found *Pueraria hirsuta* and *D. annulatum* to be the most promising. Wind erosion is another serious problem in the arid and semi-arid regions including the states of Rajasthan, Gujarat, Haryana, and Punjab. Soil and nutrient loss from different land-use systems in Shivaliks (foothills of Himalayas) has included combinations of Acacia catechu-forage grass; Leucaena-Napier grass; teak-Leucaena-Bhabar; Eucalyptus-Leucaena-Turmeric; poplar-Leucaena-Bhabar; and Sesamum-rape seed.

2.2. Salt-Affected and Desertified Soils

Out of nearly 188 million ha wastelands in India, about 9.6 million are salt-effected soils, of which a large proportion is present in the Indo-gangetic alluvial plains. Silvipasture is considered an option of great promise for the rehabilitation of such soils *(33,34)*. Salt-effected soils in the country can be grouped into two categories, sodic and saline. Sodic soils of the Indo-gangetic alluvial plain are characterized by high pH, high exchangeable sodium, low infiltration, dispersed soil, low organic matter content, and poor fertility. In most cases a precipitated $CaCO_3$ layer exists in the profile. This layer offers severe mechanical impedance for root penetration of perennial vegetation, particularly trees. Because of high sodicity, such soils do not support any kind of vegetation except the growth of some salt-tolerant indicator plants *(28)*.

The Central Arid Zone Research Institute, Jodhpur, and the Arid Forest Research Institute, Jodhpur have taken rigorous initiatives to address the problems of saline and sodic soils, and desertification. Special planting techniques have

Table 3
Species Used for Phytoremediation and Rehabilitation of Degraded Lands [Source: 28] in India

Salt-effected ravine areas	Forage and fuel species for eroded areas	Silvipasture in Bundelkhand, Uttar Pradesh	Vegetative barriers for run-off and soil loss from doon valley
Trees and shrubs	**Trees**	**Trees**	*Panicum maximum* (Guinea grass)
Acacia tortilis	*Prosopis juliflora*	*Albizia* spp.	*Eulaliopsis binata* (Bhabar grass)
Albizia amara	*Prosopis cineraria*	*Hardiwickia binata*	*Vetivera zizinoides* (Khus grass)
Dichrostachys cinerea	*A. tortilis*	*Dalbergia sissoo*	
Leucaena leucocephala	*Acacia radianoa*	*Leucaena leucocephala*	
Acacia nilotica	*Zizyphus mauritiana*	*Azadirachta indica*	
Dendrocalamus strictus		*A. nilotica*	
D. sissoo	**Shrubs**		
Albizia lebbeck	*Calligaonum polygonoides*	**Grasses**	
Prosopis juliflora	*Exotolaria burhia*	*Chrysopogon fulvus*	
Terminalia arjuna	*Aervajavanica*	*Dichanthium annulatum*	
Azadirachta indica	*Zizyphus nummularia*	*Cenchrus ciliaris*	
		P. maximum	
Grasses and legumes	**Grasses**	*Panicum pedicellatum*	
Chrysopogon fulvus	*Lasiuruss indicus*	*Heteropogon*	
C. ciliaris	*Panicum turgidum*	*Bothrochloa*, and so on	
Pennisetum pedicellatum	*P. antidotale*		
Saccharum spontanellm		**Legumes**	
D. annulatum		*Stylo*	
Phaseolus atropurpureus		*Clitaria*	
Stylosanthes species		*Sirata*	

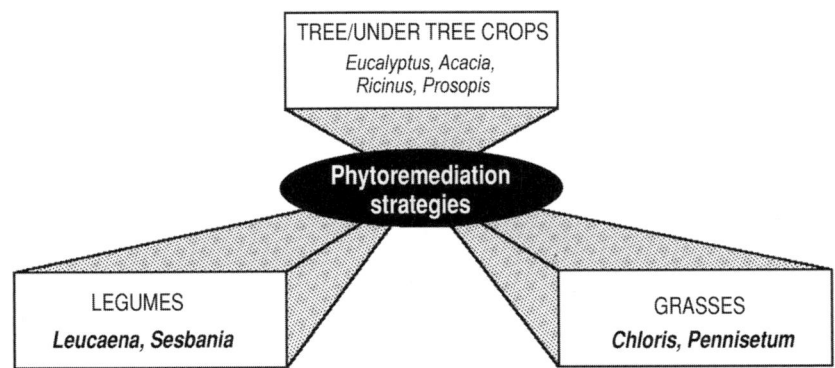

Fig. 2. Silvipastoral model of phytoremediation strategies applied to Indian ecosystems.

been developed for raising MPTS in sodic and saline soils. The planting technique for sodic soils involves digging holes of 30-cm diameter and 100- to 140-cm deep with the help of a tractor-mounted post-hole digger. These auger-holes are back-filled with a mixture of original alkali soil plus 3–4 kg gypsum plus 8–10 kg FYM + 10–15 kg river sand. This technique ensures more than 80% tree survival even after 10 yr in highly alkali soils (pH >10.0). This technology has become a common practice with the forest department, farmers, and others engaged in afforestation programs on alkali soils in the country. For saline soils, sub-surface planting gives better survival and biomass of multipurpose tree species. In this case, saplings are planted in 30-cm deep trenches.

A silvipastoral model comprising *Prosopis juliflora* and *Leptochloa fusca* has been developed and alkali soils have been standardized *(35)*. *P. juliflora* and *L. fusca* are grown for about 5 yr to produce fuel wood and fodder. Later on, when the surface soil is reclaimed, the grass is replaced with high value fodder crops such as berseem (*Trifolium alexandrinum*), shaftal (*Trifolium resupinatum*), oats (*Avena sativa*), senji (*Melilotus parviflora*), and so on. The silvipastoral model is highly suited for the development of village-community lands that have been lying abandoned because of sodicity problems (**Fig. 2**).

Overall, it has been suggested *(28)* that promising multipurpose tree plant species for salt-effected soils are (1) alkaline soils (*P. juliflora, A. nilotica, Tamarix articulata, Casurina equisetifolia, Eucalyptus tereticornis, Pithecelobium dulce, Pongamia pinnata, Terminalia arjuna, Prosopis alba, Dalbergia sissoo*), (2) saline soils (*P. juliflora, Tamarix troupii, T. articulata, Pithecellohium dulce, Acacia farnesiana, A. nilotica, Acacia tortilis, Casuarina glauca, Eucalyptus camaldulensis, Leucaena leucocephala*), and (3) promising grasses (*Leptochloa fusco, Cynodon dactylon, Braciaria mutica, Panicum* sp., *Chloris gayana*).

The species found to be most suitable for restricting the movement of sand dunes and checking the advancement of desert are *Acacia planiforms*, *Acacia albida*, *A. tortilis*, *P. juliflora*, *Prosopis cineraria*, *Teconklla undulata*, and *Zizyphus manuritiana*. Promising grasses identified for growing in association with trees are *C. ciliaris* and *Cenchrus setigaris*. For Tamil Nadu situations, species like *Acacia senegal* and *Albizia melliera* are reportedly suitable to check shifting of sand dunes. The three-step technology proposed by the Central Arid Zone Research Institute for fixation of sand dunes involves providing protection from biotic interference through the establishment of biofences, the erection or development of artificial physical barriers to minimize surface wind erosion, and revegetating the treated dunes using trees and grass species.

2.3. Waterlogged Soils

The term "waterlogging" refers to a condition of short- or long-term water stagnation caused by changes in hydrologic regime, landscape, silting-up of riverbeds, increased sedimentation, or reduced capacity of the drainage systems. The physical deterioration of soil from waterlogging or submergence/flooding has affected around 11.6 million ha land in India. Suitable trees for such situations are trees (*E. tereticornis*, *Populus deltoids*, *T. arjuna*, *Acacia auriculiformis*, *Syzigium cumini*, *Albizia lebbek*, *D. sissoo*, and *P. pinnata*) and grasses (para grass, cord grass, lemon grass, and *Setaria* grass).

2.4. Restoration of Mine Spoils

It is estimated that nearly 3000 billion tons of mine over-burden is dumped annually all over the world. At present, about 386,000 ha land per annum is disturbed by mining. Suitable plant species for mine restoration in India have been identified *(36–43)* and are listed in **Table 4**.

3. Groundwater Imbalance and Contamination

During the last two decades a major shift in groundwater level has taken place in some parts of northwestern India. A shift in cropping pattern is one of the major causes for this imbalance. For example, in southwest Punjab and Haryana the water table is rising at 0.2–0.5 mm/yr. On the other hand, in central parts of these states the water level is decreasing at the same rate, probably because of more pumping of groundwater to meet irrigation requirements of the predominant rice–wheat system. This process is adversely affecting the productivity and sustainability of agriculture.

Contamination of food and other agricultural products with pesticide residues is a widespread problem in India. Although lindane has been technically banned in many countries in recent years, γ-HCH is still in use today, especially in tropical countries where it is used for seed protection and mosquito control in fighting

Table 4
Plant Species Suitable for Revegetation of Mine Spoils in India [Adapted from 28]

Mine spoil category	Suitable plant species
Bauxite mined area of Madhya Pradesh	*Grevillea pteridifolia, Eucalyptus camaldulensis, Pinus, Shorea robusta*
Coal mine spoils of Madhya Pradesh	*Eucalyptus hybrid, Eucalyptus camaldulensis, Acacia aurifuliformis, Acacia nilotica, Dalbergia sissoo, Pongamia pinnata*
Lime stone mine spoils of outer Himalayas	*Salix tetrasperma, Leucaena lellcocephala, Bauhinia retusa, Acacia catechu, Ipomea cornea, Eulaliopsis binata, Chrysopogon fulvus, Arundo donax, Agave americana, Pennisetum purpureum, Erythrina subersosa*
Rock-phosphate mine spoils of Musoorie	*Pennisetum purpureum, Saccharum spontaneum, Vitex negundo, Rumex hastatus, Mimosa himalayana, Buddleia asiatica, Dalbergia sissoo, Acacia catechu, Leucaena leucocephela,* and *Salix letrasperma,* and so on.
Lignite mine spoils of Tamil Nadu	*Eucalyptus* species, *Leucaena leucocephala, Acacia,* and *Agave*
Mica, copper, tungesten, marble, dolomite, limestone, and mine spoils of Rajasthan	*Acacia tortilis, Prosopis juliflora, Acacia senegal, Salvadora oleodes, Tamarix articulata, Zizyphus nummularia, Grewia tenax, Cenchrus setigerus, Cymbopogon, Cynodon dactylon, Sporobollis marginatus, D. annulatum*
Iron ore wastes of Orissa	*Leucaena leucocephala*
Hematite, magnetite, manganese spoil from Karnataka	*Albizia lebbeck*

Table 5
Fly-Ash Composition

Constituent	Percentage range (%)
Silica (SiO_2)	49–67
Alumina (Al_2O_3)	16–29
Iron oxide (Fe_2O_3)	4–10
Calcium oxide (CaO)	1–4
Magnesium oxide (MgO)	0.2–2
Sulfur (SO_3)	0.1–2
Loss of ignition	0.5–3.0

malaria. Use of atrazine has been limited as well, but its residues are also still causing serious problems for groundwater and surface soil quality.

An Indo-Swiss collaboration investigated the remediation of agricultural soils contaminated with residues of γ-HCH and atrazine by means of plants, which metabolize these pesticides or stimulate their microbial degradation or inactivation in the rhizosphere. Selected native plants of India as well as common crop plants will be tested in hydroponic cultures and pot experiments. The focus of the collaboration is on the degradation of atrazine with the help of common crop plants. In the Indian partner institutes, samples from contaminated sites will be screened for atrazine- and HCH-degrading strains of rhizosphere bacteria. These will then be isolated and identified. In a further step, suitable strains will be used in greenhouse experiments to augment the rhizosphere of selected plants. In addition, one partner group will assess the risks involved in this bioremediation approach.

4. Fly-Ash Management

Nearly 73% of India's power generation capacity is thermal, of which coal-based generation is nearly 90% (diesel, wind, gas, and steam adding up to about 10%). The 85 utility thermal-power stations, in addition to several captive power plants, use bituminous or sub-bituminous coal and produce large volumes of fly ash. The high ash content (30–50%) of Indian coals contributes to these large volumes of fly ash. India's dependence on coal as a source of energy will continue in the next millennium and, therefore, fly-ash management will remain an important area of national concern.

Fly ash is the residue of the coal combustion process. Its indiscriminate disposal requires large volumes of land, water, and energy. The fine particles of fly ash, by virtue of their lightness, can become air borne if not managed well. Indian fly ashes are safer than those produced in other countries (especially on account of a lower content of sulfur, heavy/toxic elements, and radio nuclides), however, management of the large volumes produced poses a big challenge to the country. At present, nearly 90 million tons of fly ash is being generated annually in India and nearly 65,000 acres of land is presently occupied by ash ponds. It is a siliceous or aluminous material with pozzolanic properties. It is refractory and alkaline in nature, having fineness in the range of 3000–6000 cm^2/gm.

The "Mission Mode" experiment of fly-ash management has brought into focus "fly ash" as a important resource material and prompted various research projects, using fly ash in forestry systems and for the growth of *Cassia siamea (44–47)*. The integrated approach of working in 10 areas for safe disposal and utilization has led to a near doubling of fly-ash utilization in the country. A long-term perspective toward fly-ash management needs to be elucidated. More importantly, a greater participating role of thermal power plants, coal suppliers, industry, and technologies is needed on a continued basis. This is to ensure that the momentum is maintained, more so because environment issues shall be a prime concern during the coming century.

5. Oil-Slick Treatment
5.1. Progress and Challenges

Tata Energy Research Institute (TERI) developed a biological method of using micro-organisms to clean up oil-contaminated sites. More than 5000 MT of oily sludge has already been cleared and another 3500 metric trons (MT) is being cleaned up by using micro-organisms. Organisms, usually bacteria and fungi and occasionally plants, have been utilized to reduce/eliminate toxic pollutants. These micro-organisms either eat up the contaminants (mostly organic compounds) or assimilate them (heavy metals), thus cleaning up the oil-contaminated land or waters. This biotechnological intervention—bioremediation—has been in use since the 1970s. It acquired global acceptance when the US Environmental Protection Agency and the Exxon Company demonstrated its effectiveness on Alaskan beaches contaminated by the Valdez oil spill.

TERI succeeded in 1996 in developing a cost-effective method to clean up the environmental mess created by oily sludge and oil spills. "Oilzapper" is what TERI called the mixture it developed for bioremediation of oil-contaminated soils. Bioremediation is the most ecofriendly and economically viable of all the available methods of oil-sludge management. Today, TERI researchers are on firm ground after successfully demonstrating the technology in about a dozen oil refineries of the country. TERI has a ready stock of such bacteria obtained from nature that eat up the harmful compounds in oil spill sites and oily sludge.

With the help of Oilzapper, TERI has successfully biodegraded sludge sites of IOCL (Indian Oil Corporation Limited) refineries at Brauni, Mathura,

Digboi, and Guwahati, the BPCL (Bharat Petroleum Corporation Limited) refinery at Mumbai, and various other refineries. India's 15 oil refineries generate a huge amount of oily sludge annually. The cumulative sludge, generated over the decades of existence of these refineries, is life threatening in its ecological impact because it takes years for even a few hundred tons of waste to degenerate naturally. Moreover, this waste is supposed to be dumped in identified locations in secured pits. In the United States and Europe these pits are provided with a leachate-collection system and a polymer lining to prevent underground water contamination. However, the oil refineries in India do not find it a viable proposition to construct such storage pits. Moreover, storing the waste is not a sustainable approach to manage oily sludge because it exposes the local habitats to dangerous levels of toxicity through air and water pollution. This also takes a toll on the scarce soil because land requirements increase with an increase in oil-sludge generation. Besides the sludge from oil refineries, crude oil spills too are a cause of environmental degradation. Oil spills at port terminals are a frequent phenomenon, which invariably go unreported in the media. The Annual Report (1999) of the National Oil Spill Disaster Contingency Plan has reported major oil spills at the port terminals of Vadinar, Kandla, and Haldia amounting to 16,000, 4000, and 5000 m^3, respectively.

Oil spills are also common during oil explorations and at the oil well drilling sites. Oil spills also occur at oil collection centers where oil is separated from water. Scientists all over the world are battling to come up with efficient and economic solutions to combat contamination of land and water through oil sludge and crude-oil spills. All the emerging solutions indicate the use of natural (biological) processes to tackle the accompanying ecological threats. The crude oil and oily-sludge-degrading bacterial consortium (Oilzapper) developed by TERI is the only biological answer available to the oil industry in India right now, and one which is both highly efficient and cost effective. Application of a mixture of the Oilzapper to the oil-contaminated sites not only saves valuable land but also cuts down the cost of construction and maintenance of the dumping sites, besides saving the local environment from degradation.

Refineries in the country spend over 2000–3000 rupees/ton to construct pits for dumping the oil waste, while the Oilzapper solution comes for less than 800 rupees/ton. Bioremediation through this method has an additional advantage of containing the problem where it is located, thus eliminating the need for transferring large quantities of contaminated waste, which can be a potential hazard to human health during transportation. Also, Oilzapper speeds up the degradation process by three to four times the natural process. Until now, in India there has been no strict legislation against environmental pollution. However, in the near future, all industries will be required to strictly adhere to the regulatory guidelines laid down by the Central Pollution Control Board and the state regulatory authorities. Implementation of the laid guidelines is expected to rapidly increase the market of bioremediation of crude-oil spills and oily sludge in the country in coming years.

5.2. The Application of "Oilzapper"

Until the time the country's oil and petrochemicals industry resorts to natural methods of its waste degradation, India will continue to pay a very heavy price in terms of its environmental health—a price that no one cares to assess and is not possible to estimate. Oilzapper was produced in bulk and immobilized onto a carrier material (organic powder material). Carrier-based Oilzapper was used for clean up of crude-oil spills and treatment of oily sludge. Crude-oil- and oily-sludge-degradation efficiency of Oilzapper was tested under laboratory as well as under field conditions. With application of Oilzapper more than 10,000 metric tons of oily sludge have been treated, and more than 10-acre land and many lakes (northeastern part of India) contaminated as a result of oil slicks have been cleaned up in 2 yr. With application of Oilzapper, crude-oil-contaminated agricultural land near the IOCL refinery, Digboi, and Oil India Limited, Duliajan, Assam have been cleaned up. The know-how of Oilzapper technology has been transferred to Shriram Biotech Limited, Hyderabad and Bharat Petroleum Corporation Limited, Mumbai for commercialization, and this product also available at TERI, New Delhi.

Apart from accidental spills of crude oil, oily sludge, a hydrocarbon waste generated in huge quantities by oil refineries, also creates environment pollution. Oil refineries need a well-planned oily-sludge-management strategy to manage oil sludge. A straightforward approach may be to dump the oily sludge into specially constructed pits. Because the possibility of seepage cannot be ruled out, the ideal sludge pit should incorporate a leachate-collection system and a polymer lining to prevent percolation of the contaminants into the groundwater. It is possible that plants might have a role in preventing such seepage.

Sites where full-scale bioremediation has been carried out include: Barauni refinery, owned by the Indian Oil Corporation Limited, and 60 km from Patna, the capital of Bihar; Guwahati Refinery owned by Indian Oil Corporation Ltd. situated in the northeastern part of India; the Digboi Refinery owned by Indian Oil Corporation Ltd. Assam Oil Division; the Bharat Petroleum Corporation Limited refinery in Mumbai; and the Hindustan Petroleum Corporation Limited Refinery, Visakhapatnam, situated in the southern part India.

5.3. The Development and Future of "Oilzapper"

The Microbiological laboratory at TERI has, therefore, developed an efficient bacterial consortium that degrades crude oil and oily sludge very fast. This bacterial consortium was developed by mixing five bacterial strains, which could degrade aliphatic, aromatic, asphaltene and nitrogen, sulfur, and oxygen compound fractions of crude oil and oily sludge. Crude-oil and oily-sludge-degrading efficiency of the bacterial consortium was tested under laboratory conditions and field conditions. A feasibility study on the bioremediation of soil contaminated with crude oil/oily sludge was carried out at Mathura oil refinery (India). The feasibility study was carried out with six different treatments in a 25-m^2 land area contaminated with crude oil/oily sludge prior to full-scale bioremediation. The indigenous crude oil/oily-sludge-degrading bacterial population was only 104 cfu/g soil in the feasibility study. Of the six treatments, the application of the bacterial consortium and nutrients gave the maximum response, which resulted in 48.5% biodegradation of TPH in 4 mo as compared with only 17% biodegradation of TPH in soil treated with nutrients alone. Based on the feasibility study, the treatment consisted of the application of bacterial consortium, and nutrients were selected for full-scale bioremediation.

A microbial consortium was developed from five bacterial isolates. These isolates were obtained from hydrocarbon-contaminated sites using enrichment methods. The microbial consortium developed was immobilized with a suitable carrier material, namely powdered corncob, which is an environment-friendly, biodegradable product. Survivability of the consortium in the immobilized condition was determined and found to be 3 mo at ambient temperatures. The immobilized culture was put into sterile polythene bags, sealed aseptically, and transported to the place of requirement. This immobilized bacterial consortium was named "Oilzapper." The site was tilled thoroughly to mix the oily sludge uniformly with the soil and Oilzapper applied onto it. The land was tilled again and watered to maintain proper aeration and moisture levels. The land was tilled at regular intervals to facilitate faster degradation. The problem of heterogenous distribution of the oily sludge was solved by extensive tilling prior to the application of the Oilzapper.

The success of "Oilzapper" can be gauged by the tremendous response received from various oil refineries. At present, TERI is working on the bioremediation of oily sludge at the following sites as shown in **Table 6**. Prospective end-users of this technology are: Bharat Petroleum Corporation Ltd., Indian Oil Corporation Ltd., Oil and Natural Gas Corporation Ltd., Oil India Ltd., Hindustan Petroleum Corporation Ltd., Reliance Industries, Shell India, Essar Oil, Gas Authority of India Ltd., Indian Petrochemicals Ltd., Madras Refineries Ltd., Videocon Petroleum Ltd., Southern Petrochemicals and Industrial Corporation Ltd., Multinational Bioremediation companies, and the Lubricant oil manufacturing industry. Given the rapid advances currently being seen in phytoremediation of organic compounds and rhizosphere–microbe interactions,

Table 6
Oily Sludge Treatment in India

Location	Quantity of sludge under treatment (metric ton)
BPCL refinery, Mumbai	1000
BPCL terminal, Kandla	100
IOCL terminal, Rajkot	350
IOCL terminal, Kanpur	100
IOCL refinery, Barauni	1000

it is quite possible that in the future phytoremediation technologies to complement "Oilzapper" or enhance its utility will be developed *(48,49)*.

6. Conclusions

There are several advantages of phytoremediation technology. However, the government agencies in India are not coming forward for application on a large scale unlike United States, Europe, and Australia, although there is much interest in universities and research institutes in India (**Fig. 3**). Industrial crops not used for food production, e.g., fiber crops and microbes, would be the best-suited candidates for use in phytoremediation. Agricultural crops such as *Brassica juncea*, *Armoracia rusticana*, *Arabidopsis halleri*, *Gossypium hirsutum*, *Helianthus annuus*, *Eucalyptus*, *Amaranthus*, *Cannabis sativa*, and *Linum usitatissimum*-based phytoremediation systems would contribute to sustainable development as along as the produce is used for industrial products but not human or cattle consumption. Hence, fiber and energy crops and industrial crops that are amendable to genetic manipulation via in vitro culture techniques could certainly play a significant role for the success of phytoremediation technology not only in India but also for global sustainable development for which already considerable success has been attained in this direction. To enable this, sustained fundamental research to build on that already established in India *(50–52)* is likely to be necessary.

Experience with phytoremediation in India reminds us that degraded soils and marginal lands occupy a significant proportion of land in the world. Rehabilitation and management of degraded lands with appropriate agroforestry systems is a promising global opportunity to manage the buildup of greenhouse gases in the atmosphere *(53)*, which has been little exploited. For sequestering carbon through agroforestry on degraded soils, our research strategies should concentrate on: (1) the development of silvipastoral, hortipastoral, agrisilvicultural, and silvicultural models for all kinds of wastelands in different agroclimatic regions; (2) estimation of the carbon-sequestering potential of

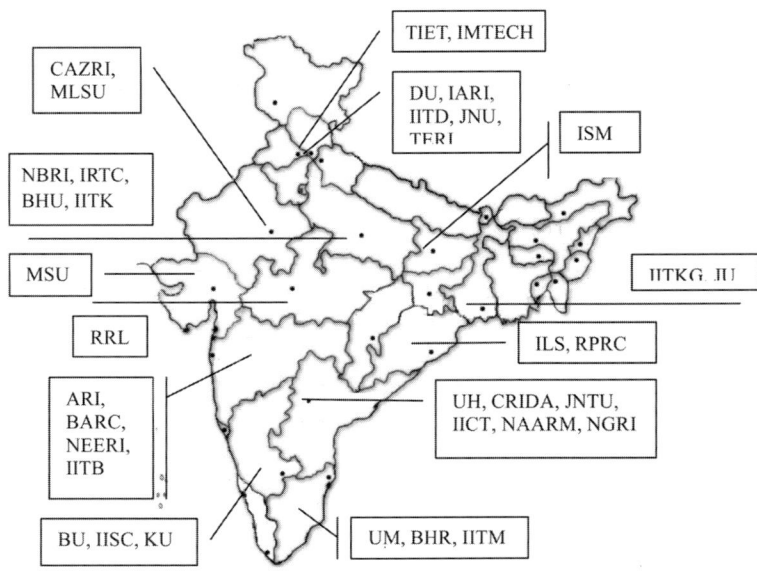

Fig. 3. Institutions and universities in different provinces of India that are involved in the phytoremediation research.

different land-use systems already in practice *viz.*, arable farming, forest plantations, and agroforestry; and (3) pilot-scale studies at selected places where the previously mentioned three systems already exist.

Acknowledgments and Disclaimer

This review has been prepared from information obtained from authentic and highly regarded sources such as internet, national and international conference deliberations; the author thankfully acknowledges these sources. The author and publisher cannot assume responsibility for any adverse consequences of the use of material inserted in this article in field of laboratory.

References

1. Verma, V. K., Chopra R., Sharma, P. K., and Singh, C. (1998) Integrated resource study for conservation and management of Ropar wetland ecosystem, Punjab. *J. Indian Soc. Remote Sensing* **26,** 85–195.
2. Singhal, V., Kumar, A., and Rai, J. P. N. (2003) Phytoremediation of pulp and paper mill and distillery effluents by channel grass (*Vallisneria spralis*). *J. Sci Industrial Research.* **62,** 319–328.
3. Pandey, J. S., Joseph, V., Shankar, R., and Kumar, R. (2000) Modelling of groundwater contamination and contextual phytoremediation: sensivity analysis for an Indian Case Study, Proceedings CSRA, Melbourne, Australia, December 04–08, pp. 545–552.
4. Prasad, M. N. V. (2004) Phytoremediation of metals in the environment for sustainable development. *Proc Indian Natl Sci Acad* **70(1)**: 71–98.
5. Prasad, M. N. V. (1997) Free floating, submerged and emergent macrophytes as biofilters of toxic trace meatls and pollutants from natural and industrially polluted aquatic systems. In: *Proceedings of the International Conference on Industrial Pollution and Control Technologies*, (Anjaneyulu, Y., ed.), Allied Publishers Ltd., Hyderabad, India, pp. 324–327.
6. Prasad, M. N. V. and Freitas, H. (1999) Feasible biotechnological and bioremediation strategies for serpentine soils and mine spoils. *Electronic J. Biotechnol.* **2,** 36–50.
7. Prasad, M. N. V., and Matsumoto, H. (2002) Bioresources for remediation and monitoring of metals in the environment. In: *Proceedings of the International Conference on Bioresources and Environmental Stress*. Research Institute for Bioresources, Okayama University, Kurashiki, Japan, pp. 7–10.
8. Prasad, M. N. V., Greger, M., and Landberg, T. (2001) *Acacia nilotica* L. bark removes toxic metals from solution: Corroboration from toxicity bioassay using *Salix viminalis* L. in hydroponic system. *Int. J. Phytorem.* **3,** 289–300.
9. Ahmed, K. S., Panwar, B. S., and Gupta, S. P. (2001) Phytoremediation of Cadmium contaminated soil by *Brassica* species. *Act Agro. Hungarica.* **49,** 351–360.
10. Kumar A., Rao N. N., and Kaul S. N. (2000) Alkali-treated straw and insoluble straw xanthate as low cost adsorbents for heavy metal removal—preparation, characterization and application. *Bioresource Technol.* **71,** 133–142.

11. Ajmal, M., Khan, R. R. A., Anwar, S., Ahmad, J., and Ahmad, R. (2003) Adsorption studies on rice husk: removal and recovery of Cd(II) from wastewater. *Bioresource Technol.* **86,** 147–149.
12. Goswami, T. and Saikia, C. N. (1994) Water hyacinth a potential source of raw material for grease proof paper. *Bioresource Technol.* **50,** 235–238.
13. Gupta, S. K., Herren, T., Wenger, K., Krebs, R., and Hari, T. (1999) In *Situ* gentle remediation measures for heavy metal-polluted soils. In: *Phytoremediation of Contaminated Soil and Water,* (Terry, N. and Banuelos, G. eds.) CRC Press, Boca Raton, pp. 303–322.
14. Kumar, S. M., Vaidya, A. N., Shivaraman, N., and Bal, A. S. (2000) Biotreatment of oil-bearing coke-oven wastewater in fixed film reactor: a viable alternative to activated sludge process. *Environmental Engineering Science* **17,** 221–226.
15. Ali, M. B., Tripathi, R. D., Rai, U. N., Pal, A., and Singh S. P. (1999) Physico-chemical characteristics and pollution level of Lake Nainital (U.P., India): role of macrophytes and phytoplankton in biomonitoring and phytoremediation of toxic metal ions. *Chemosphere.* **39,** 2171–2182.
16. Ansari, M. H., Deshkar, A. M., Dharmadhikari, D. M., Pentu Saheb, S., and Hasan, M. Z., (2000) Neem (*Azardirachta indica*) bark for removal of mercury from water. *J. Indian Assoc. Environ. Manag.* **22,** 133–137.
17. Bhati, M. and Singh, G. (2003) Growth and mineral accumulation in *Eucalyptus cameldulensis* seedlings irrigated with mixed industrial effluents. *Bioresource Technol.* **88,** 221–228.
18. Gajghate, D. G., Thakre, R., and Aggarwal, A. L. (1998) Strategic considerations for lead pollution control in Kanpur City. *J. Indian Chem. Soc.* **25,** 23–26.
19. Chandra Sekhar, K., Kamala, C. T., Chary, N. S., and Anjaneyulu, Y. (2003) Removal of heavy metals using a plant biomass with reference to environmental control. *Int. J. Min. Proc.* **68,** 37–45.
20. Chandra Sekhar, K., Rajni Supriya, K., Kamala, C. T., Chary, N. S., and Nageswara Rao, T. (2001) Speciation, accumulation of heavy metals in vegetation grown on sludge amended soils and their transfer to human food chain. *Tox. Environ. Chem.* **82,** 33–43.
21. Chandra Sekher, K., and Puvvada, G. V. K. (1997) Studies on the metal binding properties of the seeds of *Strychnos potatorum* Linn. *NML Technical J.* **39,** 239–243.
22. Dahiya, S. S., Goel, S. K., Antil, R. S., and Karwasra, S. P. S. (1987) Effect of farmyard manure and cadmium on dry matter yield and nutrients uptake by maize. *J. Indian Soc. Soil. Sci.* **35,** 460–464.
23. Manning, W. J. (2002) The ICPEP-2 meeting in India: biodiversity to the rescue! *The Sci. World J.* **2,** 1196–1197.
24. Glass, D. J. (1999) *US and International Markets for Phytoremediation, 1999–2000.* DJ Glass Associates Inc., Needham, MA. pp. 1–266.
25. Glass, D. J. (2000) *The 2000 Phytoremediation Industry.* DJ Glass Associates Inc., Needham, MA, pp. 1–100.
26. Khoosho, T. N. and Deekshatulu, B. L. (1992) *Land and Soils,* Indian National Science Academy, Har-Anand Publication, New Delhi, India.

27. Kumar, A. and Pandey, R. N. (1989) *Wasteland Management in India*, Ashish Publishing House, New Delhi, India.
28. Solanki, K. R. and Singh, G. (2000) Agroforestry technologies for wasteland development—India experience. *Proc. International Conference on Managing Natural Resources*, New Delhi, India, pp. 379–390.
29. Bhardwaj, S. P. (1990–1991) *Annual Report.* Central Soil, Water Conservation Technology Research Institute (CSWCRI), Dehra Dun, India, pp. 41–42.
30. Withington, D., MacDicken, K. G., Sastry, C. B., and Adams, N. E. (1988) *Multipurpose trees for small farm use.* Winrock International Institute for Agricultural Development and the International Research Centre of Canada, FAO Regional Office for Asia and the Pacific, p. 282.
31. Grewal, S. S. (1993) *Agroforestry in 2000 AD For the Semi Arid and Arid Tropics.* National Research Centre for Agroforestry, Jhansi, India.
32. Tejwani, K. G., Gupta, S. K., and Mathur, H. N. (1975) *ICAR Annual Report*, New Delhi, India, p. 359.
33. Bhojvaid, P. P. and Timmer, V. R. (1998) Soil dynamics in age sequence of *Prosopis juliflora* planted for sodic soil restoration in India. *Forest Eco. Manag.* **106,** 181–193.
34. Gururaja Rao, G. and Singh, R. (1997) Ecodevelopment of saline black soils—a holistic approach. *Indian J Soil Conserv.* **25,** 151–156.
35. Kikkawa, J., Dart, P., Dole, D., Ishii, K., Lamb, L., and Suzuki, K. (1988) *Overcoming Impediments to Reforestation: Tropical Forest Rehabilitation in the Asia Pacific Region. Proc. of the 6th International Workshop on Bio-Reforestation,* BIO-REFO, IURO/SPDC. pp. 1–249.
36. Aery, N. C. and Tiagi, Y. D. (1984) Studies on the reclamation of tailings dams at Zawar Mines, Udaipur, India. *Asian Mining*, IMM London, UK, 65–70.
37. Aery, N. C., Tiagi, Y. D., and Khandewal, R. (1987) Studies on the efficacy of certain plants for the stabilization of tailing dams at Zawar Mines. Rajasthan, India. In: *Proceedings of International Conference on Heavy Metals in the Environment*, New Orleans, Sept. 1987. pp. 445–447.
38. Kundu, N. K. and Ghose, M. K. (1998) Studies on the plant communities in eastern coalfield areas with a view to reclamation of mined out lands. *J. Environ. Biol.* **19,** 83–89.
39. Samantaray, S., Rout, G. R., and Das, P. (1999) Studies on the uptake of heavy metals by various plant species on chromite minespoils in sub-tropical regions of India. *Environ. Mon. Assess.* **53,** 389–399.
40. Samantaray, S., Rout, G. R., and Das, P. (2001) Heavy metal and nutrient concentration in soil and plants growing on a metalliferous chromite minespoil. *Environ. Technol.* **22,** 1147–1154.
41. Sekhar, D. M. R., Aery, N. C., and Tiagi, Y. D. (1982) Revegetation of tailing dams, in *Lead, Zinc and Cadmium at Workplace-Environment and Health Care,* New Delhi, India, pp. 569–578.
42. Sharma, A. and Aery, N. C. (2001) Phytoremediation studies on the tailings of Rajpura-Dariba, Udaipur (Raj.) lead-zinc mines. In: *Proceedings of Tenth*

National Symposium on Environment. June 4–6, 2001, BARC, Mumbai, India, pp. 244–248.
43. Sharma, A. and Aery, N. C. (2001) Studies on the phytoremediation of zinc tailing in growth performance. *Vasundhra* **6,** 21–27.
44. Bhattacharyya, K. G. and Sarma, N. (1993) Using flyash to remove mercury (II) from aqueous solution. *Indian J. Environ. Prot.* **13,** 917–920.
45. Dinesh Goyal, K., Kaur, R., Garg, V., et al. (2002) Industrial fly ash as a soil amendment agent for raising forestry plantations. In: *EPD Congress: A Symposium on Flyash, 2002 Seattle,* WA (Taylor, P. R., ed.), TMS Publication.
46. Kumar, V., Mukesh Mathur, M., and Kharia, P. S. (2001) *Fly Ash Management: Vision for the New Millenium.* TIFAC publications, DST, New Delhi, India.
47. Tripathi, R. D., Vajpayee, P., Singh, N., et al. (2004) Efficacy of various amendments for amelioration of fly-ash toxicity: growth performance and metal composition of *Cassia siamea* Lamk. *Chemosphere* **54,** 1581–1588.
48. Mishra, S., Jyot, J., Kuhad, R. C., and Lal, B. (2001) Evaluation of inoculum addition to stimulate in situ bioremediation of oily sludge contaminated soil. *App. Environ. Microbiol.* **67,** 1675–1681.
49. Mishra, S., Lal, B., Jyot, J., Rajan, S., and Khanna, S. (1999) Field study: In situ bioremediation of oily sludge contaminated land using oilzapper. In: *Proceedings of Hazardous and Industrial Wastes Symposium,* (Bishop, D., ed.), Technomic Publishing Co. Inc., Lancaster, PA, pp. 177–186.
50. Prasad, M. N. V. (ed.) (2001) *Metals in the Environment: Analysis by Biodiversity.* Marcel Dekker Inc., New York, NY. p. 504.
51. Prasad, M. N. V. (2001) Bioremediation potential of Amaranthaceae. In: *Phytoremediation, Wetlands, and Sediments: Proceedings of The 6th International In Situ and On-Site Bioremediation Symposium,* (Leeson, A., Foote, E. A., Banks, M. K., and Magar, V. S., eds.), Battelle Press, Columbus, OH, pp. 165–172.
52. Prasad, M. N. V. and Strzalka, K. (eds.) (2002) *Physiology and Biochemistry of Metal Toxicity and Tolerance in Plants.* Kluwer Academic Publishers. Dordrecht, Germany, p. 460.
53. Jha, M. N., Gupta, M. K., and Raina, A. K. (2001) Carbon sequestration: forest soil and land use management. *Annals Forestry* **9,** 249–256.
54. Prasad, M. N. V., Sajwan, K. S., and Ravi Naidu, (eds.). (2006) Trace elements in the environment: Biogeochemistry, Biotechnology and Bioremediation. CRC Press, Boca Raton pp. 1–726. Taylor and Francis Group, USA.

31

Phytoremediation in New Zealand and Australia

Brett Robinson and Chris Anderson

Summary

Phytoremediation in New Zealand and Australia stemmed from pioneering work by Professor R. R. Brooks on plants that hyperaccumulate heavy metals. Although original work focused on the extraction of heavy metals from contaminated sites, successful phytoremediation now employs plants as biopumps to reduce contaminant mobility and enhance the *in situ* degradation of some pesticides. In the first years of the 21st century, phytoremediation became established in the commercial environment with the appearance of dedicated phytoremediation companies. Phytoremediation offers a low-cost means of maintaining Australasia's "clean-green" image abroad. Use of this technology will increase because of increased pressure from regulators and future scientific achievements. In New Zealand, phytoremediation is used to improve degraded lands resulting from agricultural and silvicultural production, whereas in Australia its greatest potential is the remediation of mining-affected lands. Phytoremediation is most effective on lands where the clean-up cost of alternative technologies is greater than the land value. This reduces the importance of the longer time needed for phytoremediation. This chapter discusses, using case studies, the development of phytoremediation in Australia applied to a range of contaminated lands under various climatic conditions.

Key Words: Biopumps; biosolids; hydraulic control; mining; sheep dip; timber production.

1. Introduction

Phytoremediation is the use of plants to improve degraded environments (*1*). Pioneering work by the late Professor Robert Brooks at Massey University, Palmerston North, New Zealand, popularized the study of plants that accumulate inordinate amounts of heavy metals. Phytoremediation research in Australasia has stemmed from the investigation of these so-called "hyperaccumulator" plants. Professor Brooks was responsible for setting up a New Zealand phytoremediation program in the mid-1990s. Since these early studies on the plant extraction of heavy metals, phytoremediation has been developed for the treatment of a whole suite of contaminated sites. In Australasia, this technology has been successfully transferred to the commercial environment.

In New Zealand, HortResearch (www.hortresearch.co.nz) and the Soil and Earth Sciences Group of the Institute of Natural Resources at Massey University (soils-earth.massey.ac.nz) have active phytoremediation programs. Tiaki Resources Ltd. (projects@tiaki.co.nz) provides commercial phytoremediation. In Australia, the Botany Department at the University of Melbourne (www.botany.unimelb.edu.au), the Centre for Mined Land Rehabilitation (www.cmlr.uq.edu.au), the Commonwealth Scientific and Industrial Research Organisation (CSIRO) (www.csiro.au), and several other universities have phytoremediation programs. Phytolink Australia Pty (www.phytolink.com.au) is a dedicated phytoremediation company. As elsewhere, the commercial use of phytoremediation in Australasia is driven by pressure from regulators.

Phytoremediation in Australasia has focused on the use of plants as biopumps *(1)*. Here, plants use the sun's energy to dewater contaminated sites and control leaching, as well as enhance the organic matter and microbial activity in the rhizosphere. These root-zone processes thereby augment contaminant degradation and reduce the mobility of heavy metals. We therefore use the term phytoremediation to cover a wide-range of plant-based environmental applications, ranging from mine-site revegetation through to riparian management and phytoextraction. In Australasia, the most important role of phytoremediation is to reduce contaminant mobility and to degrade organic pollutants, rather than the phytoextraction of heavy metals. In this chapter, we discuss the most important environmental issues in New Zealand and Australia and demonstrate, using case studies, how phytoremediation can be used to address land degradation.

2. The Relevance of Phytoremediation in the Australasian Context

The New Zealand economy is underpinned by an internationally perceived "clean-green" image. Contaminant-free agricultural exports and tourism contribute 16 and 9%, respectively to New Zealand's gross domestic product *(2)*. Environmental degradation thus poses a significant risk to economic growth, and the government has consequently implemented strict environmental controls via the Resource Management Act, 1992. Australia's economy is similar to New Zealand's, although mining is now Australia's single biggest export earner. Unlike New Zealand, however, Australia has no overarching environmental legislation. Rather, disparate bills have been passed that address specific environmental issues. These may vary between states.

Most contaminated sites in New Zealand are associated with agricultural and silvicultural production: there are an estimated 50,000 disused sheep-dipping sites that may contain elevated levels of persistent pesticides such as dieldrin and sodium arsenate. Similarly, there are numerous sites contaminated with timber treatment compounds such as copper-chromium-arsenate, pentachlorophenol, and boron. In addition to agricultural- and silvicultural-contaminated sites, Australia has over 2 million ha of open-cast mining and many contaminated sites associated with smelting and processing *(3)*. Both countries face environmental issues associated with urban development, especially the disposal and treatment of sewage sludge and burgeoning landfills.

New Zealand has a temperate oceanic climate with high rainfall. Meteorological conditions seldom prohibit plant growth making phytoremediation a viable option for many contaminated sites. However, the high rainfall:evapotranspiration ratio can limit the effectiveness of phytoremediation to provide hydraulic control on contaminated sites. Australia, on the other hand, often suffers from drought and associated soil salinity, both of which can negatively affect plant growth, but render phytoremediation effective for the mitigation of leaching.

Phytoremediation is best suited to the long-term cleanup of low-value land where other remediation options are prohibitively expensive *(4)*. This technology is therefore well suited for use on contaminated sites in the extensive production systems of Australasia. The low population densities of both New Zealand (14.8 people/km^2) and Australia (2.4 people/km^2) keep land values relatively low and reduce the pressure for the rapid remediation of contaminated sites.

3. Phytoremediation Case Studies

3.1. Phytoremediation of a Timber-Industry Waste Site

New Zealand has 1.6 million ha of *Pinus radiata* plantations for pulp and timber production. Most timber products are treated with biocides to prevent decay. In the past, pentachlorophenol and boron have been used to treat timber. Currently, copper-chromium-arsenic is the treatment of choice. Treatment sites and wood-waste disposal sites have become contaminated with the aforementioned biocides and pose a risk to ground and surface waters through contaminant leaching. Here, we outline the use of phytoremediation to mitigate the environmental risk associated with a timber-industry waste site.

The Kopu timber-waste pile is located at the base of the Coromandel peninsula, North Island, New Zealand (37.2°S, 175.6°E). The pile has a surface area of 3.6 ha and an average depth of 15 m. Over a 30-yr period from 1966, sawdust and yard scrapings from timber milling in the region were dumped on the pile. Land around the pile has been engineered so that no surface or ground water enters the pile, and all leachate resulting from rainfall is collected in a small holding pond at the foot of the pile. In the past, vegetation has failed to

Fig. 1. Aerial photograph of the revegetated Kopu timber waste pile, October 2003.

establish and evaporation from the surface of the pile has been negligible, even in the summer months. This was demonstrated by the presence of saturated material at depths as shallow as 20 mm.

Leachate resulting from the annual rainfall of 1135 mm, as measured at a nearby meteorological station at Thames, regularly caused the holding pond to overflow and enter a local stream. This overflow elevated boron concentrations in the stream to levels that were in excess of 1.4 mg/L, the New Zealand drinking water standard, especially in the summer months when stream flow was low. In response to these breaches, the local environmental authority placed an order on the forestry company responsible for the site that the problem be remedied.

In July 2000, a 1-ha trial was established on the Kopu site using 10 poplar and willow clones, as well as two species of *Eucalyptus*. Two *Populus deltoides* hybrid clones were then chosen as the best candidates for phytoremediation based on survival, biomass production, and B uptake. The following year, the remainder of the pile was planted in these two clones at a density of 7000 trees/ha. Fertilizers were periodically added to the trees and a pump was installed near the holding pond at the foot of the pile for irrigation during the summer months. **Figure 1** shows tree growth on the Kopu pile after 3 yr. **Figure 1** demonstrates clearly how phytoremediation helps the contaminated site become part of the landscape by covering the bare pile with an actively growing green mantle.

Robinson et al. *(1)* calculated the monthly water balance of the pile using a computer model similar to that described in Green et al. *(5)*. The model used

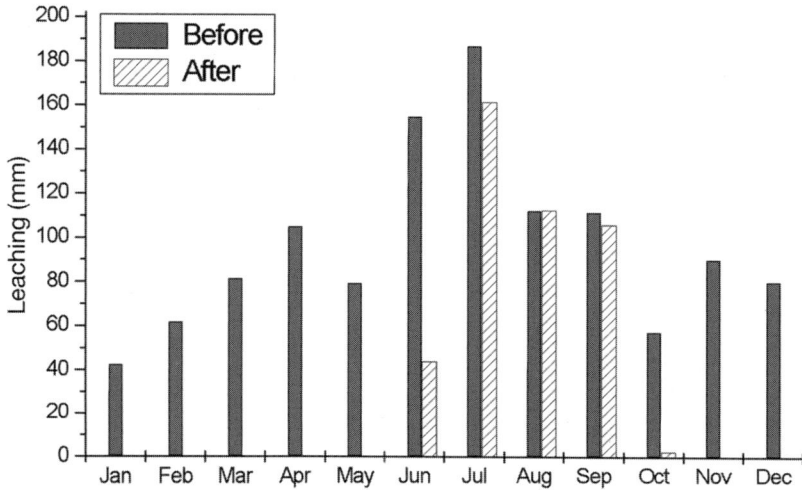

Fig. 2. Model calculations of average monthly leaching from the Kopu sawdust pile before and after phytoremediation.

daily weather data taken from a meteorological station at nearby Thames and other parameters obtained experimentally. Model calculations of leaching are shown in **Fig. 2**. As expected for such a high rainfall site, the bare pile leaches a considerable amount of drainage water through all months of the year. The impact of trees is to substantially reduce the drainage of water during the summer months when the trees are fully leafed and transpiring at their maximum. The summer months are of greatest concern for contamination of the local waterways because stream flows are lower and there is less dilution of the contaminants. The leaching that occurs during the winter months can be irrigated onto the trees in times of drought during the summer, or alternatively, released into a nearby stream at times of high flow when the risk of exceeding the New Zealand drinking water standard is minimal. Poplar leaves sampled from the sawdust pile contained Cu and Cr concentrations that were on average 6.6 and 4.9 mg/kg dry mass, respectively. Arsenic concentrations were below detection limits (1 mg/kg).

At the end of the growing season, the average leaf B concentration was nearly 700 mg/kg on a dry matter basis, over 28 times higher than the B concentration in the sawdust (40 mg/kg dry matter). Bañuelos et al. *(6)* have previously reported this B accumulation trait in poplars. The results indicate that in addition to controlling leaching at the site, poplars may also be able to reduce the B loading by phytoextraction. Unless the trees are harvested, most of the B is returned to the sawdust via leaf fall. Harvested material could, however, be used as an organic B supplement to trees in orchards that are B deficient in other parts of the country. The concentrations of other heavy metals in the leaves are unlikely to cause further environmental problems.

The cost of phytoremediation at Kopu is estimated to be New Zealand $200,000 including a site-maintenance plan more than 5 yr. Half of this total cost was taken up as site assessment, involving scientist time to conduct the plant trial and chemical analysis. The alternative cost of capping the site was estimated by the local environmental authority to be over New Zealand $1.2 million. Capping would also require ongoing maintenance to ensure its integrity.

3.2. Phytoremediation of a Disused Sheep-Dipping Site

Until 1966, there was a legal requirement that all sheep sold in New Zealand were free of pest infestations such as lice, blowflies, ticks, and mites *(7)*. The most effective means of dealing with this problem was dipping the sheep in a pesticide solution. The active ingredients of these solutions were arsenic, organochlorines, and organophosphates, the former two being persistent in the environment. Disposal of the pesticides after use resulted in areas adjacent to the sheep dip becoming contaminated. These areas pose a risk to human and animal health through groundwater contamination *(8)*, as well as direct ingestion of soil. Dipping sites were often located near wells or streams, to prepare the pesticide solution. The exact numbers and locations of historical dip sites in New Zealand are unknown, but there are probably many tens of thousands on both private and public land.

A disused sheep-dipping site in an asparagus field was discovered near the city of Hamilton, North Island, following the measurement of elevated dieldrin concentrations in a nearby well. Soil analyses revealed dieldrin concentrations from 10 to 70 mg/kg over 100 m^2. The Dutch intervention value for dieldrin in soil is 4 mg/kg. In late September 2001, the site was planted using HortResearch willow clones. In October 2003, the average height of the trees was over 3 m (**Fig. 3**). Soil collected from the site before planting was homogenized and placed in 12- to 15-L pots in HortResearch's plant-growth facilities. Willow clones were planted in eight of the pots. All pots were watered and fertilized equally. After 5 mo, soils from the pots were analyzed for dieldrin, as well as biological (dehydrogenase activity) activity. Substrate dehydrogenase activity is estimated from the rate of conversion of triphenyltetrazolium chloride to triphenylformazan (TPF). This is a measure of biological activity.

Figure 4 shows the biological activity in the root zones of grass and willows. The data shown in **Fig. 4** may approximate the surface of the site at Ngahinapouri before planting (i.e., when grass was growing on the site), and now after the planting of willows. Clearly, willows greatly enhance biological activity in the soil. Previous studies *(9,10)* have shown that biological activity

Fig. 3. Phytoremediation at the Ngahinapouri site.

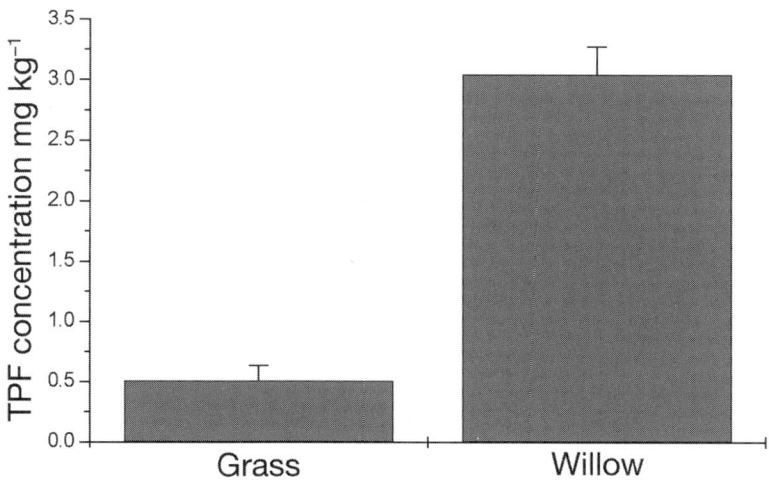

Fig. 4. Soil triphenylformazan (TPF) concentration under grass and willow vegetation.

leads to a greater rate of the decomposition of some contaminants. This increase in biological activity is caused by root exudates, such as sugars and organic acids, on which bacteria and fungi can feed. Willows have a much greater biomass production than grasses, and consequently have a greater quantity of root exudates. Willow roots also penetrate further than grass roots (up to 1 m) and improve soil aeration because of their high water use. The willows caused a significant decrease ($p < 0.05$) in the soil dieldrin concentration over the treatment

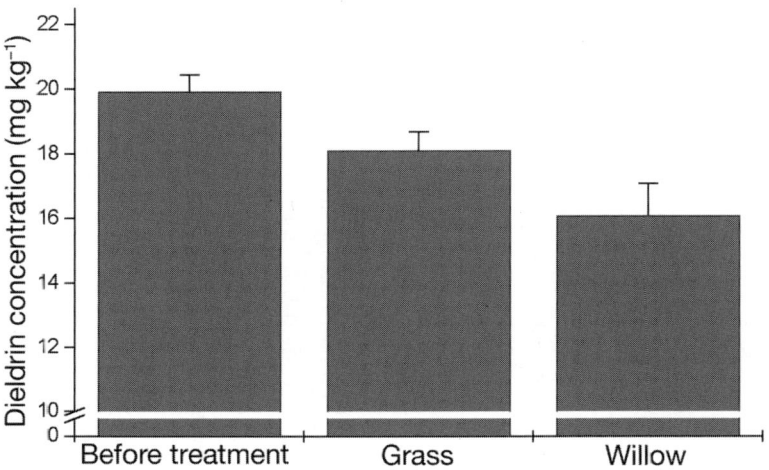

Fig. 5. Soil dieldrin concentration as affected by vegetation cover.

period (**Fig. 5**). This 20% reduction was achieved in only 5 mo of growth. The dieldrin degradation effected by the willows was greater than that by the grass species that may first colonize many disused sheep-dipping sites.

3.3. Phytoremediation of the Tui Mine Site

The Tui mine tailings near Te Aroha, is considered New Zealand's worst environmental disaster caused by mining activities *(11)*. The site consists of a 1.5-ha tailings dam containing 100,000 m^2 of toxic mining waste, principally sulfide minerals with high concentrations of lead (0.5%), cadmium (26 mg/kg), and mercury (8 mg/kg). Continual oxidation of the sulfide produces sulfuric and sulfurous acids that result in a pH <3.0 for the surface material. The low pH mobilizes heavy metals that leach out of the tailings dam into a nearby stream. Analyses of the stream water and sediments reveal that both are above the allowable limits for lead and cadmium set by the World Health Organization *(12)*.

Although the site has been abandoned for more than 30 yr, no vegetation has established itself on the tailings as a result of a low pH and the high concentration of heavy metals. During the summer months, dust containing high concentrations of heavy metals is blown around, contaminating nearby areas. Adjacent to the tailings dam is a car park used by hikers. Children have been observed playing on the tailings, their parents unaware of the risk of heavy metal poisoning. The goal of phytoremediation at Tui is to mitigate heavy metal leaching, prevent erosion and dust movement, and to return the area to native vegetation. Plant accumulation of heavy metals is not desirable because this may provide an exposure pathway for metals to enter the food chain via herbivore browsing. The pH and plant nutrient status of the tailings had to be modified so that

Fig. 6. Experimental plot on the Tui mine tailings, September 2002.

Table 1
Metal Concentrations in Tui Tailings and Supported Plant Species

	Cd	Cu	Fe	Hg	Pb	Zn
Tailings	0.7	27 (3)	1581 (180)	8 (1)	5410 (923)	83 (7)
Phormium tennax	<0.15	12 (1)	179 (6)	<0.1	102 (52)	67 (10)
Hebe stricta	<0.15	11 (1)	221 (63)	<0.1	314 (71)	32 (7)
Leptospermum scoparium	<0.15	14 (1)	285 (18)	<0.1	454 (87)	43 (13)
Populus sp.	<0.15	8 (1)	122 (15)	<0.1	148 (23)	74 (26)
Cortaderia toetoe	<0.15	14 (2)	525 (368)	<0.1	226 (94)	30 (2)

vegetation could be established. Experiments conducted at Massey University determined the optimal rate of liming and organic matter addition to permit plant growth on the Tui tailings. A 100-m^2 plot was established on the Tui tailings in April 2001. Several indigenous species were planted as well as lupin to fix nitrogen.

The plants established rapidly (**Fig. 6**), with other plants and animals colonizing the plot area over time. Chemical analyses of leaf material from the plants that were grown on the plot at the Tui mines indicated low levels of bioaccumulation (**Table 1**). Elevated lead levels may be the result of surface contamination of the leaves with dust from adjacent nonvegetated areas. Greenhouse experiments with the Tui tailings, where there is no risk of dust contamination, demonstrated that leaf lead levels never exceeded 50 mg/kg. Samples of surface material on and off the vegetated plot were analyzed for TPF, an indicator of

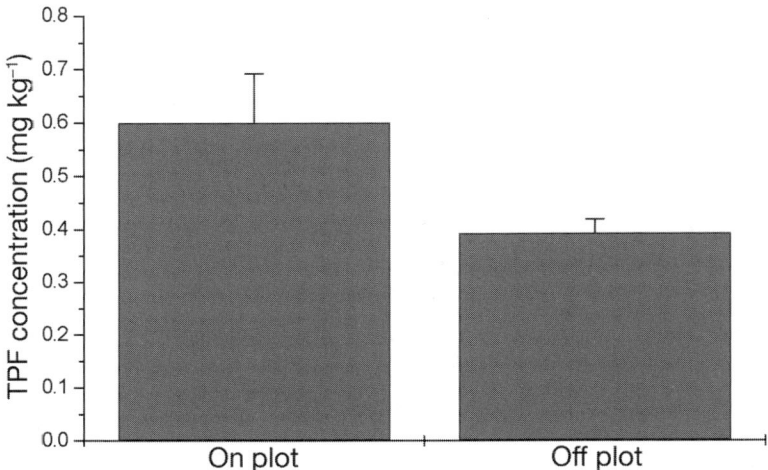

Fig. 7. Biological activity in the surface of the Tui mine tailings, as measured by triphenyl farmazan (TPF) concentration, on and off the phytoremediation plot.

microbial activity (**Fig. 7**). Microbial activity was significantly higher under the vegetated areas. The measurable TPF concentrations off the plot may be a result of the presence of *Thiobacillus* bacteria that are responsible for sulfide oxidation.

Clearly, it is possible to establish vegetation on the Tui mine tailings. Full-scale phytoremediation would eliminate lead-laden dust, reduce the visual impact of the mine, and reduce leaching by re-evaporating some of the rainfall through transpiration. Phytoremediation of the Tui tailings can be considered a revegetation operation. Quantification of the environmental benefit afforded by phytoremediation can be made through analysis of the biological activity of the substrate.

3.4. Phytostabilization of Biosolids, Western Treatment Plant, Victoria, Australia

The storage of biosolids, the solid fraction of sewerage waste, is a contentious subject for any developed area. Fresh biosolid or oxidation pond sludge waste has a water content of approx 70% by weight. When disposed of on land, natural drying will generate a crust of approx 15 cm; less than 15 cm minimal drying will occur. Mechanical drying of the fresh waste is possible, but is costly and is certainly not applicable to previously deposited volumes of waste. Every major populated area in New Zealand and Australia has large volumes of unstable, semiliquid waste stored on potentially valuable land.

Melbourne Water is the largest water utility in Melbourne, with a sewerage system of 380 km and two water treatment plants, the Western and Eastern Treatment Plants. Melbourne produces on average 900 million liters of sewage a day, 54% of which is treated by The Western Treatment Plant (WTP) situated on the western shore of Port Philip Bay near the city of Geelong *(13)*. The WTP covers 11,000 ha of valuable land, and discharges approx 600 L of treated water daily into Port Philip Bay. Three treatment methods are used for incoming sewage. An extensive lagoon system is used for peak daily and year-round wet weather flow. Land filtration is used during periods of high evaporation between October and April. Grass filtration is used during periods of low evaporation between April and October. Extensive land contamination with organic and inorganic contaminants has occurred at the WTP as a result of sewage disposal practices over the past 100 yr. Commercial and governmental groups in Melbourne have developed a "Vision for Werribee," a long-term plan that aims to turn solid and liquid sewerage waste into valuable revenue streams, and that will release decontaminated land to capital development. The Vision for Werribee brings together several environmental technologies to achieve the desired objective. One of these technologies is phytoremediation.

3.4.1. Establishing the Field Trial

The concept of plants as biopumps has been field tested at Werribee to assess the dewatering potential that phytoremediation may have on stockpiled biosolids. Biosolids have potential use as a soil amendment or energy source (as a renewable "brown coal"), however, high water content limits this potential. For the period between 1973 and 1979 (the only period when climatic data was collected), average annual rainfall at Werribee was 641 mm, whereas average annual evaporation for this period was 1386 mm (**Fig. 8**). For no month did rainfall exceed evaporation and this indicates that if water can be removed using plants then no rainfall-induced recharge should occur. An experimentally derived biosolid water-retention curve shows that, in theory, the water content could be lowered to approx 20% (10 bars) by plant transpiration. Interpretation of this physical and climatic data suggests that plants should be able to dewater the Werribee biosolids. Chemical analysis of the biosolids indicated that elevated concentrations of Cu could possibly affect plant growth, however, no signs of phytotoxicity were observed during experimental work (**Table 2**).

A 1-yr demonstration trial was initiated in May 2002 on a biosolids storage tank. The relative growth performance of 10 plant species was tested during the trial. Core samples (0–20 cm) collected from across the plot area at the time of trial setup and then again 1 yr later allowed for estimation of the level of dewatering.

3.4.2. Dewatering Potential of Eucalyptus *sp. and Vetiver Grass*

Figure 9 summarizes the end of trial performance of two of the trialed species, *Eucalyptus saligna* and vetiver grass, in dewatering the biosolids.

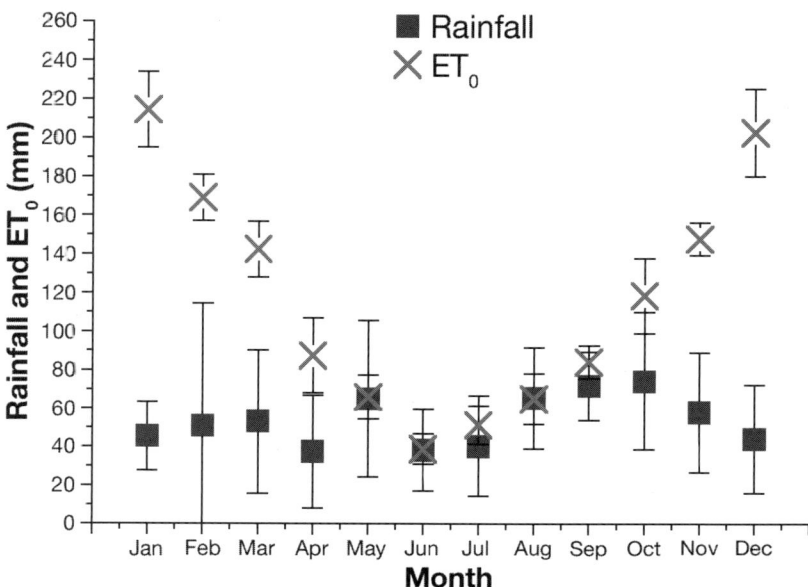

Fig. 8. Average monthly values of rainfall and potential evapotranspiration (ETo) at Werribee.

Table 2
Select Geochemical Parameters of the Werribee Biosolids

Metal	Extractable metal concentration (mg/L)	Total metal concentration (mg/kg)
As	0.04	9
Ca	464	8620
Cd	0.0008	9
Cr	0.002	646
Cu	0.16	1001
Fe	3.8	10943
K	84.6	1601
Na	272	1837
P	1.6	6724
Zn	7	1174

[a]Extractable concentrations determined by ICP analysis of 12,000g centrifuge soil solution extracts. Total concentrations determined by ICP analysis of aqua regia digest solutions.

Clearly, both species affected a significant decrease in the water content of the biosolids to 30–40% in the 5- to 10-cm sampled horizon. This is relative to a 55% water content for control sampling in May 2003 and a pretrial water content of 65% at May 2002. Dewatering was so effective that it was not possible

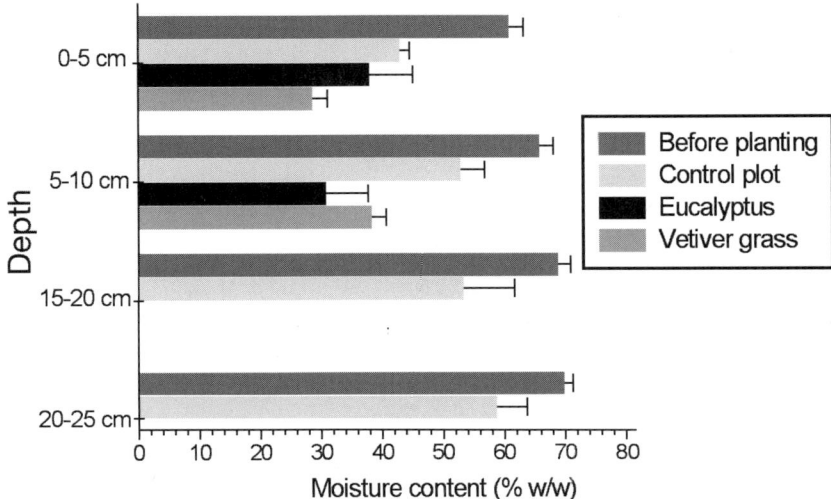

Fig. 9. Water contents of the sludge at Werribee under select plantings.

to push the soil corer into biosolids below a 10-cm depth for all cores sampled in proximity to vegetation.

There was a decrease in water content for control cores, sampled away from areas of vegetation. However, we expect that this water will have recharged because of the rainfall of late winter 2003; the timeline of the trial (May 2002 to May 2003) was particularly dry. Dewatering results indicate that in a single growing season, either *Eucalyptus* or vetiver grass could be used to dewater the top 15 cm of sludge stabilizing the material and potentially allowing reuse of the substrate. A 15-cm dewatering scenario is conservative and based on observed results; the plants had not been able to reach their maximum biomass over the trial because of cattle and sheep grazing. The true dewatering depth may be much greater, as modeled using HortResearch's Phytoextraction Decision Support System.

4. Conclusion

Ongoing research and commercial activities in New Zealand and Australia continue to apply phytoremediation to contaminated and degraded land. Four examples have been presented in this chapter; wood waste, acid mine tailings, a disused sheep-dip site, and sewage sludge. In addition, phytoremediation to promote revegetation has been successfully tested on a cyanide heap-leach pad in Victoria, Australia, and at a serpentinite waste-rock pile in the Waikato, New Zealand. Quantification of the benefits of phytoremediation is possible through analysis of variables such as biological activity and species diversity, as well as by examining the reduction in contaminant values promoted by plants. Phytoremediation is likely to remain a fertile subject for research and commercial development in New Zealand. The technology is well suited to the clean up of low-level contamination and land degradation associated with the commodity-driven economies of Australasia.

References

1. Robinson, B. H., Green, S. R., Mills, T. M., et al. (2003) Phytoremediation: using plants as biopumps to improve degraded environments. *Austr. J. Soil Res.* **41**, 599–611.
2. NZEFO (2003) New Zealand Economic and Financial Overview 2003. Treasury Report, Wellington, New Zealand.
3. Hamblin, A. (2001) *Australia State of the Environment 2001*. CSIRO Publishing, Collingwood, Vic. p. 42.
4. Schnoor, J. L. (2002) *Phytoremediation of Soil and Groundwater.* Technology Evaluation Report TE0201. Ground-water remediation technologies analysis centre, Iowa.
5. Green, S. R., Clothier, B. E., Mills, T. M., and Millar, A. (1999) Risk assessment of irrigation requirements of field crops in a maritime climate. *J. Crop Prod.* **2**, 353–377.
6. Bañuelos, G. S., Shannon, M. C., Ajwa, H., Draper, J. H., Jordahl, J., and Licht, L. (1999) Phytoextraction and accumulation of B and selenium by poplar (*Populus*) hybrid clones. *Int. J. Phytorem.* **1**, 81–96.
7. Heath, A. (1994) Ectoparasites of livestock in New Zealand. *New. Zea. J. Zool.* **21**, 23–38.
8. Robinson, P. (1995) *The Fate of Vetrazin (cyromazine) During Wool Scouring and its Effects on the Aquatic Environment,* PhD thesis. Lincoln University, Christchurch, New Zealand.
9. McLaughlin, M. J., Smolders, E., and Herckx, R. (1998) Soil: root interface: physicochemicl process. In: S., *Chemistry and Ecosystem Health* (P. M. Huang, ed.) Soil Science Society of America, Madison, WI, pp. 233–277.
10. Eriksson, M., Dalhammar, G., and Borg-Karlson, A. K. (2000) Biological degradation of selected hydrocarbons in an old PAH/creosote contaminated soil from a gas work site. *App. Micro. Biotech.* **53**, 619–626.
11. Morrell, W. J., Gregg, P. E. H., Stewart, R. B., Bolan, N. S., and Horne, D. (1995) Potential for revegetating base-metal tailings at the Tui mine site, Te Aroha, New Zealand. *Proceedings of the 1995 PACRIM Congress* (*Auckland*), 95–400.
12. Sabti, H., Hossain, M., Brooks, R., and Stewart, R. (2000) The current environmental impact of base metal mining at the Tui Mine, Te Aroha, New Zealand. *J. Royal Soc. N. Z.* **30**, 197–208.
13. Melbourne Water (2001) Infostream, public relations material. www.melbourne-water.com.au.

Index

A

Abutilon theophrastii, 238
Acacia albida, 442
Acacia auriculiformis, 442
Acacia farnesiana, 441
Acacia nilotica, 439, 441
Acacia planiformis, 442
Acacia senegal, 442
Acacia tortilis, 441, 442
Acanthaceae, 236
Achillea millefolium, 428
Aeroponics, 91, 93, 94
Agrobacterium tumefaciens, 5, 8, 11, 52, 115, 268
Agrobacterium rhizogenes, 20, 162
  cultures, 164, 167
Agropyron smithii, 75
Agrostis capillaris, 322
Air pollutants, 110
Air sparging, 101
Albizia lebbek, 442
Albizia melliera, 442
Allelochemicals, 50
Alnus glutinosa, 416
Aluminum, 272
Alyssum bertolonii, 268
Alyssum lesbiacum, 268, 395
Alyssum murale, 395
Alyssum serpyllifolium, 414
Amaranthaceae, 370
Amaranthus blitoides,
  As, 213, 215
  trace metals, 214
Amaranthus cruentus, 370
Amaranthus tricolor, 370
Aminocyclopropane-1-carboxylic acid (ACC) deaminase
  activity in plant tissues, 18, 21
  isolation of bacteria containing, 17, 19
  function, 16
  PCR, 19, 20, 23
Amplified Fragment Length Polymorphism (AFLP), 272
Andropogon gerardii, 75
Anthyllis sampaiana, 415
Antioxidants, 324
Aquatic plants, 191, 193, 223
Arabidopsis halleri, 268, 326, 328, 449
Arabidopsis lyrata, 269
Arabidopsis thaliana, 4, 7, 29, 30
  cadmium/lead detoxification, 50–51
  chimeric nitrite reductase, 110
  fer gene, 272
  floral dip transformation, 53
  growth conditions, 110
  hydroponic growth, 63, 64, 66, 69
  ILR2 gene, 271
  mineral nutrition traits, 32
  molecular markers, 31
  NO_2 uptake, 110
  phenolic compounds, 52, 54
  quantitative trait loci (QTL), 30
  recombinant inbred lines, 31, 38
  vacuum infiltration transformation, 7
Arachis hypogea, 73
Armoracia lapathifolia, 163
  hairy roots, 165–169
Armoracia rusticana, 449
Arsenic,
  arsenate reductase, 267
  chelating agents, 333
  distribution in plant, 322
  hyperaccumulators, 327, 328–329
  in irrigation water, 394

in soil, 393–394
NIRS, 208
nutrients, 334
pH changes, 333
phosphate, 320
resistance plasmid, 428
sheep dip, 460
sodium arsenite, 427, 457
toxicity, 321, 394
uptake, 322
Artemesia argii, 360
Artemesia scoparia, 360
Artemesia absinthum, 428
Arundo donax, 257
Aspergillus fumigatus, 235
Asteraceae, 370
Atomic absorption spectrophotometry, 94
Atrazine,
conjugation, 238, 239
hydroxylation, 236–238, 239
in run-off, 233
metabolites, 238–239
N-dealkylation, 235–236, 239
transpiration stream, 234
vetiver, 241
Avena barbata, 182
Avena sativa, 425, 426, 441
Azolla caroliniana, 429

B

β-glucuronidase (GUS), 4–6, 9
Basidiomycetes,
wood-rotting fungi, 4, 51
Benzoxazinones, 236, 239
Berkheya codii, 296
Beta vulgaris, 73, 133, 135, 368
Betula uliginosa, 74
Bidens frondosa, 360
Binary vector,
including ACC deaminase gene, 20, 23
pCGN, 5–7

pEGAD, 5, 11
insertion into *A. tumefaciens*, 10
pKYLX71, 23
Bioavailability, 355
Bioreactor,
ebb and flow, 222
hydroponic, 227
soil, 230
Biopumps, 456, 465
Bioremoval, 185
experiments, 198–199
Biosorption, 185
Boehmeria nivea, 366
Bouteloua curtipendula, 75
Bouteloua gracilis, 75
Braciaria mutica, 441
Brassica campestris, 239, 360, 366
Brassica juncea, 73, 127, 141, 142, 257, 449
As, 321, 323
chelates, 292, 296
NIRS metals, 213, 215
Pb, 394
Brassica napus
genetic transformation, 20
herbicide degradation, 244
POPs, 73, 82
Brassica oleracea, 394
Brassicaceae, 4, 89, 394

C

Cadmium,
accumulation, 292
cDNAs, 273
in Chinese soils, 351–353
Cajanus cajan, 73
Cannabis sativa, 449
Canna generalis, 234
Capsicum annuum, 73
Carex normalis, 75
Carthamus tinctorius, 368
Caryophyllaceae, 370
Cassia siamea, 445

Casurina equisetifolia, 441
Casurina glauca, 441
Cenchrus ciliaris, 439
Cenococcum geophilum, 414
    genetic variability, 414
Centaurea cyanus, 368
Chamaecytisus palmensis, 295
Chelates,
    application, 300
    organic acids, 292
    Pb accumulation, 292
    plant growth, 297
    synthetic, 292
Chemical warfare agents, 426–427
Chenopodium album, 239
Chlamydomonas reinhardtii, 268
Chlorella, 430
Chloris gayana, 441
Chromium, in Chinese soils, 351–352
Chrysopogon zizanoides, 241
Cirsium arvense, 428
Cladophora parraudii, 193
Commelina communis, 356
Constructed wetlands, 410–412
Cordia subcordata, 177
Coriolus versicolor, 51
Cucumis sativum, 73
Cucurbita maxima, 110
Cucurbita melo, 73
    As, 323
Cucurbita moschata, 366
Cucurbita pepo, 73
Cucurbitaceae, 74
Cyanobacteria, 429
Cynodon dactylon, 75, 368, 441
Cytisus striatus, 324
Cytochrome p450, 235, 242

D

Dahlbergia sissoo, 441, 442
Daucus carota, 73
    As, 323
Desert advancement, 442

Dewatering, 465–467
Dichanthium annulatum, 439
Dicranopteris linearis, 361
Differential display, 272
Diffuse pollution, 233
Digitaria sanguinalis, 238
Dissolved oxygen, 228
DNA repair genes, 60–61

E

E-value, 150
    equation, 151
    determination, 153–154
Echinochloa crus-galli, 238, 424
EDTA, 293, 294
    enhanced uptake, 362
    degradation, 362
E_H, 227, 228
Eichhornia crassipes, 430
Elaliopsis binata, 439
Electrodics,
    apparatus, 141
    lead (Pb), 140,
    optimization, 142–143
Elsholtzia haichowensis, 356
Elymus canadensis, 75
Elytrigia repens, 424
Enhancer trap, Ac/Ds transposon
    construction, 6
    selection of tagged lines, 8
    response to PCBs, 9
Erechtites hieracifolia, 110
Eremochloa ophiuroides, 368
Eroded soils, 438–439
Escherichia coli, 267
Ethylene, 267
    plant stress, 16
Eucalyptus camaldulensis, 441
Eucalyptus saligna, 465
Eucalyptus tereticornis, 441, 442
Eucalyptus viminalis, 111
Experimental design, 196–198
Expressed sequence tags (ESTs), 273

F

Festuca arundinacea, 75, 214
Festuca brigantine, 415
Field sites,
  Argonne National Laboratory, IL, USA, 306
  Barber Orchard, NC, USA, 140
  Bradwell Nuclear Power Station, UK, 306
  Brookhaven National Laboratory, NY, USA, 306
  Chenzhou City, Hunan, China, 398
  Hamilton, New Zealand, 460
  Hickham Air Force Base, Hawaii, USA, 176
  Kopu, North Island, New Zealand, 457
  Tui, Te Aroha, New Zealand, 460
  Werribee, Melbourne, Australia, 465
Fly ash, 436, 444–445
Frankia, 416

G

Galega officinalis, 425
Gas Chromatography, 79
  electron capture detection, 80
  mass spectrometry, 80, 181, 276, 277, 384
Genomics, 270
Glycine max, 73, 75, 110, 235
Glomeromycota, 91
Glomus intraradices, 416
Glucosyltransferases, 254
Glutathione-S-transferase, 236, 238, 239, 244, 254, 256
Gossypium arboreum, 51, 55
Gossypium hirsutum, 52, 449
Growth medium,
  Hoagland's solution, 94, 229
  Kirk, 6
  Luria-Bertani (LB), 5
  minimal medium with DF salts, 17, 22
  Murashige and Skoog (MS), 5, 21, 52, 64, 165
  Murashige minimal organic (MMO), 20
  SOC, 5
  tryptic soya bean broth, 17, 19
  yeast beef (YEB), 5
  yeast mannitol (YM), 5, 164

H

Hairy roots. See *Agrobacterium rhizogenes*
Helianthus annuus, 427, 449
Herbicides, 233
  alachlor, 235, 243
  BASTA, 11
  chloroacetanilides, 244
  chlorotoluron, 235
  conjugation, 254
  detoxification, 236, 254
  metalochlor, 235, 243
  monuron, 235
  sulfonylureas, 256
  triazines, 244
  simazine, 235
Hibiscus cannabinus, 111
Holcus lanatus, 321
  As, 322, 323
Hordeum brachyantherum, 74
Hordeum vulgare, 110, 123, 124
Hydroponics, 91
  As, 322
  atrazine, 442
  perchlorate, 221
Hyperaccumulation,
  CO_2 triggered, 363–365
Hyperaccumulators, 269, 291, 326
  characteristics, 326
  definition, 268, 325
  in China, 355, 356
  Mn, 360
  of metals, 122
  on serpentine, 414

Index

I

Inductively Coupled Plasma (ICP),
  mass spectrometry (MS), 151, 155–157, 231
  optical emission spectrometry (OES), 124, 141
Insecticides, 71–72
  organochlorines, 72
Ionomics, 277
Ipomoea batatas, 73
Iris psuedacorus, 234

J

Juncus conglomeratus, 414
Juncus effusus, 414, 415

K

K_{ow} values, 72, 253
Kochia scoparia, 244
Kochia sieversiana, 362

L

L-value 150,
  equation, 151
  determination, 154–155
Lactuca sativa, 73, 368
Land farming, 100, 102
Lead,
  accumulation, 292
  chelates, 293–294
  in Chinese soils, 351–352
  remediation, 139–140
Leachate, 457
  analysis, 124, 128
  metals, 299
  pesticides, 383
Lemna, 191, 200, 429
  L. minor, 430
  L. trisulca, 430
Lepidum virginicum, 360
Leptochloa fusca, 441
Leucaena leucocephala, 441
Linaria spartea, 414

Linum usitatissimum, 449
Lolium multiflorum, 75
Lolium perenne, 73
Lolium rigidum, 235
Lupinus albus, 73
Lupinus angustifolius, 73
Lycopersicon esculentum, 73, 74
  As, 322, 323

M

Malonyltransferases, 254
Map-based cloning, 271
Medicago sativa, 73, 75, 213
  As, 321
  PAHs, 426
Melilotus parviflora, 441
Mercuric reductase, 267
Mercury, 351
Metabolomics, 276
Metalloids, 319
Metallothionins, 16, 267, 278
Metals,
  adsorbed, 192–193
  analysis, 195, 196
  bioremoval, 186
  cadmium resistance, 267
  efflux ATPases, 267
  homeostasis, 266, 268, 270
  in batch reactor, 188
  in irrigation water, 353–354
  in plants 15–16
  in soils, 15, 89–90, 121–122, 149, 291, 351–354
  in sludge, 127
  isotope dilution, 150
  NIRS, 206, 208, 210
  resistance systems, 267
  transporters, 278
  water hyacinth, 430
Microarrays, 60, 269, 272, 273
Mine waste,
  phytostabilization, 412
  revegetation, 463

suitable species, 442
trace metals, 407
Modeling, 186
  adsorption, 188–191
  batch reactor, 187–188
  biomass growth rate, 191
  empirical, 308
Morus rubra, 74
Mutaginization, 266, 268
Mycorrhizae,
  arbuscular (AM), 89, 363
    axenic culture, 91
    inoculum, 95
    observing colonization, 94
    plant nutrition, 90
    semi-aquatics, 416
  As, 335
  Ecto (ECM), 414
  metallophytes, 413
  metals, 363
  organic pollutants, 418
  proteomics, 276
  serpentinophytes, 414
Myoporum sandwicense, 177
Myriophyllum aquaticum, 229
Near-Infrared Reflectance Spectroscopy (NIRS), 205–208
  calibration, 209–210, 212
  for arsenic, 213
  for metals, 208, 212, 213
  overtones, 207
  statistics, 211
  validation, 209, 210–211
Nerium indicum, 368
Nicotiana tabacum, 110
Nitrogen,
  Ammonium, 133
    analysis in soils, 134
    biopond removal, 431
    Cs desorption, 310
    extraction from plants, 112
    perchlorate degradation, 229
  Kjeldhal, 113, 114

nitrate,
  analysis in plants, 112, 113
  analysis in soils, 134
  in vegetables, 354
nitrite,
  in waste water, 429
  analysis in plants, 112, 113
nitrogen dioxide,
  plant metabolism, 109
  plant assimilation, 110
  plant fumigation, 112
NOx, 109
  unidentified, 110
Nitrogenous aromatics, 411–412
Nuclear magnetic resonance
  spectroscopy (NMR), 276, 277

O

Onobrychis viciifolia, 425
Operator protection, 309
Oryza sativa, 110
  As, 322
Oxidation/Reduction enzymes
  extracellular, 4
  catalase, 324
  dehalogenase, 50
  laccase, 4, 50, 51, 55, 163
  monooxygenase, 254, 258
  peroxidase, 4, 50, 163, 254, 324
    activity assay, 168
    induction, 258

P

Panicum maximum, 438
Panicum milaceum, 238
Panicum virgatum, 75
Pentachlorophenol, 49, 457
Perchlorate,
  ^{36}Cl labelled, 226, 229
  phytoaccumulation, 224
  removal kinetics, 225
  rhizodegradation, 229

Persistent organic pollutants (POPs), 71–73
  analytical procedures, 79–81
  chlordane, 73, 80
  DDE, 73, 79, 81
  DDT, 73, 386–387
  dioxins, 72, 74
  extraction from vegetation, 75, 78–79, 81–82
  extraction from soil, 75
  furans, 72, 74
  organochlorine insecticides, 72, 460
  PCBs, 71
Pesticides,
  γ-HCH, 442, 443
  DDT, 386–387
  dieldrin, 456, 460
  in China, 383
  in Portugal, 406
  organo-phosphorus, 386, 460
  riparian corridors
Petroleum hydrocarbons (PHCs, TPHs), 99, 176
  analysis, 180–181
  biodegradation, 181, 445–448
  bioremediation, 101–102
  cleanup options, 101
  crude oil, 387
  degradation by grasses, 102–103
  in Russia, 424–426
  physicochemical properties, 100
  phytoremediation strategy, 104
  Phytopet database, 105
  phytotoxicity, 101, 103
  plant sensitivity, 425
  plant tolerance, 425
  standards, 177, 179
Phalaris arundinacea, 75
Phalaris paradoxa, 239
Phanerochaete chrysosporium,
  culture, 9
  degradation of PCBs, 5
  purchase, 5

Phaseolus vulgaris, 73
Phaseolus aureus, 360
Phenol,
  assay, 167–168, 169
Phragmites australis, 234, 412
  oil degradation, 426
Phylogeny,
  of ferns, 329
  of fungi, 91
  of plants, 29, 71, 233
Phytochelatins, 16, 270, 276, 324
Phytoextraction,
  As, 394
  choice of species, 135
  heavy metals, 122, 123
  radionuclides, 132
Phytolacca acinosa, 360
Phytolaccaceae, 370
Phytosiderophores, 312
Phytostabilization,
  biosolids, 464–465
  heavy metals, 122
  mine wastes, 412
Pichia pastoris, 52
Pinus halepensis and As, 323
Pinus pinea, and As, 323
Pinus radiata, 295, 457
  As, 323
Pinus taeda, 229
Pisum sativum, 73, 235
  As, 323
Pithecelobium dulce, 441
Pittosporum tobira, 111
Pityrogramma calomelanos, 328
Plantago majus, 257
Plantago virfinica, 360
Poa alpina, 75
Poa annua, 239, 360
Poa trivialis, 425
Poaceae, 236
Polychlorinated biphenyls (PCBs), 3, 74, 252
  congeners, 4

Polychlorinated biphenyls *(cont.)*
  coplanar, 4
  in China, 384
Polychlorinated phenols, 49
  trichlorophenol degradation, 51
Polycyclic aromatic hydrocarbons
    (PAHs), 72, 75, 176
  degradation, 252, 258, 426
  in water, 383
Polygonaceae, 370
Polygonum hydropiper, 360
Polygonum microcephalum, 356, 360
Polygonum sibiricum, 362
Polymerase chain reaction (PCR),
  ACC deaminase, 19, 23
  fungal identification, 91
  Lac, Lip, Mnp, 10
  RT-PCR
    general methods, 61–63, 67–68
    primers, 69
  *Sna*BI, GUS, 9
Pongomia pinnata, 441
Pontaderia cordata, 234
Pontederiaceae, 430
Populus deltoides, 229, 235, 442, 458
Prosopis cineraria, 442
Prosopis juliflora, 441, 442
Proteomics, 274
  effects of metals, 276
  subcellular fractions, 275
Pteridium aquilinum, 368
Pteris cretica, 328, 361
Pteris longifolia, 328
Pteris umbrosa, 328
Pteris vittata
  As concentration, 396
  As detoxification, 332
  As distribution, 330, 395
  As hyperaccumulator, 327, 395–396
  As speciation, 331
  As uptake, 329, 361
  calcium, 397
  field trials, 398–401

heavy metals, 398
N fertilizers, 397
P fertilizers, 397, 399
Pueraria hirsuta, 439

Q

Quantitative trait loci (QTL), 268, 269,
    271
  Al resistance, 272
  Cs, 31–32, 313.
  fine mapping, 35
  LOD scores, 34
  mapping software, 32, 33
Quercus ilex, 414

R

Radioactive contamination, 28, 59, 131
Radionuclides,
  Cl, 310
  ^{134}Cs, 355, 368–370
  ^{137}Cs, 131–132, 306, 313, 365
    accumulation by plants, 28–30,
      132, 430
    analysis, 134, 136
    clays, 310–311
    exclusion from food chains, 29
    exposing plants to, 65, 66
    gene expression, 61
    in soils, 28, 131
    natural genetic variation, 29
    shoot concentrations, 38, 39, 44
    soil activity, 133, 136
    soil amendments, 133
    transport proteins, 29, 44
  in fertilizers, 355
  in fly ash, 444
  in soils, 305
  I, 355
  ^{32}P
  Pu, 311
  soil decontamination, 307
  S, 310
  Sr, 131, 305, 313, 365, 430

Index

Tc, 309–310
Th, 355
transfer factors, 308
U, 131, 292, 355, 366
Ranunculaceae, 236
Raphanus sativus, 395
Rare Earth Elements,
 uptake by ferns, 361
Research Technologies Development Forum, 103, 104
Rhaphiolepsis umbellate, 115
Rhizobiaceae, 162
Rhizofiltration, 306
Rhizophere, 227
 designing, 260
 heavy metals, 356
 herbicides, 244
 oil degradation, 425
 Phragmites australis, 415
Rhododendron mucronatum, 111
Riparian corridors, 412
RNAi, 278
Root concentration factor (RCF), 253
Root exudates, 50, 277, 311, 427
 As, 335
Rumex acetosa, 356, 360
Rumex hastatus, 356, 360

S

Saccharomyces cerevisiae, 270
Safe crops, 28, 30
Salix atrocinerea, 416
Salix babylonica, 75
Salix nigra, 229
Salt-affected soils, 43
 species for, 441
Salvia anthemifolia, 360
Salsola collina, 366
Schima nervosum, 438
Schizachyrium scoparius, 75
Schoenoplectus lacustris, 234
Scirpus holoschoenus, 414
Scrophulariaceae, 236

Secale cereale, 426
Sectored planters, 175–176
 construction of, 179
 uses of, 182
Serpentine, 267, 414, 417
Sedum alfredii, 360
Senecio vulgaris, 239
Setaria adherens, 239
Setaria verticillata, 239
Setaria viridis, 239
Serpentine soils, 267
Sewage sludge, 457
 in Chinese soils, 352–354
Silene cucubalus, 276
Silene vulgaris, 268, 324
Silybum marianum, 368
Soil amendments,
 ^{137}Cs, 132, 133, 135
 EDTA, 122, 123, 125, 127, 140–143
 HEDTA, 122
 DTPA, 122
Soil columns, 124
Solanum melongena, 73
Solanum nigrum, 74, 239
Solanum tuberosum, 73, 235
Sorghastrum nutans, 75
Sorghum saccharatum, 427
Sorghum vulgare, 75, 426
Spartina alterniflora, 235
 As, 322
Spinacia oleracea, 73, 110
Spirodela polyrhiza, 229
Stellaria media, 424
Suaeda corniculata, 362
Suaeda glauca, 362
Subtractive hybridization, 273
Superoxide dismutase, 16, 324
Syzigium cimini, 442

T

Tamarix articulata, 441
Tamarix troupii, 441

Taraxacum officinale, 73, 428
Teconklla undulata, 442
Terminalia arjuna, 441, 442
Thermal Ionisation Mass Spectrometry (TIMS), 151, 155
Thespesia populnea, 177
Thlapsi caerulescens, 268, 326, 362
    Ni/Zn phytoextraction
    Zn transporter, 269
Thlapsi goesingense, 268
Thymus mastichina, 414
Timber-Industry waste, 457
Toxicity test,
    Vibrio fischeri, 170
Trametes versicolor
    culture, 9
    purchase, 5
Transcriptomics, 273
Trifolium alexandrum, 441
Trifolium pratense, 73
Trifolium resupinatum, 441
Triticum aestivum, 74, 110, 235, 360
Typha latifolia, 229, 233

V

Vetiver, *See Chrysopogon zizanoides*
    eroded soils, 439
    dewatering, 465–467
Veronica perefina, 360
Vicia villosa, 73

W

Waste disposal, 309
Waterlogged soils, 442
Wind erosion, 439

Z

Zea mays, 73, 425
    As, 323
Zinc,
    in Chinese soils, 352–354
    uptake by plants, 360
ZIP transporters, 274
    Cd uptake, 270
    Fe transport, 270
    Zn transport, 270
Zizyphus mauritiana, 442